THE OXFORD AUTHORS

General Editor: Frank Kermode

MATTHEW ARNOLD was born at Laleham-on-Thames on 24 December 1822 as the eldest son of Dr Thomas Arnold and his wife Mary. He was educated at Winchester College, his father's old school; Rugby, where his father was headmaster; and Oxford. In 1851 he was appointed Inspector of Schools, pursuing this taxing career to support his wife and family until his retirement in 1886. He published his first volume of verse, *The Strayed Reveller, and other Poems*, in 1849 followed by *Empedocles on Etna, and other Poems* (1852) and five further collections which appeared, with a diminishing number of new poems in each, between 1853 and 1867, after which his creative gift appeared to dwindle still further and he published little poetry. His career as a writer of prose began to take over after his election to the Professorship of Poetry at Oxford in 1857. Stimulated by preparing his lectures, many of the earliest published in 1865 as *Essays in Criticism (First Series)*, he turned increasingly to the vigorous and widely ranging polemical commentaries on culture, religion, and society which were to make him known at home and abroad as the foremost critic of his day. He died suddenly of heart failure on 15 April 1888 while awaiting at Liverpool the arrival of his married daughter from America.

MIRIAM ALLOTT is Emeritus Professor of English in the University of London. She has edited *The Poems of John Keats* for Longman's Annotated English Poets, and prepared the revised edition of *The Poems of Matthew Arnold* for the same series, first edited by Kenneth Allott in 1965. She is currently working on a biography of Arthur Hugh Clough and a critical study of Victorian poetry. She has also worked on the nineteenth- and twentieth-century novels and is studying aspects of English fiction since the 1960s.

ROBERT H. SUPER is Professor of English at the University of Michigan, editor of Arnold's *Complete Prose Works*, and author of *The Time-Spirit of Matthew Arnold* (1970).

THE OXFORD AUTHORS

MATTHEW ARNOLD

EDITED BY
MIRIAM ALLOTT
AND
ROBERT H. SUPER

Oxford   New York
OXFORD UNIVERSITY PRESS
1986

Oxford University Press, Walton Street, Oxford OX2 6DP

Oxford New York Toronto
Delhi Bombay Calcutta Madras Karachi
Petaling Jaya Singapore Hong Kong Tokyo
Nairobi Dar es Salaam Cape Town
Melbourne Auckland
and associated companies in
Beirut Berlin Ibadan Nicosia

Oxford is a trade mark of Oxford University Press

British Library Cataloguing in Publication Data
Arnold, Matthew
Matthew Arnold.—(The Oxford authors)
I. Title   II. Allott, Miriam   III. Super, Robert H.
828'.809   PR4020.A1
ISBN 0–19–254187–0
ISBN 0–19–281376–5 Pbk

Library of Congress Cataloging in Publication Data
Arnold, Matthew, 1822–1888.
Matthew Arnold.
(The Oxford authors)
Bibliography: p.
Includes index.
1. Allott, Miriam Farris.   II. Super, R. H.
(Robert Henry), 1914–   .   III. Title.   IV. Series.
PR4021.A46 1986   828'.809   85–18866
ISBN 0–19–254187–0
ISBN 0–19–281376–5 (pbk.)

Set by Wyvern Typesetting Ltd.
Printed in Great Britain by

Cox & Wyman Ltd.
Reading, Berks.

CONTENTS

PROSE

ABBREVIATIONS

The following abbreviations have been used in the Introduction and explanatory notes:

1849	*The Strayed Reveller, and Other Poems* (1849).
1852	*Empedocles on Etna, and Other Poems* (1852).
1853	*Poems.* A new edition (1853).
1854	*Poems.* Second edition (1854).
1855	*Poems.* Second series (1855).
1857	*Poems.* Third edition (1857).
1867	*New Poems* (1867).
1868	*New Poems.* Second edition (1868).
1869	*Poems* (1869). First collected edition (2 volumes).
1877	*Poems* (1877). Collected edition (2 volumes).
1881	*Poems* (1881). Collected edition (2 volumes).
1885	*Poems* (1885). Collected Library edition (3 volumes).
1890	*Poetical Works of Matthew Arnold* (1890). Globe edition.
1950	*Poetical Works of Matthew Arnold*, ed. C. B. Tinker and H. F. Lowry (London and New York, 1950).
Allott	*Matthew Arnold*, 'Writers and Their Background' series. ed. Kenneth Allott (London, 1975).
Baum	*Ten Studies in the Poetry of Matthew Arnold*, by P. F. Baum (Durham, NC, 1958).
Bonnerot	*Matthew Arnold—Poète: Essai de biographie psychologique*, by L. Bonnerot (Paris, 1947).
Buckler	*Matthew Arnold's Books: Toward a Publishing Diary*, written and edited by W. E. Buckler (Geneva and Paris, 1958).
CL	*The Letters of Matthew Arnold to Arthur Hugh Clough*, ed. H. F. Lowry (London and New York, 1932).
Commentary	*The Poetry of Matthew Arnold: A Commentary*, by C. B. Tinker and H. F. Lowry (London and New York, 1940).
Conversations	*Conversations with Goethe*, by J. P. Eckermann, tr. J. Oxenford (London, 1850).
Correspondence of AHC	*Correspondence of Arthur Hugh Clough*, ed. F. L. Mulhauser in 2 volumes (Oxford, 1957).
CPW	*The Complete Prose Works of Matthew Arnold*, ed. R. H. Super in 11 volumes. (Ann Arbor, Mich., 1960–77).
E in C I	*Essays in Criticism* (1865).
E in C II	*Essays in Criticism.* Second series (1888).
L	*Letters of Matthew Arnold, 1848–88*, ed. G. W. E. Russell in 2 volumes (London and New York, 1895).
Life of JDC	*Life and Correspondence of John Duke, Lord Coleridge . . .*, written and edited by E. H. Coleridge in 2 volumes (London and New York, 1904).

Neiman	*Essays, Letters, and Reviews by Matthew Arnold*, ed. F. Neiman (Cambridge, Mass., 1960).
Note-books	*The Note-books of Matthew Arnold* ed. H. F. Lowry, K. Young, and W. H. Dunn (London and New York, 1952).
Obermann	*Obermann*, by E. P. de Senancour, ed. G. Michaut, in 2 volumes (Paris, 1912–13).
PMLA	*Publications of the Modern Language Association of America.*
Parrish	*A Concordance to the Poems of Matthew Arnold*, ed. S. M. Parrish (Ithaca. NY, 1959).
Poems	*Arnold. The Complete Poems*, ed. Kenneth Allott (London, 1965); Second edition by Miriam Allott (1979).
Remains of AHC	*The Poems and Prose Remains of Arthur Hugh Clough, with a selection from his letters and a memoir*, edited by his wife in 2 volumes (London, 1869)
Russell	*Matthew Arnold*, by G. W. E. Russell (Literary Lives series, second edition, London, 1904).
Sells	*Matthew Arnold and France: The Poet*, by I. E. Sells (Cambridge, 1935; rev. edn. 1970).
TLS	*The Times Literary Supplement.*
Traill	*The Works of Thomas Carlyle*, ed. H. D. Traill in 30 volumes (London and New York, 1896–9).
Trilling	*Matthew Arnold*, by Lionel Trilling (New York and London, 1939).
UL	*Unpublished Letters of Matthew Arnold*, ed. A. Whitridge (New Haven, Conn., 1923).
Yale MS	A Collection of notes, poetic drafts, etc. by Matthew Arnold which are preserved as a notebook at Yale.
Yale Papers	The larger collection of Arnold's poetic manuscripts, note-books, diaries, letters, etc. preserved at Yale.

INTRODUCTION

(i) Poetry

WHEN the little volume *The Strayed Reveller, and Other Poems*, containing twenty-seven poems by 'A', first appeared in 1849, those who knew Arnold were as much taken aback as they were pleased. 'It is the moral strength, or at any rate, the *moral consciousness*, which struck and surprised me so much', wrote his sister Mary: 'I could have been prepared for any degree of poetical power ... but there is something which such a man as Clough has, for instance, which I did not expect to find in Matt.' In another letter she says: 'there was so much more practical questioning ... than I was prepared for ... it showed a knowledge of life and conflict which was *strangely like experience*'.[1] The surprise was not caused by the author's youth—he was by now, after all, two months into his twenty-seventh year—so much as the distance between the apparent gaiety and insouciance of the young man many of his friends thought they knew and the serious themes, anxious self-debate, and struggle to overcome melancholy and self-doubt expressed in his poems. These are the qualities ensuring that plangent, elegiac note which for most of his admirers remains the sign of his authentic poetic gift, however firmly in later years he tried to subdue it by writing a more public poetry designed to 'animate' and 'ennoble' troubled fellow-sufferers from the 'strange disease of modern life' with its 'doubts, disputes, distractions, fears', its 'confused alarms of struggle and flight', and its 'complaining millions of men' who 'Darken in labour and pain'.[2]

Lines such as these from some of Arnold's best-known poems have passed into the language, but before 1849 his family and friends were accustomed to—and in some cases concerned by—the bantering style often enlivening his letters, his pranks as a schoolboy and undergraduate, and the Olympian manners and dandyism of his enthusiastic Francophile days in the later 1840s. 'But, my dear Clough,' he writes, after attending a stuffily pietistic Masters' Meeting during the brief period in 1845 when he was teaching classics at Rugby, 'have you a great Force of Character? ... I am a reed, a very whoreson Bullrush'; 'Tell Edward', he says in 1848, the year of revolutions abroad and fear of

---

[1] Mrs Humphry Ward, *A Writer's Recollections* (London, 1918), pp. 44–5.
[2] 'The Scholar-Gipsy', l. 203; 'Memorial Verses', l. 44; 'Dover Beach', l. 36; 'The Youth of Nature', ll. 51–2.

French invasion at home, 'I have engaged a Hansom to convey us both from the possible scene of carnage', and am prepared to leave at once; in February 1847, after returning from Paris where he had spent more than two months going to balls and parties and attending every one of Rachel's performances at the Théâtre Français, Clough records in a famous passage:

Matt is full of Parisianism, theatres in general, and Rachel in special: he enters the room with a chanson of Béranger's on his lips—for the sake of French words almost conscious of tune: his carriage shows him in fancy parading the rue de Rivoli—and his hair is guiltless of English scissors; he breakfasts at twelve, and never dines in Hall, and in the week or 8 days rather (for 2 Sundays must be included) he has been to Chapel *once* . . .[3]

Charlotte Brontë, who with Harriet Martineau met him in the Lake District in 1850, was among those who felt 'the shade of Dr Arnold seemed to frown upon his young representative', and his letter at the time ('sent the lions roaring to their dens at half-past nine') might have confirmed her misgivings; yet she was sharp enough to perceive 'intellectual aspiration' and 'a real modesty . . . under his assumed conceit'.[4]

In these early poems the 'aspiration' is found to be a restless yearning for knowledge and truth, and for the calm these might bring; and the 'modesty' emerges as the troubled self-questioning and uncertainty which was at once a symptom and cause of the poet's restlessness and troubled sense of the 'divided self'. The 'Youth' of his misleadingly named title poem 'The Strayed Reveller', endowed with the qualities he found seductive when reading Maurice de Guérin in 1847 ('extraordinary delicacy of organisation and susceptibility to impressions . . . To assist at the evolution of the whole life of the world is his craving'),[5] contrasts his feverish creativity with calm breadth of vision:

> The Gods are happy.
> They turn on all sides
> Their shining eyes,
> And see below them
> The earth and men.

'Wise bards' are granted the power to 'behold and sing' what the gods calmly survey:

> But oh, what labour!
> O prince, what pain!

[3] *CL*, pp. 56, 66; *Correspondence of AHC*, i. 178–9.
[4] Letter to James Taylor, 15 January 1851, in *The Brontës: Their Lives, Friendships and Correspondence* (ed. Wise and Symons, 1932), iii. 199.
[5] *E in C* I (1863; *CPW* iii. 30).

The 'prince' is Ulysses and Arnold has made his reveller 'stray' into Circe's temple, but he uses his setting from the *Odyssey* to establish the balancing of different visions which is habitual to his creative temper from first to last. His dualities helped to make him very much a 'dramatic' poet, which means that he was also a representative one, as he recognized in his pronouncement of 1869: 'My poems represent, on the whole, the main movement of mind of the last quarter of a century, and thus they will probably have their day as people become conscious to themselves of what that movement of mind is, and interested in the literary productions which reflect it'.[6] We have indeed become conscious of 'that movement of mind' for we can now recognize the manner in which the mind debated its pressing concerns as well as the nature of the concerns themselves. It would be difficult not to do so when confronted with the extraordinarily varied range of dramatic procedures employed in Victorian poems as different in temper as 'Maud' and 'Dipsychus', 'Caliban upon Setebos' and 'Amours de Voyage', or 'St Simeon Stylites' and 'Mr Sludge, "the Medium"'. The 'modern spirit' as Arnold saw it— he is our real inventor of 'the modern', being the first to attempt, if with varying success, a systematic analysis of its complicated properties—is introspective, questioning, perplexed: 'the dialogue of the mind with itself has commenced', as he says in his 1853 Preface (p. 172 below). We have to be alert to his habit of self-debate if we are to understand him properly because his constant movement between, and attempt to reconcile, opposites—whether Lockian reason and Kantian intuition, Hellenism and Hebraism, classical *architectonicē* and imaginative energy, or Teutonic stringency and Celtic natural magic—governs not only the formal characteristics of most of his poems and the relationship between one poetic statement and another, but also the connection between the poetry and the prose. The latter are 'inseparable', as H. F. Lowry emphasizes: 'Voices reverberate back and forth between the verse and the essays, the questions raised in one are answered in the other.'[7]

This habit of self-debate is not perhaps a final safeguard against the risks facing Arnold's distinguished and finely tempered but not naturally robust or incandescent poetic gift. He said himself that he lacked Browning's 'vigour and abundance' and Tennyson's 'poetical senti-ment', though adding with prescience that since he had 'perhaps more of a fusion of the two than either of them, and had more regularly applied that fusion to the main line of modern development, I am likely enough

[6] *L* ii. 9.   [7] *CL*, p. 36.

to have my turn, as they have had theirs'.[8] The risks include the occasional stiffness that comes from the missionary side of his nature and is linked with the passing of the 'open and liberal state of our youth', a loss, sensed early, which he understood well and wrote of with some bitterness in the later stages of his brief poetic career, whose most prolific period lasted little more than a decade before the 'man of morality and character' took over and he turned to prose. But his most important poems find a strong imaginative stimulus in that native habit, including 'Thyrsis' and even the second 'Obermann' poem, both written in the 1860s when the stream had almost dried up. The habit is responsible among other things for the shaping of symbolic landscapes whose contrasted features—cool moonlit shores, high mountains and steadfast stars set against hot and dizzying 'cities of the plain', centres of 'sick hurry and divided aims'—compose a region as powerfully marked with the impress of individual idiosyncrasy as Hardy's Wessex or 'Greeneland'.

The two most impressive poems in the 1849 volume, 'Resignation. To Fausta' and 'The Forsaken Merman' (a much loved poem from the time of the first reviews), in spite of stylistic differences testifying to the technical resources of this new talent, both move characteristically between contrasted scenes. In 'Resignation' the two walks separated by ten years are taken through the same Lakeland countryside, but the walkers have changed with time so that nature, like the deceptively serene 'moonblanch'd' shore of 'Dover Beach', now 'seems to bear rather than rejoice'. The poem, certainly a 'dialogue of the mind with itself', uses Fausta to represent an aspect of the 'divided self', the much quoted line *Not deep the poet sees but wide* (in context italicized because attributed to her unspoken thoughts) being part of an argument balancing unreflecting 'torpor', focused in the gipsies, and the restless impatience focused in 'Fausta'. It also engages with the line of thought which produced the youthful poet's feverish creativity in 'The Strayed Reveller', the procession 'Of eddying forms' sweeping through his brain corresponding with 'action's dizzying eddy' in 'Resignation', from which refuge is sought in stoical quietism, though not, it seems, with total 'resignation'. At the close, 'the something that infects the world' is not restless longing but life's seemingly inevitable frustration of ardent expectancy.

'Resignation' is explicitly an exploratory debating poem, as with all poems Arnold addresses to others, among them the two 'Obermann'

---

[8] *L* ii. 9.

poems; the 'Stanzas from the Grande Chartreuse', which directs its communings about modern faith and doubt to the inhabitants of the monastery; the 'Switzerland' sequence for 'Marguerite'; and the group of poems written to Clough in the later 1840s and arguing for the values of quietism over those of involvement in the world of decision and action. Even the group of poems combining elegy and literary criticism—for instance the various tributes to Wordsworth and 'Heine's Grave'—carry on a running debate with their subjects (intermittently they do so with whoever may be reading, in which case the tone becomes more conventionally that of the explicator and mentor). 'The Forsaken Merman', on the other hand, employs more indirect and dynamic dramatic and lyrical procedures, balancing 'inner' themes of loss and separation caused by incompatible longings with the 'outer' world of its chosen legend. The romantic, pagan submarine world and the austere white-walled town on the shore, regions respectively of passionate desire and spiritual peace, achieve an effect of immediacy and of 'voices reverberating to and fro' not unlike the impression made by the heights and the valley, regions of storm and calm, in *Wuthering Heights*, whose author Arnold singles out enthusiastically in his elegy 'Haworth Churchyard'. 'The Forsaken Merman' springs from his myth-making side, which in the same early volume uses the ancient Egyptian legend of Mycerinus to project painful loss of faith in a divinely sanctioned order, and fastens on the Islamic story related in 'The Sick King in Bokhara' because it exemplifies human need for submission to moral law against all claims of reason and expediency. In 'Tristram and Iseult', the earliest modern treatment of the legend in English (published in the second collection of 1852), Arnold's portrayal of the two Iseults—the first passionately loved and lost, the second calm, gentle and stoical—again shows this creative imagination taking over and transforming its source under the pressure of personal idiosyncrasy. The dualities governing Joseph Glanvill's story of the poor scholar who abandons one kind of existence to pursue his quest for truth in another, totally different, world, seized upon in the early 1850s when shaping 'The Scholar-Gipsy', are still potent thirteen years later in the companion piece 'Thyrsis', Arnold's elegy for Clough. These major elegiac poems show, if nothing else, how long established and deeply felt was the sense of differences and alternatives experienced in those Oxford days, themselves lamented as a time of irrecoverable freedom and happiness, past and present, youth and disenchanted middle age thus adding their weight to abiding preoccupations.

All these characteristics are found purposefully at work in

'Empedocles on Etna', Arnold's—and arguably the age's—major long poem, notoriously dropped from the 1853 collection for its alleged 'morbidity' and introspection. It is, as the subheading tells us, a 'dramatic poem', setting the youthful singer Callicles against the ageing Empedocles, an Ancient Greek philosopher remodelled for the sake of this modern myth, whose poetic gift has left him. His recognition that 'The brave impetuous heart yields everywhere / To the subtle, contriving head' places him with the dwellers in the Waste Land of modern life, sufferers from inner division and imbalance, with 'heads o'ertaxed' and 'palsied hearts'. The qualities dramatized in the contrasted figures are imaged in the changing features of Etna's landscape—the green forest region is the natural home of the young Callicles, the region's outskirts on the upper slopes signal Empedocles' climb to bleak maturity, and the charred summit is the place where the fount 'shall not flow again', as Arnold puts it in his comfortless late poem 'The Progress of Poesy' (p. 268). The work also displays the metrical resource heralded in the 1849 poems, which vary prosodic measures to accommodate, underline, and also harmonize the different terms of the debate—though such imaginative reconciliation of opposites is ignored by the 'man of morality and character' who disconcertingly takes over in the dismissive 1853 Preface. Callicles sings of traditional wisdom in the flexible two-stressed and three-stressed lines gathered in irregular verse paragraphs, like the 'pindarics' of 'The Youth of Nature' and 'The Youth of Man' (see head-note on pp. 532–3 below), and used earlier to accommodate the forsaken Merman's opening 'lyric cry' as well as his role as narrator and his final stoicism, associated with the closing moonlit scene on the shore, Arnold's favourite image of calm. Callicles uses this irregular chant to sing of balance and harmony, closing with celebration of 'The night in her silence, / The stars in their calm'. Empedocles' formal philosophical analysis of 'modern thought' in Act I is cast in studiedly regular hexameters, his reflections on 'modern feeling' in Act II in a more relaxed but still elevated metrical style, and his final exultant utterances, recapturing the note of his early power at the very moment of renouncing it, echo the lyrical chanting of his youthful counterpart. Arnold's metrical ear cannot rival Tennyson's, but these skills ensure the poignant lyricism come upon from time to time in his poems when the movement of thought and the object contemplated are brought together and unified. The movement mimics the action in 'Dover Beach' where

> the light
> Gleams and is gone

and the pulse of the waves reminds the observer of time and change as they

> Begin, and cease, and then again begin,
> With tremulous cadence slow, and bring
> The eternal note of sadness in.

The ebbing Sea of Faith is the central emblem of 'Dover Beach', as 'the unplumbed, salt, estranging sea' is the central motif in the 'Marguerite' poems, which lament a general sense of human isolation not escaped even by lovers ('what heart knows another? / Ah! who knows his own?'). As might be expected from the poet who confessed himself 'one who looks upon water as the Mediator between the inanimate and man',[9] images of seas and rivers figure prominently in his landscapes (which also include 'high places' from which 'to see life steadily and see it whole', especially the Alpine heights in Switzerland, that country carrying a high emotional charge whenever it appears in the poems). Linked with the idea of voyages and journeying, the images of water, seas, and rivers come to express the mysterious courses of life and also the hidden currents of the 'buried self' whose true nature, obscured by other warring 'selves', once known and understood will make for clear vision and firm purpose. On 'life's incognisable sea' in 'Human Life' the baffled traveller finds himself 'charter'd by some unknown Powers' even when thinking himself free, and in 'A Summer Night' as the 'tempest strikes' the lightning flashes reveal

> the pale master on his spar-strewn deck
> With anguish'd face and flying hair,
> Grasping the rudder hard,
> Still bent to make some port he knows not where,
> Still standing for some false, impossible shore.

'The Buried Life' permits a brief respite at rare moments of happiness in love when—as not in the 'Marguerite' poems—the longing to understand 'the mystery of this heart which beats / So wild, so deep in us' is momentarily fulfilled: 'a lost pulse of feeling stirs again', 'A man becomes aware of his life's flow',

> And then he thinks he knows
> The hills where his life rose,
> And the sea where it goes.

This longed-for calm is pictured again at the close of 'The Future',

---

[9] Letter of 29 September 1848 (*CL*, p. 92).

where the river flowing out to sea brings together ideas of the furthest reaches of human destiny and the sense of an individual life's ending:

> the banks fade dimmer away,
> As the stars come out, and the night-wind
> Brings up the stream
> Murmurs and scents of the infinite sea.

The most widely known expression of this feeling is of course the coda to 'Sohrab and Rustum' where the Oxus, flowing by the tragic father and his dead son on the shore, is gradually freed from 'foiled and circuitous wanderings' and at last

> The long'd-for dash of waves is heard, and wide
> His luminous home of waters opens, bright
> And tranquil, from whose floor the new-bathed stars
> Emerge, and shine upon the Aral Sea.

All this is a long way from the young man of the 1840s who sang little French tunes while 'in fancy parading the rue de Rivoli' and played impish tricks on his friends. We have to wait till his last years and the composition of affectionate elegies for beloved family pets before discovering in his poetry any trace of his native playfulness and wit. His vivacity, fortunately, remains engagingly alive for us in his prose, but his belief that wit and humour were inappropriate for serious poetry was a damaging limitation. Clough's brilliant and original satirical style in his major work made him uneasy; and his celebratory poem for Heine is carefully silent about the wit and irony he salutes in his prose essay (see headnote on p. 552 below). It is tempting to think that the gaiety which quickens and delights us in his prose might if drawn on have helped to prolong his poetic life. But the reasons for his poetic decline are complex: his taxing work as a school inspector, his settlement to a contented life as a family man remote from the wilder shores of love, his assumption, once he took the Oxford Professorship, of a missionary role as the major polemicist of his day, all played their part. The fact remains that of the five collections published between 1849 and 1867, only the first two reflect any degree of prolificity, with some thirty-five poems in 1852 to add to the twenty-seven in 1849. New poems dwindle in 1853, 1855, and 1867, all these collections needing selections from earlier work to fill them out. The four remaining volumes published in Arnold's lifetime are Collected Poems, or selections from published work, the arrange-ment in the two-volume edition of 1877 representing his final classifica-tion under the headings Early Poems, Narrative Poems, Sonnets, Lyric,

Dramatic, and Elegiac Poems, settled upon after severe taxonomic struggles—unsurprisingly since one kind runs into another, with the interrogatory to-and-fro of different ways of seeing and feeling, and the effort to unite them, as the highest common factor. The creative tide had already begun to ebb by 1853, when Arnold dropped 'Empedocles on Etna', though there was still sufficient *élan* to complete its substitute 'Sohrab and Rustum', which he liked because 'in its poor way I think [it] *animates*'; 'The Scholar-Gipsy', which, perversely, he did not, because it 'at best awakens a pleasing melancholy';[10] and appealing shorter pieces such as 'The Neckan' and 'Philomela'. These helped to make up the half-dozen or so new poems among the total of thirty-six. The 'classical' principles announced in 1853 saw to it that lengthy 'elevating' poems should now occupy time and space. 'Sohrab and Rustum' was followed in 1855 by the more frigid if honourably conceived 'Balder Dead' and by the classical tragedy *Merope*, that misguided undertaking published in 1858. Apart from 'Thyrsis', the high points of the 1867 volume are 'Dover Beach', 'Stanzas from the Grande Chartreuse', and 'Calais Sands', all written in the early 1850s. Even the handful of more recent poems are inspired by early experience, outstandingly in 'Thyrsis', the companion poem to 'The Scholar-Gipsy', which celebrates those youthful Oxford days shared with Clough, and two poems poignantly recalling 'Youth's Agitations' (the title of a poem of 1855 reprinted here), namely 'The Terrace at Berne', Arnold's last poem for 'Marguerite', and 'Obermann Once More', his last major poem.

For such reasons, then, the poems in this selection are not printed in the order in which they appeared in individual collections, nor according to the classifications of 1877. They are set out in chronological order of composition in the hope that, together with their principles of selection, they will demonstrate that 'movement of mind', at once highly individual and highly representative, of which Arnold spoke in 1869 when his career as a master of prose was already well under way.

M.A.

(ii) Prose

Matthew Arnold was from birth a member of the intellectual élite; as he touchingly, if indirectly, reminds us in more than one of his essays, he was an Oxford man (and the son of an Oxford man). His father, Thomas, was an undergraduate at Corpus Christi College, took holy orders in the Church of England, was Fellow of Oriel College when (as Newman

[10] *CL*, p. 146.

makes us aware in his *Apologia*) some of the most important theological minds of that Church were his colleagues, then moved into secondary education and became Headmaster of Rugby School. He was granted a doctorate by his university in 1828, and the year before his death was appointed Regius Professor of Modern History at Oxford (1841). His friend John Keble, later one of the founders of the Tractarian movement, was Matthew's godfather; another Oriel associate, Richard Whately, later Archbishop of Dublin, was godfather to one of Matthew's sons. Samuel Taylor Coleridge's nephew, John Taylor Coleridge, was Thomas Arnold's close friend at Oxford; John Taylor's son, John Duke Coleridge (later Lord Chief Justice of England) was Matthew's lifelong friend. But Thomas Arnold added one non-English dimension to this background—a great interest in German thought (especially theological) and a personal friendship with a number of important German writers.

Matthew, then, was brought up in an atmosphere of serious concern for religious questions, for education, for the impact of history on modern life. His father was not a poet, but the tradition of prize competitions in poetry both at school and the university spurred Matthew in that direction and his success in the competitions did much to confirm his vocation. (One evidence of how his father's interests influenced him is that Thomas set 'Mycerinus' as subject of a prize poem in 1831 at Rugby, and Matthew took it as the subject of one of his earliest mature poems.) His father's connection with some of England's great poets of the era also helped him: that with the Coleridge family has been mentioned; Wordsworth became a neighbour and close friend when Dr Arnold built a summer home in the Lake District, and Southey's son-in-law was for a time the boy Matthew's private tutor.

The lad followed in his father's footsteps, through Winchester College (the school Thomas had attended), Rugby School, a scholarship at Oxford (Balliol College) and election to a fellowship at Oriel. In due course he received doctorates from both Oxford and Cambridge. (He stopped short of ordination, however.) He early learned to admire Goethe, and owned the 60-volume edition of Goethe's *Works* (1827–42). His first significant prose writing was the Preface to his *Poems* (1853), deriving principally from Aristotle and from recent German literary criticism. But the real turning point in his literary career was his election to the Professorship of Poetry at Oxford in 1857, when he was thirty-four—a post he held for ten years. The principal duty of the professor was to lecture three times a year on a subject of his choosing, and Arnold first determined upon a series of lectures on the concept of 'modernism', following a pattern his father had set in the interpretation of history. Only one of this group of lectures survives, the first, 'On the Modern Element in Literature', in which the modern spirit is represented by Thucydides (whom Thomas Arnold had edited) and by Lucretius (whose *De Rerum Natura* had strongly influenced Matthew's 'Empedocles on Etna'). The series floundered, apparently through Arnold's lack of conceptual vigour; luckily two thoroughly inept new translations of Homer moved him to consider both Homer and the theory of translation in three lectures, and when one of the inept translators was foolish enough to make an even more inept reply, Arnold had the material for yet another lecture. By this time it was becoming apparent that he could not resume his original course. Moreover, he had by now travelled extensively in France on an official mission to study the State-supported school system there and had come to know personally some of the leading French intellectuals, including Sainte-Beuve. And so began a series of lectures aimed at introducing to his audience some of the lesser-known contemporary Continental literary figures, principally but not exclusively French. The founding of the *Cornhill Magazine* about this time gave him a means of publishing his lectures, and thereafter nearly everything he wrote was published in a periodical before it found its way into a book. When enough lectures had appeared to justify collecting them into a volume, he composed two on more general topics, which should serve as introduction to the book—'The Literary Influence of Academies', contrasting French urbanity with British provincialism in letters (as recently as 1937 'provincialism' was set as the subject of a prize essay at Oxford), and 'The Function of Criticism at the Present Time', to show the fruitful interplay between criticism and the creative process.

If what has been said thus far seems to paint too solemn a picture of a young man with a mission, urged by awe of his father's ghost (Thomas Arnold died at the end of Matthew's first year at Oxford), an important aspect of Matthew's character has been neglected—his playfulness. This was one who as a lad could slip up behind his father while Thomas was teaching and make faces at the class, putting them on their mettle not to burst out laughing; who not many years later, travelling with a friend in a public coach, could intimate to fellow-passengers that the poor friend was mentally afflicted and that Matthew was his keeper, transporting him to an asylum. His love of fun remained with him through life, going hand in hand with a quickness of wit and perception of anomalies that frequently peeks out from his prose, as in the Preface to *Essays in Criticism*, in *Culture and Anarchy* and *Friendship's Garland*, and much later in 'Civilisation in the United States'.

While the Professorship of Poetry gave Arnold an audience in a public he never thereafter lost, his appointment as Inspector of Schools in 1851 at the age of twenty-eight showed him an aspect of English life Oxford hardly recognized. The schools were elementary schools assisted by government grants, and conducted for the most part by the Protestant Nonconformists—Congregationalists, Presbyterians, Methodists, Unitarians (Church of England schools had their own, clerical, inspectors). This was quite a different world from the Winchester and Rugby he had known. If there is some touch of youthful superciliousness in the comment he made soon after his appointment—'I think I shall get interested in the schools after a little time; their effects on the children are so immense and their future effects in civilizing the next generation of the lower classes, who, as things are going, will have most of the political power of the country in their hands, may be so important'—the tone quickly disappeared. He saw now a whole new aspect of English society, and it became a principal motive of his life to improve and elevate the tone of those common people over whose schools he exercised some supervision. Brought up a Liberal, he was more than ever convinced that the French Revolution had been the great turning point in modern history and that England must follow the lead of that nation against whom she had fought so bitterly. His French mission of 1859 confirmed with careful observation the rightness of his conclusion, and when he republished his long official report for the edification of the general reader, he prefaced it with an essay on 'Democracy'—a word not in good odour in mid-century England. A second official mission to the Continent in 1865, this time to study higher schools and universities, took him to Germany and refreshed his early respect for German thought; thereafter he regarded the German concept of 'Culture' as the path to the successful operation of the inevitable democracy of the future, the corrective to the 'anarchy' which many Englishmen were certain was the more descriptive synonym of 'democracy'. Arnold's last lecture from the Chair of Poetry at Oxford was 'Culture and Its Enemies'; when this lecture was criticized severely (and sometimes, to his delight, wittily) he replied in a series of articles which, combined with the lecture, became his best-known statement of his doctrine for society, *Culture and Anarchy*.

The tour of France had one other, rather personal, consequence for his lectures. Arnold's mother was a Cornish woman, and Arnold took considerable pride in his ethnic origins; he christened one of his sons with the Cornish family name Trevenen. This was the country of romance, of Tristram and Iseult, of King Arthur and the Round Table. When Arnold reached Brittany in the spring of 1859, he found Cornwall in a sense reduplicated; in the French Cornouaille Mont-St-Michel faced the English St Michael's Mount, and Finisterre corresponded to Land's End. 'I could not but think of you in Brittany,' he wrote to his mother, 'with Cranics and Trevenecs all about me, and the peasantry with their expressive, rather mournful faces, long noses, and dark eyes, reminding me perpetually of dear Tom [his brother] and Uncle Trevenen, and utterly unlike the French.' He turned to Ernest Renan's essay 'Sur la poésie des races celtiques'—'I have read few things for a long time with more pleasure'—and designed an Oxford lecture on 'The Claim of the Celtic Race, and the Claim of the Christian Religion, to Have Originated Chivalrous Sentiment'. The lecture has vanished, but it was in fact only preliminary to the series of four 'On the Study of Celtic Literature' that he delivered in 1865–6 after he had completed the series that became *Essays in Criticism*; the Celtic lectures made one of his most interesting books, and their influence on Yeats and the poets of the 'Celtic Twilight' is marked. Thereafter also Arnold took a lively interest in the politics of the 'Irish question'.

Arnold's father, as we have noted, had been closely associated with the undogmatic theologians, the 'Noetics', at Oriel College, and Matthew's interest in theological questions was constant throughout his life. From Goethe he early learned to read and value the Dutch Jewish philosopher Spinoza, whose naturalism went hand in hand with a deep sense of spiritual values not unlike those of the Stoics. The course of nature is not set aside by special dispensations; as Arnold put it later, simply, 'Miracles do not happen.' Occasionally a public discussion of the nature of scriptural inspiration like Bishop Colenso's solemn, humourless revelation of mathematical inconsistencies in the Pentateuch moved him to witty reply. 'A Zulu hut in Natal contains on an average only 3½ [persons]', wrote Colenso; 'Half a Zulu?' asked Arnold. The constant association with the Nonconformists in their schools, reading their journals, hearing them speak, led him increasingly to reflect on the nature of Dissent and the validity of its doctrinal justifications, especially in the writings of St Paul on which they professed to base their arguments. When the Oxford lectures came to an end, and the essays on modern English society that were the immediate sequel of the last of them, he felt the need to examine one of the great barriers that divided Englishmen, their disparate concepts of Christianity. Like Thomas Arnold and like Coleridge before him, Matthew believed that the spiritual well-being of the people was as much the business of the State as their intellectual and physical well-being, and that a unified national church was the proper instrument for State action in that direction. But the essays he collected as *St Paul and Protestantism* were only a beginning; St Paul's writings could not really be grasped without an understanding of the gospel preached by Jesus, and the gospel of Jesus led directly to the notion of divine inspiration and the question of the nature of God. (God, in Arnold's view, is not a 'person'; he does not feel 'wrath' or 'love', he does not speak, he does not have sons. God is a transcendental moral ideal that dominates the universe, a force, 'the stream of tendency by which all things seek to fulfil the law of their being'.) *Literature and Dogma* (which started as a series of essays in the *Cornhill* but was halted there by the protests of the subscribers and had to be published entire as a book) and *God and the Bible* occupied Arnold for nearly a decade. Nor was it a wasted decade; *Literature and Dogma* had by far the greatest sale of any of his books, and in many ways presages modern developments in the Christian Church. In fact it completed a course he had marked out for himself intellectually in his earlier poetry: 'Empedocles on Etna' was a metaphoric treatment of the nature of Christ's divinity and of his miracles, founded on the same Stoic philosophy. In a few later essays Arnold amplified ideas he perhaps had not made clear in these longer works. 'A Psychological Parallel' shows how belief in miracles can be so much a part of the intellectual climate of an age that even honest and intelligent people may genuinely believe in them; but the believers' integrity does not make the miracles true. And 'The Church of England' is a persuasive discourse on the State's responsibilities for the spiritual life of its citizens—persuasive, yet perhaps not compatible with the temper of our own times.

The works of Arnold's final decade are quite miscellaneous, as indicated by the titles of the books into which they were collected— *Mixed Essays*, *Irish Essays and Others*, *Discourses in America*. They commonly dealt with aspects of matters he had treated earlier, and often are the fruit of more mature reflection, more polished and assured in style than the earlier work. As frequently happens with literary reputations, the writings with which Arnold first caught the public eye remain the best known, the phrases he coined in his earlier works—and he loved to deal in catch-phrases—are often taken to be the essence of all he had to say. But the maturity and balance of the later essays, generally without the old catchwords, are usually more pleasing and more convincing. 'Equality', a maturer sequel to 'Democracy', reminds us with its title of Arnold's continued conviction that the ideals of the French Revolution, 'Liberty, Equality, Fraternity', showed the direction in which the world was moving, and must move. The ideas of 'The Function of Criticism' retain their value, but the approach to a practical criticism of literature in 'A

French Critic on Milton' comes as a pleasant surprise: we learn how to judge critics as well as literature, and if the varieties of critical approach are not now quite what they were when Arnold wrote, the ideal criticism for lovers of literature is not very different.

One kind of enterprise occupied Arnold's attention in this last decade that brought together his love of poetry and his pedagogical instincts—the editing of selections from some of his favourite authors in order to give them a wider audience both amongst the general public and in the schools. His editions of the two Isaiahs reflected his religious concern, of some of Burke's letters and speeches on Irish affairs his political concern, of six of Dr Johnson's *Lives of the Poets* his critical concern (forecast in 'A French Critic on Milton'), but best known are the prefaces he wrote for his selections from the poetry of Wordsworth and of Byron. When his niece's husband, Humphry Ward, conceived a four-volume, easily readable, selection from the works of the English poets at large, Arnold wrote brief introductions to the selections from Gray and Keats, but most notably the general introduction to the anthology, in which he attempted to face the question the layman always asks—'How can I tell the good from the bad?' Arnold devised his now famous touchstone method (which he had foreshadowed in his Oxford lectures on translating Homer): 'Have in your mind passages from some of the world's greatest poetry, and see if the work you are judging has the ring of the same quality.' Though intended as advice for the general reader, the touchstone method was not easy to apply, since at its fullest it required a grasp not of the line or two Arnold quoted, but of a whole context in a great (and usually long) poem. But it had the virtue of directing attention to the poetry itself—its language, its concept, its emotion. This Introduction, 'The Study of Poetry', with the introductions to Wordsworth, Byron, Gray, and Keats, made the principal contents of Arnold's last (indeed, as it befell, posthumous) book, the second series of *Essays in Criticism*.

There was also a retrospective aspect about Arnold's later work. As a schoolboy he won praise for reciting a passage from Byron's *Marino Faliero*; he knew Wordsworth personally; he talked passionately of Keats with Clough at Oxford and some of his best earlier poetry was Keatsian. One author he came close to editing (but the scheme fell through) was Emerson, whom he had met and whom he regarded (with Carlyle, Newman, and Goethe) as one of the great formative influences on his life during his Oxford years. His lecture tour of America in the year after Emerson's death gave him the opportunity to repay that debt and to look back somewhat nostalgically on those years when he was learning to be a poet, when 'life ran gaily as the sparkling Thames'. Goethe, whose published conversations, letters, apophthegms, and autobiography had been of such importance to the young Arnold, was called back in an essay that approached its subject indirectly, with the modern French critic Edmond Scherer (whom Arnold knew) as intermediary.

It was Arnold's love of debate, his witty polemicism, that led him to pick up the challenge thrown out by his friend Thomas Henry Huxley that the movement of the age was in the direction of science, not 'culture'; Arnold took the challenge as an invitation to formulate in a single lecture his theory of the nature and value of education. It is today no longer 'science' in Huxley's sense that threatens humane learning: 'science' has become too difficult and too specialized. But one might read this essay on 'Literature and Science' as a comparison of the humanities with quantified 'social science', or even more currently, remembering Arnold's and Carlyle's hatred of 'machinery', with our devotion to an all-too-frequently unthinking use of the computer.

The lecture tour of America gave Arnold the chance he had long wanted—the chance to see how a nation founded a century earlier on the principles of the French Enlightenment was faring. His experiences in the United States were sometimes amusing, but not unpleasant, and the last lecture/essay he wrote, 'Civilisation in the United States', was a kind of field report on the workings of the sociological ideas that had been at the heart of 'Democracy', *Culture and Anarchy*, and 'Equality'. It was an appropriate close to a career abruptly curtailed at the age of sixty-five by a heart attack in Liverpool, where he had gone to greet the arrival of the daughter who had married an American.

The essays and lectures collected here are intended to show both the range of Arnold's interests (education, politics, religion, and literary criticism) and the interrelatedness of his treatments of these subjects. Indeed, the frequent repetition of key phrases and allusions in different contexts suggests how much of a piece Arnold's writing on these matters was. Each essay in a sense epitomizes his mature thinking on its subject; most of them date from the last dozen years of his life. And each of them has an intellectual value as great, or nearly as great, in the late twentieth century as it had when first published a century or more ago. Though (at least in America) he had the reputation of being a conspicuously bad lecturer, more than half the essays included here were first delivered from the lecture platform.

R.H.S.

CHRONOLOGY

1822   Born (24 December) at Laleham-on-Thames as the eldest son of the Reverend Thomas Arnold and Mary Arnold (née Penrose).

1828   The family move to Rugby on Dr Arnold's appointment as Headmaster of Rugby School.

1831   Arnold is tutored by his uncle, the Reverend John Buckland, at Laleham (January to December) and makes his first visit to the Lake District (August).

1832   Tennyson's *Poems* published.

1833   Southey's cousin, Herbert Hill, engaged by Dr Arnold as tutor for his sons (May). John Keble preaches his 'National Apostasy' sermon; beginning of the Oxford Movement. Carlyle's *Sartor Resartus* published.

1834   Fox How, near Ambleside, completed (July) and from now on becomes the holiday home of the Arnolds in the Lakes. Wordsworth is a neighbour and frequent visitor.

1836   Arnold's first attempts at writing verse, including 'The First Sight of Italy' (March) and 'Lines written on the seashore at Eaglehurst' (July). With his brother Tom enters Winchester College (his father's old school) as a Commoner.

1837   Arnold wins a school verse-speaking prize with a speech from Byron's *Marino Faliero* and makes his first visit to France (August); leaves Winchester and enters the Fifth Form at Rugby. Publication of Carlyle's *The French Revolution* and *Lectures on German Literature*.

1838   First number of the manuscript *Fox How Magazine* (January); brought out twice yearly by Arnold with Tom's help until January 1842. He wins the Fifth Form prize for Latin verse and removes to the Sixth Form under his father. Publication of Dr Arnold's *Early History of Rome* (concluded 1843).

1839   Walter Pater born. Publication of Carlyle's *Chartism*.

1840   Arnold wins school prize for English essay and English verse (June) and his prize poem, 'Alaric at Rome', printed at Rugby; gains open scholarship to Balliol College, Oxford.

1841   Arnold shares school prizes for Latin essay and Latin verse (June); his father appointed Regius Professor of Modern History at Oxford (August). He goes into residence at Oxford, when his close friendship with Arthur Hugh Clough begins. He deeply admires Newman's preaching at St Mary's but is not drawn into the Oxford Movement. Publication of Newman's 'Tract 90'; Carlyle's

On Heroes, Hero-Worship and the Heroic in History; Emerson's *Essays: First Series* (Second Series 1844).

1842
Arnold *proxime accessit* for Hertford Latin Scholarship (March). Sudden death of his father of heart disease (12 June). Publication of Dr Arnold's *Study of Modern History*.

1842–5
Arnold reads and is influenced by Carlyle, Emerson, George Sand, Goethe and Spinoza. Member of the 'Decade' undergraduate society.

1843
Arnold's Newdigate prize poem *Cromwell* printed. Wordsworth Poet Laureate. Publication of Carlyle's *Past and Present*.

1844
A. P. Stanley's *Life and Correspondence of Thomas Arnold* published. Arnold obtains BA Second Class in 'Greats' (November).

1845
Arnold appointed temporary assistant master at Rugby (February–April) and elected Fellow of Oriel College, Oxford (March). Publication of Carlyle's *Cromwell*.

1846
Arnold visits France; meets George Sand at Nohant (July) and sees Rachel act in Paris (December), where he stays until February 1847. Probably begins his close reading of Senancour and Sainte-Beuve at this time.

1847
Arnold becomes Private Secretary to Lord Lansdowne, Lord President of the Council and Whig elder stateman (April). Tom Arnold emigrates to New Zealand in search of 'Liberty, Equality, and Fraternity' (November).

1848
Arnold's brother William Delafield leaves Oxford for India as ensign in the Bengal Army of the East India Company (February). Arnold himself visits Switzerland and meets 'Marguerite' at Thun (September). Clough writes 'The Bothie of Toper-na-Fuosich' (November), after having resigned his fellowship at Oriel College because of scruples about religious subscription.

1849
Arnold publishes his first volume, *The Strayed Reveller, and Other Poems* (February). He visits Switzerland and meets 'Marguerite' for the second and last time (September).

1850
Death of Wordsworth (23 April); Arnold publishes 'Memorial Verses' in *Fraser's Magazine*. His favourite sister Jane marries W. E. Forster (August), but his own engagement to Frances Lucy Wightman is delayed because of his need to find a settled source of livelihood. Meets Charlotte Brontë at Fox How. Tennyson publishes *In Memoriam* and made Poet Laureate. Publication of Wordsworth's *The Prelude*; Carlyle's *Latter-Day Pamphlets*; Emerson's *Representative Men*.

1851
Arnold appointed Inspector of Schools (15 April), marries Frances Lucy, daughter of Sir William Wightman, Justice of the Queen's Bench, at Hampton (10 June) and goes on a delayed honeymoon journey in France, Italy, and Switzerland, during which he visits

the Grande Chartreuse (September–October). Begins work as a school inspector (11 October), and from now on is committed to a heavy programme of work and constant travelling as an inspector and, for some years, marshal to his father-in-law on circuit.

1852   Arnold publishes *Empedocles on Etna, and Other Poems* (October).

1853   Arnold publishes *Poems. A New Edition*, a selection of his poems, excluding 'Empedocles on Etna' and including among new poems 'Sohrab and Rustum' and 'The Scholar-Gipsy'.

1854   Arnold publishes *Poems. Second Series*, a further selection from his two earlier volumes, with 'Balder Dead' as the single important new poem. (Title-page dated 1855.)

1855   'Stanzas from the Grande Chartreuse' and 'Haworth Churchyard' published in *Fraser's Magazine* (April, May).

1856   Publication of Carlyle's *Collected Works*; Emerson's *English Traits*.

1857   Arnold elected Professor of Poetry at Oxford (May) and delivers his Inaugural Lecture 'On the Modern Element in Literature' (14 November; creates a precedent by lecturing in English instead of Latin; re-elected (1862) at the end of his first term of five years); publishes *Merope* (December).

1858   Arnold settles in London at 2 Chester Square (February: 'it will be something to unpack one's portmanteau for the first time since I was married, now nearly seven years ago'). Takes a walking holiday in Switzerland with Theodore Walrond (August–September). Publication of Carlyle's *Frederick the Great*; Goethe's *Poems and Ballads*, translated by Aytoun and Martin; Clough's 'Amours de Voyage'.

1859   Arnold visits France, Holland, and Switzerland as Foreign Assistant Commissioner to the Newcastle Commission on Elementary Education (March–August); his brother William dies at Gibraltar (April); his *England and the Italian Question* published (August).

1860   *Cornhill Magazine* started, with Thackeray as editor (until 1862).

1861   Arnold publishes *On Translating Homer* (January) and *The Popular Education of France* with the introductory essay 'Democracy' (November). Clough dies at Florence. Publication of Palgrave's *Golden Treasury* (revised 1896).

1862   Arnold risks official hostility by publishing in *Fraser's Magazine* 'The Twice-Revised Code', attacking Robert Lowe's 'Payment by Results' as a method of distributing government grants for education (March); also publishes *On Translating Homer: Last Words*. Publication of Clough's *Collected Poems* with memoir by Palgrave.

1864   Arnold publishes *A French Eton*. From now on most of his work appears in periodicals before being published in book form.

1865   Arnold publishes *Essays in Criticism: First Series* (February); visits France, Italy, Germany, and Switzerland as Foreign Assistant Commissioner to the Taunton Commission (Schools Inquiry).

1866   Arnold applies unsuccessfully for the post of Charity Commissioner (March); publishes 'Thyrsis', his elegy on Clough, in *Macmillan's Magazine*.

1867   Arnold applies unsuccessfully for the Librarianship of the House of Commons (April). Publishes *Celtic Literature, New Poems* (July; restoring 'Empedocles on Etna' at Browning's request). From now on writes little verse, and is increasingly known for his controversial social and religious writings. Publication of Carlyle's *Shooting Niagara (Macmillan's Magazine)*; Pater's *Essay on Winckelmann (Westminster Review)*.

1868   Arnold loses two of his sons, his infant son Basil (January) and his eldest son Thomas, aged sixteen and a Harrow schoolboy (November). Moves to Byron House, Harrow (March).

1869   Arnold publishes *Culture and Anarchy*, his major work of social criticism (January), the first collected edition of his *Poems* (two volumes, June); his essay on 'Obermann' in the *Academy* (October), and, after the death of Sainte-Beuve, his commemorative essay also in the *Academy* (November). Applies unsuccessfully for appointment as one of the three commissioners under the Endowed Schools Act. Publication of Clough's *Poems and Prose Remains, with Selection from his Letters*, ed. by his wife.

1870   Arnold publishes *St Paul and Protestantism* (May), receives the Honorary Degree of DCL at Oxford (June) and is promoted Senior Inspector of Schools.

1871   Arnold publishes *Friendship's Garland*, his 'half serious, half playful' letters on English life and culture (February) and visits France and Switzerland with his wife and his son Richard (August).

1872   Arnold loses a third son, Trevenen William, aged eighteen (February).

1873   Arnold publishes *Literature and Dogma*, his most important work on religion (February), takes a holiday leave from school inspection with his wife in Italy (February–May) and moves to Pains Hill Cottage, Cobham, Surrey. His mother dies at Fox How (September). Publication of Pater's *Studies in the Renaissance*.

1875   Arnold publishes *God and the Bible*, reviewing objections to *Literature and Dogma* (November).

1876   George Sand dies in June; Arnold reprints 'The New Sirens' in *Macmillan's Magazine* (December). Publication of Thomas Arnold's edition of *Beowulf*.

1877   Arnold declines renomination for the Professorship of Poetry at Oxford (February) and nomination for the Lord Rectorship of St

Andrews University (November). He publishes *Last Essays on Church and Religion* and 'George Sand' in *The Fortnightly Review* (June). W. H. Mallock portrays Arnold as 'Mr Luke' in his *The New Republic* (April–May).

1878   *Selected Poems of Matthew Arnold* (Golden Treasury Series) published (June).

1879   Arnold publishes *Mixed Essays* (*c*. March) and his selected *Poems of Wordsworth* (August).

1880   Arnold attends the reception in London given in honour of Cardinal Newman by the Duke of Norfolk 'because I wanted to have spoken once in my life to Newman' (12 May). He contributes three essays to T. H. Ward's *The English Poets*: Introduction (later called 'On the Study of Poetry'), 'Thomas Gray', and 'John Keats'. Holiday in Switzerland and Italy (September). His brother-in-law, W. E. Forster, appointed Chief Secretary for Ireland by Gladstone (resigning 1882, shortly before the Phoenix Park Murders).

1881   Arnold publishes his selected *Poetry of Byron* (June).

1882   Arnold publishes 'Westminster Abbey', his elegy on Stanley, in the *Nineteenth Century* (January), and *Irish Essays*.

1883   Arnold accepts Civil List Pension of £250 a year 'in public recognition of service to the poetry and literature of England' (August). Begins his lecture tour of the USA (October to March 1884).

1884   Arnold becomes Chief Inspector of Schools.

1885   Arnold publishes *Discourses in America* (June) and his three-volume collected edition of poems (Library Edition *c*. August). He again declines renomination for the Professorship of Poetry in spite of a memorial from Oxford heads of colleges and another from four hundred undergraduates (October: 'Everyone is very kind as one grows old'). Visits Germany for the Education Department (November–December). Publication of Pater's *Marius the Epicurean*.

1886   Arnold abroad again in France, Switzerland and Germany for the Education Department (February–March). He retires from Inspectorship of Schools (30 April) and makes his second visit to USA (May–August).

1887   Publication of Carlyle's *Early Letters* and *Correspondence with Goethe*, ed. Norton.

1888   Arnold dies suddenly of heart failure at Liverpool while awaiting the arrival of his married daughter from America (15 April). His *Essays in Criticism. Second Series* published posthumously (November).

NOTE ON THE TEXT

The text of the poetry is from *Poetical Works of Matthew Arnold*, edited by C. B. Tinker and H. F. Lowry (Oxford, 1950).

M.A.

The text of the prose is that of my edition of Arnold's *Complete Prose Works* (11 vols., Ann Arbor, Mich., 1960–77: referred to in the notes as *CPW*), used by kind permission of the University of Michigan Press. Footnotes cued by superior figures and printed at the bottom of a page of text are Arnold's own.

R.H.S.

The degree sign (°) indicates a note at the end of the book. More general notes and headnotes are not cued.

Mycerinus

'Not by the justice that my father spurn'd,
Not for the thousands whom my father slew,
Altars unfed and temples overturn'd,
Cold hearts and thankless tongues, where thanks are due;
Fell this dread voice from lips that cannot lie,
Stern sentence of the Powers of Destiny.

'I will unfold my sentence and my crime.
My crime—that, rapt in reverential awe,
I sate obedient, in the fiery prime
Of youth, self-govern'd, at the feet of Law;°   10
Ennobling this dull pomp, the life of kings,
By contemplation of diviner things.

'My father loved injustice, and lived long;
Crown'd with gray hairs he died, and full of sway.
I loved the good he scorn'd, and hated wrong—
The Gods declare my recompense to-day.
I look'd for life more lasting, rule more high;
And when six years are measured, lo, I die!

'Yet surely, O my people, did I deem
Man's justice from the all-just Gods was given;   20
A light that from some upper fount did beam,
Some better archetype, whose seat was heaven;
A light that, shining from the blest abodes,
Did shadow somewhat of the life of Gods.

'Mere phantoms of man's self-tormenting heart,
Which on the sweets that woo it dares not feed!
Vain dreams, which quench our pleasures, then depart,
When the duped soul, self-master'd, claims its meed;
When, on the strenuous just man, Heaven bestows,
Crown of his struggling life, an unjust close!   30

'Seems it so light a thing, then, austere Powers,
To spurn man's common lure, life's pleasant things?
Seems there no joy in dances crown'd with flowers,
Love, free to range, and regal banquetings?

Bend ye on these, indeed, an unmoved eye,
Not Gods but ghosts, in frozen apathy?

'Or is it that some Force, too wise, too strong,
Even for yourselves to conquer or beguile,
Sweeps earth, and heaven, and men, and gods along,
Like the broad volume of the insurgent Nile?   40
And the great powers we serve, themselves may be
Slaves of a tyrannous necessity?

'Or in mid-heaven, perhaps, your golden cars,
Where earthly voice climbs never, wing their flight,
And in wild hunt, through mazy tracts of stars,
Sweep in the sounding stillness of the night?
Or in deaf ease, on thrones of dazzling sheen,
Drinking deep draughts of joy, ye dwell serene?

'Oh, wherefore cheat our youth, if thus it be,
Of one short joy, one lust, one pleasant dream?   50
Stringing vain words of powers we cannot see,
Blind divinations of a will supreme;
Lost labour! when the circumambient gloom
But hides, if Gods, Gods careless of our doom?

'The rest I give to joy. Even while I speak,
My sand runs short; and—as yon star-shot ray,
Hemm'd by two banks of cloud, peers pale and weak,
Now, as the barrier closes, dies away—
Even so do past and future intertwine,
Blotting this six years' space, which yet is mine.   60

'Six years—six little years—six drops of time!
Yet suns shall rise, and many moons shall wane,
And old men die, and young men pass their prime,
And languid pleasure fade and flower again,
And the dull Gods behold, ere these are flown,
Revels more deep, joy keener than their own.

'Into the silence of the groves and woods
I will go forth; though something would I say—
Something—yet what, I know not; for the Gods
The doom they pass revoke not, nor delay;   70

And prayers, and gifts, and tears, are fruitless all,
And the night waxes, and the shadows fall.

'Ye men of Egypt, ye have heard your king!
I go, and I return not. But the will
Of the great Gods is plain; and ye must bring
Ill deeds, ill passions, zealous to fulfil
Their pleasure, to their feet; and reap their praise,
The praise of Gods, rich boon! and length of days.'

—So spake he, half in anger, half in scorn;
And one loud cry of grief and of amaze   80
Broke from his sorrowing people; so he spake,
And turning, left them there; and with brief pause,
Girt with a throng of revellers, bent his way
To the cool region of the groves he loved.
There by the river-banks he wander'd on,
From palm-grove on to palm-grove, happy trees,
Their smooth tops shining sunward, and beneath
Burying their unsunn'd stems in grass and flowers;
Where in one dream the feverish time of youth
Might fade in slumber, and the feet of joy   90
Might wander all day long and never tire.
Here came the king, holding high feast, at morn,
Rose-crown'd; and ever, when the sun went down,
A hundred lamps beam'd in the tranquil gloom,
From tree to tree all through the twinkling grove,
Revealing all the tumult of the feast—
Flush'd guests, and golden goblets foam'd with wine;
While the deep-burnish'd foliage overhead
Splinter'd the silver arrows of the moon.
    It may be that sometimes his wondering soul   100
From the loud joyful laughter of his lips
Might shrink half startled, like a guilty man°
Who wrestles with his dream; as some pale shape
Gliding half hidden through the dusky stems,
Would thrust a hand before the lifted bowl,
Whispering: *A little space, and thou art mine!*
It may be on that joyless feast his eye°
Dwelt with mere outward seeming; he, within,
Took measure of his soul, and knew its strength,
And by that silent knowledge, day by day,   110

Was calm'd, ennobled, comforted, sustain'd.
It may be; but not less his brow was smooth,
And his clear laugh fled ringing through the gloom,
And his mirth quail'd not at the mild reproof
Sigh'd out by winter's sad tranquillity;
Nor, pall'd with its own fulness, ebb'd and died
In the rich languor of long summer-days;
Nor wither'd when the palm-tree plumes, that roof'd
With their mild dark his grassy banquet-hall,
Bent to the cold winds of the showerless spring;   120
No, nor grew dark when autumn brought the clouds.
   So six long years he revell'd, night and day.
And when the mirth wax'd loudest, with dull sound
Sometimes from the grove's centre echoes came,
To tell his wondering people of their king;
In the still night, across the steaming flats,
Mix'd with the murmur of the moving Nile.

Stagirius

Thou, who dost dwell alone—
Thou, who dost know thine own—
Thou, to whom all are known
From the cradle to the grave—
      Save, oh! save.
From the world's temptations,
   From tribulations,
From that fierce anguish
Wherein we languish,
From that torpor deep   10
Wherein we lie asleep,
Heavy as death, cold as the grave,
      Save, oh! save.

When the soul, growing clearer,
   Sees God no nearer;
When the soul, mounting higher,
   To God comes no nigher;
But the arch-fiend Pride
Mounts at her side,

Foiling her high emprise,   20
Sealing her eagle eyes,
And, when she fain would soar,
Makes idols to adore,
Changing the pure emotion
Of her high devotion,
To a skin-deep sense
Of her own eloquence;
Strong to deceive, strong to enslave—
    Save, oh! save.

From the ingrain'd fashion   30
Of this earthly nature
That mars thy creature;
From grief that is but passion,
From mirth that is but feigning,
From tears that bring no healing,
From wild and weak complaining,
    Thine old strength revealing,
        Save, oh! save.
From doubt, where all is double;
Where wise men are not strong,   40
Where comfort turns to trouble,
Where just men suffer wrong;
Where sorrow treads on joy,
Where sweet things soonest cloy,
Where faiths are built on dust,
Where love is half mistrust,
Hungry, and barren, and sharp as the sea—
    Oh! set us free.
O let the false dream fly,
Where our sick souls do lie   50
    Tossing continually!
        O where thy voice doth come
            Let all doubts be dumb,
            Let all worlds be mild,
            All strifes be reconciled,
            All pains beguiled!
            Light bring no blindness,
            Love no unkindness,

Knowledge no ruin,
Fear no undoing!   60
From the cradle to the grave,
Save, oh! save.

The Voice

As the kindling glances,
    Queen-like and clear,
Which the bright moon lances
    From her tranquil sphere
At the sleepless waters
    Of a lonely mere,
On the wild whirling waves, mournfully, mournfully,
        Shiver and die.

As the tears of sorrow
    Mothers have shed—   10
Prayers that to-morrow
    Shall in vain be sped
When the flower they flow for
    Lies frozen and dead—
Fall on the throbbing brow, fall on the burning breast,
        Bringing no rest.

Like bright waves that fall
    With a lifelike motion
On the lifeless margin of the sparkling Ocean;
A wild rose climbing up a mouldering wall—   20
A gush of sunbeams through a ruin'd hall—
Strains of glad music at a funeral—
        So sad, and with so wild a start
        To this deep-sober'd heart,
        So anxiously and painfully,
        So drearily and doubtfully,
And oh, with such intolerable change
        Of thought, such contrast strange,
O unforgotten voice, thy accents come,
Like wanderers from the world's extremity,   30
        Unto their ancient home!

In vain, all, all in vain,
They beat upon mine ear again,
Those melancholy tones so sweet and still.
Those lute-like tones which in the bygone year
    Did steal into mine ear—
Blew such a thrilling summons to my will,
    Yet could not shake it;
Made my tost heart its very life-blood spill,
    Yet could not break it.   40

A Question

TO FAUSTA

Joy comes and goes, hope ebbs and flows
    Like the wave;
Change doth unknit the tranquil strength of men.
    Love lends life a little grace,
    A few sad smiles; and then,
    Both are laid in one cold place,
      In the grave.

Dreams dawn and fly, friends smile and die
    Like spring flowers;
Our vaunted life is one long funeral.   10
    Men dig graves with bitter tears
    For their dead hopes; and all,
    Mazed with doubts and sick with fears,
      Count the hours.

We count the hours! These dreams of ours,
    False and hollow,
Do we go hence and find they are not dead?
    Joys we dimly apprehend,
    Faces that smiled and fled,
    Hopes born here, and born to end,   20
      Shall we follow?

Shakespeare

Others abide our question. Thou art free.
We ask and ask—Thou smilest and art still,
Out-topping knowledge. For the loftiest hill,
Who to the stars uncrowns his majesty,

Planting his steadfast footsteps in the sea,°
Making the heaven of heavens his dwelling-place,
Spares but the cloudy border of his base
To the foil'd searching of mortality;

And thou, who didst the stars and sunbeams know,
Self-school'd, self-scann'd, self-honour'd, self-secure,   10
Didst tread on earth unguess'd at.—Better so!

All pains the immortal spirit must endure,
All weakness which impairs, all griefs which bow,
Find their sole speech in that victorious brow.

Written in Emerson's Essays

'O monstrous, dead, unprofitable world,°
That thou canst hear, and hearing, hold thy way!
A voice oracular hath peal'd to-day,
To-day a hero's banner is unfurl'd;

'Hast thou no lip for welcome?'—So I said.
Man after man, the world smiled and pass'd by;
A smile of wistful incredulity
As though one spake of life unto the dead—

Scornful, and strange, and sorrowful, and full
Of bitter knowledge. Yet the will is free;   10
Strong is the soul, and wise, and beautiful;

The seeds of godlike power are in us still;
Gods are we, bards, saints, heroes, if we will!—
Dumb judges, answer, truth or mockery?°

In Harmony with Nature

TO A PREACHER

'In harmony with Nature?' Restless fool,
Who with such heat dost preach what were to thee,
When true, the last impossibility—
To be like Nature strong, like Nature cool!

Know, man hath all which Nature hath, but more,
And in that *more* lie all his hopes of good.
Nature is cruel, man is sick of blood;
Nature is stubborn, man would fain adore;

Nature is fickle, man hath need of rest;
Nature forgives no debt, and fears no grave;   10
Man would be mild, and with safe conscience blest.

Man must begin, know this, where Nature ends;
Nature and man can never be fast friends.°
Fool, if thou canst not pass her, rest her slave!

In Utrumque Paratus

If, in the silent mind of One all-pure,
    At first imagined lay
The sacred world; and by procession sure
From those still deeps, in form and colour drest,
Seasons alternating, and night and day,
The long-mused thought to north, south, east, and west,
    Took then its all-seen way;

O waking on a world which thus-wise springs!
    Whether it needs thee count
Betwixt thy waking and the birth of things   10
Ages or hours—O waking on life's stream!
By lonely pureness to the all-pure fount
(Only by this thou canst) the colour'd dream
    Of life remount!

Thin, thin the pleasant human noises grow,
    And faint the city gleams;
Rare the lone pastoral huts—marvel not thou!
The solemn peaks but to the stars are known,
But to the stars, and the cold lunar beams;
Alone the sun arises, and alone   20
    Spring the great streams.

But, if the wild unfather'd mass no birth
    In divine seats hath known;
In the blank, echoing solitude if Earth,
Rocking her obscure body to and fro,
Ceases not from all time to heave and groan,
Unfruitful oft, and at her happiest throe
    Forms, what she forms, alone;

O seeming sole to awake, thy sun-bathed head
    Piercing the solemn cloud   30
Round thy still dreaming brother-world outspread!°
O man, whom Earth, thy long-vext mother, bare
Not without joy—so radiant, so endow'd
(Such happy issue crown'd her painful care)—
    Be not too proud!

Oh when most self-exalted, most alone,
    Chief dreamer, own thy dream!
Thy brother-world stirs at thy feet unknown,
Who hath a monarch's hath no brother's part;
Yet doth thine inmost soul with yearning teem.   40
—Oh, what a spasm shakes the dreamer's heart!
    'I, too, but seem.'

The New Sirens

In the cedarn shadow sleeping,
Where cool grass and fragrant glooms
Forth at noon had lured me, creeping
From your darken'd palace rooms—

I, who in your train at morning
Stroll'd and sang with joyful mind,
Heard, in slumber, sounds of warning;
Heard the hoarse boughs labour in the wind.

Who are they, O pensive Graces,
—For I dream'd they wore your forms—   10
Who on shores and sea-wash'd places
Scoop the shelves and fret the storms?
Who, when ships are that way tending,
Troop across the flushing sands,
To all reefs and narrows wending,
With blown tresses, and with beckoning hands?

Yet I see, the howling levels
Of the deep are not your lair;
And your tragic-vaunted revels
Are less lonely than they were.   20
Like those Kings with treasure steering
From the jewell'd lands of dawn,
Troops, with gold and gifts, appearing,
Stream all day through your enchanted lawn.

And we too, from upland valleys,
Where some Muse with half-curved frown
Leans her ear to your mad sallies
Which the charm'd winds never drown;
By faint music guided, ranging
The scared glens, we wander'd on,   30
Left our awful laurels hanging,°
And came heap'd with myrtles to your throne.

From the dragon-warder'd fountains
Where the springs of knowledge are,
From the watchers on the mountains,
And the bright and morning star;
We are exiles, we are falling,
We have lost them at your call—
O ye false ones, at your calling
Seeking ceiled chambers and a palace-hall!°   40

Are the accents of your luring
More melodious than of yore?
Are those frail forms more enduring
Than the charms Ulysses bore?°
That we sought you with rejoicings,
Till at evening we descry
At a pause of Siren voicings
These vext branches and this howling sky? . . .

*   *   *   *

Oh, your pardon! The uncouthness
Of that primal age is gone,   50
And the skin of dazzling smoothness
Screen not now a heart of stone.
Love has flush'd those cruel faces;
And those slacken'd arms forgo
The delight of death-embraces,
And yon whitening bone-mounds do not grow.

'Ah,' you say; 'the large appearance
Of man's labour is but vain,
And we plead as staunch adherence
Due to pleasure as to pain.'   60
Pointing to earth's careworn creatures,
'Come,' you murmur with a sigh:
'Ah! we own diviner features,
Loftier bearing, and a prouder eye.

'Come,' you say, 'the hours were dreary;
Dull did life in torpor fade;
Time is lame, and we grew weary
In the slumbrous cedarn shade.
Round our hearts with long caresses,
With low sighings, Silence stole,   70
And her load of steaming tresses
Fell, like Ossa, on the climbing soul.

'Come,' you say, 'the soul is fainting
Till she search and learn her own,
And the wisdom of man's painting
Leaves her riddle half unknown.

Come,' you say, 'the brain is seeking,
While the sovran heart is dead;
Yet this glean'd, when Gods were speaking,
Rarer secrets than the toiling head.   80

'Come,' you say, 'opinion trembles,
Judgment shifts, convictions go;
Life dries up, the heart dissembles—
Only, what we feel, we know.
Hath your wisdom felt emotions?
Will it weep our burning tears?
Hath it drunk of our love-potions
Crowning moments with the wealth of years?'

—I am dumb. Alas, too soon all
Man's grave reasons disappear!   90
Yet, I think, at God's tribunal
Some large answer you shall hear.
But, for me, my thoughts are straying
Where at sunrise, through your vines,
On these lawns I saw you playing,
Hanging garlands on your odorous pines;

When your showering locks enwound you,
And your heavenly eyes shone through;
When the pine-boughs yielded round you,
And your brows were starr'd with dew;   100
And immortal forms, to meet you,
Down the statued alleys came,
And through golden horns, to greet you,
Blew such music as a God may frame.

Yes, I muse! And if the dawning
Into daylight never grew,
If the glistering wings of morning
On the dry noon shook their dew,
If the fits of joy were longer,
Or the day were sooner done,   110
Or, perhaps, if hope were stronger,
No weak nursling of an earthly sun . . .

Pluck, pluck cypress, O pale maidens,
    Dusk the hall with yew!

      *   *   *   *

For a bound was set to meetings,
And the sombre day dragg'd on;
And the burst of joyful greetings,
And the joyful dawn, were gone.
For the eye grows fill'd with gazing,
And on raptures follow calms;   120
And those warm locks men were praising,
Droop'd, unbraided, on your listless arms.

Storms unsmooth'd your folded valleys,
And made all your cedars frown;
Leaves were whirling in the alleys
Which your lovers wander'd down.
—Sitting cheerless in your bowers,
The hands propping the sunk head,
Still they gall you, the long hours,
And the hungry thought, that must be fed!   130

Is the pleasure that is tasted
Patient of a long review?
Will the fire joy hath wasted,
Mused on, warm the heart anew?
—Or, are those old thoughts returning,
Guests the dull sense never knew,
Stars, set deep, yet inly burning,
Germs, your untrimm'd passion overgrew?

Once, like us, you took your station
Watchers for a purer fire;   140
But you droop'd in expectation,
And you wearied in desire.
When the first rose flush was steeping
All the frore peak's awful crown,
Shepherds say, they found you sleeping
In some windless valley, farther down.

Then you wept, and slowly raising
Your dozed eyelids, sought again,
Half in doubt, they say, and gazing
Sadly back, the seats of men;—   150
Snatch'd a turbid inspiration
From some transient earthly sun,
And proclaim'd your vain ovation
For those mimic raptures you had won . . .

*   *   *   *

With a sad, majestic motion,
With a stately, slow surprise,
From their earthward-bound devotion
Lifting up your languid eyes—
Would you freeze my too loud boldness,
Dumbly smiling as you go,   160
One faint frown of distant coldness
Flitting fast across each marble brow?

Do I brighten at your sorrow,
O sweet Pleaders?—doth my lot
Find assurance in to-morrow
Of one joy, which you have not?
O, speak once, and shame my sadness!
Let this sobbing, Phrygian strain,
Mock'd and baffled by your gladness,
Mar the music of your feasts in vain!   170

*   *   *   *

Scent, and song, and light, and flowers!
Gust on gust, the harsh winds blow—
Come, bind up those ringlet showers!
Roses for that dreaming brow!
Come, once more that ancient lightness,
Glancing feet, and eager eyes!
Let your broad lamps flash the brightness
Which the sorrow-stricken day denies!

Through black depths of serried shadows,
　Up cold aisles of buried glade;　　　　　　　　180
In the midst of river-meadows
　Where the looming kine are laid;
　From your dazzled windows streaming,
　　From your humming festal room,
　Deep and far, a broken gleaming
Reels and shivers on the ruffled gloom.

Where I stand, the grass is glowing;
　Doubtless you are passing fair!
But I hear the north wind blowing,
　And I feel the cold night-air.　　　　　　　　190
　Can I look on your sweet faces,
　　And your proud heads backward thrown,
　From this dusk of leaf-strewn places
With the dumb woods and the night alone?

Yet, indeed, this flux of guesses—
　Mad delight, and frozen calms—
Mirth to-day and vine-bound tresses,
　And to-morrow—folded palms;
　Is this all? this balanced measure?
　　Could life run no happier way?　　　　　　200
　Joyous, at the height of pleasure,
Passive at the nadir of dismay?

But, indeed, this proud possession,
　This far-reaching, magic chain,
Linking in a mad succession
　Fits of joy and fits of pain—
　Have you seen it at the closing?
　　Have you track'd its clouded ways?
　Can your eyes, while fools are dozing,
Drop, with mine, adown life's latter days?　　210

When a dreary dawn is wading
　Through this waste of sunless greens,
When the flushing hues are fading
　On the peerless cheek of queens;

When the mean shall no more sorrow,
And the proudest no more smile;
As old age, youth's fatal morrow,
Spreads its cold light wider all that while?

Then, when change itself is over,
When the slow tide sets one way,   220
Shall you find the radiant lover,
Even by moments, of to-day?
The eye wanders, faith is failing—
O, loose hands, and let it be!
Proudly, like a king bewailing,
O, let fall one tear, and set us free!

All true speech and large avowal
Which the jealous soul concedes;
All man's heart which brooks bestowal,
All frank faith which passion breeds—   230
These we had, and we gave truly;
Doubt not, what we had, we gave!
False we were not, nor unruly;
Lodgers in the forest and the cave.

Long we wander'd with you, feeding
Our rapt souls on your replies,
In a wistful silence reading
All the meaning of your eyes.
By moss-border'd statues sitting,
By well-heads, in summer days.   240
But we turn, our eyes are flitting—
See, the white east, and the morning rays!

And you too, O worshipp'd Graces,
Sylvan Gods of this fair shade!
Is there doubt on divine faces?
Are the blessed Gods dismay'd?
Can men worship the wan features,
The sunk eyes, the wailing tone,
Of unsphered, discrownéd creatures,
Souls as little godlike as their own?   250

Come, loose hands! The wingéd fleetness
Of immortal feet is gone;
And your scents have shed their sweetness,
And your flowers are overblown.
And your jewell'd gauds surrender
Half their glories to the day;
Freely did they flash their splendour,
Freely gave it—but it dies away.

In the pines the thrush is waking—
Lo, yon orient hill in flames!   260
Scores of true love knots are breaking
At divorce which it proclaims.
When the lamps are paled at morning,
Heart quits heart and hand quits hand.
Cold in that unlovely dawning,
Loveless, rayless, joyless you shall stand!

Pluck no more red roses, maidens,
Leave the lilies in their dew—
Pluck, pluck cypress, O pale maidens,
Dusk, oh, dusk the hall with yew!   270
—Shall I seek, that I may scorn her,
Her I loved at eventide?
Shall I ask, what faded mourner
Stands, at daybreak, weeping by my side?
Pluck, pluck cypress, O pale maidens!
Dusk the hall with yew!

Horatian Echo

(TO AN AMBITIOUS FRIEND)

Omit, omit, my simple friend,°
Still to enquire how parties tend,
Or what we fix with foreign powers.
If France and we are really friends,
And what the Russian Czar intends,
    Is no concern of ours.

Us not the daily quickening race°
Of the invading populace
Shall draw to swell that shouldering herd.
Mourn will we not your closing hour,
Ye imbeciles in present power,   10
     Doom'd, pompous, and absurd!

And let us bear, that they debate
Of all the engine-work of state,
Of commerce, laws, and policy,
The secrets of the world's machine,
And what the rights of man may mean,
     With readier tongue than we.

Only, that with no finer art
They cloak the troubles of the heart
With pleasant smile, let us take care;   20
Nor with a lighter hand dispose
Fresh garlands of this dewy rose,
     To crown Eugenia's hair.°

Of little threads our life is spun,
And he spins ill, who misses one.
But is thy fair Eugenia cold?
Yet Helen had an equal grace,
And Juliet's was as fair a face,
     And now their years are told.   30

The day approaches, when we must
Be crumbling bones and windy dust;
And scorn us as our mistress may,
Her beauty will no better be
Than the poor face she slights in thee,
     When dawns that day, that day.

Fragment of an 'Antigone'

THE CHORUS

Well hath he done who hath seized happiness!
For little do the all-containing hours,
    Though opulent, freely give.
    Who, weighing that life well
    Fortune presents unpray'd,
Declines her ministry, and carves his own;
    And, justice not infringed,
Makes his own welfare his unswerved-from law.

He does well too, who keeps that clue the mild
Birth-Goddess and the austere Fates first gave.   10
    For from the day when these
    Bring him, a weeping child,
    First to the light, and mark
A country for him, kinsfolk, and a home,
    Unguided he remains,
Till the Fates come again, this time with death.

    In little companies,
    And, our own place once left,
Ignorant where to stand, or whom to avoid,
By city and household group'd, we live; and many shocks   20
    Our order heaven-ordain'd
    Must every day endure:
Voyages, exiles, hates, dissensions, wars.
    Besides what waste *he* makes,
    The all-hated, order-breaking,
    Without friend, city, or home,
    Death, who dissevers all.
    Him then I praise, who dares
    To self-selected good
Prefer obedience to the primal law,   30
Which consecrates the ties of blood; for these, indeed,
    Are to the Gods a care;
    That touches but himself.
For every day man may be link'd and loosed

With strangers; but the bond
Original, deep-inwound,
Of blood, can he not bind,
Nor, if Fate binds, not bear.

But hush! Haemon, whom Antigone,
Robbing herself of life in burying,　　　40
Against Creon's law, Polynices,
Robs of a loved bride—pale, imploring,
　　Waiting her passage,
Forth from the palace hitherward comes.

HAEMON

No, no, old men, Creon I curse not!
　　I weep, Thebans,
　　One than Creon crueller far!
For he, he, at least, by slaying her,
August laws doth mightily vindicate;
But thou, too-bold, headstrong, pitiless!　　　50
Ah me!—honourest more than thy lover,
　　O Antigone!
A dead, ignorant, thankless corpse.

THE CHORUS

　　Nor was the love untrue
　　Which the Dawn-Goddess bore
　　To that fair youth she erst,
　　Leaving the salt sea-beds
And coming flush'd over the stormy frith
　　Of loud Euripus, saw—°
　　Saw and snatch'd, wild with love,　　　60
　　From the pine-dotted spurs
　　Of Parnes, where thy waves,
　　Asopus! gleam rock-hemm'd—
The Hunter of the Tanagraean Field.°

　　But him, in his sweet prime,
　　By severance immature,
　　By Artemis' soft shafts,
　　She, though a Goddess born,
Saw in the rocky isle of Delos die.

Such end o'ertook that love.   70
For she desired to make
Immortal mortal man,
And blend his happy life,
Far from the Gods, with hers;
To him postponing an eternal law.

HAEMON

But like me, she, wroth, complaining,
Succumb'd to the envy of unkind Gods;
And, her beautiful arms unclasping,
Her fair youth unwillingly gave.

THE CHORUS

Nor, though enthroned too high   80
To fear assault of envious Gods,
His beloved Argive seer would Zeus retain
From his appointed end

In this our Thebes; but when
His flying steeds came near
To cross the steep Ismenian glen,°
The broad earth open'd, and whelm'd them and him;
And through the void air sang
At large his enemy's spear.

And fain would Zeus have saved his tired son°   90
Beholding him where the Two Pillars stand
O'er the sun-redden'd western straits,°
Or at his work in that dim lower world.
Fain would he have recall'd
The fraudulent oath which bound
To a much feebler wight the heroic man.

But he preferr'd Fate to his strong desire.
Nor did there need less than the burning pile°
Under the towering Trachis crags,
And the Spercheios vale, shaken with groans,   100
And the roused Maliac gulph,
And scared Oetaean snows,
To achieve his son's deliverance, O my child!

Fragment of Chorus of a 'Dejaneira'

O frivolous mind of man,
Light ignorance, and hurrying, unsure thoughts!
Though man bewails you not,
How *I* bewail you!

Little in your prosperity
Do you seek counsel of the Gods.·
Proud, ignorant, self-adored, you live alone.
In profound silence stern,
Among their savage gorges and cold springs,
Unvisited remain   10
The great oracular shrines.

Thither in your adversity
Do you betake yourselves for light,
But strangely misinterpret all you hear.
For you will not put on
New hearts with the enquirer's holy robe,°
And purged, considerate minds.

And him on whom, at the end
Of toil and dolour untold,
The Gods have said that repose   20
At last shall descend undisturb'd—
Him you expect to behold
In an easy old age, in a happy home;
No end but this you praise.

But him, on whom, in the prime
Of life, with vigour undimm'd,
With unspent mind, and a soul
Unworn, undebased, undecay'd,
Mournfully grating, the gates
Of the city of death have for ever closed—   30
Him, I count *him*, well-starr'd.

The Strayed Reveller

THE PORTICO OF CIRCE'S PALACE. EVENING
A Youth. Circe

THE YOUTH

Faster, faster,
O Circe, Goddess,
Let the wild, thronging train,
The bright procession
Of eddying forms,
Sweep through my soul!

Thou standest, smiling
Down on me! thy right arm,
Lean'd up against the column there,
Props thy soft cheek;   10
Thy left holds, hanging loosely,
The deep cup, ivy-cinctured,
I held but now.

Is it, then, evening
So soon? I see, the night-dews,
Cluster'd in thick beads, dim
The agate brooch-stones
On thy white shoulder;
The cool night-wind, too,
Blows through the portico,   20
Stirs thy hair, Goddess,
Waves thy white robe!

CIRCE

Whence art thou, sleeper?

THE YOUTH

When the white dawn first
Through the rough fir-planks
Of my hut, by the chestnuts,
Up at the valley-head,
Came breaking, Goddess!

I sprang up, I threw round me
My dappled fawn-skin;   30
Passing out, from the wet turf,
Where they lay, by the hut door,
I snatch'd up my vine-crown, my fir-staff,
All drench'd in dew—
Came swift down to join
The rout early gather'd
In the town, round the temple,
Iacchus' white fane°
On yonder hill.
Quick I pass'd, following   40
The wood-cutters' cart-track
Down the dark valley;—I saw
On my left, through the beeches,
Thy palace, Goddess,
Smokeless, empty!
Trembling, I enter'd; beheld
The court all silent,
The lions sleeping,
On the altar this bowl.
I drank, Goddess!   50
And sank down here, sleeping,
On the steps of thy portico.

CIRCE

Foolish boy! Why tremblest thou?
Thou lovest it, then, my wine?
Wouldst more of it? See, how glows,
Through the delicate, flush'd marble,
The red, creaming liquor,
Strown with dark seeds!
Drink, then! I chide thee not,
Deny thee not my bowl.   60
Come, stretch forth thy hand, then—so!
Drink—drink again!

THE YOUTH

Thanks, gracious one!
Ah, the sweet fumes again!
More soft, ah me,

More subtle-winding
Than Pan's flute-music!
Faint—faint! Ah me,
Again the sweet sleep!

CIRCE

Hist! Thou—within there!   70
Come forth, Ulysses!
Art tired with hunting?
While we range the woodland,
See what the day brings.

ULYSSES

Ever new magic!
Hast thou then lured hither,
Wonderful Goddess, by thy art,
The young, languid-eyed Ampelus,°
Iacchus' darling—
Or some youth beloved of Pan,   80
Of Pan and the Nymphs?
That he sits, bending downward
His white, delicate neck
To the ivy-wreathed marge
Of thy cup; the bright, glancing vine-leaves
That crown his hair,
Falling forward, mingling
With the dark ivy-plants—
His fawn-skin, half untied,
Smear'd with red wine-stains? Who is he,   90
That he sits, overweigh'd
By fumes of wine and sleep,
So late, in thy portico?
What youth, Goddess,—what guest
Of Gods or mortals?

CIRCE

Hist! he wakes!
I lured him not hither, Ulysses.
Nay, ask him!

THE YOUTH

Who speaks? Ah, who comes forth
To thy side, Goddess, from within?   100
How shall I name him?
This spare, dark-featured,
Quick-eyed stranger?
Ah, and I see too
His sailor's bonnet,
His short coat, travel-tarnish'd,
With one arm bare!—
Art thou not he, whom fame
This long time rumours
The favour'd guest of Circe, brought by the waves?
Art thou he, stranger?   111
The wise Ulysses,
Laertes' son?

ULYSSES

I am Ulysses.
And thou, too, sleeper?
Thy voice is sweet.
It may be thou hast follow'd
Through the islands some divine bard,
By age taught many things,
Age and the Muses;   120
And heard him delighting
The chiefs and people
In the banquet, and learn'd his songs,
Of Gods and Heroes,
Of war and arts,
And peopled cities,
Inland, or built
By the grey sea.—If so, then hail!
I honour and welcome thee.

THE YOUTH

The Gods are happy.   130
They turn on all sides
Their shining eyes,
And see below them
The earth and men.

They see Tiresias°
Sitting, staff in hand,
On the warm, grassy
Asopus bank,
His robe drawn over
His old, sightless head,   140
Revolving inly
The doom of Thebes.

They see the Centaurs
In the upper glens
Of Pelion, in the streams,
Where red-berried ashes fringe
The clear-brown shallow pools,
With streaming flanks, and heads
Rear'd proudly, snuffling
The mountain wind.   150

They see the Indian
Drifting, knife in hand,
His frail boat moor'd to
A floating isle thick-matted
With large-leaved, low-creeping melon-plants,
And the dark cucumber.
He reaps, and stows them,
Drifting—drifting;—round him,
Round his green harvest-plot,
Flow the cool lake-waves,   160
The mountains ring them.

They see the Scythian
On the wide steppe, unharnessing
His wheel'd house at noon.
He tethers his beast down, and makes his meal—
Mares' milk, and bread
Baked on the embers;—all around
The boundless, waving grass-plains stretch, thick-starr'd
With saffron and the yellow hollyhock
And flag-leaved iris-flowers.   170
Sitting in his cart
He makes his meal; before him, for long miles,

Alive with bright green lizards,
And the springing bustard-fowl,
The track, a straight black line,
Furrows the rich soil; here and there
Clusters of lonely mounds
Topp'd with rough-hewn,
Grey, rain-blear'd statues, overpeer
The sunny waste.   180

They see the ferry
On the broad, clay-laden
Lone Chorasmian stream;—thereon,°
With snort and strain,
Two horses, strongly swimming, tow
The ferry-boat, with woven ropes
To either bow
Firm harness'd by the mane; a chief,
With shout and shaken spear,
Stands at the prow, and guides them; but astern   190
The cowering merchants, in long robes,
Sit pale beside their wealth
Of silk-bales and of balsam-drops,
Of gold and ivory,
Of turquoise-earth and amethyst,
Jasper and chalcedony,
And milk-barr'd onyx-stones.
The loaded boat swings groaning
In the yellow eddies;
The Gods behold them.   200
They see the Heroes
Sitting in the dark ship
On the foamless, long-heaving
Violet sea,
At sunset nearing
The Happy Islands.°

   These things, Ulysses,
The wise bards also
Behold and sing.
But oh, what labour!   210
O prince, what pain!

They too can see
Tiresias;—but the Gods,
Who give them vision,
Added this law:
That they should bear too
His groping blindness,
His dark foreboding,
His scorn'd white hairs;
Bear Hera's anger°   220
Through a life lengthen'd
To seven ages.

They see the Centaurs°
On Pelion;—then they feel,
They too, the maddening wine
Swell their large veins to bursting; in wild pain
They feel the biting spears
Of the grim Lapithae, and Theseus, drive,
Drive crashing through their bones; they feel
High on a jutting rock in the red stream   230
Alcmena's dreadful son
Ply his bow;—such a price
The Gods exact for song:
To become what we sing.

They see the Indian
On his mountain lake; but squalls
Make their skiff reel, and worms
In the unkind spring have gnawn
Their melon-harvest to the heart.—They see
The Scythian; but long frosts   240
Parch them in winter-time on the bare stepp,
Till they too fade like grass; they crawl
Like shadows forth in spring.

They see the merchants
On the Oxus stream;—but care
Must visit first them too, and make them pale.
Whether, through whirling sand,
A cloud of desert robber-horse have burst
Upon their caravan; or greedy kings,

In the wall'd cities the way passes through,   250
Crush'd them with tolls; or fever-airs,
Or some great river's marge,
Mown them down, far from home.

They see the Heroes
Near harbour;—but they share
Their lives, and former violent toil in Thebes,
Seven-gated Thebes, or Troy;
Or where the echoing oars°
Of Argo first
Startled the unknown sea.°   260

The old Silenus°
Came, lolling in the sunshine,
From the dewy forest-coverts,
This way, at noon.
Sitting by me, while his Fauns
Down at the water-side
Sprinkled and smoothed
His drooping garland,
He told me these things.

But I, Ulysses,   270
Sitting on the warm steps,
Looking over the valley,
All day long, have seen,
Without pain, without labour,
Sometimes a wild-hair'd Maenad—
Sometimes a Faun with torches—
And sometimes, for a moment,
Passing through the dark stems
Flowing-robed, the beloved,
The desired, the divine,   280
Beloved Iacchus.

Ah, cool night-wind, tremulous stars!
Ah, glimmering water,
Fitful earth-murmur,
Dreaming woods!
Ah, golden-hair'd, strangely smiling, Goddess,

And thou, proved, much enduring,°
Wave-toss'd Wanderer!
Who can stand still?
Ye fade, ye swim, ye waver before me—   290
The cup again!

Faster, faster,
O Circe, Goddess,
Let the wild, thronging train,
The bright procession
Of eddying forms,
Sweep through my soul!

The Sick King in Bokhara

HUSSEIN

O most just Vizier, send away
The cloth-merchants, and let them be,
Them and their dues, this day! the King
Is ill at ease, and calls for thee.

THE VIZIER

O merchants, tarry yet a day
Here in Bokhara! but at noon,
To-morrow, come, and ye shall pay
Each fortieth web of cloth to me,
As the law is, and go your way.
O Hussein, lead me to the King!   10
Thou teller of sweet tales, thine own,
Ferdousi's, and the others', lead!°
How is it with my lord?

HUSSEIN

                    Alone,
Ever since prayer-time, he doth wait,
O Vizier! without lying down,
In the great window of the gate,

Looking into the Registàn,
Where through the sellers' booths the slaves
Are this way bringing the dead man.—
O Vizier, here is the King's door!

THE KING

O Vizier, I may bury him?

THE VIZIER

O King, thou know'st, I have been sick
These many days, and heard no thing
(For Allah shut my ears and mind),
Not even what thou dost, O King!
Wherefore, that I may counsel thee,
Let Hussein, if thou wilt, make haste
To speak in order what hath chanced.

THE KING

O Vizier, be it as thou say'st!

HUSSEIN

Three days since, at the time of prayer   30
A certain Moollah, with his robe°
All rent, and dust upon his hair,
Watch'd my lord's coming forth, and push'd
The golden mace-bearers aside,
And fell at the King's feet, and cried:
'Justice, O King, and on myself!
On this great sinner, who did break
The law, and by the law must die!
Vengeance, O King!'

                    But the King spake:
'What fool is this, that hurts our ears   40
With folly? or what drunken slave?
My guards, what, prick him with your spears!
Prick me the fellow from the path!'
As the King said, so it was done,
And to the mosque my lord pass'd on.

But on the morrow, when the King
Went forth again, the holy book
Carried before him, as is right,
And through the square his way he took;
My man comes running, fleck'd with blood   50
From yesterday, and falling down
Cries out most earnestly: 'O King,
My lord, O King, do right, I pray!

'How canst thou, ere thou hear, discern
If I speak folly? but a king,
Whether a thing be great or small,
Like Allah, hears and judges all.

'Wherefore hear thou! Thou know'st, how fierce
In these last days the sun hath burn'd;
That the green water in the tanks   60
Is to a putrid puddle turn'd;
And the canal, which from the stream
Of Samarcand is brought this way,
Wastes, and runs thinner every day.

'Now I at nightfall had gone forth
Alone, and in a darksome place
Under some mulberry-trees I found
A little pool; and in short space,
With all the water that was there
I fill'd my pitcher, and stole home   70
Unseen; and having drink to spare,
I hid the can behind the door,
And went up on the roof to sleep.

'But in the night, which was with wind
And burning dust, again I creep
Down, having fever, for a drink.

'Now meanwhile had my brethren found
The water-pitcher, where it stood
Behind the door upon the ground,
And call'd my mother; and they all,   80

As they were thirsty, and the night
Most sultry, drain'd the pitcher there;
That they sate with it, in my sight,
Their lips still wet, when I came down.

'Now mark! I, being fever'd, sick
(Most unblest also), at that sight
Brake forth, and cursed them—dost thou hear?—
One was my mother—Now, do right!'

But my lord mused a space, and said:°
'Send him away, Sirs, and make on!
It is some madman!' the King said.
As the King bade, so was it done.

The morrow, at the self-same hour,
In the King's path, behold, the man,
Not kneeling, sternly fix'd! he stood
Right opposite, and thus began,
Frowning grim down: 'Thou wicked King,
Most deaf where thou shouldst most give ear!
What, must I howl in the next world,
Because thou wilt not listen here?   100

'What, wilt thou pray, and get thee grace,
And all grace shall to me be grudged?
Nay but, I swear, from this thy path
I will not stir till I be judged!'

Then they who stood about the King
Drew close together and conferr'd;
Till that the King stood forth and said:
'Before the priests thou shalt be heard.'

But when the Ulemas were met,
And the thing heard, they doubted not;   110
But sentenced him, as the law is,
To die by stoning on the spot.

Now the King charged us secretly:
'Stoned must he be, the law stands so.
Yet, if he seek to fly, give way;
Hinder him not, but let him go.'

So saying, the King took a stone,
And cast it softly;—but the man,
With a great joy upon his face,
Kneel'd down, and cried not, neither ran.   120

So they, whose lot it was, cast stones,
That they flew thick and bruised him sore.
But he praised Allah with loud voice,
And remain'd kneeling as before.

My lord had cover'd up his face;
But when one told him, 'He is dead,'
Turning him quickly to go in,
'Bring thou to me his corpse', he said.

And truly, while I speak, O King,
I hear the bearers on the stair;   130
Wilt thou they straightway bring him in?
—Ho! enter ye who tarry there!

THE VIZIER

O King, in this I praise thee not!
Now must I call thy grief not wise.
Is he thy friend, or of thy blood,
To find such favour in thine eyes?

Nay, were he thine own mother's son,
Still, thou art king, and the law stands.
It were not meet the balance swerved,
The sword were broken in thy hands.   140

But being nothing, as he is,
Why for no cause make sad thy face?—
Lo, I am old! three kings, ere thee,
Have I seen reigning in this place.

But who, through all this length of time,
Could bear the burden of his years,
If he for strangers pain'd his heart
Not less than those who merit tears?

Fathers we *must* have, wife and child,
And grievous is the grief for these;   150
This pain alone, which *must* be borne,
Makes the head white, and bows the knees.

But other loads than this his own
One man is not well made to bear.
Besides, to each are his own friends,
To mourn with him, and show him care.

Look, this is but one single place,
Though it be great; all the earth round,
If a man bear to have it so,
Things which might vex him shall be found.   160

Upon the Russian frontier, where
The watchers of two armies stand
Near one another, many a man,
Seeking a prey unto his hand,

Hath snatch'd a little fair-hair'd slave;
They snatch also, towards Mervè,
The Shiah dogs, who pasture sheep,°
And up from thence to Orgunjè.

And these all, labouring for a lord,
Eat not the fruit of their own hands;   170
Which is the heaviest of all plagues,
To that man's mind, who understands.

The kaffirs also (whom God curse!)°
Vex one another, night and day;
There are the lepers, and all sick;
There are the poor, who faint alway.

All these have sorrow, and keep still,
Whilst other men make cheer, and sing.
Wilt thou have pity on all these?
No, nor on this dead dog, O King!   180

THE KING

O Vizier, thou art old, I young!
Clear in these things I cannot see.
My head is burning, and a heat
Is in my skin which angers me.

But hear ye this, ye sons of men!
They that bear rule, and are obey'd,°
Unto a rule more strong than theirs
Are in their turn obedient made.

In vain therefore, with wistful eyes
Gazing up hither, the poor man,   190
Who loiters by the high-heap'd booths,
Below there, in the Registàn,

Says: 'Happy he, who lodges there!
With silken raiment, store of rice,
And for this drought, all kinds of fruits,
Grape-syrup, squares of colour'd ice,

'With cherries serv'd in drifts of snow.'
In vain hath a king power to build
Houses, arcades, enamell'd mosques;
And to make orchard-closes, fill'd   200

With curious fruit-trees brought from far;
With cisterns for the winter-rain,
And, in the desert, spacious inns
In divers places—if that pain

Is not more lighten'd, which he feels,
If his will be not satisfied;
And that it be not, from all time
The law is planted, to abide.

Thou wast a sinner, thou poor man!
Thou wast athirst; and didst not see, 210
That, though we take what we desire,
We must not snatch it eagerly.

And I have meat and drink at will,
And rooms of treasures, not a few.
But I am sick, nor heed I these;
And what I would, I cannot do.

Even the great honour which I have,
When I am dead, will soon grow still;
So have I neither joy, nor fame.
But what I can do, that I will. 220

I have a fretted brick-work tomb
Upon a hill on the right hand,
Hard by a close of apricots,
Upon the road of Samarcand;

Thither, O Vizier, will I bear
This man my pity could not save,
And, plucking up the marble flags,
There lay his body in my grave.

Bring water, nard, and linen rolls!
Wash off all blood, set smooth each limb! 230
Then say: 'He was not wholly vile,
Because a king shall bury him.'

Resignation

TO FAUSTA

To die be given us, or attain!°
Fierce work it were, to do again.
So pilgrims, bound for Mecca, pray'd
At burning noon; so warriors said,
Scarf'd with the cross, who watch'd the miles
Of dust which wreathed their struggling files

Down Lydian mountains; so, when snows
Round Alpine summits, eddying rose,
The Goth, bound Rome-wards; so the Hun,
Crouch'd on his saddle, while the sun   10
Went lurid down o'er flooded plains
Through which the groaning Danube strains
To the drear Euxine;—so pray all,
Whom labours, self-ordain'd, enthrall;
Because they to themselves propose
On this side the all-common close
A goal which, gain'd, may give repose.
So pray they; and to stand again
Where they stood once, to them were pain;
Pain to thread back and to renew   20
Past straits, and currents long steer'd through.°

But milder natures, and more free—
Whom an unblamed serenity°
Hath freed from passions, and the state
Of struggle these necessitate;
Whom schooling of the stubborn mind
Hath made, or birth hath found, resign'd—
These mourn not, that their goings pay
Obedience to the passing day.
These claim not every laughing Hour   30
For handmaid to their striding power;
Each in her turn, with torch uprear'd,
To await their march; and when appear'd,
Through the cold gloom, with measured race,
To usher for a destined space
(Her own sweet errands all forgone)
The too imperious traveller on.
These, Fausta, ask not this; nor thou,
Time's chafing prisoner, ask it now!
    We left, just ten years since, you say,°   40
That wayside inn we left to-day.
Our jovial host, as forth we fare,
Shouts greeting from his easy chair.
High on a bank our leader stands,
Reviews and ranks his motley bands,
Makes clear our goal to every eye—

The valley's western boundary.
A gate swings to! our tide hath flow'd
Already from the silent road.
The valley-pastures, one by one,   50
Are threaded, quiet in the sun;
And now beyond the rude stone bridge
Slopes gracious up the western ridge.
Its woody border, and the last
Of its dark upland farms is past—
Cool farms, with open-lying stores,
Under their burnish'd sycamores;
All past! and through the trees we glide,
Emerging on the green hill-side.
There climbing hangs, a far-seen sign,   60
Our wavering, many-colour'd line;
There winds, upstreaming slowly still
Over the summit of the hill.
And now, in front, behold outspread
Those upper regions we must tread!
Mild hollows, and clear heathy swells,
The cheerful silence of the fells.
Some two hours' march with serious air,
Through the deep noontide heats we fare;
The red-grouse, springing at our sound,   70
Skims, now and then, the shining ground;
No life, save his and ours, intrudes
Upon these breathless solitudes.
O joy! again the farms appear.
Cool shade is there, and rustic cheer;
There springs the brook will guide us down,
Bright comrade, to the noisy town.
Lingering, we follow down; we gain
The town, the highway, and the plain.
And many a mile of dusty way,   80
Parch'd and road-worn, we made that day;
But, Fausta, I remember well,
That as the balmy darkness fell
We bathed our hands with speechless glee,
That night, in the wide-glimmering sea.
Once more we tread this self-same road,°
Fausta, which ten years since we trod;

Alone we tread it, you and I,
Ghosts of that boisterous company.
Here, where the brook shines, near its head,   90
In its clear, shallow, turf-fringed bed;
Here, whence the eye first sees, far down,
Capp'd with faint smoke, the noisy town;
Here sit we, and again unroll,
Though slowly, the familiar whole.
The solemn wastes of heathy hill
Sleep in the July sunshine still;
The self-same shadows now, as then,
Play through this grassy upland glen;
The loose dark stones on the green way   100
Lie strewn, it seems, where then they lay;
On this mild bank above the stream,
(You crush them!) the blue gentians gleam.
Still this wild brook, the rushes cool,
The sailing foam, the shining pool!
These are not changed; and we, you say,
Are scarce more changed, in truth, than they.

The gipsies, whom we met below,
They, too, have long roam'd to and fro;
They ramble, leaving, where they pass,   110
Their fragments on the cumber'd grass.
And often to some kindly place
Chance guides the migratory race,
Where, though long wanderings intervene,
They recognise a former scene.
The dingy tents are pitch'd; the fires°
Give to the wind their wavering spires;
In dark knots crouch round the wild flame
Their children, as when first they came;
They see their shackled beasts again   120
Move, browsing, up the gray-wall'd lane.
Signs are not wanting, which might raise
The ghost in them of former days—
Signs are not wanting, if they would;
Suggestions to disquietude.
For them, for all, time's busy touch,
While it mends little, troubles much.

Their joints grow stiffer—but the year
Runs his old round of dubious cheer;
Chilly they grow—yet winds in March,   130
Still, sharp as ever, freeze and parch;
They must live still—and yet, God knows,
Crowded and keen the country grows;
It seems as if, in their decay,
The law grew stronger every day.
So might they reason, so compare,
Fausta, times past with times that are.
But no!—they rubb'd through yesterday
In their hereditary way,
And they will rub through, if they can,   140
To-morrow on the self-same plan,
Till death arrive to supersede,
For them, vicissitude and need.

The poet, to whose mighty heart°
Heaven doth a quicker pulse impart,
Subdues that energy to scan
Not his own course, but that of man.
Though he move mountains, though his day
Be pass'd on the proud heights of sway,
Though he hath loosed a thousand chains,   150
Though he hath borne immortal pains,
Action and suffering though he know—
He hath not lived, if he lives so.
He sees, in some great-historied land,
A ruler of the people stand,
Sees his strong thought in fiery flood
Roll through the heaving multitude;
Exults—yet for no moment's space
Envies the all-regarded place.
Beautiful eyes meet his—and he   160
Bears to admire uncravingly;
They pass—he, mingled with the crowd,
Is in their far-off triumphs proud.
From some high station he looks down,
At sunset, on a populous town;
Surveys each happy group, which fleets,
Toil ended, through the shining streets,

Each with some errand of its own—
And does not say; *I am alone.*
He sees the gentle stir of birth   170
When morning purifies the earth;
He leans upon a gate and sees
The pastures, and the quiet trees.
Low, woody hill, with gracious bound,
Folds the still valley almost round;
The cuckoo, loud on some high lawn,
Is answer'd from the depth of dawn;
In the hedge straggling to the stream,
Pale, dew-drench'd, half-shut roses gleam;
But, where the farther side slopes down,   180
He sees the drowsy new-waked clown
In his white quaint-embroider'd frock
Make, whistling, tow'rd his mist-wreathed flock—
Slowly, behind his heavy tread,
The wet, flower'd grass heaves up its head.
Lean'd on his gate, he gazes—tears
And in his eyes, and in his ears
The murmur of a thousand years.
Before him he sees life unroll,
A placid and continuous whole—   190
That general life, which does not cease,
Whose secret is not joy, but peace;
That life, whose dumb wish is not miss'd
If birth proceeds, if things subsist;
The life of plants, and stones, and rain,
The life he craves—if not in vain
Fate gave, what chance shall not control,
His sad lucidity of soul.

You listen—but that wandering smile,
Fausta, betrays you cold the while!   200
Your eyes pursue the bells of foam
Wash'd, eddying, from this bank, their home.
Those gipsies, so your thoughts I scan,
Are less, the poet more, than man.
They feel not, though they move and see;
Deeper the poet feels; but he
Breathes, when he will, immortal air,

Where Orpheus and where Homer are.
In the day's life, whose iron round
Hems us all in, he is not bound;   210
He leaves his kind, o'erleaps their pen,
And flees the common life of men.
He escapes thence, but we abide—
Not deep the poet sees, but wide.

The world in which we live and move
Outlasts aversion, outlasts love,
Outlasts each effort, interest, hope,
Remorse, grief, joy;—and were the scope
Of these affections wider made,
Man still would see, and see dismay'd,   220
Beyond his passion's widest range,
Far regions of eternal change.
Nay, and since death, which wipes out man,
Finds him with many an unsolved plan,
With much unknown, and much untried,
Wonder not dead, and thirst not dried,
Still gazing on the ever full
Eternal mundane spectacle—
This world in which we draw our breath,
In some sense, Fausta, outlasts death.   230

Blame thou not, therefore, him who dares
Judge vain beforehand human cares;
Whose natural insight can discern
What through experience others learn;
Who needs not love and power, to know
Love transient, power an unreal show;
Who treads at ease life's uncheer'd ways—
Him blame not, Fausta, rather praise!
Rather thyself for some aim pray
Nobler than this, to fill the day;   240
Rather that heart, which burns in thee,
Ask, not to amuse, but to set free;
Be passionate hopes not ill resign'd
For quiet, and a fearless mind.
And though Fate grudge to thee and me
The poet's rapt security,

Yet they, believe me, who await
No gifts from chance, have conquer'd fate.
They, winning room to see and hear,
And to men's business not too near,   250
Through clouds of individual strife
Draw homeward to the general life.
Like leaves by suns not yet uncurl'd;°
To the wise, foolish; to the world,
Weak;—yet not weak, I might reply,
Not foolish, Fausta, in His eye,
To whom each moment in its race,
Crowd as we will its neutral space,
Is but a quiet watershed
Whence, equally, the seas of life and death are fed.   260

Enough, we live!—and if a life,
With large results so little rife,
Though bearable, seem hardly worth
This pomp of worlds, this pain of birth;
Yet, Fausta, the mute turf we tread,°
The solemn hills around us spread,
This stream which falls incessantly,
The strange-scrawl'd rocks, the lonely sky,°
If I might lend their life a voice,
Seem to bear rather than rejoice.   270
And even could the intemperate prayer
Man iterates, while these forbear,
For movement, for an ampler sphere,
Pierce Fate's impenetrable ear;
Not milder is the general lot
Because our spirits have forgot,
In action's dizzying eddy whirl'd,
The something that infects the world.°

The Forsaken Merman

Come, dear children, let us away;
Down and away below!
Now my brothers call from the bay,
Now the great winds shoreward blow,

Now the salt tides seaward flow;
Now the wild white horses play,
Champ and chafe and toss in the spray.
Children dear, let us away!
This way, this way!

Call her once before you go—   10
Call once yet!
In a voice that she will know:
'Margaret! Margaret!'
Children's voices should be dear
(Call once more) to a mother's ear;

Children's voices, wild with pain—
Surely she will come again!
Call her once and come away;
This way, this way!
'Mother dear, we cannot stay!   20
The wild white horses foam and fret.'
Margaret! Margaret!

Come, dear children, come away down;
Call no more!
One last look at the white-wall'd town,
And the little grey church on the windy shore,
Then come down!
She will not come though you call all day;
Come away, come away!

Children dear, was it yesterday   30
We heard the sweet bells over the bay?
In the caverns where we lay,
Through the surf and through the swell,
The far-off sound of a silver bell?
Sand-strewn caverns, cool and deep,
Where the winds are all asleep;
Where the spent lights quiver and gleam,
Where the salt weed sways in the stream,
Where the sea-beasts, ranged all round,
Feed in the ooze of their pasture-ground;   40
Where the sea-snakes coil and twine,
Dry their mail and bask in the brine;

Where great whales come sailing by,
Sail and sail, with unshut eye,
Round the world for ever and aye?
When did music come this way?
Children dear, was it yesterday?

Children dear, was it yesterday
(Call yet once) that she went away?
Once she sate with you and me,   50
On a red gold throne in the heart of the sea,
And the youngest sate on her knee.
She comb'd its bright hair, and she tended it well,
When down swung the sound of a far-off bell.
She sigh'd, she look'd up through the clear green sea;
She said: 'I must go, for my kinsfolk pray
In the little grey church on the shore to-day.
'Twill be Easter-time in the world—ah me!
And I lose my poor soul, Merman! here with thee.'
I said: 'Go up, dear heart, through the waves;   60
Say thy prayer, and come back to the kind sea-caves!'
She smiled, she went up through the surf in the bay.
Children dear, was it yesterday?

    Children dear, were we long alone?
'The sea grows stormy, the little ones moan;
Long prayers,' I said, 'in the world they say;
Come!' I said; and we rose through the surf in the bay.
We went up the beach, by the sandy down
Where the sea-stocks bloom, to the white-wall'd town;
Through the narrow paved streets, where all was still,   70
To the little grey church on the windy hill.
From the church came a murmur of folk at their prayers,
But we stood without in the cold blowing airs.
We climb'd on the graves, on the stones worn with rains,
And we gazed up the aisle through the small leaded panes.
She sate by the pillar; we saw her clear:
'Margaret, hist! come quick, we are here!
Dear heart,' I said, 'we are long alone;
The sea grows stormy, the little ones moan.'
But, ah, she gave me never a look,   80
For her eyes were seal'd to the holy book!

Loud prays the priest; shut stands the door.

Come away, children, call no more!
Come away, come down, call no more!

   Down, down, down!
Down to the depths of the sea!
She sits at her wheel in the humming town,
Singing most joyfully.
Hark what she sings: 'O joy, O joy,
For the humming street, and the child with its toy!   90
For the priest, and the bell, and the holy well;
For the wheel where I spun,
And the blessed light of the sun!'
And so she sings her fill,
Singing most joyfully,
Till the spindle drops from her hand,°
And the whizzing wheel stands still.
She steals to the window, and looks at the sand,
And over the sand at the sea;
And her eyes are set in a stare;   100
And anon there breaks a sigh,
And anon there drops a tear,
From a sorrow-clouded eye,
And a heart sorrow-laden,
A long, long sigh;
For the cold strange eyes of a little Mermaiden
And the gleam of her golden hair.

   Come away, away children;
Come children, come down!
The hoarse wind blows coldly;   110
Lights shine in the town.
She will start from her slumber
When gusts shake the door;
She will hear the winds howling,
Will hear the waves roar.
We shall see, while above us
The waves roar and whirl,
A ceiling of amber,
A pavement of pearl.

Singing: 'Here came a mortal,   120
But faithless was she!
And alone dwell for ever
The kings of the sea.'

But, children, at midnight,
When soft the winds blow,
When clear falls the moonlight,
When spring-tides are low;
When sweet airs come seaward
From heaths starr'd with broom,
And high rocks throw mildly   130
On the blanch'd sands a gloom;
Up the still, glistening beaches,
Up the creeks we will hie,
Over banks of bright seaweed
The ebb-tide leaves dry.
We will gaze, from the sand-hills,
At the white, sleeping town;
At the church on the hill-side—
And then come back down.
Singing: 'There dwells a loved one,   140
But cruel is she!
She left lonely for ever
The kings of the sea.'

The World and the Quietist

TO CRITIAS

'Why, when the world's great mind
Hath finally inclined,
Why,' you say, Critias, 'be debating still?
Why, with these mournful rhymes
Learn'd in more languid climes,
Blame our activity
Who, with such passionate will,
Are what we mean to be?'

Critias, long since, I know
(For Fate decreed it so),
Long since the world hath set its heart to live;   10
   Long since, with credulous zeal
     It turns life's mighty wheel,
   Still doth for labourers send
     Who still their labour give,
   And still expects an end.

   Yet, as the wheel flies round,
   With no ungrateful sound
Do adverse voices fall on the world's ear.
   Deafen'd by his own stir   20
   The rugged labourer
   Caught not till then a sense
   So glowing and so near
     Of his omnipotence.

   So, when the feast grew loud°
   In Susa's palace proud,
A white-robed slave stole to the Great King's side.
   He spake—the Great King heard;
   Felt the slow-rolling word
   Swell his attentive soul;   30
   Breathed deeply as it died,
     And drain'd his mighty bowl.

To a Republican Friend, 1848

God knows it, I am with you. If to prize
Those virtues, prized and practised by too few,
But prized, but loved, but eminent in you,
Man's fundamental life; if to despise

The barren optimistic sophistries
Of comfortable moles, whom what they do
Teaches the limit of the just and true
(And for such doing they require not eyes);

If sadness at the long heart-wasting show
Wherein earth's great ones are disquieted;   10
If thoughts, not idle, while before me flow

The armies of the homeless and unfed—
If these are yours, if this is what you are,
Then am I yours, and what you feel, I share.

To a Republican Friend, Continued

Yet, when I muse on what life is, I seem
Rather to patience prompted, than that proud
Prospect of hope which France proclaims so loud—
France, famed in all great arts, in none supreme;°

Seeing this vale, this earth, whereon we dream,
Is on all sides o'ershadow'd by the high
Uno'erleap'd Mountains of Necessity,
Sparing us narrower margin than we deem.

Nor will that day dawn at a human nod,
When, bursting through the network superposed   10
By selfish occupation—plot and plan,

Lust, avarice, envy—liberated man,
All difference with his fellow-mortal closed,
Shall be left standing face to face with God.

Religious Isolation

TO THE SAME FRIEND

Children (as such forgive them) have I known,
Ever in their own eager pastime bent
To make the incurious bystander, intent
On his own swarming thoughts, an interest own—

Too fearful or too fond to play alone.
Do thou, whom light in thine own inmost soul
(Not less thy boast) illuminates, control
Wishes unworthy of a man full-grown.

What though the holy secret, which moulds thee,°
Mould not the solid earth? though never winds   10
Have whisper'd it to the complaining sea,

Nature's great law, and law of all men's minds?—
To its own impulse every creature stirs;
Live by thy light, and earth will live by hers!

To a Friend

Who prop, thou ask'st, in these bad days, my mind?—°
He much, the old man, who, clearest-soul'd of men,°
Saw The Wide Prospect, and the Asian Fen,°
And Tmolus hill, and Smyrna bay, though blind.°

Much he, whose friendship I not long since won,
That halting slave, who in Nicopolis°
Taught Arrian, when Vespasian's brutal son
Clear'd Rome of what most shamed him. But be his

My special thanks, whose even-balanced soul,°
From first youth tested up to extreme old age,   10
Business could not make dull, nor passion wild;

Who saw life steadily, and saw it whole;
The mellow glory of the Attic stage,
Singer of sweet Colonus, and its child.

Quiet Work

One lesson, Nature, let me learn of thee,
One lesson which in every wind is blown,
One lesson of two duties kept at one
Though the loud world proclaim their enmity—

Of toil unsever'd from tranquillity!
Of labour, that in lasting fruit outgrows
Far noisier schemes, accomplish'd in repose,
Too great for haste, too high for rivalry!

Yes, while on earth a thousand discords ring,
Man's fitful uproar mingling with his toil,   10
Still do thy sleepless ministers move on,°

Their glorious tasks in silence perfecting;
Still working, blaming still our vain turmoil,
Labourers that shall not fail, when man is gone.

Switzerland

A MEMORY-PICTURE

Laugh, my friends, and without blame
Lightly quit what lightly came;
Rich to-morrow as to-day,
Spend as madly as you may!
I, with little land to stir,
Am the exacter labourer.
   Ere the parting hour go by,
   Quick, thy tablets, Memory!

Once I said: 'A face is gone
If too hotly mused upon; 10
And our best impressions are
Those that do themselves repair.'
Many a face I so let flee,
Ah! is faded utterly.
   Ere the parting hour go by,
   Quick, thy tablets, Memory!

Marguerite says: 'As last year went,
So the coming year'll be spent;
Some day next year, I shall be,
Entering heedless, kiss'd by thee.' 20
Ah, I hope!—yet, once away,
What may chain us, who can say?
   Ere the parting hour go by,
   Quick, thy tablets, Memory!

Paint that lilac kerchief, bound
Her soft face, her hair around;
Tied under the archest chin
Mockery ever ambush'd in.
Let the fluttering fringes streak
All her pale, sweet-rounded cheek. 30
   Ere the parting hour go by,
   Quick, thy tablets, Memory!

Paint that figure's pliant grace
As she tow'rd me lean'd her face,
Half refused and half resign'd
Murmuring: 'Art thou still unkind?'
Many a broken promise then
Was new made—to break again.
   Ere the parting hour go by,
     Quick, thy tablets, Memory!   40

Paint those eyes, so blue, so kind,
Eager tell-tales of her mind;
Paint, with their impetuous stress
Of inquiring tenderness,
Those frank eyes, where deep I see
An angelic gravity.
   Ere the parting hour go by,
     Quick, thy tablets, Memory!

What, my friends, these feeble lines
Show, you say, my love declines?   50
To paint ill as I have done,
Proves forgetfulness begun?
Time's gay minions, pleased you see,
Time, your master, governs me;
   Pleased, you mock the fruitless cry:
     'Quick, thy tablets, Memory!'

Ah, too true! Time's current strong
Leaves us fixt to nothing long.
Yet, if little stays with man,
Ah, retain we all we can!   60
If the clear impression dies,
Ah, the dim remembrance prize!
   Ere the parting hour go by,
     Quick, thy tablets, Memory!

MEETING

Again I see my bliss at hand,
The town, the lake are here;°
My Marguerite smiles upon the strand,
Unalter'd with the year.

I know that graceful figure fair,
That cheek of languid hue;
I know that soft, enkerchief'd hair,
And those sweet eyes of blue.

Again I spring to make my choice;
Again in tones of ire   10
I hear a God's tremendous voice:°
'Be counsell'd, and retire.'

Ye guiding Powers who join and part,
What would ye have with me?
Ah, warn some more ambitious heart,
And let the peaceful be!

PARTING

Ye storm-winds of Autumn!
Who rush by, who shake
The window, and ruffle
The gleam-lighted lake;
Who cross to the hill-side
Thin-sprinkled with farms,
Where the high woods strip sadly
Their yellowing arms—
Ye are bound for the mountains!
Ah! with you let me go   10
Where your cold, distant barrier,
The vast range of snow,
Through the loose clouds lifts dimly
Its white peaks in air—
How deep is their stillness!
Ah, would I were there!

But on the stairs what voice is this I hear,
Buoyant as morning, and as morning clear?
Say, has some wet bird-haunted English lawn
Lent it the music of its trees at dawn?   20
Or was it from some sun-fleck'd mountain-brook
That the sweet voice its upland clearness took?
            Ah! it comes nearer—
            Sweet notes, this way!

Hark! fast by the window
The rushing winds go,
To the ice-cumber'd gorges,
The vast seas of snow!
There the torrents drive upward
Their rock-strangled hum;   30
There the avalanche thunders
The hoarse torrent dumb.
—I come, O ye mountains!
Ye torrents, I come!

But who is this, by the half-open'd door,
Whose figure casts a shadow on the floor?
The sweet blue eyes—the soft, ash-colour'd hair—
The cheeks that still their gentle paleness wear—
The lovely lips, with their arch smile that tells
The unconquer'd joy in which her spirit dwells—   40
Ah! they bend nearer—
Sweet lips, this way!

Hark! the wind rushes past us!
Ah! with that let me go
To the clear, waning hill-side,
Unspotted by snow,
There to watch, o'er the sunk vale,
The frore mountain-wall,
Where the niched snow-bed sprays down
Its powdery fall.   50
There its dusky blue clusters
The aconite spreads;
There the pines slope, the cloud-strips
Hung soft in their heads.
No life but, at moments,
The mountain-bee's hum.
—I come, O ye mountains!
Ye pine-woods, I come!

Forgive me! forgive me!
Ah, Marguerite, fain   60
Would these arms reach to clasp thee!
But see! 'tis in vain.

In the void air, towards thee,
  My stretch'd arms are cast;
But a sea rolls between us—
  Our different past!

To the lips, ah! of others
  Those lips have been prest,
And others, ere I was,
  Were strain'd to that breast;   70

Far, far from each other
  Our spirits have grown;
And what heart knows another?
  Ah! who knows his own?

Blow, ye winds! lift me with you!
  I come to the wild.
Fold closely, O Nature!
  Thine arms round thy child.

To thee only God granted
  A heart ever new—   80
To all always open,
  To all always true.

Ah! calm me, restore me;
  And dry up my tears
On thy high mountain-platforms,
  Where morn first appears;

Where the white mists, for ever,
  Are spread and upfurl'd—
In the stir of the forces
  Whence issued the world   90

ISOLATION. TO MARGUERITE

We were apart; yet, day by day,
I bade my heart more constant be.
I bade it keep the world away,
And grow a home for only thee;
Nor fear'd but thy love likewise grew,
Like mine, each day, more tried, more true.

The fault was grave! I might have known,
What far too soon, alas! I learn'd—
The heart can bind itself alone,
And faith may oft be unreturn'd.
Self-sway'd our feelings ebb and swell—°   10
Thou lov'st no more;—Farewell! Farewell!

Farewell!—and thou, thou lonely heart,
Which never yet without remorse
Even for a moment didst depart
From thy remote and spheréd course
To haunt the place where passions reign—
Back to thy solitude again!

Back! with the conscious thrill of shame
Which Luna felt, that summer-night,   20
Flash through her pure immortal frame,
When she forsook the starry height
To hang over Endymion's sleep
Upon the pine-grown Latmian steep.

Yet she, chaste queen, had never proved
How vain a thing is mortal love,
Wandering in Heaven, far removed.
But thou hast long had place to prove
This truth—to prove, and make thine own:
'Thou hast been, shalt be, art, alone.'   30

Or, if not quite alone, yet they
Which touch thee are unmating things—
Ocean and clouds and night and day;
Lorn autumns and triumphant springs;
And life, and others' joy and pain,
And love, if love, of happier men.

Of happier men—for they, at least,
Have *dream'd* two human hearts might blend
In one, and were through faith released
From isolation without end   40
Prolong'd; nor knew, although not less
Alone than thou, their loneliness.

TO MARGUERITE—CONTINUED

Yes! in the sea of life enisled,
With echoing straits between us thrown,°
Dotting the shoreless watery wild,
We mortal millions live *alone*.
The islands feel the enclasping flow,
And then their endless bounds they know.°

But when the moon their hollows lights,°
And they are swept by balms of spring,
And in their glens, on starry nights,
The nightingales divinely sing;   10
And lovely notes, from shore to shore,
Across the sounds and channels pour—

Oh! then a longing like despair
Is to their farthest caverns sent;
For surely once, they feel, we were
Parts of a single continent!
Now round us spreads the watery plain—
Oh might our marges meet again!

Who order'd, that their longing's fire
Should be, as soon as kindled, cool'd?   20
Who renders vain their deep desire?—
A God, a God their severance ruled!°
And bade betwixt their shores to be
The unplumb'd, salt, estranging sea.

A FAREWELL

My horse's feet beside the lake,
Where sweet the unbroken moonbeams lay,
Sent echoes through the night to wake
Each glistening strand, each heath-fringed bay.

The poplar avenue was pass'd,°
And the roof'd bridge that spans the stream;
Up the steep street I hurried fast,
Led by thy taper's starlike beam.

I came! I saw thee rise!—the blood
Pour'd flushing to thy languid cheek.   10
Lock'd in each other's arms we stood,
In tears, with hearts too full to speak.

Days flew;—ah, soon I could discern
A trouble in thine alter'd air!
Thy hand lay languidly in mine,
Thy cheek was grave, thy speech grew rare.

I blame thee not!—this heart, I know,
To be long loved was never framed;
For something in its depths doth glow
Too strange, too restless, too untamed.   20

And women—things that live and move
Mined by the fever of the soul—
They seek to find in those they love
Stern strength, and promise of control.

They ask not kindness, gentle ways—
These they themselves have tried and known;
They ask a soul which never sways
With the blind gusts that shake their own.

I too have felt the load I bore
In a too strong emotion's sway;   30
I too have wish'd, no woman more,
This starting, feverish heart away.

I too have long'd for trenchant force,
And will like a dividing spear;
Have praised the keen, unscrupulous course,
Which knows no doubt, which feels no fear.

But in the world I learnt, what there
Thou too wilt surely one day prove,
That will, that energy, though rare,
Are yet far, far less rare than love.   40

Go, then!—till time and fate impress
This truth on thee, be mine no more!
They will!—for thou, I feel, not less
Than I, wast destined to this lore.

We school our manners, act our parts—
But He, who sees us through and through,
Knows that the bent of both our hearts
Was to be gentle, tranquil, true.

And though we wear out life, alas!
Distracted as a homeless wind,   50
In beating where we must not pass,
In seeking what we shall not find;

Yet we shall one day gain, life past,
Clear prospect o'er our being's whole;
Shall see ourselves, and learn at last
Our true affinities of soul.°

We shall not then deny a course
To every thought the mass ignore;
We shall not then call hardness force,
Nor lightness wisdom any more.   60

Then, in the eternal Father's smile,
Our soothed, encouraged souls will dare
To seem as free from pride and guile,
As good, as generous, as they are.

Then we shall know our friends!—though much
Will have been lost—the help in strife,
The thousand sweet, still joys of such
As hand in hand face earthly life—

Though these be lost, there will be yet
A sympathy august and pure;   70
Ennobled by a vast regret,
And by contrition seal'd thrice sure.

And we, whose ways were unlike here,°
May then more neighbouring courses ply;
May to each other be brought near,
And greet across infinity.

How sweet, unreach'd by earthly jars,
My sister! to maintain with thee
The hush among the shining stars,
The calm upon the moonlit sea!   80

How sweet to feel, on the boon air,°
All our unquiet pulses cease!
To feel that nothing can impair
The gentleness, the thirst for peace—

The gentleness too rudely hurl'd
On this wild earth of hate and fear;
The thirst for peace a raving world
Would never let us satiate here.

ABSENCE

In this fair stranger's eyes of grey
Thine eyes, my love! I see.
I shiver; for the passing day
Had borne me far from thee.

This is the curse of life! that not
A nobler, calmer train
Of wiser thoughts and feelings blot
Our passions from our brain;

But each day brings its petty dust
Our soon-choked souls to fill,   10
And we forget because we must
And not because we will.

I struggle towards the light; and ye,
Once-long'd-for storms of love!
If with the light ye cannot be,
I bear that ye remove.

I struggle towards the light—but oh,
While yet the night is chill,
Upon time's barren, stormy flow,
Stay with me, Marguerite, still!   20

A DREAM

Was it a dream? We sail'd, I thought we sail'd,
Martin and I, down a green Alpine stream,
Border'd, each bank, with pines; the morning sun,
On the wet umbrage of their glossy tops,
On the red pinings of their forest-floor,
Drew a warm scent abroad; behind the pines
The mountain-skirts, with all their sylvan change
Of bright-leaf'd chestnuts and moss'd walnut-trees
And the frail scarlet-berried ash, began.
Swiss chalets glitter'd on the dewy slopes,   10
And from some swarded shelf, high up, there came
Notes of wild pastoral music—over all
Ranged, diamond-bright, the eternal wall of snow.
Upon the mossy rocks at the stream's edge,
Back'd by the pines, a plank-built cottage stood,
Bright in the sun; the climbing gourd-plant's leaves
Muffled its walls, and on the stone-strewn roof
Lay the warm golden gourds; golden, within,
Under the eaves, peer'd rows of Indian corn.
We shot beneath the cottage with the stream.   20
On the brown, rude-carved balcony, two forms
Came forth—Olivia's, Marguerite! and thine.
Clad were they both in white, flowers in their breast;
Straw hats bedeck'd their heads, with ribbons blue,
Which danced, and on their shoulders, fluttering, play'd.
They saw us, they conferr'd; their bosoms heaved,°
And more than mortal impulse fill'd their eyes.
Their lips moved; their white arms, waved eagerly,
Flash'd once, like falling streams; we rose, we gazed.
One moment, on the rapid's top, our boat   30
Hung poised—and then the darting river of Life
(Such now, methought, it was), the river of Life,

Loud thundering, bore us by; swift, swift it foam'd,
Black under cliffs it raced, round headlands shone.
Soon the plank'd cottage by the sun-warm'd pines
Faded—the moss—the rocks; us burning plains,°
Bristled with cities, us the sea received.

Stanzas in Memory of the Author of 'Obermann'

NOVEMBER, 1849

In front the awful Alpine track
Crawls up its rocky stair;
The autumn storm-winds drive the rack,°
Close o'er it, in the air.

Behind are the abandon'd baths°
Mute in their meadows lone;
The leaves are on the valley-paths,
The mists are on the Rhone—

The white mists rolling like a sea!
I hear the torrents roar.   10
—Yes, Obermann, all speaks of thee;
I feel thee near once more!

I turn thy leaves! I feel their breath
Once more upon me roll;
That air of languor, cold, and death,
Which brooded o'er thy soul.

Fly hence, poor wretch, whoe'er thou art,
Condemn'd to cast about,°
All shipwreck in thy own weak heart,
For comfort from without!   20

A fever in these pages burns
Beneath the calm they feign;
A wounded human spirit turns,
Here, on its bed of pain.

Yes, though the virgin mountain-air
Fresh through these pages blows;
Though to these leaves the glaciers spare°
The soul of their white snows;

Though here a mountain-murmur swells
Of many a dark-bough'd pine;   30
Though, as you read, you hear the bells
Of the high-pasturing kine—

Yet, through the hum of torrent lone,
And brooding mountain-bee,
There sobs I know not what ground-tone
Of human agony.

Is it for this, because the sound
Is fraught too deep with pain,
That, Obermann! the world around
So little loves thy strain?   40

Some secrets may the poet tell,
For the world loves new ways;
To tell too deep ones is not well—
It knows not what he says.°

Yet, of the spirits who have reign'd
In this our troubled day,
I know but two, who have attain'd,
Save thee, to see their way.

By England's lakes, in grey old age,°
His quiet home one keeps;   50
And one, the strong much-toiling sage°
In German Weimar sleeps.

But Wordsworth's eyes avert their ken
From half of human fate;
And Goethe's course few sons of men
May think to emulate.

For he pursued a lonely road,
His eyes on Nature's plan;
Neither made man too much a God,
Nor God too much a man.   60

Strong was he, with a spirit free
From mists, and sane, and clear;
Clearer, how much! than ours—yet we
Have a worse course to steer.

For though his manhood bore the blast°
Of a tremendous time,
Yet in a tranquil world was pass'd
His tenderer youthful prime.

But we, brought forth and rear'd in hours
Of change, alarm, surprise—   70
What shelter to grow ripe is ours?
What leisure to grow wise?

Like children bathing on the shore,
Buried a wave beneath,
The second wave succeeds, before
We have had time to breathe.

Too fast we live, too much are tried,
Too harass'd, to attain
Wordsworth's sweet calm, or Goethe's wide
And luminous view to gain.   80

And then we turn, thou sadder sage,
To thee! we feel thy spell!
—The hopeless tangle of our age,
Thou too hast scann'd it well!

Immoveable thou sittest, still
As death, composed to bear!
Thy head is clear, thy feeling chill,
And icy thy despair.

Yes, as the son of Thetis said,
I hear thee saying now: 90
Greater by far than thou are dead;°
Strive not! die also thou!

Ah! two desires toss about
The poet's feverish blood.
One drives him to the world without,
And one to solitude.

The glow, he cries, *the thrill of life*,
Where, where do these abound?—
Not in the world, not in the strife
Of men, shall they be found.   100

He who hath watch'd, not shared, the strife,
Knows how the day hath gone.
He only lives with the world's life,
Who hath renounced his own.

To thee we come, then! Clouds are roll'd
Where thou, O seer! art set;
Thy realm of thought is drear and cold—
The world is colder yet!

And thou hast pleasures, too, to share
With those who come to thee—   110
Balms floating on thy mountain-air,
And healing sights to see.

How often, where the slopes are green
On Jaman, hast thou sate°
By some high chalet-door, and seen
The summer-day grow late;

And darkness steal o'er the wet grass
With the pale crocus starr'd,
And reach that glimmering sheet of glass
Beneath the piny sward,   120

Lake Leman's waters, far below!
And watch'd the rosy light
Fade from the distant peaks of snow;
And on the air of night

Heard accents of the eternal tongue
Through the pine branches play—
Listen'd, and felt thyself grow young!
Listen'd and wept——Away!

Away the dreams that but deceive
And thou, sad guide, adieu!   130
I go, fate drives me; but I leave
Half of my life with you.°

We, in some unknown Power's employ,
Move on a rigorous line;
Can neither, when we will, enjoy,
Nor, when we will, resign.

I in the world must live; but thou,
Thou melancholy shade!
Wilt not, if thou canst see me now,
Condemn me, nor upbraid.   140

For thou art gone away from earth,
And place with those dost claim,
The Children of the Second Birth,°
Whom the world could not tame;

And with that small, transfigured band,
Whom many a different way
Conducted to their common land,
Thou learn'st to think as they.

Christian and pagan, king and slave,
Soldier and anchorite,   150
Distinctions we esteem so grave,
Are nothing in their sight.

They do not ask, who pined unseen,
Who was on action hurl'd,
Whose one bond is, that all have been
Unspotted by the world.°

There without anger thou wilt see
Him who obeys thy spell
No more, so he but rest, like thee,
Unsoil'd!—and so, farewell.   160

Farewell!—Whether thou now liest near
That much-loved inland sea,
The ripples of whose blue waves cheer
Vevey and Meillerie:

And in that gracious region bland,
Where with clear-rustling wave
The scented pines of Switzerland
Stand dark round thy green grave,

Between the dusty vineyard-walls
Issuing on that green place   170
The early peasant still recalls
The pensive stranger's face,

And stoops to clear thy moss-grown date
Ere he plods on again;—
Or whether, by maligner fate,
Among the swarms of men,

Where between granite terraces
The blue Seine rolls her wave,
The Capital of Pleasure sees
The hardly-heard-of grave;—°   180

Farewell! Under the sky we part,
In this stern Alpine dell.
O unstrung will! O broken heart!°
A last, a last farewell!

Human Life

What mortal, when he saw,
Life's voyage done, his heavenly Friend,
Could ever yet dare tell him fearlessly:
'I have kept uninfringed my nature's law;
The inly-written chart thou gavest me,°
To guide me, I have steer'd by to the end'?

Ah! let us make no claim,
On life's incognisable sea,
To too exact a steering of our way;
Let us not fret and fear to miss our aim,   10
If some fair coast have lured us to make stay,
Or some friend hail'd us to keep company.

Ay! we would each fain drive
At random, and not steer by rule.
Weakness! and worse, weakness bestow'd in vain!
Winds from our side the unsuiting consort rive,
We rush by coasts where we had lief remain;
Man cannot, though he would, live chance's fool.

No! as the foaming swath
Of torn-up water, on the main,   20
Falls heavily away with long-drawn roar
On either side the black deep-furrow'd path
Cut by an onward-labouring vessel's prore,°
And never touches the ship-side again;

Even so we leave behind,
As, charter'd by some unknown Powers,
We stem across the sea of life by night,°
The joys which were not for our use design'd;—
The friends to whom we had no natural right,
The homes that were not destined to be ours.   30

Self-Dependence

Weary of myself, and sick of asking
What I am, and what I ought to be,
At this vessel's prow I stand, which bears me
Forwards, forwards, o'er the starlit sea.

And a look of passionate desire
O'er the sea and to the stars I send:
'Ye who from my childhood up have calm'd me,
Calm me, ah, compose me to the end!

'Ah, once more,' I cried, 'ye stars, ye waters,
On my heart your mighty charm renew;   10
Still, still let me, as I gaze upon you,
Feel my soul becoming vast like you!'

From the intense, clear, star-sown vault of heaven,
Over the lit sea's unquiet way,
In the rustling night-air came the answer:
'Wouldst thou *be* as these are? *Live* as they.

'Unaffrighted by the silence round them,°
Undistracted by the sights they see,
These demand not that the things without them
Yield them love, amusement, sympathy.   20

'And with joy the stars perform their shining,
And the sea its long moon-silver'd roll;
For self-poised they live, nor pine with noting
All the fever of some differing soul.

'Bounded by themselves, and unregardful
In what state God's other works may be,
In their own tasks all their powers pouring,
These attain the mighty life you see.'

O air-born voice! long since, severely clear,
A cry like thine in mine own heart I hear: 30
'Resolve to be thyself; and know that he,
Who finds himself, loses his misery!'

Destiny

Why each is striving, from of old,
To love more deeply than he can?
Still would be true, yet still grows cold?
—Ask of the Powers that sport with man!

They yok'd in him, for endless strife,
A heart of ice, a soul of fire;
And hurl'd him on the Field of Life,
An aimless unallay'd Desire.

Youth's Agitations

When I shall be divorced, some ten years hence,
From this poor present self which I am now;
When youth has done its tedious vain expense°
Of passions that for ever ebb and flow;

Shall I not joy youth's heats are left behind,
And breathe more happy in an even clime?—
Ah no, for then I shall begin to find
A thousand virtues in this hated time!

Then I shall wish its agitations back,
And all its thwarting currents of desire;   10
Then I shall praise the heat which then I lack,
And call this hurrying fever, generous fire;

And sigh that one thing only has been lent
To youth and age in common—discontent.

The World's Triumphs

So far as I conceive the world's rebuke
To him address'd who would recast her new,
Not from herself her fame of strength she took,
But from their weakness who would work her rue.

'Behold,' she cries, 'so many rages lull'd,
So many fiery spirits quite cool'd down;
Look how so many valours, long undull'd,
After short commerce with me, fear my frown!

'Thou too, when thou against my crimes wouldst cry,
Let thy foreboded homage check thy tongue!'—   10
The world speaks well; yet might her foe reply:
'Are wills so weak?—then let not mine wait long!

'Hast thou so rare a poison?—let me be
Keener to slay thee, lest thou poison me!'

Empedocles on Etna

A DRAMATIC POEM

PERSONS

EMPEDOCLES.
PAUSANIAS, *a Physician.*
CALLICLES, *a young Harp-player.*

The Scene of the Poem is on Mount Etna; at first in the forest region, afterwards on the summit of the mountain.

ACT I. SCENE I

Morning. A Pass in the forest region of Etna.

CALLICLES.
(Alone, resting on a rock by the path.)

The mules, I think, will not be here this hour;
They feel the cool wet turf under their feet
By the stream-side, after the dusty lanes
In which they have toil'd all night from Catana,
And scarcely will they budge a yard. O Pan,
How gracious is the mountain at this hour!
A thousand times have I been here alone,
Or with the revellers from the mountain-towns,
But never on so fair a morn;—the sun
Is shining on the brilliant mountain-crests,   10
And on the highest pines; but farther down,
Here in the valley, is in shade; the sward
Is dark, and on the stream the mist still hangs;
One sees one's footprints crush'd in the wet grass,
One's breath curls in the air; and on these pines
That climb from the stream's edge, the long grey tufts,
Which the goats love, are jewell'd thick with dew.
Here will I stay till the slow litter comes.
I have my harp too—that is well.—Apollo!
What mortal could be sick or sorry here?°   20
I know not in what mind Empedocles,
Whose mules I follow'd, may be coming up,
But if, as most men say, he is half mad
With exile, and with brooding on his wrongs,

Pausanias, his sage friend, who mounts with him,
Could scarce have lighted on a lovelier cure.
The mules must be below, far down. I hear
Their tinkling bells, mix'd with the song of birds,
Rise faintly to me—now it stops!—Who's here?
Pausanias! and on foot? alone?

PAUSANIAS

                    And thou, then?   30
I left thee supping with Peisianax,°
With thy head full of wine, and thy hair crown'd,
Touching thy harp as the whim came on thee,
And praised and spoil'd by master and by guests
Almost as much as the new dancing-girl.
Why hast thou follow'd us?

CALLICLES

                    The night was hot,
And the feast past its prime; so we slipp'd out,
Some of us, to the portico to breathe;—
Peisianax, thou know'st, drinks late;—and then,
As I was lifting my soil'd garland off,   40
I saw the mules and litter in the court,
And in the litter sate Empedocles;
Thou, too, wast with him. Straightway I sped home;
I saddled my white mule, and all night long
Through the cool lovely country follow'd you,
Pass'd you a little since as morning dawn'd,
And have this hour sate by the torrent here,
Till the slow mules should climb in sight again.
And now?

PAUSANIAS

             And now, back to the town with speed!
Crouch in the wood first, till the mules have pass'd; 50
They do but halt, they will be here anon.
Thou must be viewless to Empedocles;
Save mine, he must not meet a human eye.
One of his moods is on him that thou know'st;
I think, thou wouldst not vex him.

CALLICLES

No—and yet
I would fain stay, and help thee tend him. Once
He knew me well, and would oft notice me;
And still, I know not how, he draws me to him,
And I could watch him with his proud sad face,
His flowing locks and gold-encircled brow   60
And kingly gait, for ever; such a spell
In his severe looks, such a majesty
As drew of old the people after him,
In Agrigentum and Olympia,
When his star reign'd, before his banishment,
Is potent still on me in his decline.
But oh! Pausanias, he is changed of late;
There is a settled trouble in his air
Admits no momentary brightening now,
And when he comes among his friends at feasts,   70
'Tis as an orphan among prosperous boys.
Thou know'st of old he loved this harp of mine,
When first he sojourn'd with Peisianax;
He is now always moody, and I fear him;
But I would serve him, soothe him, if I could,
Dared one but try.

PAUSANIAS

Thou wast a kind child ever!
He loves thee, but he must not see thee now.
Thou hast indeed a rare touch on thy harp,
He loves that in thee, too;—there was a time
(But that is pass'd), he would have paid thy strain   80
With music to have drawn the stars from heaven.
He hath his harp and laurel with him still,
But he has laid the use of music by,
And all which might relax his settled gloom.
Yet thou may'st try thy playing, if thou wilt—°
But thou must keep unseen; follow us on,
But at a distance! in these solitudes,
In this clear mountain-air, a voice will rise,
Though from afar, distinctly; it may soothe him.
Play when we halt, and, when the evening comes   90

And I must leave him (for his pleasure is
To be left musing these soft nights alone
In the high unfrequented mountain-spots),
Then watch him, for he ranges swift and far,
Sometimes to Etna's top, and to the cone;
But hide thee in the rocks a great way down,
And try thy noblest strains, my Callicles,
With the sweet night to help thy harmony!
Thou wilt earn my thanks sure, and perhaps his.

CALLICLES

More than a day and night, Pausanias,   100
Of this fair summer-weather, on these hills,
Would I bestow to help Empedocles.
That needs no thanks; one is far better here
Than in the broiling city in these heats.
But tell me, how hast thou persuaded him
In this his present fierce, man-hating mood,
To bring thee out with him alone on Etna?

PAUSANIAS

Thou hast heard all men speaking of Pantheia
The woman who at Agrigentum lay
Thirty long days in a cold trance of death,   110
And whom Empedocles call'd back to life.
Thou art too young to note it, but his power
Swells with the swelling evil of this time,
And holds men mute to see where it will rise.
He could stay swift diseases in old days,
Chain madmen by the music of his lyre,
Cleanse to sweet airs the breath of poisonous streams,
And in the mountain-chinks inter the winds.
This he could do of old; but now, since all
Clouds and grows daily worse in Sicily,   120
Since broils tear us in twain, since this new swarm
Of sophists has got empire in our schools
Where he was paramount, since he is banish'd
And lives a lonely man in triple gloom—
He grasps the very reins of life and death.
I ask'd him of Pantheia yesterday,
When we were gather'd with Peisianax,

And he made answer, I should come at night
On Etna here, and be alone with him,
And he would tell me, as his old, tried friend,   130
Who still was faithful, what might profit me;
That is, the secret of this miracle.

CALLICLES

Bah! Thou a doctor! Thou art superstitious.
Simple Pausanias, 'twas no miracle!
Pantheia, for I know her kinsmen well,
Was subject to these trances from a girl.
Empedocles would say so, did he deign;
But he still lets the people, whom he scorns,
Gape and cry *wizard* at him, if they list.
But thou, thou art no company for him!   140
Thou art as cross, as sour'd as himself!
Thou hast some wrong from thine own citizens,
And then thy friend is banish'd, and on that,
Straightway thou fallest to arraign the times,
As if the sky was impious not to fall.
The sophists are no enemies of his;
I hear, Gorgias, their chief, speaks nobly of him,
As of his gifted master, and once friend.
He is too scornful, too high-wrought, too bitter.
'Tis not the times, 'tis not the sophists vex him;   150
There is some root of suffering in himself,°
Some secret and unfollow'd vein of woe,
Which makes the time look black and sad to him.
Pester him not in this his sombre mood
With questionings about an idle tale,
But lead him through the lovely mountain-paths,
And keep his mind from preying on itself,
And talk to him of things at hand and common,
Not miracles! thou art a learned man,
But credulous of fables as a girl.   160

PAUSANIAS

And thou, a boy whose tongue outruns his knowledge,
And on whose lightness blame is thrown away.
Enough of this! I see the litter wind
Up by the torrent-side, under the pines.

I must rejoin Empedocles. Do thou
Crouch in the brushwood till the mules have pass'd;
Then play thy kind part well. Farewell till night!

SCENE II

Noon. A Glen on the highest skirts of the woody region of Etna.

EMPEDOCLES—PAUSANIAS

PAUSANIAS

The noon is hot. When we have cross'd the stream,
We shall have left the woody tract, and come
Upon the open shoulder of the hill.
See how the giant spires of yellow bloom°
Of the sun-loving gentian, in the heat,
Are shining on those naked slopes like flame!
Let us rest here; and now, Empedocles,
Pantheia's history!

                                              [A harp-note below is heard.°

EMPEDOCLES

               Hark! what sound was that
Rose from below? If it were possible,
And we were not so far from human haunt,   10
I should have said that someone touch'd a harp.
Hark! there again!

PAUSANIAS

              'Tis the boy Callicles,
The sweetest harp-player in Catana.
He is for ever coming on these hills,
In summer, to all country-festivals,
With a gay revelling band; he breaks from them
Sometimes, and wanders far among the glens.
But heed him not, he will not mount to us;
I spoke with him this morning. Once more, therefore,
Instruct me of Pantheia's story, Master,   20
As I have pray'd thee.

EMPEDOCLES

            That? and to what end?

PAUSANIAS

It is enough that all men speak of it.
But I will also say, that when the Gods
Visit us as they do with sign and plague,
To know those spells of thine which stay their hand
Were to live free from terror.

EMPEDOCLES

Spells? Mistrust them!
Mind is the spell which governs earth and heaven.
Man has a mind with which to plan his safety;
Know that, and help thyself!

PAUSANIAS

But thine own words?
'The wit and counsel of man was never clear,　　　　30
Troubles confound the little wit he has.'
Mind is a light which the Gods mock us with,
To lead those false who trust it.

[The harp sounds again.

EMPEDOCLES

Hist! once more!
Listen, Pausanias!—Ay, 'tis Callicles;
I know these notes among a thousand. Hark!

CALLICLES

(*Sings unseen, from below*)
The track winds down to the clear stream,°
To cross the sparkling shallows; there
The cattle love to gather, on their way
To the high mountain-pastures, and to stay,
Till the rough cow-herds drive them past,　　　　40
Knee-deep in the cool ford; for 'tis the last
Of all the woody, high, well-water'd dells
On Etna; and the beam
Of noon is broken there by chestnut-boughs
Down its steep verdant sides; the air

Is freshen'd by the leaping stream, which throws
Eternal showers of spray on the moss'd roots
Of trees, and veins of turf, and long dark shoots
Of ivy-plants, and fragrant hanging bells
Of hyacinths, and on late anemonies,   50
That muffle its wet banks; but glade,°
And stream, and sward, and chestnut-trees,
End here; Etna beyond, in the broad glare
Of the hot noon, without a shade,
Slope behind slope, up to the peak, lies bare;
The peak, round which the white clouds play.

    In such a glen, on such a day,°
On Pelion, on the grassy ground,
Chiron, the aged Centaur lay,
The young Achilles standing by.   60
The Centaur taught him to explore
The mountains; where the glens are dry
And the tired Centaurs come to rest,
And where the soaking springs abound
And the straight ashes grow for spears,
And where the hill-goats come to feed,
And the sea-eagles build their nest.
He show'd him Phthia far away,
And said: O boy, I taught this lore
To Peleus, in long distant years!   70
He told him of the Gods, the stars,
The tides;—and then of mortal wars,
And of the life which heroes lead
Before they reach the Elysian place
And rest in the immortal mead;
And all the wisdom of his race.

The music below ceases, and EMPEDOCLES *speaks, accompanying himself in a
solemn manner on his harp.*

    The out-spread world to span°
A cord the Gods first slung,
And then the soul of man
There, like a mirror, hung,   80
And bade the winds through space impel the gusty toy.

Hither and thither spins
The wind-borne, mirroring soul,
A thousand glimpses wins,
And never sees a whole;
Looks once, and drives elsewhere, and leaves its last employ.

The Gods laugh in their sleeve
To watch man doubt and fear,
Who knows not what to believe
Since he sees nothing clear,   90
And dares stamp nothing false where he finds nothing sure.

Is this, Pausanias, so?
And can our souls not strive,
But with the winds must go,
And hurry where they drive?
Is fate indeed so strong, man's strength indeed so poor?

I will not judge. That man,
Howbeit, I judge as lost,
Whose mind allows a plan,
Which would degrade it most;   100
And he treats doubt the best who tries to see least ill.°

Be not, then, fear's blind slave!
Thou art my friend; to thee,
All knowledge that I have,
All skill I wield, are free.
Ask not the latest news of the last miracle,

Ask not what days and nights
In trance Pantheia lay,
But ask how thou such sights
May'st see without dismay;   110
Ask what most helps when known, thou son of Anchitus!°

What? hate, and awe, and shame
Fill thee to see our time;
Thou feelest thy soul's frame
Shaken and out of chime?
What? life and chance go hard with thee too, as with us;

Thy citizens, 'tis said,
Envy thee and oppress,
Thy goodness no men aid,
All strive to make it less;   120
Tyranny, pride, and lust, fill Sicily's abodes;

Heaven is with earth at strife,°
Signs make thy soul afraid,
The dead return to life,
Rivers are dried, winds stay'd;
Scarce can one think in calm, so threatening are the Gods;

And we feel, day and night,
The burden of ourselves—°
Well, then, the wiser wight
In his own bosom delves,   130
And asks what ails him so, and gets what cure he can.

The sophist sneers: Fool, take
Thy pleasure, right or wrong.
The pious wail: Forsake
A world these sophists throng.
Be neither saint- nor sophist-led, but be a man!

These hundred doctors try
To preach thee to their school.
We have the truth! they cry;
And yet their oracle,   140
Trumpet it as they will, is but the same as thine.

Once read thy own breast right,
And thou hast done with fears;
Man gets no other light,
Search he a thousand years.
Sink in thyself! there ask what ails thee, at that shrine!

What makes thee struggle and rave?
Why are men ill at ease?—
'Tis that the lot they have
Fails their own will to please;   150
For man would make no murmuring, were his will obey'd.

And why is it, that still°
Man with his lot thus fights?—
'Tis that he makes this *will*
The measure of his *rights*,
And believes Nature outraged if his will's gainsaid.

Couldst thou, Pausanias, learn
How deep a fault is this;
Couldst thou but once discern
Thou hast no *right* to bliss,   160
No title from the Gods to welfare and repose;

Then thou wouldst look less mazed
Whene'er of bliss debarr'd,
Nor think the Gods were crazed
When thy own lot went hard.
But we are all the same—the fools of our own woes!

For, from the first faint morn
Of life, the thirst for bliss
Deep in man's heart is born;
And, sceptic as he is,   170
He fails not to judge clear if this be quench'd or no.

Nor is the thirst to blame.
Man errs not that he deems
His welfare his true aim,
He errs because he dreams
The world does but exist that welfare to bestow.

We mortals are no kings
For each of whom to sway
A new-made world up-springs,
Meant merely for his play;   180
No, we are strangers here; the world is from of old.

In vain our pent wills fret,
And would the world subdue.
Limits we did not set
Condition all we do;
Born into life we are, and life must be our mould.

Born into life!—man grows
Forth from his parents' stem,
And blends their bloods, as those
Of theirs are blent in them;   190
So each new man strikes root into a far fore-time.

Born into life!—we bring
A bias with us here,
And, when here, each new thing
Affects us we come near;
To tunes we did not call our being must keep chime.

Born into life!—in vain,°
Opinions, those or these,
Unalter'd to retain
The obstinate mind decrees;   200
Experience, like a sea, soaks all-effacing in.

Born into life!—who lists
May what is false hold dear,
And for himself make mists
Through which to see less clear;
The world is what it is, for all our dust and din.

Born into life!—'tis we,
And not the world, are new;
Our cry for bliss, our plea,
Others have urged it too—   210
Our wants have all been felt, our errors made before.

No eye could be too sound
To observe a world so vast,
No patience too profound
To sort what's here amass'd;
How man may here best live no care too great to explore.

But we—as some rude guest°
Would change, where'er he roam,
The manners there profess'd
To those he brings from home—   220
We mark not the world's course, but would have *it* take *ours*.

The world's course proves the terms
On which man wins content;
Reason the proof confirms—
We spurn it, and invent
A false course for the world, and for ourselves, false powers.

Riches we wish to get,
Yet remain spendthrifts still;
We would have health, and yet
Still use our bodies ill;   230
Bafflers of our own prayers, from youth to life's last scenes.

We would have inward peace,
Yet will not look within;
We would have misery cease,
Yet will not cease from sin;
We want all pleasant ends, but will use no harsh means;

We do not what we ought,°
What we ought not, we do,
And lean upon the thought
That chance will bring us through;   240
But our own acts, for good or ill, are mightier powers.°

Yet, even when man forsakes
All sin,—is just, is pure,
Abandons all which makes
His welfare insecure,—
Other existences there are, that clash with ours.

Like us, the lightning-fires
Love to have scope and play;
The stream, like us, desires
An unimpeded way;   250
Like us, the Libyan wind delights to roam at large.

Streams will not curb their pride
The just man not to entomb,
Nor lightnings go aside
To give his virtues room;
Nor is that wind less rough which blows a good man's barge.

Nature, with equal mind,
Sees all her sons at play;
Sees man control the wind,
The wind sweep man away;   260
Allows the proudly-riding and the foundering bark.

And, lastly, though of ours
No weakness spoil our lot,
Though the non-human powers
Of Nature harm us not,
The ill deeds of other men make often *our* life dark.°

What were the wise man's plan?—
Through this sharp, toil-set life,
To work as best he can,
And win what's won by strife.—   270
But we an easier way to cheat our pains have found.

Scratch'd by a fall, with moans
As children of weak age
Lend life to the dumb stones
Whereon to vent their rage,
And bend their little fists, and rate the senseless ground;

So, loath to suffer mute,
We, peopling the void air,
Make Gods to whom to impute
The ills we ought to bear;   280
With God and Fate to rail at, suffering easily.

Yet grant—as sense long miss'd
Things that are now perceived,
And much may still exist
Which is not yet believed—
Grant that the world were full of Gods we cannot see;

All things the world which fill°
Of but one stuff are spun,
That we who rail are still,
With what we rail at, one;   290
One with the o'erlabour'd Power that through the breadth and length

Of earth, and air, and sea,
In men, and plants, and stones,
Hath toil perpetually,
And travails, pants, and moans;
Fain would do all things well, but sometimes fails in strength.

And patiently exact
This universal God
Alike to any act
Proceeds at any nod,   300
And quietly declaims the cursings of himself.

This is not what man hates,
Yet he can curse but this.
Harsh Gods and hostile Fates
Are dreams! this only *is*—
Is everywhere; sustains the wise, the foolish elf.

Nor only, in the intent
To attach blame elsewhere,
Do we at will invent
Stern Powers who make their care   310
To embitter human life, malignant Deities;

But, next, we would reverse
The scheme ourselves have spun,
And what we made to curse
We now would lean upon,
And feign kind Gods who perfect what man vainly tries.

Look, the world tempts our eye,
And we would know it all!
We map the starry sky,
We mine this earthen ball,   320
We measure the sea-tides, we number the sea-sands;

We scrutinise the dates
Of long-past human things,
The bounds of effaced states,
The lines of deceased kings;
We search out dead men's words, and works of dead men's hands;

We shut our eyes, and muse
How our own minds are made,
What springs of thought they use,
How righten'd, how betray'd—   330
And spend our wit to name what most employ unnamed.

But still, as we proceed°
The mass swells more and more
Of volumes yet to read,
Of secrets yet to explore.
Our hair grows grey, our eyes are dimm'd, our heat is tamed;

We rest our faculties,
And thus address the Gods:
'True science if there is,
It stays in your abodes!   340
Man's measures cannot mete the immeasurable All.

'You only can take in
The world's immense design.
Our desperate search was sin,
Which henceforth we resign,
Sure only that your mind sees all things which befall.'

Fools! That in man's brief term
He cannot all things view,
Affords no ground to affirm
That there are Gods who do;   350
Nor does being weary prove that he has where to rest.

Again.—Our youthful blood
Claims rapture as its right;
The world, a rolling flood
Of newness and delight,
Draws in the enamour'd gazer to its shining breast;

Pleasure, to our hot grasp,
Gives flowers after flowers;
With passionate warmth we clasp
Hand after hand in ours;   360
Now do we soon perceive how fast our youth is spent.

At once our eyes grow clear!°
We see, in blank dismay,
Year posting after year,
Sense after sense decay;
Our shivering heart is mined by secret discontent;

Yet still, in spite of truth,
In spite of hopes entomb'd,
That longing of our youth
Burns ever unconsumed,   370
Still hungrier for delight as delights grow more rare.

We pause; we hush our heart,
And thus address the Gods:
'The world hath fail'd to impart
The joy our youth forebodes,
Fail'd to fill up the void which in our breasts we bear.

'Changeful till now, we still
Look'd on to something new;
Let us, with changeless will,
Henceforth look on to you,   380
To find with you the joy we in vain here require!'

Fools! That so often here
Happiness mock'd our prayer,
I think, might make us fear
A like event elsewhere;
Make us, not fly to dreams, but moderate desire.°

And yet, for those who know
Themselves, who wisely take
Their way through life, and bow
To what they cannot break,   390
Why should I say that life need yield but *moderate* bliss?

Shall we, with temper spoil'd,
Health sapp'd by living ill,
And judgment all embroil'd
By sadness and self-will,
Shall *we* judge what for man is not true bliss or is?

Is it so small a thing
To have enjoy'd the sun,
To have lived light in the spring,
To have loved, to have thought, to have done;   400
To have advanced true friends, and beat down baffling foes—

That we must feign a bliss
Of doubtful future date,
And, while we dream on this,
Lose all our present state,
And relegate to worlds yet distant our repose?

Not much, I know, you prize
What pleasures may be had,
Who look on life with eyes
Estranged, like mine, and sad;   410
And yet the village-churl feels the truth more than you,

Who's loath to leave this life
Which to him little yields—
His hard-task'd sunburnt wife,
His often-labour'd fields,
The boors with whom he talk'd, the country-spots he knew.

But thou, because thou hear'st
Men scoff at Heaven and Fate,
Because the Gods thou fear'st
Fail to make blest thy state,   420
Tremblest, and wilt not dare to trust the joys there are!

I say: Fear not! Life still°
Leaves human effort scope.
But, since life teems with ill,
Nurse no extravagant hope;
Because thou must not dream, thou need'st not then despair!

*A long pause. At the end of it the notes of a harp below are again
heard, and* CALLICLES *sings:—*

Far, far from here,°
The Adriatic breaks in a warm bay
Among the green Illyrian hills; and there

The sunshine in the happy glens is fair,   430
And by the sea, and in the brakes.
The grass is cool, the sea-side air
Buoyant and fresh, the mountain flowers
More virginal and sweet than ours.

And there, they say, two bright and aged snakes,
Who once were Cadmus and Harmonia,
Bask in the glens or on the warm sea-shore,
In breathless quiet, after all their ills;
Nor do they see their country, nor the place
Where the Sphinx lived among the frowning hills,°   440
Nor the unhappy palace of their race,
Nor Thebes, nor the Ismenus, any more

   There those two live, far in the Illyrian brakes!
They had stay'd long enough to see,
In Thebes, the billow of calamity°
Over their own dear children roll'd,°
Curse upon curse, pang upon pang,
For years, they sitting helpless in their home,
A grey old man and woman; yet of old
The Gods had to their marriage come,   450
And at the banquet all the Muses sang.

Therefore they did not end their days
In sight of blood; but were rapt, far away,
To where the west-wind plays,
And murmurs of the Adriatic come
To those untrodden mountain-lawns; and there
Placed safely in changed forms, the pair°
Wholly forgot their first sad life, and home,
And all that Theban woe, and stray
For ever through the glens, placid and dumb.   460

EMPEDOCLES

That was my harp-player again!—where is he?
Down by the stream?

PAUSANIAS

          Yes, Master, in the wood.

EMPEDOCLES

He ever loved the Theban story well!
But the day wears. Go now, Pausanias,
For I must be alone. Leave me one mule;
Take down with thee the rest to Catana.
And for young Callicles, thank him from me;
Tell him, I never fail'd to love his lyre—
But he must follow me no more to-night.

PAUSANIAS

Thou wilt return to-morrow to the city?   470

EMPEDOCLES

Either to-morrow or some other day,
In the sure revolutions of the world,
Good friend, I shall revisit Catana.
I have seen many cities in my time,
Till mine eyes ache with the long spectacle,
And I shall doubtless see them all again;
Thou know'st me for a wanderer from of old.
Meanwhile, stay me not now. Farewell, Pausanias!
                    He departs on his way up the mountain.

PAUSANIAS (*alone*)

I dare not urge him further—he must go;
But he is strangely wrought!—I will speed back   480
And bring Peisianax to him from the city;
His counsel could once soothe him. But, Apollo!
How his brow lighten'd as the music rose!
Callicles must wait here, and play to him;
I saw him through the chestnuts far below,
Just since, down at the stream.—Ho! Callicles!
                    He descends, calling.

ACT II

Evening. The Summit of Etna.

EMPEDOCLES

                              Alone!—
On this charr'd, blacken'd, melancholy waste,

Crown'd by the awful peak, Etna's great mouth.
Round which the sullen vapour rolls—alone!
Pausanias is far hence, and that is well,
For I must henceforth speak no more with man.
He hath his lesson too, and that debt's paid;
And the good, learned, friendly, quiet man
May bravelier front his life, and in himself
Find henceforth energy and heart. But I—   10
The weary man, the banish'd citizen,°
Whose banishment is not his greatest ill,
Whose weariness no energy can reach,
And for whose hurt courage is not the cure—
What should I do with life and living more?

No, thou art come too late, Empedocles!
And the world hath the day, and must break thee,
Not thou the world. With men thou canst not live,
Their thoughts, their ways, their wishes, are not thine;
And being lonely thou art miserable,   20
For something has impair'd thy spirit's strength,°
And dried its self-sufficing fount of joy.
Thou canst not live with men nor with thyself—
O sage! O sage!—Take then the one way left;
And turn thee to the elements, thy friends,°
Thy well-tried friends, thy willing ministers,
And say: Ye helpers, hear Empedocles,
Who asks this final service at your hands!
Before the sophist-brood hath overlaid°
The last spark of man's consciousness with words—   30
Ere quite the being of man, ere quite the world
Be disarray'd of their divinity—
Before the soul lose all her solemn joys,
And awe be dead, and hope impossible,
And the soul's deep eternal night come on—
Receive me, hide me, quench me, take me home!

*He advances to the edge of the crater. Smoke and fire break forth with a loud
noise, and* CALLICLES *is heard below singing:—*

The lyre's voice is lovely everywhere;°
In the court of Gods, in the city of men,
And in the lonely rock-strewn mountain-glen,
In the still mountain air.   40

Only to Typho it sounds hatefully;
To Typho only, the rebel o'erthrown,
Through whose heart Etna drives her roots of stone
To imbed them in the sea.

Wherefore dost thou groan so loud?
Wherefore do thy nostrils flash,
Through the dark night, suddenly,
Typho, such red jets of flame?—
Is thy tortured heart still proud?
Is thy fire-scathed arm still rash?   50
Still alert thy stone-crush'd frame?
Doth thy fierce soul still deplore
Thine ancient rout by the Cilician hills,
And that curst treachery on the Mount of Gore?°
Do thy bloodshot eyes still weep
The fight which crown'd thine ills,
Thy last mischance on this Sicilian deep?
Hast thou sworn, in thy sad lair,
Where erst the strong sea-currents suck'd thee down,
Never to cease to writhe, and try to rest,   60
Letting the sea-stream wander through thy hair?
That thy groans, like thunder prest,
Begin to roll, and almost drown
The sweet notes whose lulling spell
Gods and the race of mortals love so well,
When through thy caves thou hearest music swell?

But an awful pleasure bland°
Spreading o'er the Thunderer's face,
When the sound climbs near his seat,
The Olympian council sees;   70
As he lets his lax right hand,
Which the lightnings doth embrace,
Sink upon his mighty knees.
And the eagle, at the beck
Of the appeasing, gracious harmony,
Droops all his sheeny, brown, deep-feather'd neck,
Nestling nearer to Jove's feet;
While o'er his sovran eye
The curtains of the blue films slowly meet

And the white Olympus-peaks   80
Rosily brighten, and the soothed Gods smile
At one another from their golden chairs,
And no one round the charmed circle speaks.
Only the loved Hebe bears
The cup about, whose draughts beguile
Pain and care, with a dark store
Of fresh-pull'd violets wreathed and nodding o'er;
And her flush'd feet glow on the marble floor.

EMPEDOCLES

He fables, yet speaks truth!
The brave, impetuous heart yields everywhere°   90
To the subtle, contriving head;
Great qualities are trodden down,°
And littleness united
Is become invincible.

These rumblings are not Typho's groans, I know!°
These angry smoke-bursts
Are not the passionate breath
Of the mountain-crush'd, tortured, intractable Titan king—
But over all the world
What suffering is there not seen   100
Of plainness oppress'd by cunning,
As the well-counsell'd Zeus oppress'd
That self-helping son of earth!
What anguish of greatness,
Rail'd and hunted from the world,
Because its simplicity rebukes
This envious, miserable age!

I am weary of it.
—Lie there, ye ensigns°
Of my unloved preëminence   110
In an age like this!
Among a people of children,
Who throng'd me in their cities,
Who worshipp'd me in their houses,
And ask'd, not wisdom,
But drugs to charm with,

But spells to mutter—
All the fool's-armoury of magic!—Lie there,
My golden circlet,
My purple robe!   120

CALLICLES (*from below*)

As the sky-brightening south-wind clears the day,°
And makes the mass'd clouds roll,
The music of the lyre blows away
The clouds which wrap the soul.

Oh! that Fate had let me see
That triumph of the sweet persuasive lyre,
That famous, final victory,
When jealous Pan with Marsyas did conspire;°
When, from far Parnassus' side,
Young Apollo, all the pride   130
Of the Phrygian flutes to tame,
To the Phrygian highlands came;
Where the long green reed-beds sway
In the rippled waters grey
Of that solitary lake
Where Maeander's springs are born;
Whence the ridged pine-wooded roots
Of Messogis westward break,
Mounting westward, high and higher.
There was held the famous strife;   140
There the Phrygian brought his flutes,
And Apollo brought his lyre;
And, when now the westering sun
Touch'd the hills, the strife was done,
And the attentive Muses said;
'Marsyas, thou art vanquishéd!'
Then Apollo's minister
Hang'd upon a branching fir
Marsyas, that unhappy Faun,
And began to whet his knife.   150
But the Maenads, who were there,
Left their friend, and with robes flowing
In the wind, and loose dark hair
O'er their polish'd bosoms blowing,

Each her ribbon'd tambourine
Flinging on the mountain-sod,
With a lovely frighten'd mien
Came about the youthful God.
But he turn'd his beauteous face
Haughtily another way,   160
From the grassy sun-warm'd place
Where in proud repose he lay,
With one arm over his head,
Watching how the whetting sped.

    But aloof, on the lake-strand,
Did the young Olympus stand,
Weeping at his master's end;
For the Faun had been his friend.
For he taught him how to sing,
And he taught him flute-playing.   170
Many a morning had they gone
To the glimmering mountain-lakes,
And had torn up by the roots
The tall crested water-reeds
With long plumes and soft brown seeds,
And had carved them into flutes,
Sitting on a tabled stone
Where the shoreward ripple breaks.
And he taught him how to please
The red-snooded Phrygian girls,   180
Whom the summer evening sees
Flashing in the dance's whirls
Underneath the starlit trees
In the mountain-villages.
Therefore now Olympus stands,
At his master's piteous cries
Pressing fast with both his hands
His white garment to his eyes,
Not to see Apollo's scorn;—
Ah, poor Faun, poor Faun! ah, poor Faun!   190

EMPEDOCLES

And lie thou there,°
My laurel bough!

Scornful Apollo's ensign, lie thou there!
Though thou hast been my shade in the world's heat—
Though I have loved thee, lived in honouring thee—
Yet lie thou there,
My laurel bough!

I am weary of thee.
I am weary of the solitude
Where he who bears thee must abide—   200
Of the rocks of Parnassus,
Of the gorge of Delphi,
Of the moonlit peaks, and the caves.
Thou guardest them, Apollo!
Over the grave of the slain Pytho,
Though young, intolerably severe!
Thou keepest aloof the profane,
But the solitude oppresses thy votary!
The jars of men reach him not in thy valley—
But can life reach him?   210
Thou fencest him from the multitude—
Who will fence him from himself?
He hears nothing but the cry of the torrents,
And the beating of his own heart.
The air is thin, the veins swell,
The temples tighten and throb there—
Air! air!

Take thy bough, set me free from my solitude;
I have been enough alone!

Where shall thy votary fly then? back to men?—°   220
But they will gladly welcome him once more,
And help him to unbend his too tense thought,
And rid him of the presence of himself,
And keep their friendly chatter at his ear,
And haunt him, till the absence from himself,
That other torment, grow unbearable;
And he will fly to solitude again,
And he will find its air too keen for him,
And so change back; and many thousand times
Be miserably bandied to and fro   230

Like a sea-wave, betwixt the world and thee,
Thou young, implacable God! and only death
Can cut his oscillations short, and so
Bring him to poise. There is no other way.

And yet what days were those, Parmenides!
When we were young, when we could number friends
In all the Italian cities like ourselves,
When with elated hearts we join'd your train,
Ye Sun-born Virgins! on the road of truth.
Then we could still enjoy, then neither thought   240
Nor outward things were closed and dead to us;
But we received the shock of mighty thoughts
On simple minds with a pure natural joy;
And if the sacred load oppress'd our brain,
We had the power to feel the pressure eased,
The brow unbound, the thoughts flow free again,
In the delightful commerce of the world.
We had not lost our balance then, nor grown
Thought's slaves, and dead to every natural joy.
The smallest thing could give us pleasure then—   250
The sports of the country-people,
A flute-note from the woods,
Sunset over the sea;
Seed-time and harvest,
The reapers in the corn,
The vinedresser in his vineyard,
The village-girl at her wheel.

Fulness of life and power of feeling, ye
Are for the happy, for the souls at ease,
Who dwell on a firm basis of content!   260
But he, who has outlived his prosperous days—
But he, whose youth fell on a different world
From that on which his exiled age is thrown—
Whose mind was fed on other food, was train'd
By other rules than are in vogue to-day—
Whose habit of thought is fix'd, who will not change,
But, in a world he loves not, must subsist
In ceaseless opposition, be the guard
Of his own breast, fetter'd to what he guards,

That the world win no mastery over him—   270
Who has no friend, no fellow left, not one;
Who has no minute's breathing space allow'd
To nurse his dwindling faculty of joy——
Joy and the outward world must die to him,
As they are dead to me.

> *A long pause, during which* EMPEDOCLES *remains motion-*
> *less, plunged in thought. The night deepens. He moves*
> *forward and gazes round him, and proceeds:—*

And you, ye stars,
Who slowly begin to marshal,
As of old, in the fields of heaven,
Your distant, melancholy lines!
Have you, too, survived yourselves?   280
Are you, too, what I fear to become?
You, too, once lived;
You, too, moved joyfully
Among august companions,
In an older world, peopled by Gods,
In a mightier order,
The radiant, rejoicing, intelligent Sons of Heaven.
But now, ye kindle
Your lonely, cold-shining lights,
Unwilling lingerers   290
In the heavenly wilderness,
For a younger, ignoble world;
And renew, by necessity,
Night after night your courses,
In echoing, unnear'd silence,
Above a race you know not—
Uncaring and undelighted,
Without friend and without home;
Weary like us, though not
Weary with our weariness.   300

No, no, ye stars! there is no death with you,
No languor, no decay! languor and death,
They are with me, not you! ye are alive—
Ye, and the pure dark ether where ye ride
Brilliant above me! And thou, fiery world,

That sapp'st the vitals of this terrible mount
Upon whose charr'd and quaking crust I stand—
Thou, too, brimmest with life!—the sea of cloud,
That heaves its white and billowy vapours up
To moat this isle of ashes from the world,   310
Lives; and that other fainter sea, far down,
O'er whose lit floor a road of moonbeams leads
To Etna's Liparëan sister-fires°
And the long dusky line of Italy—
That mild and luminous floor of waters lives,
With held-in joy swelling its heart; I only,
Whose spring of hope is dried, whose spirit has fail'd,
I, who have not, like these, in solitude
Maintain'd courage and force, and in myself
Nursed an immortal vigour—I alone   320
Am dead to life and joy, therefore I read
In all things my own deadness.

A long silence. He continues:—

Oh, that I could glow like this mountain!
Oh, that my heart bounded with the swell of the sea!
Oh, that my soul were full of light as the stars!
Oh, that it brooded over the world like the air!

But no, this heart will glow no more; thou art
A living man no more, Empedocles!
Nothing but a devouring flame of thought—
But a naked, eternally restless mind!   330

After a pause:—

To the elements it came from
Everything will return—
Our bodies to earth,
Our blood to water,
Heat to fire,
Breath to air.
They were well born, they will be well entomb'd—
But mind?. . .

And we might gladly share the fruitful stir
Down in our mother earth's miraculous womb;   340
Well would it be

With what roll'd of us in the stormy main;
We might have joy, blent with the all-bathing air,
Or with the nimble, radiant life of fire.

But mind, but thought—
If these have been the master part of us—
Where will *they* find their parent element?
What will receive *them*, who will call *them* home?
But we shall still be in them, and they in us,
And we shall be the strangers of the world,   350
And they will be our lords, as they are now;
And keep us prisoners of our consciousness,
And never let us clasp and feel the All°
But through their forms, and modes, and stifling veils.
And we shall be unsatisfied as now;
And we shall feel the agony of thirst,
The ineffable longing for the life of life
Baffled for ever; and still thought and mind
Will hurry us with them on their homeless march,
Over the unallied unopening earth,   360
Over the unrecognising sea; while air
Will blow us fiercely back to sea and earth,
And fire repel us from its living waves.
And then we shall unwillingly return
Back to this meadow of calamity,
This uncongenial place, this human life;
And in our individual human state
Go through the sad probation all again,°
To see if we will poise our life at last,
To see if we will now at last be true   370
To our own only true, deep-buried selves,°
Being one with which we are one with the whole world;
Or whether we will once more fall away
Into some bondage of the flesh or mind,
Some slough of sense, or some fantastic maze
Forged by the imperious lonely thinking-power.
And each succeeding age in which we are born
Will have more peril for us than the last;
Will goad our senses with a sharper spur,
Will fret our minds to an intenser play,   380
Will make ourselves harder to be discern'd.

And we shall struggle awhile, gasp and rebel—
And we shall fly for refuge to past times,
Their soul of unworn youth, their breath of greatness;
And the reality will pluck us back,
Knead us in its hot hand, and change our nature
And we shall feel our powers of effort flag,
And rally them for one last fight—and fail;
And we shall sink in the impossible strife,
And be astray for ever.

              Slave of sense   390
I have in no wise been;—but slave of thought?. . .
And who can say: I have been always free,
Lived ever in the light of my own soul?—
I cannot; I have lived in wrath and gloom,
Fierce, disputatious, ever at war with man,
Far from my own soul, far from warmth and light.
But I have not grown easy in these bonds—
But I have not denied what bonds these were.
Yea, I take myself to witness,
That I have loved no darkness,   400
Sophisticated no truth,
Nursed no delusion,
Allow'd no fear!

    And therefore, O ye elements! I know—°
Ye know it too—it hath been granted me
Not to die wholly, not to be all enslaved.
I feel it in this hour. The numbing cloud
Mounts off my soul; I feel it, I breathe free.

Is it but for a moment?
—Ah, boil up, ye vapours!   410
Leap and roar, thou sea of fire!
My soul glows to meet you.
Ere it flag, ere the mists
Of despondency and gloom
Rush over it again,
Receive me, save me!

                              [He plunges into the crater.

CALLICLES

(from below)

Through the black, rushing smoke-bursts,°
Thick breaks the red flame;
All Etna heaves fiercely
Her forest-clothed frame.   420

Not here, O Apollo!
Are haunts meet for thee.
But, where Helicon breaks down
In cliff to the sea,

Where the moon-silver'd inlets
Send far their light voice
Up the still vale of Thisbe,°
O speed, and rejoice!

On the sward at the cliff-top
Lie strewn the white flocks,   430
On the cliff-side the pigeons
Roost deep in the rocks.

In the moonlight the shepherds,
Soft lull'd by the rills,
Lie wrapt in their blankets
Asleep on the hills.

—What forms are these coming
So white through the gloom?
What garments out-glistening
The gold-flower'd broom?   440

What sweet-breathing presence
Out-perfumes the thyme?
What voices enrapture
The night's balmy prime?—

'Tis Apollo comes leading
His choir, the Nine.
—The leader is fairest,
But all are divine.

They are lost in the hollows!
They stream up again!   450
What seeks on this mountain
The glorified train?—

They bathe on this mountain,
In the spring by their road;
Then on to Olympus,
Their endless abode.

—Whose praise do they mention?°
Of what is it told?—
What will be for ever;
What was from of old.   460

First hymn they the Father
Of all things; and then,
The rest of immortals,
The action of men.

The day in his hotness,
The strife with the palm;
The night in her silence,
The stars in their calm.

Tristram and Iseult

I

𝕿ristram

TRISTRAM

Is she not come? The messenger was sure.
Prop me upon the pillows once again—
Raise me, my page! this cannot long endure.
—Christ, what a night! how the sleet whips the pane!°
What lights will those out to the northward be?

THE PAGE

The lanterns of the fishing-boats at sea.

TRISTRAM

Soft—who is that, stands by the dying fire?

THE PAGE

Iseult.

TRISTRAM

Ah! not the Iseult I desire.

*   *   *   *

What Knight is this so weak and pale,
Through the locks are yet brown on his noble head,   10
Propt on pillows in his bed,
Gazing seaward for the light
Of some ship that fights the gale
On this wild December night?
Over the sick man's feet is spread
A dark green forest-dress;
A gold harp leans against the bed,
Ruddy in the fire's light.
I know him by his harp of gold,
Famous in Arthur's court of old;   20
I know him by his forest-dress—
The peerless hunter, harper, knight,
Tristram of Lyoness.

What Lady is this, whose silk attire
Gleams so rich in the light of the fire?
The ringlets on her shoulders lying
In their flitting lustre vying
With the clasp of burnish'd gold
Which her heavy robe doth hold.
Her looks are mild, her fingers slight   30
As the driven snow are white;
But her cheeks are sunk and pale.
Is it that the bleak sea-gale
Beating from the Atlantic sea
On this coast of Brittany,
Nips too keenly the sweet flower?
Is it that a deep fatigue
Hath come on her, a chilly fear,
Passing all her youthful hour
Spinning with her maidens here,   40
Listlessly through the window-bars
Gazing seawards many a league,
From her lonely shore-built tower,
While the knights are at the wars?
Or, perhaps, has her young heart
Felt already some deeper smart,
Of those that in secret the heart-strings rive,
Leaving her sunk and pale, though fair?
Who is this snowdrop by the sea?—
I know her by her mildness rare,   50
Her snow-white hands, her golden hair;
I know her by her rich silk dress,
And her fragile loveliness—
The sweetest Christian soul alive,
Iseult of Brittany.

Iseult of Brittany?—but where
Is that other Iseult fair,
That proud, first Iseult, Cornwall's queen?
She, whom Tristram's ship of yore
From Ireland to Cornwall bore,   60
To Tyntagel, to the side
Of King Marc, to be his bride?
She who, as they voyaged, quaff'd

With Tristram that spiced magic draught,
Which since then for ever rolls
Through their blood, and binds their souls,
Working love, but working teen?—
There were two Iseults who did sway
Each her hour of Tristram's day;
But one possess'd his waning time,   70
The other his resplendent prime.
Behold her here, the patient flower,
Who possess'd his darker hour!
Iseult of the Snow-White Hand°
Watches pale by Tristram's bed.
She is here who had his gloom,
Where art thou who hadst his bloom?
One such kiss as those of yore
Might thy dying knight restore!
Does the love-draught work no more?   80
Art thou cold, or false, or dead,
Iseult of Ireland?

* * * *

Loud howls the wind, sharp patters the rain,
And the knight sinks back on his pillows again.
He is weak with fever and pain,
And his spirit is not clear.
Hark! he mutters in his sleep,
As he wanders far from here,
Changes place and time of year,
And his closéd eye doth sweep   90
O'er some fair unwintry sea,
Not this fierce Atlantic deep,
While he mutters brokenly:—

TRISTRAM

The calm sea shines, loose hang the vessel's sails;
Before us are the sweet green fields of Wales,
And overhead the cloudless sky of May.—
'Ah, would I were in those green fields at play,
Not pent on ship-board this delicious day!
Tristram, I pray thee, of thy courtesy,
Reach me my golden phial stands by thee,   100

But pledge me in it first for courtesy.—'
Ha! dost thou start? are thy lips blanch'd like mine?
Child, 'tis no true draught this, 'tis poison'd wine!
Iseult!. . . .

*   *   *   *

Ah, sweet angels, let him dream!
Keep his eyelids! let him seem
Not this fever-wasted wight
Thinn'd and paled before his time,
But the brilliant youthful knight
In the glory of his prime,   110
Sitting in the gilded barge,
At thy side, thou lovely charge,
Bending gaily o'er thy hand,
Iseult of Ireland!
And she too, that princess fair,
If her bloom be now less rare,
Let her have her youth again—
Let her be as she was then!
Let her have her proud dark eyes,
And her petulant quick replies—   120
Let her sweep her dazzling hand
With its gesture of command,
And shake back her raven hair
With the old imperious air!
As of old, so let her be,
That first Iseult, princess bright,
Chatting with her youthful knight
As he steers her o'er the sea,
Quitting at her father's will
The green isle where she was bred,   130
And her bower in Ireland,
For the surge-beat Cornish strand;
Where the prince whom she must wed
Dwells on loud Tyntagel's hill,
High above the sounding sea.
And that potion rare her mother
Gave her, that her future lord,
Gave her, that King Marc and she,
Might drink it on their marriage-day,

And for ever love each other— 140
Let her, as she sits on board,
Ah, sweet saints, unwittingly!
See it shine, and take it up,
And to Tristram laughing say:
'Sir Tristram, of thy courtesy,
Pledge me in my golden cup!'
Let them drink it—let their hands
Tremble, and their cheeks be flame,
As they feel the fatal bands
Of a love they dare not name, 150
With a wild delicious pain,
Twine about their hearts again!
Let the early summer be
Once more round them, and the sea
Blue, and o'er its mirror kind
Let the breath of the May-wind,
Wandering through their drooping sails,
Die on the green fields of Wales!
Let a dream like this restore
What his eye must see no more! 160

TRISTRAM

Chill blows the wind, the pleasaunce-walks are drear—
Madcap, what jest was this, to meet me here?
Were feet like those made for so wild a way?
The southern winter-parlour, by my fay,
Had been the likeliest trysting-place to-day!
'Tristram!—nay, nay—thou must not take my hand!—
Tristram!—sweet love!—we are betray'd—out-plann'd.
Fly—save thyself—save me!—I dare not stay.'—
One last kiss first!— ''Tis vain—to horse—away!'

*   *   *   *

Ah! sweet saints, his dream doth move 170
Faster surely than it should,
From the fever in his blood!
All the spring-time of his love
Is already gone and past,
And instead thereof is seen
Its winter, which endureth still—

Tyntagel on its surge-beat hill,
The pleasaunce-walks, the weeping queen,
The flying leaves, the straining blast,
And that long, wild kiss—their last.　　　　　180
And this rough December-night,
And his burning fever-pain,
Mingle with his hurrying dream,
Till they rule it, till he seem
The press'd fugitive again,
The love-desperate banish'd knight
With a fire in his brain
Flying o'er the stormy main.
—Whither does he wander now?
Haply in his dreams the wind　　　　　190
Wafts him here, and lets him find
The lovely orphan child again
In her castle by the coast;
The youngest, fairest chatelaine,
Whom this realm of France can boast,
Our snowdrop by the Atlantic sea,
Iseult of Brittany.
And—for through the haggard air,
The stain'd arms, the matted hair
Of that stranger-knight ill-starr'd,　　　　　200
There gleam'd something, which recall'd
The Tristram who in better days
Was Launcelot's guest at Joyous Gard—°
Welcomed here, and here install'd,
Tended of his fever here,
Haply he seems again to move
His young guardian's heart with love;
In his exiled loneliness,
In his stately, deep distress,
Without a word, without a tear.　　　　　210
—Ah! 'tis well he should retrace
His tranquil life in this lone place:
His gentle bearing at the side
Of his timid youthful bride;
His long rambles by the shore
On winter-evenings, when the roar
Of the near waves came, sadly grand,

Through the dark, up the drown'd sand,
Or his endless reveries
In the woods, where the gleams play   220
On the grass under the trees,
Passing the long summer's day
Idle as a mossy stone
In the forest-depths alone,
The chase neglected, and his hound
Couch'd beside him on the ground.
—Ah! what trouble's on his brow?
Hither let him wander now;
Hither, to the quiet hours
Pass'd among these heaths of ours   230
By the grey Atlantic sea;
Hours, if not of ecstasy,°
From violent anguish surely free!

TRISTRAM

All red with blood the whirling river flows,°
The wide plain rings, the dazed air throbs with blows.
Upon us are the chivalry of Rome—
Their spears are down, their steeds are bathed in foam.
'Up, Tristram, up,' men cry, 'thou moonstruck knight!
What foul fiend rides thee? On into the fight!'
—Above the din her voice is in my ears;   240
I see her form glide through the crossing spears.—
Iseult!. . . .

                *   *   *   *

Ah! he wanders forth again;
We cannot keep him; now, as then,
There's a secret in his breast
Which will never let him rest.
These musing fits in the green wood
They cloud the brain, they dull the blood!
—His sword is sharp, his horse is good;
Beyond the mountains will he see   250
The famous towns of Italy,
And label with the blessed sign
The heathen Saxons on the Rhine.
At Arthur's side he fights once more

With the Roman Emperor.
There's many a gay knight where he goes
Will help him to forget his care;
The march, the leaguer, Heaven's blithe air,
The neighing steeds, the ringing blows—
Sick pining comes not where these are.   260
Ah! what boots it, that the jest
Lightens every other brow,
What, that every other breast
Dances as the trumpets blow,
If one's own heart beats not light
On the waves of the toss'd fight,
If oneself cannot get free
From the clog of misery?
Thy lovely youthful wife grows pale
Watching by the salt sea-tide   270
With her children at her side
For the gleam of thy white sail.
Home, Tristram, to thy halls again!
To our lonely sea complain,
To our forests tell thy pain!

TRISTRAM

All round the forest sweeps off, black in shade,
But it is moonlight in the open glade;
And in the bottom of the glade shine clear
The forest-chapel and the fountain near.
—I think, I have a fever in my blood;   280
Come, let me leave the shadow of this wood,
Ride down, and bathe my hot brow in the flood.
—Mild shines the cold spring in the moon's clear light;
God! 'tis *her* face plays in the waters bright.
'Fair love,' she says, 'canst thou forget so soon,
At this soft hour, under this sweet moon?'—
Iseult!. . . .

     *   *   *   *

Ah, poor soul! if this be so,
Only death can balm thy woe.
The solitudes of the green wood   290
Had no medicine for thy mood;

The rushing battle clear'd thy blood
As little as did solitude.
—Ah! his eyelids slowly break
Their hot seals, and let him wake;
What new change shall we now see?
A happier? Worse it cannot be.

TRISTRAM

Is my page here! Come, turn me to the fire!
Upon the window-panes the moon shines bright;
The wind is down—but she'll not come to-night.   300
Ah no! she is asleep in Cornwall now,
Far hence; her dreams are fair—smooth is her brow.
Of me she recks not, nor my vain desire.
—I have had dreams, I have had dreams, my page,
Would take a score years from a strong man's age;
And with a blood like mine, will leave, I fear,
Scant leisure for a second messenger.
—My princess, art thou there? Sweet, do not wait!
To bed, and sleep! my fever is gone by;
To-night my page shall keep me company.   310
Where do the children sleep? kiss them for me!
Poor child, thou art almost as pale as I;
This comes of nursing long and watching late.
To bed—good night!

     *   *   *   *

She left the gleam-lit fireplace,
She came to the bed-side;
She took his hands in hers—her tears
Down on his wasted fingers rain'd.
She raised her eyes upon his face—
Not with a look of wounded pride,   320
A look as if the heart complained—
Her look was like a sad embrace;
The gaze of one who can divine
A grief, and sympathise.
Sweet flower! thy children's eyes
Are not more innocent than thine.
   But they sleep in shelter'd rest,
Like helpless birds in the warm nest,

On the castle's southern side;
Where feebly comes the mournful roar   330
Of buffeting wind and surging tide
Through many a room and corridor.
—Full on their window the moon's ray
Makes their chamber as bright as day.
It shines upon the blank white walls,
And on the snowy pillow falls,
And on two angel-heads doth play
Turn'd to each other—the eyes closed,
The lashes on the cheeks reposed.
Round each sweet brow the cap close-set   340
Hardly lets peep the golden hair;
Through the soft-open'd lips the air
Scarcely moves the coverlet.
One little wandering arm is thrown
At random on the counterpane,
And often the fingers close in haste
As if their baby-owner chased
The butterflies again.
This stir they have, and this alone;
But else they are so still!   350
—Ah, tired madcaps! you lie still;
But were you at the window now,
To look forth on the fairy sight
Of your illumined haunts by night,
To see the park-glades where you play
Far lovelier than they are by day,
To see the sparkle on the eaves,
And upon every giant-bough
Of those old oaks, whose wet red leaves
Are jewell'd with bright drops of rain—   360
How would your voices run again!
And far beyond the sparkling trees
Of the castle-park one sees
The bare heaths spreading, clear as day,
Moor behind moor, far, far away,
Into the heart of Brittany.
And here and there, lock'd by the land,
Long inlets of smooth glittering sea,
And many a stretch of watery sand

All shining in the white moon-beams—   370
But you see fairer in your dreams!
What voices are these on the clear night-air?
What lights in the court—what steps on the stair?

II

Iseult of Ireland

TRISTRAM

Raise the light, my page! that I may see her.—
    Thou art come at last, then, haughty Queen!
Long I've waited, long I've fought my fever;
    Late thou comest, cruel thou hast been.

ISEULT

Blame me not, poor sufferer! that I tarried;
    Bound I was, I could not break the band.
Chide not with the past, but feel the present!
    I am here—we meet—I hold thy hand.

TRISTRAM

Thou art come, indeed—thou hast rejoin'd me;
    Thou hast dared it—but too late to save.   10
Fear not now that men should tax thine honour!
    I am dying: build—(thou may'st)—my grave!

ISEULT

Tristram, ah, for love of Heaven, speak kindly!
    What, I hear these bitter words from thee?
Sick with grief I am, and faint with travel—
    Take my hand—dear Tristram, look on me!

TRISTRAM

I forgot, thou comest from thy voyage—
    Yes, the spray is on thy cloak and hair.
But thy dark eyes are not dimm'd, proud Iseult!
    And thy beauty never was more fair.   20

ISEULT

Ah, harsh flatterer! let alone my beauty!
　I, like thee, have left my youth afar.
Take my hand, and touch these wasted fingers—
　See my cheek and lips, how white they are!

TRISTRAM

Thou art paler—but thy sweet charm, Iseult!
　Would not fade with the dull years away.
Ah, how fair thou standest in the moonlight!
　I forgive thee, Iseult!—thou wilt stay?

ISEULT

Fear me not, I will be always with thee;
　I will watch thee, tend thee, soothe thy pain;   30
Sing thee tales of true, long-parted lovers,
　Join'd at evening of their days again.

TRISTRAM

No, thou shalt not speak! I should be finding
　Something alter'd in thy courtly tone.
Sit—sit by me! I will think we've lived so
　In the green wood, all our lives, alone.

ISEULT

Alter'd, Tristram? Not in courts, believe me,
　Love like mine is alter'd in the breast;
Courtly life is light and cannot reach it—
　Ah! it lives, because so deep-suppress'd!   40

What, thou think'st men speak in courtly chambers
　Words by which the wretched are consoled?
What, thou think'st this aching brow was cooler,
　Circled, Tristram, by a band of gold?

Royal state with Marc, my deep-wrong'd husband—
　That was bliss to make my sorrows flee!
Silken courtiers whispering honied nothings—
　Those were friends to make me false to thee!

Ah, on which, if both our lots were balanced,
  Was indeed the heaviest burden thrown—   50
Thee, a pining exile in thy forest,
  Me, a smiling queen upon my throne?

Vain and strange debate, where both have suffer'd,
  Both have pass'd a youth consumed and sad,
Both have brought their anxious day to evening,
  And have now short space for being glad!

Join'd we are henceforth; nor will thy people,
  Nor thy younger Iseult take it ill,
That a former rival shares her office,
  When she sees her humbled, pale, and still.   60

I, a faded watcher by thy pillow,
  I, a statue on thy chapel-floor,
Pour'd in prayer before the Virgin-Mother,
  Rouse no anger, make no rivals more.

She will cry: 'Is this the foe I dreaded?
  This his idol? this that royal bride?
Ah, an hour of health would purge his eyesight!
  Stay, pale queen! for ever by my side.'

Hush, no words! that smile, I see, forgives me.
  I am now thy nurse, I bid thee sleep.   70
Close thine eyes—this flooding moonlight blinds them!—
  Nay, all's well again! thou must not weep.

TRISTRAM

I am happy! yet I feel, there's something
  Swells my heart, and takes my breath away.
Through a mist I see thee; near—come nearer!
  Bend—bend down!—I yet have much to say.

ISEULT

Heaven! his head sinks back upon the pillow—
  Tristram! Tristram! let thy heart not fail!
Call on God and on the holy angels!
  What, love, courage!—Christ! he is so pale.   80

TRISTRAM

Hush, 'tis vain, I feel my end approaching!
  This is what my mother said should be
When the fierce pains took her in the forest,
  The deep draughts of death, in bearing me.°

'Son,' she said, 'thy name shall be of sorrow;
  Tristram art thou call'd for my death's sake.'
So she said, and died in the drear forest.
  Grief since then his home with me doth make.

I am dying.—Start not, nor look wildly!
  Me, thy living friend, thou canst not save.   90
But, since living we were ununited,
  Go not far, O Iseult! from my grave.

Close mine eyes, then seek the princess Iseult;
  Speak her fair, she is of royal blood!
Say, I will'd so, that thou stay beside me—
  She will grant it; she is kind and good.

Now to sail the seas of death I leave thee—
  One last kiss upon the living shore!

ISEULT

Tristram!—Tristram!—stay—receive me with thee!
  Iseult leaves thee, Tristram! never more.   100

                *   *   *   *

You see them clear—the moon shines bright.°
Slow, slow and softly, where she stood,
She sinks upon the ground;—her hood
Had fallen back; her arms outspread
Still hold her lover's hand; her head
Is bow'd, half-buried, on the bed.
O'er the blanch'd sheet her raven hair
Lies in disorder'd streams; and there,
Strung like white stars, the pearls still are,
And the golden bracelets, heavy and rare,   110
Flash on her white arms still.

The very same which yesternight
Flash'd in the silver sconces' light,
When the feast was gay and the laughter loud
In Tyntagel's palace proud.
But then they deck'd a restless ghost
With hot-flush'd cheeks and brilliant eyes,
And quivering lips on which the tide
Of courtly speech abruptly died,
And a glance which over the crowded floor,   120
The dancers, and the festive host,
Flew ever to the door.
That the knights eyed her in surprise,
And the dames whispered scoffingly:
'Her moods, good lack, they pass like showers!
But yesternight and she would be
As pale and still as wither'd flowers,
And now to-night she laughs and speaks
And has a colour in her cheeks;
Christ keep us from such fantasy!'—   130

Yes, now the longing is o'erpast,°
Which, dogg'd by fear and fought by shame
Shook her weak bosom day and night,
Consumed her beauty like a flame,
And dimm'd it like the desert-blast.
And though the bed-clothes hide her face,
Yet were it lifted to the light,
The sweet expression of her brow
Would charm the gazer, till his thought
Erased the ravages of time,   140
Fill'd up the hollow cheek, and brought
A freshness back as of her prime—
So healing is her quiet now.
So perfectly the lines express
A tranquil, settled loveliness,
Her younger rival's purest grace.

The air of the December-night°
Steals coldly around the chamber bright,
Where those lifeless lovers be;
Swinging with it, in the light   150

Flaps the ghostlike tapestry.
And on the arras wrought you see°
A stately Huntsman, clad in green,
And round him a fresh forest-scene.
On that clear forest-knoll he stays,
With his pack round him, and delays.
He stares and stares, with troubled face,
At this huge, gleam-lit fireplace,
At that bright, iron-figured door,
And those blown rushes on the floor.   160
He gazes down into the room
With heated cheeks and flurried air,
And to himself he seems to say:°
'What place is this, and who are they?°
Who is that kneeling Lady fair?
And on his pillows that pale Knight
Who seems of marble on a tomb?
How comes it here, this chamber bright,
Through whose mullion'd windows clear
The castle-court all wet with rain,   170
The drawbridge and the moat appear,
And then the beach, and, mark'd with spray,
The sunken reefs, and far away
The unquiet bright Atlantic plain?°
—What, has some glamour made me sleep,
And sent me with my dogs to sweep,
By night, with boisterous bugle-peal,
Through some old, sea-side, knightly hall,
Not in the free green wood at all?
That Knight's asleep, and at her prayer   180
That Lady by the bed doth kneel—
Then hush, thou boisterous bugle-peal!'
—The wild boar rustles in his lair;
The fierce hounds snuff the tainted air;
But lord and hounds keep rooted there.

Cheer, cheer thy dogs into the brake,
O Hunter! and without a fear
Thy golden-tassell'd bugle blow,
And through the glades thy pastime take—
For thou wilt rouse no sleepers here!   190

For these thou seest are unmoved;
Cold, cold as those who lived and loved°
A thousand years ago.

III
Iseult of Brittany

A year had flown, and o'er the sea away,
In Cornwall, Tristram and Queen Iseult lay;
In King Marc's chapel, in Tyntagel old—
There in a ship they bore those lovers cold.

The young surviving Iseult, one bright day,
Had wander'd forth. Her children were at play
In a green circular hollow in the heath°
Which borders the sea-shore—a country path
Creeps over it from the till'd fields behind.
The hollow's grassy banks are soft-inclined,   10
And to one standing on them, far and near
The lone unbroken view spreads bright and clear
Over the waste. This cirque of open ground
Is light and green; the heather, which all round
Creeps thickly, grows not here; but the pale grass
Is strewn with rocks, and many a shiver'd mass
Of vein'd white-gleaming quartz, and here and there
Dotted with holly-trees and juniper.
In the smooth centre of the opening stood
Three hollies side by side, and made a screen,   20
Warm with the winter-sun, of burnish'd green
With scarlet berries gemm'd, the fell-fare's food.°
Under the glittering hollies Iseult stands,
Watching her children play; their little hands
Are busy gathering spars of quartz, and streams
Of stagshorn for their hats; anon, with screams
Of mad delight they drop their spoils, and bound
Among the holly-clumps and broken ground,
Racing full speed, and startling in their rush
The fell-fares and the speckled missel-thrush   30
Out of their glossy coverts;—but when now
Their cheeks were flush'd, and over each hot brow,

Under the feather'd hats of the sweet pair,
In blinding masses shower'd the golden hair—
Then Iseult call'd them to her, and the three
Cluster'd under the holly-screen, and she
Told them an old-world Breton history.

Warm in their mantles wrapt the three stood there,
Under the hollies, in the clear still air—
Mantles with those rich furs deep glistering   40
Which Venice ships do from swart Egypt bring.
Long they stay'd still—then, pacing at their ease,
Moved up and down under the glossy trees.
But still, as they pursued their warm dry road,
From Iseult's lips the unbroken story flow'd,
And still the children listen'd, their blue eyes
Fix'd on their mother's face in wide surprise;
Nor did their looks stray once to the sea-side,
Nor to the brown heaths round them, bright and wide,
Nor to the snow, which, though 't was all away   50
From the open heath, still by the hedgerows lay,
Nor to the shining sea-fowl, that with screams
Bore up from where the bright Atlantic gleams,
Swooping to landward; nor to where, quite clear,
The fell-fares settled on the thickets near.
And they would still have listen'd, till dark night
Came keen and chill down on the heather bright;
But, when the red glow on the sea grew cold,
And the grey turrets of the castle old
Look'd sternly through the frosty evening-air,   60
Then Iseult took by the hand those children fair,
And brought her tale to an end, and found the path,
And led them home over the darkening heath.

And is she happy? Does she see unmoved
The days in which she might have lived and loved
Slip without bringing bliss slowly away,
One after one, to-morrow like to-day?
Joy has not found her yet, nor ever will—
Is it this thought which makes her mien so still,
Her features so fatigued, her eyes, though sweet,   70
So sunk, so rarely lifted save to meet

Her children's? She moves slow; her voice alone
Hath yet an infantine and silver tone,
But even that comes languidly; in truth,
She seems one dying in a mask of youth.
And now she will go home, and softly lay
Her laughing children in their beds, and play
Awhile with them before they sleep; and then
She'll light her silver lamp, which fishermen
Dragging their nets through the rough waves, afar,   80
Along this iron coast, know like a star,
And take her broidery-frame, and there she'll sit
Hour after hour, her gold curls sweeping it;
Lifting her soft-bent head only to mind
Her children, or to listen to the wind.
And when the clock peals midnight, she will move
Her work away, and let her fingers rove
Across the shaggy brows of Tristram's hound
Who lies, guarding her feet, along the ground;
Or else she will fall musing her blue eyes   90
Fixt, her slight hands clasp'd on her lap; then rise,
And at her prie-dieu kneel, until she have told
Her rosary-beads of ebony tipp'd with gold,
Then to her soft sleep—and to-morrow'll be
To-day's exact repeated effigy.

Yes, it is lonely for her in her hall.
The children, and the grey-hair'd seneschal,
Her women, and Sir Tristram's aged hound,
Are there the sole companions to be found.
But these she loves; and noisier life than this   100
She would find ill to bear, weak as she is.
She has her children, too, and night and day
Is with them; and the wide heaths where they play,
The hollies, and the cliff, and the sea-shore,
The sand, the sea-birds, and the distant sails,
These are to her dear as to them; the tales
With which this day the children she beguiled
She gleaned from Breton grandames, when a child,
In every hut along this sea-coast wild.
She herself loves them still, and, when they are told,   110
Can forget all to hear them, as of old.

Dear saints, it is not sorrow, as I hear.°
Not suffering, which shuts up eye and ear
To all that has delighted them before,
And lets us be what we were once no more.
No, we may suffer deeply, yet retain
Power to be moved and soothed, for all our pain,
By what of old pleased us, and will again.
No, 'tis the gradual furnace of the world,
In whose hot air our spirits are upcurl'd   120
Until they crumble, or else grow like steel—
Which kills in us the bloom, the youth, the spring—°
Which leaves the fierce necessity to feel,
But takes away the power—this can avail,
By drying up our joy in everything,
To make our former pleasures all seem stale.
This, or some tyrannous single thought, some fit
Of passion, which subdues our souls to it,
Till for its sake alone we live and move—
Call it ambition, or remorse, or love—   130
This too can change us wholly, and make seem
All which we did before, shadow and dream.

    And yet, I swear, it angers me to see
How this fool passion gulls men potently;
Being, in truth, but a diseased unrest,
And an unnatural overheat at best.
How they are full of languor and distress
Not having it; which when they do possess,
They straightway are burnt up with fume and care,
And spend their lives in posting here and there   140
Where this plague drives them; and have little ease,
Are furious with themselves, and hard to please.
Like that bald Caesar, the famed Roman wight,°
Who wept at reading of a Grecian knight
Who made a name at younger years than he;
Or that renown'd mirror of chivalry,
Prince Alexander, Philip's peerless son,
Who carried the great war from Macedon
Into the Soudan's realm, and thundered on°
To die at thirty-five in Babylon.   150

What tale did Iseult to the children say,
Under the hollies, that bright winter's day?

She told them of the fairy-haunted land
Away the other side of Brittany,
Beyond the heaths, edged by the lonely sea;
Of the deep forest-glades of Broce-liande,°
Through whose green boughs the golden sunshine creeps,
Where Merlin by the enchanted thorn-tree sleeps.
For here he came with the fay Vivian,
One April, when the warm days first began.   160
He was on foot, and that false fay, his friend,
On her white palfrey; here he met his end,
In these lone sylvan glades, that April-day.
This tale of Merlin and the lovely fay
Was the one Iseult chose, and she brought clear
Before the children's fancy him and her.

Blowing between the stems, the forest-air
Had loosen'd the brown locks of Vivian's hair,
Which play'd on her flush'd cheek, and her blue eyes
Sparkled with mocking glee and exercise.   170
Her palfrey's flanks were mired and bathed in sweat,
For they had travell'd far and not stopp'd yet.
A brier in that tangled wilderness
Had scored her white right hand, which she allows
To rest ungloved on her green riding-dress;
The other warded off the drooping boughs.
But still she chatted on, with her blue eyes
Fix'd full on Merlin's face, her stately prize.
Her 'haviour had the morning's fresh clear grace,°
The spirit of the woods was in her face.   180
She look'd so witching fair, that learned wight
Forgot his craft, and his best wits took flight;
And he grew fond, and eager to obey
His mistress, use her empire as she may.

They came to where the brushwood ceased, and day
Peer'd 'twixt the stems; and the ground broke away,
In a sloped sward down to a brawling brook;
And up as high as where they stood to look

On the brook's farther side was clear, but then
The underwood and trees began again.   190
This open glen was studded thick with thorns
Then white with blossom; and you saw the horns,
Through last year's fern, of the shy fallow-deer
Who come at noon down to the water here.
You saw the bright-eyed squirrels dart along
Under the thorns on the green sward; and strong
The blackbird whistled from the dingles near,
And the weird chipping of the woodpecker
Rang lonelily and sharp; the sky was fair,
And a fresh breath of spring stirr'd everywhere.   200
Merlin and Vivian stopp'd on the slope's brow,
To gaze on the light sea of leaf and bough
Which glistering plays all round them, lone and mild,
As if to itself the quiet forest smiled.
Upon the brow-top grew a thorn, and here
The grass was dry and moss'd, and you saw clear
Across the hollow; white anemonies
Starr'd the cool turf, and clumps of primroses
Ran out from the dark underwood behind.
No fairer resting-place a man could find.   210
'Here let us halt,' said Merlin then; and she
Nodded, and tied her palfrey to a tree.

They sate them down together, and a sleep
Fell upon Merlin, more like death, so deep.

Her finger on her lips, then Vivian rose,
And from her brown-lock'd head the wimple throws,
And takes it in her hand, and waves it over
The blossom'd thorn-tree and her sleeping lover.
Nine times she waved the fluttering wimple round,
And made a little plot of magic ground.   220
And in that daisied circle, as men say,
Is Merlin prisoner till the judgment-day;
But she herself whither she will can rove—
For she was passing weary of his love.°

Youth and Calm

'Tis death! and peace, indeed, is here,
And ease from shame, and rest from fear.
There's nothing can dismarble now
The smoothness of that limpid brow.
But is a calm like this, in truth,
The crowning end of life and youth,
And when this boon rewards the dead,
Are all debts paid, has all been said?
And is the heart of youth so light,
Its step so firm, its eye so bright,   10
Because on its hot brow there blows
A wind of promise and repose
From the far grave, to which it goes;
Because it hath the hope to come,

One day, to harbour in the tomb?
Ah no, the bliss youth dreams is one
For daylight, for the cheerful sun,
For feeling nerves and living breath—
Youth dreams a bliss on this side death.
It dreams a rest, if not more deep,   20
More grateful than this marble sleep;°
It hears a voice within it tell:
Calm's not life's crown, though calm is well.°
'Tis all perhaps which man acquires,
But 'tis not what our youth desires.

Faded Leaves

I. THE RIVER

Still glides the stream, slow drops the boat
Under the rustling poplars' shade;
Silent the swans beside us float—
None speaks, none heeds; ah, turn thy head!

Let those arch eyes now softly shine,
That mocking mouth grow sweetly bland;
Ah, let them rest, those eyes, on mine!
On mine let rest that lovely hand!

My pent-up tears oppress my brain,
My heart is swoln with love unsaid.   10
Ah, let me weep, and tell my pain,
And on thy shoulder rest my head!

Before I die—before the soul,
Which now is mine, must re-attain
Immunity from my control,
And wander round the world again;

Before this teased o'erlabour'd heart
For ever leaves its vain employ,
Dead to its deep habitual smart,
And dead to hopes of future joy.   20

II. TOO LATE

Each on his own strict line we move,
And some find death ere they find love;
So far apart their lives are thrown
From the twin soul which halves their own.

And sometimes, by still harder fate,
The lovers meet, but meet too late.
—Thy heart is mine!—*True, true! ah, true!*
—Then, love, thy hand!—*Ah no! adieu!*

III. SEPARATION

Stop!—not to me, at this bitter departing,
    Speak of the sure consolations of time!
Fresh be the wound, still-renew'd be its smarting,
    So but thy image endure in its prime.

But, if the stedfast commandment of Nature
    Wills that remembrance should always decay—
If the loved form and the deep-cherish'd feature
    Must, when unseen, from the soul fade away—

Me let no half-effaced memories cumber!
    Fled, fled at once, be all vestige of thee!   10
Deep be the darkness and still be the slumber—
    Dead be the past and its phantoms to me!

Then, when we meet, and thy look strays toward me,
    Scanning my face and the changes wrought there:
Who, let me say, *is this stranger regards me,*°
    With the grey eyes, and the lovely brown hair?

IV. ON THE RHINE

Vain is the effort to forget.
Some day I shall be cold, I know,
As is the eternal moonlit snow
Of the high Alps, to which I go—
But ah, not yet, not yet!

Vain is the agony of grief.
'Tis true, indeed, an iron knot
Ties straitly up from mine thy lot,
And were it snapt—thou lov'st me not!
But is despair relief?   10

Awhile let me with thought have done.
And as this brimm'd unwrinkled Rhine,
And that far purple mountain-line,
Lie sweetly in the look divine
Of the slow-sinking sun;

So let me lie, and, calm as they,
Let beam upon my inward view
Those eyes of deep, soft, lucent hue—
Eyes too expressive to be blue,
Too lovely to be grey.   20

Ah, Quiet, all things feel thy balm!
Those blue hills too, this river's flow,
Were restless once, but long ago.
Tamed is their turbulent youthful glow;
Their joy is in their calm.

V. LONGING

Come to me in my dreams, and then
By day I shall be well again!
For then the night will more than pay
The hopeless longing of the day.

Come, as thou cam'st a thousand times,
A messenger from radiant climes,
And smile on thy new world, and be
As kind to others as to me!

Or, as thou never cam'st in sooth,
Come now, and let me dream it truth;   10
And part my hair, and kiss my brow,
And say: *My love! why sufferest thou?*

Come to me in my dreams, and then
By day I shall be well again!
For then the night will more than pay
The hopeless longing of the day.

Calais Sands

A thousand knights have rein'd their steeds
To watch this line of sand-hills run,
Along the never-silent Strait,
To Calais glittering in the sun;

To look tow'rd Ardres' Golden Field°
Across this wide aërial plain,
Which glows as if the Middle Age
Were gorgeous upon earth again.

Oh, that to share this famous scene,
I saw, upon the open sand,   10
Thy lovely presence at my side,
Thy shawl, thy look, thy smile, thy hand!

How exquisite thy voice would come,
My darling, on this lonely air!
How sweetly would the fresh sea-breeze
Shake loose some band of soft brown hair!

Yet now my glance but once hath roved
O'er Calais and its famous plain;
To England's cliffs my gaze is turn'd,
On the blue strait mine eyes I strain.   20

Thou comest! Yes! the vessel's cloud
Hangs dark upon the rolling sea.
Oh, that yon sea-bird's wings were mine,
To win one instant's glimpse of thee!

I must not spring to grasp thy hand,
To woo thy smile, to seek thine eye;
But I may stand far off, and gaze,
And watch thee pass unconscious by,

And spell thy looks, and guess thy thoughts,
Mixt with the idlers on the pier.—   30
Ah, might I always rest unseen,
So I might have thee always near!

To-morrow hurry through the fields
Of Flanders to the storied Rhine!
To-night those soft-fringed eyes shall close
Beneath one roof, my queen! with mine.

Dover Beach

The sea is calm to-night.°
The tide is full, the moon lies fair
Upon the straits;—on the French coast the light°
Gleams and is gone; the cliffs of England stand,
Glimmering and vast, out in the tranquil bay.
Come to the window, sweet is the night-air!

Only, from the long line of spray
Where the sea meets the moon-blanch'd land,
Listen! you hear the grating roar
Of pebbles which the waves draw back, and fling,   10
At their return, up the high strand,
Begin, and cease, and then again begin,
With tremulous cadence slow, and bring°
The eternal note of sadness in.°

Sophocles long ago
Heard it on the Aegean, and it brought
Into his mind the turbid ebb and flow
Of human misery; we
Find also in the sound a thought,
Hearing it by this distant northern sea.°   20

The Sea of Faith°
Was once, too, at the full, and round earth's shore
Lay like the folds of a bright girdle furl'd.°
But now I only hear°
Its melancholy, long, withdrawing roar,
Retreating, to the breath
Of the night-wind, down the vast edges drear
And naked shingles of the world.

Ah, love, let us be true°
To one another! for the world, which seems   30
To lie before us like a land of dreams,
So various, so beautiful, so new,
Hath really neither joy, nor love, nor light,
Nor certitude, nor peace, nor help for pain;
And we are here as on a darkling plain°
Swept with confused alarms of struggle and flight,
Where ignorant armies clash by night.

Stanzas in Memory of
Edward Quillinan

I saw him sensitive in frame,
  I knew his spirits low;°
And wish'd him health, success, and fame—
  I do not wish it now.

For these are all their own reward,
  And leave no good behind;
They try us, oftenest make us hard,
  Less modest, pure, and kind.

Alas! yet to the suffering man,
  In this his mortal state,   10
Friends could not give what fortune can—
  Health, ease, a heart elate.

But he is now by fortune foil'd
  No more; and we retain
The memory of a man unspoil'd,
  Sweet, generous, and humane—

With all the fortunate have not,
  With gentle voice and brow.
—Alive, we would have changed his lot,
  We would not change it now.   20

Memorial Verses

APRIL 1850

Goethe in Weimar sleeps, and Greece,°
Long since, saw Byron's struggle cease.
But one such death remain'd to come;
The last poetic voice is dumb—
We stand to-day by Wordsworth's tomb.

When Byron's eyes were shut in death,
We bow'd our head and held our breath.
He taught us little; but our soul°
Had *felt* him like the thunder's roll.
With shivering heart the strife we saw   10
Of passion with eternal law;
And yet with reverential awe
We watch'd the fount of fiery life°
Which served for that Titanic strife.
    When Goethe's death was told, we said:
Sunk, then, is Europe's sagest head.
Physician of the iron age,
Goethe has done his pilgrimage.
He took the suffering human race,°
He read each wound, each weakness clear;   20
And struck his finger on the place,
And said: *Thou ailest here, and here!*
He look'd on Europe's dying hour
Of fitful dream and feverish power;
His eye plunged down the weltering strife,
The turmoil of expiring life—
He said: *The end is everywhere,*
Art still has truth, take refuge there!
And he was happy, if to know°
Causes of things, and far below   30
His feet to see the lurid flow
Of terror, and insane distress,
And headlong fate, be happiness.

And Wordsworth!—Ah, pale ghosts, rejoice!
For never has such soothing voice
Been to your shadowy world convey'd,
Since erst, at morn, some wandering shade
Heard the clear song of Orpheus come
Through Hades, and the mournful gloom.
Wordsworth has gone from us—and ye,   40
Ah, may ye feel his voice as we!
He too upon a wintry clime
Had fallen—on this iron time°
Of doubts, disputes, distractions, fears.
He found us when the age had bound

Our souls in its benumbing round;
He spoke, and loosed our heart in tears.
He laid us as we lay at birth
On the cool flowery lap of earth,
Smiles broke from us and we had ease;   50
The hills were round us, and the breeze
Went o'er the sun-lit fields again;
Our foreheads felt the wind and rain.
Our youth return'd; for there was shed
On spirits that had long been dead,
Spirits dried up and closely furl'd,°
The freshness of the early world.

Ah! since dark days still bring to light
Man's prudence and man's fiery might,
Time may restore us in his course   60
Goethe's sage mind and Byron's force;
But where will Europe's latter hour
Again find Wordsworth's healing power?
Others will teach us how to dare,
And against fear our breast to steel;
Others will strengthen us to bear—
But who, ah! who, will make us feel?
The cloud of mortal destiny,
Others will front it fearlessly—
But who, like him, will put it by?   70

Keep fresh the grass upon his grave
O Rotha, with thy living wave!°
Sing him thy best! for few or none
Hears thy voice right, now he is gone.

The Youth of Nature

Raised are the dripping oars,
Silent the boat! the lake,
Lovely and soft as a dream,
Swims in the sheen of the moon.
The mountains stand at its head

Clear in the pure June-night,
But the valleys are flooded with haze.
Rydal and Fairfield are there;
In the shadow Wordsworth lies dead.
So it is, so it will be for aye.   10
Nature is fresh as of old,
Is lovely; a mortal is dead.

The spots which recall him survive,
For he lent a new life to these hills.
The Pillar still broods o'er the fields°
Which border Ennerdale Lake,
And Egremont sleeps by the sea.
The gleam of The Evening Star°
Twinkles on Grasmere no more,
But ruin'd and solemn and grey   20
The sheepfold of Michael survives;
And, far to the south, the heath

Still blows in the Quantock coombs,
By the favourite waters of Ruth.°
These survive!—yet not without pain,
Pain and dejection to-night,
Can I feel that their poet is gone.

He grew old in an age he condemn'd.°
He look'd on the rushing decay
Of the times which had shelter'd his youth;   30
Felt the dissolving throes
Of a social order he loved;
Outlived his brethren, his peers,
And, like the Theban seer,
Died in his enemies' day.

Cold bubbled the spring of Tilphusa,°
Copais lay bright in the moon,
Helicon glass'd in the lake
Its firs, and afar rose the peaks
Of Parnassus, snowily clear;   40
Thebes was behind him in flames,°
And the clang of arms in his ear,

When his awe-struck captors led
The Theban seer to the spring.
Tiresias drank and died.
Nor did reviving Thebes
See such a prophet again.

Well may we mourn, when the head
Of a sacred poet lies low
In an age which can rear them no more!   50
The complaining millions of men
Darken in labour and pain;
But he was a priest to us all°
Of the wonder and bloom of the world,
Which we saw with his eyes, and were glad.
He is dead, and the fruit-bearing day
Of his race is past on the earth;
And darkness returns to our eyes.

For, oh! is it you, is it you,°
Moonlight, and shadow, and lake,   60
And mountains, that fill us with joy,
Or the poet who sings you so well?
Is it you, O beauty, O grace,
O charm, O romance, that we feel,
Or the voice which reveals what you are?
Are ye, like daylight and sun,
Shared and rejoiced in by all?
Or are ye immersed in the mass
Of matter, and hard to extract,
Or sunk at the core of the world   70
Too deep for the most to discern?
Like stars in the deep of the sky,
Which arise on the glass of the sage,
But are lost when their watcher is gone.

'They are here'—I heard, as men heard
In Mysian Ida the voice°
Of the Mighty Mother, or Crete,
The murmur of Nature reply—
'Loveliness, magic, and grace,
They are here! they are set in the world,   80

They abide; and the finest of souls
Hath not been thrill'd by them all,
Nor the dullest been dead to them quite.
The poet who sings them may die,
But they are immortal and live,
For they are the life of the world.
Will ye not learn it, and know,
When ye mourn that a poet is dead,
That the singer was less than his themes,
Life, and emotion, and I?   90

'More than the singer are these.
Weak is the tremor of pain
That thrills in his mournfullest chord
To that which once ran through his soul.
Cold the elation of joy
In his gladdest, airiest song,
To that which of old in his youth
Fill'd him and made him divine.
Hardly his voice at its best
Gives us a sense of the awe,   100
The vastness, the grandeur, the gloom
Of the unlit gulph of himself.

'Ye know not yourselves; and your bards—
The clearest, the best, who have read
Most in themselves—have beheld
Less than they left unreveal'd.
Ye express not yourselves;—can you make
With marble, with colour, with word,
What charm'd you in others re-live?
Can thy pencil, O artist! restore°   110
The figure, the bloom of thy love,
As she was in her morning of spring?
Canst thou paint the ineffable smile
Of her eyes as they rested on thine?
Can the image of life have the glow,°
The motion of life itself?

'Yourselves and your fellows ye know not; and me,
The mateless, the one, will ye know?

Will ye scan me, and read me, and tell
Of the thoughts that ferment in my breast,   120
My longing, my sadness, my joy?
Will ye claim for your great ones the gift
To have render'd the gleam of my skies,
To have echoed the moan of my seas,
Utter'd the voice of my hills?
When your great ones depart, will ye say:
All things have suffer'd a loss,
Nature is hid in their grave?

'Race after race, man after man,
Have thought that my secret was theirs,   130
Have dream'd that I lived but for them,
That they were my glory and joy.
—They are dust, they are changed, they are gone!°
I remain.'

The Youth of Man

We, O Nature, depart,
Thou survivest us! this,
This, I know, is the law.
Yes? but more than this,
Thou who seest us die
Seest us change while we live;
Seest our dreams, one by one,
Seest our errors depart;
Watchest us, Nature! throughout,
Mild and inscrutably calm.   10

Well for us that we change!
Well for us that the power
Which in our morning-prime
Saw the mistakes of our youth,
Sweet, and forgiving, and good,
Sees the contrition of age!

Behold, O Nature, this pair!
See them to-night where they stand,
Not with the halo of youth
Crowning their brows with its light,   20
Not with the sunshine of hope,
Not with the rapture of spring,
Which they had of old, when they stood
Years ago at my side
In this self-same garden, and said:
'We are young, and the world is ours;
Man, man is the king of the world!
Fools that these mystics are
Who prate of Nature! for she
Hath neither beauty, nor warmth,   30
Nor life, nor emotion, nor power.
But man has a thousand gifts,
And the generous dreamer invests
The senseless world with them all.
Nature is nothing; her charm
Lives in our eyes which can paint,
Lives in our hearts which can feel.'

Thou, O Nature, wast mute,
Mute as of old! days flew,
Days and years; and Time   40
With the ceaseless stroke of his wings
Brush'd off the bloom from their soul.
Clouded and dim grew their eye,
Languid their heart—for youth
Quicken'd its pulses no more.
Slowly, within the walls
Of an ever-narrowing world,
They droop'd, they grew blind, they grew old.
Thee and their youth in thee,
Nature! they saw no more.   50

Murmur of living,°
Stir of existence,
Soul of the world!
Make, oh, make yourselves felt
To the dying spirit of youth!

Come, like the breath of the spring!
Leave not a human soul
To grow old in darkness and pain!
Only the living can feel you,
But leave us not while we live!   60

Here they stand to-night—
Here, where this grey balustrade
Crowns the still valley; behind
Is the castled house, with its woods,
Which shelter'd their childhood—the sun
On its ivied windows; a scent
From the grey-wall'd gardens, a breath
Of the fragrant stock and the pink,
Perfumes the evening air.
Their children play on the lawns.   70
They stand and listen; they hear
The children's shouts, and at times,
Faintly, the bark of a dog
From a distant farm in the hills.
Nothing besides! in front
The wide, wide valley outspreads
To the dim horizon, reposed
In the twilight, and bathed in dew,
Corn-field and hamlet and copse
Darkening fast; but a light,   80
Far off, a glory of day,
Still plays on the city spires;
And there in the dusk by the walls,
With the grey mist marking its course
Through the silent, flowery land,
On, to the plains, to the sea,
Floats the imperial stream.

    Well I know what they feel!
They gaze, and the evening wind
Plays on their faces; they gaze—   90
Airs from the Eden of youth
Awake and stir in their soul;
The past returns—they feel
What they are, alas! what they were.

They, not Nature, are changed.
Well I know what they feel!

Hush, for tears
Begin to steal to their eyes!
Hush, for fruit
Grows from such sorrow as theirs!   100

And they remember,
With piercing, untold anguish,
The proud boasting of their youth.
And they feel how Nature was fair.
And the mists of delusion,
And the scales of habit,
Fall away from their eyes;
And they see, for a moment,
Stretching out, like the desert
In its weary, unprofitable length,   110
Their faded, ignoble lives.

While the locks are yet brown on thy head,°
While the soul still looks through thine eyes,
While the heart still pours
The mantling blood to thy cheek,
Sink, O youth, in thy soul!
Yearn to the greatness of Nature;
Rally the good in the depths of thyself!

Lines Written in Kensington Gardens

In this lone, open glade I lie,
Screen'd by deep boughs on either hand;
And at its end, to stay the eye,
Those black-crown'd, red-boled pine-trees stand!

Birds here make song, each bird has his,
Across the girdling city's hum.
How green under the boughs it is!
How thick the tremulous sheep-cries come!

Sometimes a child will cross the glade
To take his nurse his broken toy;   10
Sometimes a thrush flit overhead
Deep in her unknown day's employ.

Here at my feet what wonders pass,°
What endless, active life is here!
What blowing daisies, fragrant grass!
An air-stirr'd forest, fresh and clear.

Scarce fresher is the mountain-sod
Where the tired angler lies, stretch'd out,
And, eased of basket and of rod,
Counts his day's spoil, the spotted trout.   20

In the huge world, which roars hard by,
Be others happy if they can!
But in my helpless cradle I
Was breathed on by the rural Pan.

I, on men's impious uproar hurl'd,
Think often, as I hear them rave,
That peace has left the upper world
And now keeps only in the grave.

Yet here is peace for ever new!
When I who watch them am away,   30
Still all things in this glade go through
The changes of their quiet day.

Then to their happy rest they pass!
The flowers upclose, the birds are fed,
The night comes down upon the grass,
The child sleeps warmly in his bed.°

Calm soul of all things! make it mine
To feel, amid the city's jar,
That there abides a peace of thine,
Man did not make, and cannot mar.   40

The will to neither strive nor cry,
The power to feel with others give!
Calm, calm me more! nor let me die°
Before I have begun to live.

The Future

A wanderer is man from his birth.
He was born in a ship
On the breast of the river of Time;
Brimming with wonder and joy
He spreads out his arms to the light,
Rivets his gaze on the banks of the stream.

As what he sees is, so have his thoughts been.
Whether he wakes,
Where the snowy mountainous pass,
Echoing the screams of the eagles,   10
Hems in its gorges the bed
Of the new-born clear-flowing stream;
Whether he first sees light
Where the river in gleaming rings
Sluggishly winds through the plain;
Whether in sound of the swallowing sea—
As is the world on the banks,
So is the mind of the man.

  Vainly does each, as he glides,
Fable and dream   20
Of the lands which the river of Time
Had left ere he woke on its breast,
Or shall reach when his eyes have been closed.
Only the tract where he sails
He wots of; only the thoughts,
Raised by the objects he passes, are his.

Who can see the green earth any more
As she was by the sources of Time?
Who imagines her fields as they lay
In the sunshine, unworn by the plough?   30
Who thinks as they thought,
The tribes who then roam'd on her breast,
Her vigorous, primitive sons?

What girl
Now reads in her bosom as clear
As Rebekah read, when she sate°
At eve by the palm-shaded well?
Who guards in her breast
As deep, as pellucid a spring
Of feeling, as tranquil, as sure?   40

   What bard,°
At the height of his vision, can deem
Of God, of the world, of the soul,
With a plainness as near,
As flashing as Moses felt
When he lay in the night by his flock
On the starlit Arabian waste?
Can rise and obey
The beck of the Spirit like him?

This tract which the river of Time°   50
Now flows through with us, is the plain.
Gone is the calm of its earlier shore.
Border'd by cities and hoarse
With a thousand cries is its stream.
And we on its breast, our minds
Are confused as the cries which we hear,
Changing and shot as the sights which we see.°

And we say that repose has fled
For ever the course of the river of Time.
That cities will crowd to its edge   60
In a blacker, incessanter line;
That the din will be more on its banks,
Denser the trade on its stream,
Flatter the plain where it flows,
Fiercer the sun overhead.
That never will those on its breast
See an ennobling sight,
Drink of the feeling of quiet again.

But what was before us we know not,
And we know not what shall succeed.   70

Haply, the river of Time—
As it grows, as the towns on its marge
Fling their wavering lights
On a wider, statelier stream—
May acquire, if not the calm
Of its early mountainous shore,
Yet a solemn peace of its own.

And the width of the waters, the hush
Of the grey expanse where he floats,
Freshening its current and spotted with foam   80
As it draws to the Ocean, may strike
Peace to the soul of the man on its breast—°
As the pale waste widens around him,
As the banks fade dimmer away,
As the stars come out, and the night-wind
Brings up the stream
Murmurs and scents of the infinite sea.

A Summer Night

In the deserted, moon-blanch'd street,
How lonely rings the echo of my feet!
Those windows, which I gaze at, frown,
Silent and white, unopening down,
Repellent as the world;—but see,
A break between the housetops shows
The moon! and, lost behind her, fading dim
Into the dewy dark obscurity
Down at the far horizon's rim,
Doth a whole tract of heaven disclose!   10

And to my mind the thought
Is on a sudden brought
Of a past night, and a far different scene.
Headlands stood out into the moonlit deep
As clearly as at noon;
The spring-tide's brimming flow
Heaved dazzlingly between;

Houses, with long white sweep,
Girdled the glistening bay;
Behind, through the soft air,   20
The blue haze-cradled mountains spread away,
That night was far more fair—
But the same restless pacings to and fro,
And the same vainly throbbing heart was there,
And the same bright, calm moon.
And the calm moonlight seems to say:
Hast thou then still the old unquiet breast,
Which neither deadens into rest,
Nor ever feels the fiery glow
That whirls the spirit from itself away,   30
But fluctuates to and fro,
Never by passion quite possess'd
And never quite benumb'd by the world's sway?—
And I, I know not if to pray
Still to be what I am, or yield and be
Like all the other men I see.

For most men in a brazen prison live,°
Where, in the sun's hot eye,
With heads bent o'er their toil, they languidly
Their lives to some unmeaning taskwork give,   40
Dreaming of nought beyond their prison-wall.
And as, year after year,
Fresh products of their barren labour fall
From their tired hands, and rest
Never yet comes more near,
Gloom settles slowly down over their breast;
And while they try to stem
The waves of mournful thought by which they are prest,
Death in their prison reaches them,
Unfreed, having seen nothing, still unblest.   50

And the rest, a few,
Escape their prison and depart
On the wide ocean of life anew.
There the freed prisoner, where'er his heart
Listeth, will sail;
Nor doth he know how there prevail,

Despotic on that sea,
Trade-winds which cross it from eternity.
Awhile he holds some false way, undebarr'd
By thwarting signs, and braves   60
The freshening wind and blackening waves.
And then the tempest strikes him; and between
The lightning-bursts is seen
Only a driving wreck,
And the pale master on his spar-strewn deck
With anguish'd face and flying hair
Grasping the rudder hard,
Still bent to make some port he knows not where,
Still standing for some false, impossible shore.
And sterner comes the roar   70
Of sea and wind, and through the deepening gloom
Fainter and fainter wreck and helmsman loom,
And he too disappears, and comes no more.

Is there no life, but these alone?
Madman or slave, must man be one?

Plainness and clearness without shadow of stain!
Clearness divine!
Ye heavens, whose pure dark regions have no sign
Of languor, though so calm, and, though so great,
Are yet untroubled and unpassionate;   80
Who, though so noble, share in the world's toil,
And, though so task'd, keep free from dust and soil!
I will not say that your mild deeps retain
A tinge, it may be, of their silent pain
Who have long'd deeply once, and long'd in vain—
But I will rather say that you remain
A world above man's head, to let him see
How boundless might his soul's horizons be,
How vast, yet of what clear transparency!
How it were good to abide there, and breathe free;   90
How fair a lot to fill
Is left to each man still!

The Buried Life

Light flows our war of mocking words, and yet,
Behold, with tears mine eyes are wet!
I feel a nameless sadness o'er me roll.
'Yes, yes, we know that we can jest,
We know, we know that we can smile!
But there's a something in this breast,
To which thy light words bring no rest,
And thy gay smiles no anodyne.
Give me thy hand, and hush awhile,
And turn those limpid eyes on mine,   10
And let me read there, love! thy inmost soul.

Alas! is even love too weak
To unlock the heart, and let it speak?
Are even lovers powerless to reveal
To one another what indeed they feel?
I knew the mass of men conceal'd
Their thoughts, for fear that if reveal'd
They would by other men be met
With blank indifference, or with blame reproved;
I knew they lived and moved   20
Trick'd in disguises, alien to the rest
Of men, and alien to themselves—and yet
The same heart beats in every human breast!

But we, my love!—doth a like spell benumb
Our hearts, our voices?—must we too be dumb?

Ah! well for us, if even we,
Even for a moment, can get free
Our heart, and have our lips unchain'd;
For that which seals them hath been deep-ordain'd!
Fate, which foresaw°   30
How frivolous a baby man would be—
By what distractions he would be possess'd,
How he would pour himself in every strife,
And well-nigh change his own identity—
That it might keep from his capricious play
His genuine self, and force him to obey

Even in his own despite his being's law,
Bade through the deep recesses of our breast
The unregarded river of our life
Pursue with indiscernible flow its way;   40
And that we should not see
The buried stream, and seem to be
Eddying at large in blind uncertainty,
Though driving on with it eternally.

But often, in the world's most crowded streets,
But often, in the din of strife,
There rises an unspeakable desire
After the knowledge of our buried life;
A thirst to spend our fire and restless force
In tracking out our true, original course;   50
A longing to inquire
Into the mystery of this heart which beats
So wild, so deep in us—to know
Whence our lives come and where they go.
And many a man in his own breast then delves,°
But deep enough, alas! none ever mines.
And we have been on many thousand lines°
And we have shown, on each, spirit and power;
But hardly have we, for one little hour,
Been on our own line, have we been ourselves—   60
Hardly had skill to utter one of all
The nameless feelings that course through our breast,
But they course on for ever unexpress'd.
And long we try in vain to speak and act
Our hidden self, and what we say and do
Is eloquent, is well—but 'tis not true!
And then we will no more be rack'd°
With inward striving, and demand
Of all the thousand nothings of the hour
Their stupefying power;   70
Ah yes, and they benumb us at our call!
Yet still, from time to time, vague and forlorn,
From the soul's subterranean depth upborne
As from an infinitely distant land,
Come airs, and floating echoes, and convey
A melancholy into all our day.

Only—but this is rare—
When a belovéd hand is laid in ours,
When, jaded with the rush and glare
Of the interminable hours,   80
Our eyes can in another's eyes read clear,
When our world-deafen'd ear
Is by the tones of a loved voice caress'd—
A bolt is shot back somewhere in our breast,
And a lost pulse of feeling stirs again.
The eye sinks inward, and the heart lies plain,
And what we mean, we say, and what we would, we know.
A man becomes aware of his life's flow,°
And hears its winding murmur; and he sees
The meadows where it glides, the sun, the breeze.   90

And there arrives a lull in the hot race
Wherein he doth for ever chase
That flying and elusive shadow, rest.
An air of coolness plays upon his face,
And an unwonted calm pervades his breast.
And then he thinks he knows
The hills where his life rose,
And the sea where it goes.

Self-Deception

Say, what blinds us, that we claim the glory
Of possessing powers not our share?
—Since man woke on earth, he knows his story,
But, before we woke on earth, we were.

Long, long since, undower'd yet, our spirit
Roam'd, ere birth, the treasuries of God;
Saw the gifts, the powers it might inherit,
Ask'd an outfit for its earthly road.

Then, as now, this tremulous, eager being
Strain'd and long'd and grasp'd each gift it saw;   10
Then, as now, a Power beyond our seeing
Staved us back, and gave our choice the law.

Ah, whose hand that day through Heaven guided
Man's new spirit, since it was not we?
Ah, who sway'd our choice, and who decided
What our gifts, and what our wants should be?

For, alas! he left us each retaining
Shreds of gifts which he refused in full.
Still these waste us with their hopeless straining,
Still the attempt to use them proves them null.   20

And on earth we wander, groping, reeling;
Powers stir in us, stir and disappear.
Ah! and he, who placed our master-feeling,
Fail'd to place that master-feeling clear.

We but dream we have our wish'd-for powers,
Ends we seek we never shall attain.
Ah! *some* power exists there, which is ours?
Some end is there, we indeed may gain?

Despondency

The thoughts that rain their steady glow
Like stars on life's cold sea,
Which others know, or say they know—
They never shone for me.

Thoughts light, like gleams, my spirit's sky,
But they will not remain.
They light me once, they hurry by;
And never come again.

The Neckan

In summer, on the headlands,
    The Baltic Sea along,
Sits Neckan with his harp of gold,
    And sings his plaintive song.

Green rolls beneath the headlands,
   Green rolls the Baltic Sea;
And there, below the Neckan's feet,
   His wife and children be.

He sings not of the ocean,
   Its shells and roses pale;   10
Of earth, of earth the Neckan sings,
   He hath no other tale.

He sits upon the headlands,
   And sings a mournful stave
Of all he saw and felt on earth
   Far from the kind sea-wave.

Sings how, a knight, he wander'd
   By castle, field, and town—
But earthly knights have harder hearts
   Than the sea-children own.   20

Sings of his earthly bridal—
   Priest, knights, and ladies gay.
'—And who art thou,' the priest began,
   'Sir Knight, who wedd'st to-day?'—

'—I am no knight,' he answered;
   'From the sea-waves I come.'—
The knights drew sword, the ladies scream'd,
   The surpliced priest stood dumb.

He sings how from the chapel
   He vanish'd with his bride,   30
And bore her down to the sea-halls,
   Beneath the salt sea-tide.

He sings how she sits weeping
   'Mid shells that round her lie.
'—False Neckan shares my bed,' she weeps;
   'No Christian mate have I.'—

He sings how through the billows
    He rose to earth again,
And sought a priest to sign the cross,
    That Neckan Heaven might gain.   40

He sings how, on an evening,
    Beneath the birch-trees cool,
He sate and play'd his harp of gold,
    Beside the river-pool.

Beside the pool sate Neckan—
    Tears fill'd his mild blue eye.
On his white mule, across the bridge,
    A cassock'd priest rode by.

'—Why sitt'st thou there, O Neckan,
    And play'st thy harp of gold?   50
Sooner shall this my staff bear leaves,
    Than thou shalt Heaven behold.'—

But, lo, the staff, it budded!
    It green'd, it branch'd, it waved.
'—O ruth of God,' the priest cried out,
    'This lost sea-creature saved!'

The cassock'd priest rode onwards,
    And vanished with his mule;
But Neckan in the twilight grey
    Wept by the river-pool.   60

He wept: 'The earth hath kindness,
    The sea, the starry poles;
Earth, sea, and sky, and God above—
    But, ah, not human souls!'

In summer, on the headlands,
    The Baltic Sea along,
Sits Neckan with his harp of gold,
    And sings this plaintive song.

A Caution to Poets

What poets feel not, when they make,
    A pleasure in creating,
The world, in *its* turn, will not take
    Pleasure in contemplating.

Stanzas from the Grande Chartreuse

Through Alpine meadows soft-suffused
With rain, where thick the crocus blows,°
Past the dark forges long disused,
The mule-track from Saint Laurent goes.
The bridge is cross'd, and slow we ride,
Through forest, up the mountain-side.

The autumnal evening darkens round,
The wind is up, and drives the rain;
While, hark! far down, with strangled sound
Doth the Dead Guier's stream complain,°   10
Where that wet smoke, among the woods,
Over his boiling cauldron broods.

Swift rush the spectral vapours white
Past limestone scars with ragged pines,
Showing—then blotting from our sight!—
Halt—through the cloud-drift something shines!
High in the valley, wet and drear,
The huts of Courrerie appear.

Strike leftward! cries our guide; and higher
Mounts up the stony forest-way.   20
At last the encircling trees retire;
Look! through the showery twilight grey
What pointed roofs are these advance?—
A palace of the Kings of France?

Approach, for what we seek is here!
Alight, and sparely sup, and wait
For rest in this outbuilding near;°
Then cross the sward and reach that gate.
Knock; pass the wicket! Thou art come
To the Carthusians' world-famed home.°   30

The silent courts, where night and day
Into their stone-carved basins cold
The splashing icy fountains play—
The humid corridors behold!
Where, ghostlike in the deepening night,
Cowl'd forms brush by in gleaming white.

The chapel, where no organ's peal
Invests the stern and naked prayer—
With penitential cries they kneel
And wrestle; rising then, with bare°   40
And white uplifted faces stand,
Passing the Host from hand to hand;

Each takes, and then his visage wan
Is buried in his cowl once more.
The cells!—the suffering Son of Man
Upon the wall—the knee-worn floor—
And where they sleep, that wooden bed,°
Which shall their coffin be, when dead!

The library, where tract and tome
Not to feed priestly pride are there,   50
To hymn the conquering march of Rome,
Nor yet to amuse, as ours are!
They paint of souls the inner strife,
Their drops of blood, their death in life.

The garden, overgrown—yet mild,
See, fragrant herbs are flowering there!°
Strong children of the Alpine wild
Whose culture is the brethren's care;
Of human tasks their only one,
And cheerful works beneath the sun.   60

Those halls, too, destined to contain
Each its own pilgrim-host of old,
From England, Germany, or Spain—
All are before me! I behold
The House, the Brotherhood austere!
—And what am I, that I am here?

For rigorous teachers seized my youth,°
And purged its faith, and trimm'd its fire,
Show'd me the high, white star of Truth,
There bade me gaze, and there aspire.   70
Even now their whispers pierce the gloom:
What dost thou in this living tomb?

Forgive me, masters of the mind!
At whose behest I long ago
So much unlearnt, so much resign'd—
I come not here to be your foe!
I seek these anchorites, not in ruth,
To curse and to deny your truth;

Not as their friend, or child, I speak!°
But as, on some far northern strand,   80
Thinking of his own Gods, a Greek
In pity and mournful awe might stand
Before some fallen Runic stone—
For both were faiths, and both are gone.

Wandering between two worlds, one dead,
The other powerless to be born,
With nowhere yet to rest my head,°
Like these, on earth I wait forlorn.
Their faith, my tears, the world deride—
I come to shed them at their side.   90

Oh, hide me in your gloom profound,
Ye solemn seats of holy pain!
Take me, cowl'd forms, and fence me round,
Till I possess my soul again;
Till free my thoughts before me roll,
Not chafed by hourly false control!

For the world cries your faith is now
But a dead time's exploded dream;
My melancholy, sciolists say,°
Is a pass'd mode, an outworn theme—   100
As if the world had ever had
A faith, or sciolists been sad!

Ah, if it *be* pass'd, take away,
At least, the restlessness, the pain;
Be man henceforth no more a prey
To these out-dated stings again!
The nobleness of grief is gone—
Ah, leave us not the fret alone!

But—if you cannot give us ease—°
Last of the race of them who grieve   110
Here leave us to die out with these
Last of the people who believe!
Silent, while years engrave the brow;
Silent—the best are silent now.

Achilles ponders in his tent,
The kings of modern thought are dumb;
Silent they are, though not content,
And wait to see the future come.
They have the grief men had of yore,
But they contend and cry no more.   120

Our fathers water'd with their tears
This sea of time whereon we sail,
Their voices were in all men's ears
Who pass'd within their puissant hail.
Still the same ocean round us raves,
But we stand mute, and watch the waves.

For what avail'd it, all the noise
And outcry of the former men?—
Say, have their sons achieved more joys,
Say, is life lighter now than then?   130
The sufferers died, they left their pain—
The pangs which tortured them remain.

What helps it now, that Byron bore,°
With haughty scorn which mock'd the smart,
Through Europe to the Aetolian shore
The pageant of his bleeding heart?
That thousands counted every groan,
And Europe made his woe her own?

What boots it, Shelley! that the breeze
Carried thy lovely wail away,   140
Musical through Italian trees
Which fringe thy soft blue Spezzian bay?
Inheritors of thy distress
Have restless hearts one throb the less?

Or are we easier, to have read,°
O Obermann! the sad, stern page,
Which tells us how thou hidd'st thy head
From the fierce tempest of thine age
In the lone brakes of Fontainebleau,
Or chalets near the Alpine snow?   150

Ye slumber in your silent grave!—
The world, which for an idle day
Grace to your mood of sadness gave,
Long since hath flung her weeds away.
The eternal trifler breaks your spell;
But we—we learnt your lore too well!

Years hence, perhaps, may dawn an age,
More fortunate, alas! than we,
Which without hardness will be sage,
And gay without frivolity.   160
Sons of the world, oh, speed those years;
But, while we wait, allow our tears!

Allow them! We admire with awe°
The exulting thunder of your race;
You give the universe your law,
You triumph over time and space!
Your pride of life, your tireless powers,
We laud them, but they are not ours.

We are like children rear'd in shade°
Beneath some old-world abbey wall,   170
Forgotten in a forest-glade,
And secret from the eyes of all.
Deep, deep the greenwood round them waves,
Their abbey, and its close of graves!

But, where the road runs near the stream,
Oft through the trees they catch a glance
Of passing troops in the sun's beam—
Pennon, and plume, and flashing lance!
Forth to the world those soldiers fare,
To life, to cities, and to war!   180

And through the wood, another way,
Faint bugle-notes from far are borne,
Where hunters gather, staghounds bay,
Round some fair forest-lodge at morn.
Gay dames are there, in sylvan green;
Laughter and cries—those notes between!

The banners flashing through the trees
Make their blood dance and chain their eyes;
That bugle-music on the breeze
Arrests them with a charm'd surprise.   190
Banner by turns and bugle woo:
Ye shy recluses, follow too!

O children, what do ye reply?—
'Action and pleasure, will ye roam
Through these secluded dells to cry
And call us?—but too late ye come!
Too late for us your call ye blow,
Whose bent was taken long ago.

'Long since we pace this shadow'd nave;
We watch those yellow tapers shine,   200
Emblems of hope over the grave,
In the high altar's depth divine;
The organ carries to our ear
Its accents of another sphere.

'Fenced early in this cloistral round
Of reverie, of shade, of prayer,
How should we grow in other ground?
How can we flower in foreign air?
—Pass, banners, pass, and bugles, cease;
And leave our desert to its peace!'°   210

The Church of Brou

I

The Castle

Down the Savoy valleys sounding,
    Echoing round this castle old,
'Mid the distant mountain-chalets
    Hark! what bell for church is toll'd?

In the bright October morning
    Savoy's Duke had left his bride.°
From the castle, past the drawbridge,
    Flow'd the hunters' merry tide.

Steeds are neighing, gallants glittering;
    Gay, her smiling lord to greet,   10
From her mullion'd chamber-casement
    Smiles the Duchess Marguerite.°

From Vienna, by the Danube,
    Here she came, a bride, in spring.
Now the autumn crisps the forest;°
    Hunters gather, bugles ring.

Hounds are pulling, prickers swearing,°
    Horses fret, and boar-spears glance.
Off!—They sweep the marshy forests,
    Westward, on the side of France.   20

Hark! the game's on foot; they scatter!—
  Down the forest-ridings lone,
Furious, single horsemen gallop—
  Hark! a shout—a crash—a groan!°

Pale and breathless, came the hunters;
  On the turf dead lies the boar—
God! the Duke lies stretch'd beside him,
  Senseless, weltering in his gore.

     *   *   *   *

In the dull October evening,
  Down the leaf-strewn forest-road,   30
To the castle, past the drawbridge,
  Came the hunters with their load.

In the hall, with sconces blazing,
  Ladies waiting round her seat,
Clothed in smiles, beneath the daïs
  Sate the Duchess Marguerite.

Hark! below the gates unbarring!
  Tramp of men and quick commands!
'—'Tis my lord come back from hunting—'
  And the Duchess claps her hands.   40

Slow and tired, came the hunters—
  Stopp'd in darkness in the court.
'—Ho, this way, ye laggard hunters!
  To the hall! What sport? What sport?'—

Slow they enter'd with their master;
  In the hall they laid him down.
On his coat were leaves and blood-stains,
  On his brow an angry frown.

Dead her princely youthful husband
  Lay before his youthful wife,   50
Bloody, 'neath the flaring sconces—
  And the sight froze all her life.

     *   *   *   *

In Vienna, by the Danube,
  Kings hold revel, gallants meet.
Gay of old amid the gayest
  Was the Duchess Marguerite.

In Vienna, by the Danube,
  Feast and dance her youth beguiled.
Till that hour she never sorrow'd;
  But from then she never smiled.   60

'Mid the Savoy mountain valleys°
  Far from town or haunt of man,
Stands a lonely church, unfinish'd,
  Which the Duchess Maud began;

Old, that Duchess stern began it,
  In gray age, with palsied hands;
But she died while it was building,
  And the Church unfinish'd stands—

Stands as erst the builders left it,
  When she sank into her grave;   70
Mountain greensward paves the chancel,
  Harebells flower in the nave.

'—In my castle all is sorrow,'
  Said the Duchess Marguerite then;
'Guide me, some one, to the mountain!
  We will build the Church again.'—

Sandall'd palmers, faring homeward,
  Austrian knights from Syria came.
'—Austrian wanderers bring, O warders!
  Homage to your Austrian dame.'—   80

From the gate the warders answer'd:
  '—Gone, O knights, is she you knew!
Dead our Duke, and gone his Duchess;
  Seek her at the Church of Brou!'—

Austrian knights and march-worn palmers
  Climb the winding mountain-way—
Reach the valley, where the Fabric
  Rises higher day by day.

Stones are sawing, hammers ringing;
  On the work the bright sun shines,   90
In the Savoy mountain-meadows,
  By the stream, below the pines.

On her palfrey white the Duchess
  Sate and watch'd her working train—
Flemish carvers, Lombard gilders,°
  German masons, smiths from Spain.

Clad in black, on her white palfrey,
  Her old architect beside—°
There they found her in the mountains,
  Morn and noon and eventide.   100

There she sate, and watch'd the builders,
  Till the Church was roof'd and done.
Last of all the builders rear'd her
  In the nave a tomb of stone.°

On the tomb two forms they sculptured,
  Lifelike in the marble pale—
One, the Duke in helm and armour;
  One, the Duchess in her veil.

Round the tomb the carved stone fretwork
  Was at Easter-tide put on.   110
Then the Duchess closed her labours;
  And she died at the St John.

II

The Church

Upon the glistening leaden roof°
Of the new Pile, the sunlight shines;
    The stream goes leaping by.
The hills are clothed with pines sun-proof;
'Mid bright green fields, below the pines,
    Stands the Church on high.
What Church is this, from men aloof?—
'Tis the Church of Brou.

At sunrise, from their dewy lair
Crossing the stream, the kine are seen   10
    Round the wall to stray—
The churchyard wall that clips the square
Of open hill-sward fresh and green
    Where last year they lay.
But all things now are order'd fair
Round the Church of Brou.

On Sundays, at the matin-chime,
The Alpine peasants, two and three,
    Climb up here to pray;
Burghers and dames, at summer's prime,   20
Ride out to church from Chambery,
    Dight with mantles gay.
But else it is a lonely time
Round the Church of Brou.

On Sundays, too, a priest doth come
From the wall'd town beyond the pass,
    Down the mountain-way;
And then you hear the organ's hum,
You hear the white-robed priest say mass,
    And the people pray.   30
But else the woods and fields are dumb
Round the Church of Brou.

And after church, when mass is done,
The people to the nave repair
  Round the tomb to stray;
And marvel at the Forms of stone,
And praise the chisell'd broideries rare—°
  Then they drop away.
The princely Pair are left alone
In the Church of Brou.   40

III

𝕿𝖍𝖊 𝕿𝖔𝖒𝖇

So rest, for ever rest, O princely Pair!°
In your high church, 'mid the still mountain-air,
Where horn, and hound, and vassals, never come.
Only the blessed Saints are smiling dumb,
From the rich painted windows of the nave,
On aisle, and transept, and your marble grave;
Where thou, young Prince! shalt never more arise
From the fringed mattress where thy Duchess lies,
On autumn-mornings, when the bugle sounds,
And ride across the drawbridge with thy hounds 10
To hunt the boar in the crisp woods till eve;
And thou, O Princess! shalt no more receive,
Thou and thy ladies, in the hall of state,
The jaded hunters with their bloody freight,
Coming benighted to the castle-gate.
 So sleep, for ever sleep, O marble Pair!°
Or, if ye wake, let it be then, when fair
On the carved western front a flood of light
Streams from the setting sun, and colours bright
Prophets, transfigured Saints, and Martyrs brave, 20
In the vast western window of the nave;
And on the pavement round the Tomb there glints
A chequer-work of glowing sapphire-tints,
And amethyst, and ruby—then unclose
Your eyelids on the stone where ye repose,
And from your broider'd pillows lift your heads,
And rise upon your cold white marble beds;

And, looking down on the warm rosy tints,
Which chequer, at your feet, the illumined flints,
Say: *What is this? we are in bliss—forgiven—*°   30
Behold the pavement of the courts of Heaven!
Or let it be on autumn nights, when rain
Doth rustlingly above your heads complain
On the smooth leaden roof, and on the walls
Shedding her pensive light at intervals
The moon through the clere-story windows shines,
And the wind washes through the mountain-pines.
Then, gazing up 'mid the dim pillars high,
The foliaged marble forest where ye lie,
Hush, ye will say, *it is eternity!*   40
This is the glimmering verge of Heaven, and these
The columns of the heavenly palaces!
And, in the sweeping of the wind, your ear
The passage of the Angels' wings will hear,
And on the lichen-crusted leads above
The rustle of the eternal rain of love.°

Preface to First Edition of Poems (*1853*)

IN two small volumes of Poems, published anonymously, one in 1849, the other in 1852, many of the poems which compose the present volume have already appeared. The rest are now published for the first time.

I have, in the present collection, omitted the poem from which the volume published in 1852 took its title. I have done so, not because the subject of it was a Sicilian Greek born between two and three thousand years ago, although many persons would think this a sufficient reason. Neither have I done so because I had, in my own opinion, failed in the delineation which I intended to effect. I intended to delineate the feelings of one of the last of the Greek religious philosophers, one of the family of Orpheus and Musaeus, having survived his fellows, living on into a time when the habits of Greek thought and feeling had begun fast to change, character to dwindle, the influence of the Sophists to prevail. Into the feelings of a man so situated there entered much that we are accustomed to consider as exclusively modern; how much, the fragments of Empedocles himself which remain to us are sufficient at least to indicate. What those who are familiar only with the great monuments of early Greek genius suppose to be its exclusive characteristics, have disappeared: the calm, the cheerfulness, the disinterested objectivity have disappeared; the dialogue of the mind with itself has commenced; modern problems have presented themselves; we hear already the doubts, we witness the discouragement, of Hamlet and of Faust.

The representation of such a man's feelings must be interesting, if consistently drawn. We all naturally take pleasure, says Aristotle,° in any imitation or representation whatever: this is the basis of our love of poetry; and we take pleasure in them, he adds, because all knowledge is naturally agreeable to us; not to the philosopher only, but to mankind at large. Every representation, therefore, which is consistently drawn may be supposed to be interesting, inasmuch as it gratifies this natural interest in knowledge of all kinds. What is *not* interesting, is that which does not add to our knowledge of any kind; that which is vaguely conceived and loosely drawn; a representation which is general, indeterminate, and faint, instead of being particular, precise, and firm.

Any accurate representation may therefore be expected to be interesting; but, if the representation be a poetical one, more than this is demanded. It is demanded, not only that it shall interest, but also that it shall inspirit and rejoice the reader; that it shall convey a charm, and infuse delight. For the Muses, as Hesiod says,° were born that they might

be 'a forgetfulness of evils, and a truce from cares': and it is not enough°
that the poet should add to the knowledge of men, it is required of him
also that he should add to their happiness. 'All art', says Schiller,° 'is
dedicated to Joy, and there is no higher and no more serious problem,
than how to make men happy. The right art is that alone, which creates
the highest enjoyment.'

A poetical work, therefore, is not yet justified when it has been shown
to be an accurate, and therefore interesting representation; it has to be
shown also that it is a representation from which men can derive
enjoyment.° In presence of the most tragic circumstances, represented in
a work of art, the feeling of enjoyment, as is well known, may still subsist;
the representation of the most utter calamity, of the liveliest anguish, is
not sufficient to destroy it; the more tragic the situation, the deeper
becomes the enjoyment; and the situation is more tragic in proportion as
it becomes more terrible.

What then are the situations, from the representation of which,
though accurate, no poetical enjoyment can be derived? They are those
in which the suffering finds no vent in action; in which a continuous state
of mental distress is prolonged, unrelieved by incident, hope, or
resistance; in which there is everything to be endured, nothing to be
done. In such situations there is inevitably something morbid, in the
description of them something monotonous. When they occur in actual
life, they are painful, not tragic; the representation of them in poetry is
painful also.

To this class of situations, poetically faulty as it appears to me, that of
Empedocles, as I have endeavoured to represent him, belongs; and I
have therefore excluded the poem from the present collection.

And why, it may be asked, have I entered into this explanation
respecting a matter so unimportant as the admission or exclusion of the
poem in question? I have done so, because I was anxious to avow that the
sole reason for its exclusion was that which has been stated above; and
that it has not been excluded in deference to the opinion which many
critics of the present day appear to entertain against subjects chosen
from distant times and countries: against the choice, in short, of any
subjects but modern ones.

'The poet', it is said,[1] and by an intelligent critic,° 'the poet who
would really fix the public attention must leave the exhausted past, and
draw his subjects from matters of present import, and *therefore* both of
interest and novelty.'

[1] In the *Spectator* of April 2, 1853. The words quoted were not used with
reference to poems of mine.

Now this view I believe to be completely false. It is worth examining, inasmuch as it is a fair sample of a class of critical dicta everywhere current at the present day, having a philosophical form and air, but no real basis in fact; and which are calculated to vitiate the judgment of readers of poetry, while they exert, so far as they are adopted, a misleading influence on the practice of those who make it.

What are the eternal objects of poetry, among all nations, and at all times? They are actions;° human actions; possessing an inherent interest in themselves, and which are to be communicated in an interesting manner by the art of the poet. Vainly will the latter imagine° that he has everything in his own power; that he can make an intrinsically inferior action equally delightful with a more excellent one by his treatment of it. He may indeed compel us to admire his skill, but his work will possess, within itself, an incurable defect.

The poet, then, has in the first place to select an excellent action;° and what actions are the most excellent? Those, certainly, which most powerfully appeal to the great primary human affections:° to those elementary feelings which subsist permanently in the race, and which are independent of time. These feelings are permanent and the same; that which interests them is permanent and the same also. The modernness or antiquity of an action, therefore, has nothing to do with its fitness for poetical representation; this depends upon its inherent qualities. To the elementary part of our nature, to our passions, that which is great and passionate is eternally interesting; and interesting solely in proportion to its greatness and to its passion. A great human action of a thousand years ago is more interesting to it than a smaller human action of to-day, even though upon the representation of this last the most consummate skill may have been expended, and though it has the advantage of appealing by its modern language, familiar manners, and contemporary allusions, to all our transient feelings and interests. These, however, have no right to demand of a poetical work that it shall satisfy them; their claims are to be directed elsewhere. Poetical works belong to the domain of our permanent passions; let them interest these, and the voice of all subordinate claims upon them is at once silenced.

Achilles, Prometheus, Clytemnestra, Dido,°—what modern poem presents personages as interesting, even to us moderns, as these personages of an 'exhausted past'? We have the domestic epic dealing with the details of modern life which pass daily under our eyes; we have poems representing modern personages in contact with the problems of modern life, moral, intellectual, and social; these works have been produced by poets the most distinguished of their nation and time; yet I fearlessly assert that *Hermann and Dorothea, Childe Harold, Jocelyn, The Excursion*,° leave the reader cold in comparison with the effect produced upon him by the latter books of the *Iliad*, by the *Oresteia*, or by the episode of Dido. And why is this? Simply because in the three last-named cases the action is greater, the personages nobler, the situations more intense: and this is the true basis of the interest in a poetical work, and this alone.

It may be urged, however, that past actions may be interesting in themselves, but that they are not to be adopted by the modern poet, because it is impossible for him to have them clearly present to his own mind, and he cannot therefore feel them deeply, nor represent them forcibly. But this is not necessarily the case. The externals of a past action, indeed, he cannot know with the precision of a contemporary; but his business is with its essentials. The outward man of Oedipus or of Macbeth, the houses in which they lived, the ceremonies of their courts, he cannot accurately figure to himself; but neither do they essentially concern him. His business is with their inward man; with their feelings and behaviour in certain tragic situations, which engage their passions as men; these have in them nothing local and casual; they are as accessible to the modern poet as to a contemporary.

The date of an action, then, signifies nothing: the action itself, its selection and construction, this is what is all-important. This the Greeks understood far more clearly than we do. The radical difference between their poetical theory and ours consists, as it appears to me, in this: that, with them, the poetical character of the action in itself, and the conduct of it, was the first consideration; with us, attention is fixed mainly on the value of the separate thoughts and images which occur in the treatment of an action. They regarded the whole; we regard the parts. With them, the action predominated over the expression of it; with us, the expression predominates over the action. Not that they failed in expression, or were inattentive to it; on the contrary, they are the highest models of expression, the unapproached masters of the *grand style*.° But their expression is so excellent because it is so admirably kept in its right degree of prominence; because it is so simple and so well subordinated; because it draws its force directly from the pregnancy of the matter which it conveys. For what reason was the Greek tragic poet confined to so limited a range of subjects? Because there are so few actions which unite in themselves, in the highest degree, the conditions of excellence: and it was not thought that on any but an excellent subject could an excellent poem be constructed. A few actions, therefore, eminently adapted for tragedy, maintained almost exclusive possession of the Greek tragic

stage. Their significance appeared inexhaustible; they were as permanent problems, perpetually offered to the genius of every fresh poet. This
too is the reason of what appears to us moderns a certain baldness of
expression in Greek tragedy; of the triviality with which we often
reproach the remarks of the chorus, where it takes part in the dialogue:
that the action itself, the situation of Orestes, or Merope, or Alcmaeon,°
was to stand the central point of interest, unforgotten, absorbing,
principal; that no accessories were for a moment to distract the spectator's attention from this; that the tone of the parts was to be perpetually
kept down, in order not to impair the grandiose effect of the whole. The
terrible old mythic story on which the drama was founded stood, before
he entered the theatre, traced in its bare outlines upon the spectator's
mind; it stood in his memory,° as a group of statuary, faintly seen,
at the end of a long and dark vista: then came the poet, embodying
outlines, developing situations, not a word wasted, not a sentiment
capriciously thrown in: stroke upon stroke, the drama proceeded: the
light deepened upon the group; more and more it revealed itself to the
riveted gaze of the spectator: until at last, when the final words were
spoken, it stood before him in broad sunlight, a model of immortal
beauty.

This was what a Greek critic demanded; this was what a Greek poet
endeavoured to effect. It signified nothing to what time an action
belonged. We do not find that the *Persae*° occupied a particularly high
rank among the dramas of Aeschylus, because it represented a matter of
contemporary interest; this was not what a cultivated Athenian required.
He required that the permanent elements of his nature should be moved;
and dramas of which the action, though taken from a long-distant mythic
time, yet was calculated to accomplish this in a higher degree than that of
the *Persae*, stood higher in his estimation accordingly. The Greeks felt,
no doubt, with their exquisite sagacity of taste, that an action of present
times was too near them, too much mixed up with what was accidental
and passing, to form a sufficiently grand, detached, and self-subsistent
object for a tragic poem. Such objects belonged to the domain of the
comic poet, and of the lighter kinds of poetry. For the more serious
kinds, for *pragmatic* poetry, to use an excellent expression of Polybius,°
they were more difficult and severe in the range of subjects which they
permitted. Their theory and practice alike, the admirable treatise of
Aristotle,° and the unrivalled works of their poets, exclaim with a
thousand tongues—'All depends upon the subject; choose a fitting
action, penetrate yourself with the feeling of its situations; this done,
everything else will follow.'

But for all kinds of poetry alike there was one point on which they were rigidly exacting: the adaptability of the subject to the kind of poetry selected, and the careful construction of the poem.

How different a way of thinking from this is ours! We can hardly at the present day understand what Menander meant,° when he told a man who enquired as to the progress of his comedy that he had finished it, not having yet written a single line, because he had constructed the action of it in his mind. A modern critic would have assured him that the merit of his piece depended on the brilliant things which arose under his pen as he went along. We have poems which seem to exist merely for the sake of single lines and passages; not for the sake of producing any total impression. We have critics who seem to direct their attention merely to detached expressions, to the language about the action, not to the action itself. I verily think that the majority of them do not in their hearts believe that there is such a thing as a total impression to be derived from a poem at all, or to be demanded from a poet; they think the term a commonplace of metaphysical criticism. They will permit the poet to select any action he pleases, and to suffer that action to go as it will, provided he gratifies them with occasional bursts of fine writing, and with a shower of isolated thoughts and images. That is, they permit him to leave their poetical sense ungratified, provided that he gratifies their rhetorical sense and their curiosity. Of his neglecting to gratify these, there is little danger. He needs rather to be warned against the danger of attempting to gratify these alone; he needs rather to be perpetually reminded to prefer his action to everything else; so to treat this, as to permit its inherent excellences to develop themselves, without interruption from the intrusion of his personal peculiarities; most fortunate, when he most entirely succeeds in effacing himself, and in enabling a noble action to subsist as it did in nature.

But the modern critic not only permits a false practice; he absolutely prescribes false aims.—'A true allegory of the state of one's own mind in a representative history',° the poet is told, 'is perhaps the highest thing that one can attempt in the way of poetry.' And accordingly he attempts it. An allegory of the state of one's own mind, the highest problem of an art which imitates actions! No assuredly, it is not, it never can be so: no great poetical work has ever been produced with such an aim. *Faust* itself, in which something of the kind is attempted, wonderful passages as it contains, and in spite of the unsurpassed beauty of the scenes which relate to Margaret, *Faust* itself, judged as a whole, and judged strictly as a poetical work, is defective: its illustrious author, the greatest poet of modern times, the greatest critic of all times, would have been the first to acknowledge it; he only defended his work, indeed, by asserting it to be 'something incommensurable'.°

The confusion of the present times is great, the multitude of voices counselling different things bewildering, the number of existing works capable of attracting a young writer's attention and of becoming his models, immense. What he wants is a hand to guide him through the confusion, a voice to prescribe to him the aim which he should keep in view, and to explain to him that the value of the literary works which offer themselves to his attention is relative to their power of helping him forward on his road towards this aim. Such a guide the English writer at the present day will nowhere find. Failing this, all that can be looked for, all indeed that can be desired, is, that his attention should be fixed on excellent models; that he may reproduce, at any rate, something of their excellence, by penetrating himself with their works and by catching their spirit, if he cannot be taught to produce what is excellent independently.

Foremost among these models for the English writer stands Shakspeare: a name the greatest perhaps of all poetical names; a name never to be mentioned without reverence. I will venture, however, to express a doubt, whether the influence of his works, excellent and fruitful for the readers of poetry, for the great majority, has been of unmixed advantage to the writers of it. Shakspeare indeed chose excellent subjects; the world could afford no better than Macbeth, or Romeo and Juliet, or Othello; he had no theory respecting the necessity of choosing subjects of present import, or the paramount interest attaching to allegories of the state of one's own mind; like all great poets, he knew well what constituted a poetical action; like them, wherever he found such an action, he took it; like them, too, he found his best in past times. But to these general characteristics of all great poets he added a special one of his own; a gift, namely, of happy, abundant, and ingenious expression, eminent and unrivalled: so eminent as irresistibly to strike the attention first in him, and even to throw into comparative shade his other excellences as a poet. Here has been the mischief. These other excellences were his fundamental excellences *as a poet*; what distinguishes the artist from the mere amateur, says Goethe,° is *Architectonicē* in the highest sense; that power of execution, which creates, forms, and constitutes: not the profoundness of single thoughts, not the richness of imagery, not the abundance of illustration. But these attractive accessories of a poetical work being more easily seized than the spirit of the whole, and these accessories being possessed by Shakspeare in an unequalled degree, a young writer having recourse to Shakspeare as his model runs great risk of being vanquished and absorbed by them, and, in consequence, of reproducing, according to the measure of his power, these, and these alone. Of this preponderating quality of Shakspeare's genius, accordingly, almost the whole of modern English poetry has, it appears to me, felt the influence. To the exclusive attention on the part of his imitators to this it is in a great degree owing, that of the majority of modern poetical works the details alone are valuable, the composition worthless. In reading them one is perpetually reminded of that terrible sentence on a modern French poet:—*Il dit tout ce qu'il veut, mais malheureusement il n'a rien à dire.*°

Let me give an instance of what I mean. I will take it from the works of the very chief among those who seem to have been formed in the school of Shakspeare:° of one whose exquisite genius and pathetic death render him for ever interesting. I will take the poem of *Isabella, or the Pot of Basil*, by Keats. I choose this rather than the *Endymion*, because the latter work (which a modern critic has classed with the *Fairy Queen*!), although undoubtedly there blows through it the breath of genius, is yet as a whole so utterly incoherent, as not strictly to merit the name of a poem at all. The poem of *Isabella*, then, is a perfect treasure-house of graceful and felicitous words and images: almost in every stanza there occurs one of those vivid and picturesque turns of expression, by which the object is made to flash upon the eye of the mind,° and which thrill the reader with a sudden delight. This one short poem contains, perhaps, a greater number of happy single expressions which one could quote than all the extant tragedies of Sophocles. But the action, the story? The action in itself is an excellent one; but so feebly is it conceived by the poet, so loosely constructed, that the effect produced by it, in and for itself, is absolutely null. Let the reader, after he has finished the poem of Keats, turn to the same story in the *Decameron*:° he will then feel how pregnant and interesting the same action has become in the hands of a great artist, who above all things delineates his object; who subordinates expression to that which it is designed to express.

I have said that the imitators of Shakspeare, fixing their attention on his wonderful gift of expression, have directed their imitation to this, neglecting his other excellences. These excellences, the fundamental excellences of poetical art, Shakspeare no doubt possessed them,—possessed many of them in a splendid degree; but it may perhaps be doubted whether even he himself did not sometimes give scope to his faculty of expression to the prejudice of a higher poetical duty. For we must never forget that Shakspeare is the great poet he is from his skill in discerning and firmly conceiving an excellent action, from his power of intensely feeling a situation, of intimately associating himself with a character; not from his gift of expression, which rather even leads him astray, degenerating sometimes into a fondness for curiosity of expression, into an irritability of fancy, which seems to make it impossible for him to say a thing plainly, even when the press of the action demands the very directest language, or its level character the very simplest. Mr Hallam,° than whom it is impossible to find a saner and more judicious critic, has had the courage (for at the present day it needs courage) to remark, how extremely and faultily difficult Shakspeare's language often is. It is so: you may find main scenes in some of his greatest tragedies, *King Lear* for instance, where the language is so artificial, so curiously tortured, and so difficult, that every speech has to be read two or three times before its meaning can be comprehended. This over-curiousness of expression is indeed but the excessive employment of a wonderful gift,—of the power of saying a thing in a happier way than any other man; nevertheless, it is carried so far that one understands what M. Guizot meant,° when he said that Shakspeare appears in his language to have tried all styles except that of simplicity. He has not the severe and scrupulous self-restraint of the ancients, partly, no doubt, because he had a far less cultivated and exacting audience. He has indeed a far wider range than they had, a far richer fertility of thought; in this respect he rises above them. In his strong conception of his subject, in the genuine way in which he is penetrated with it, he resembles them, and is unlike the moderns. But in the accurate limitation of it, the conscientious rejection of superfluities, the simple and rigorous development of it from the first line of his work to the last, he falls below them, and comes nearer to the moderns. In his chief works, besides what he has of his own, he has the elementary soundness of the ancients; he has their important action and their large and broad manner; but he has not their purity of method. He is therefore a less safe model; for what he has of his own is personal, and inseparable from his own rich nature; it may be imitated and exaggerated, it cannot be learned or applied as an art. He is above all suggestive; more valuable, therefore, to young writers as men than as artists. But clearness of arrangement, rigour of development, simplicity of style,—these may to a certain extent be learned; and these may, I am convinced, be learned best from the ancients, who, although infinitely less suggestive than Shakspeare, are thus, to the artist, more instructive.

What then, it will be asked, are the ancients to be our sole models? the ancients with their comparatively narrow range of experience, and their widely different circumstances? Not, certainly, that which is narrow in the ancients, nor that in which we can no longer sympathise. An action like the action of the *Antigone* of Sophocles, which turns upon the conflict between the heroine's duty to her brother's corpse and that to the laws of her country,° is no longer one in which it is possible that we should feel a deep interest. I am speaking too, it will be remembered, not of the best sources of intellectual stimulus for the general reader, but of the best models of instruction for the individual writer. This last may certainly learn of the ancients, better than anywhere else, three things which it is vitally important for him to know:—the all-importance of the choice of a subject; the necessity of accurate construction; and the subordinate character of expression. He will learn from them how unspeakably superior is the effect of the one moral impression left by a great action treated as a whole, to the effect produced by the most striking single thought or by the happiest image. As he penetrates into the spirit of the great classical works, as he becomes gradually aware of their intense significance, their noble simplicity, and their calm pathos, he will be convinced that it is this effect, unity and profoundness of moral impression, at which the ancient poets aimed; that it is this which constitutes the grandeur of their works, and which makes them immortal. He will desire to direct his own efforts towards producing the same effect. Above all, he will deliver himself from the jargon of modern criticism, and escape the danger of producing poetical works conceived in the spirit of the passing time, and which partake of its transitoriness.

The present age makes great claims upon us: we owe it service, it will not be satisfied without our admiration. I know not how it is, but their commerce with the ancients appears to me to produce, in those who constantly practise it, a steadying and composing effect upon their judgment, not of literary works only, but of men and events in general. They are like persons who have had a very weighty and impressive experience: they are more truly than others under the empire of facts, and more independent of the language current among those with whom they live. They wish neither to applaud nor to revile their age; they wish to know what it is, what it can give them, and whether this is what they want. What they want, they know very well; they want to educe and cultivate what is best and noblest in themselves; they know, too, that this is no easy task—χαλεπὸν as Pittacus said,° χαλεπὸν ἐσθλὸν ἔμμεναι—and they ask themselves sincerely whether their age and its literature can assist them in the attempt. If they are endeavouring to practise any art, they remember the plain and simple proceedings of the old artists, who attained their grand results by penetrating themselves with some noble and significant action, not by inflating themselves with a belief in the pre-eminent importance and greatness of their own times. They do not talk of their mission, nor of interpreting their age,° nor of the coming poet; all this, they know, is the mere delirium of vanity; their business is not to praise their age, but to afford to the men who live in it the highest pleasure which they are capable of feeling. If asked to afford this by means of subjects drawn from the age itself, they ask what special fitness the present age has for supplying them. They are told that it is an era of progress, an age commissioned to carry out the great ideas of industrial development and social amelioration. They reply that with all this they can do nothing; that the elements they need for the exercise of their art are great actions, calculated powerfully and delightfully to affect what is permanent in the human soul; that so far as the present age can supply such actions, they will gladly make use of them; but that an age wanting in moral grandeur can with difficulty supply such, and an age of spiritual discomfort with difficulty be powerfully and delightfully affected by them.

A host of voices will indignantly rejoin that the present age is inferior to the past neither in moral grandeur nor in spiritual health. He who possesses the discipline I speak of will content himself with remembering the judgments passed upon the present age, in this respect, by the men of strongest head and widest culture whom it has produced; by Goethe and by Niebuhr.° It will be sufficient for him that he knows the opinions held by these two great men respecting the present age and its literature; and that he feels assured in his own mind that their aims and demands upon life were such as he would wish, at any rate, his own to be; and their judgment as to what is impeding and disabling such as he may safely follow. He will not, however, maintain a hostile attitude towards the false pretensions of his age: he will content himself with not being overwhelmed by them. He will esteem himself fortunate if he can succeed in banishing from his mind all feelings of contradiction, and irritation, and impatience; in order to delight himself with the contemplation of some noble action of a heroic time, and to enable others, through his representation of it, to delight in it also.

I am far indeed from making any claim, for myself, that I possess this discipline; or for the following poems, that they breathe its spirit. But I say, that in the sincere endeavour to learn and practise, amid the bewildering confusion of our times, what is sound and true in poetical art, I seemed to myself to find the only sure guidance, the only solid footing, among the ancients. They, at any rate, knew what they wanted in art, and we do not. It is this uncertainty which is disheartening, and not hostile criticism. How often have I felt this when reading words of disparagement or of cavil: that it is the uncertainty as to what is really to be aimed at which makes our difficulty, not the dissatisfaction of the critic, who himself suffers from the same uncertainty! *Non me tua fervida terrent Dicta*; . . . *Dii me terrent, et Jupiter hostis.*°

Two kinds of *dilettanti*, says Goethe,° there are in poetry: he who neglects the indispensable mechanical part, and thinks he has done enough if he shows spirituality and feeling; and he who seeks to arrive at poetry merely by mechanism, in which he can acquire an artisan's readiness, and is without soul and matter. And he adds, that the first does most harm to art, and the last to himself. If we must be *dilettanti*: if it is impossible for us, under the circumstances amidst which we live, to think clearly, to feel nobly, and to delineate firmly: if we cannot attain to the mastery of the great artists;—let us, at least, have so much respect for our art as to prefer it to ourselves. Let us not bewilder our successors; let us transmit to them the practice of poetry, with its boundaries and whole-some regulative laws, under which excellent works may again, perhaps, at some future time, be produced, not yet fallen into oblivion through our neglect, not yet condemned and cancelled by the influence of their eternal enemy, caprice.

Preface to Second Edition of Poems (*1854*)

I HAVE allowed the Preface to the former edition of these Poems to stand almost without change, because I still believe it to be, in the main, true. I must not, however, be supposed insensible to the force of much that has been alleged against portions of it, or unaware that it contains many things incompletely stated, many things which need limitation. It leaves, too, untouched the question, how far and in what manner the opinions there expressed respecting the choice of subjects apply to lyric poetry,— that region of the poetical field which is chiefly cultivated at present. But neither do I propose at the present time to supply these deficiencies, nor, indeed, would this be the proper place for attempting it. On one or two points alone I wish to offer, in the briefest possible way, some explanation.

An objection° has been warmly urged to the classing together, as subjects equally belonging to a past time, Oedipus and Macbeth. And it is no doubt true that to Shakspeare, standing on the verge of the middle ages, the epoch of Macbeth was more familiar than that of Oedipus. But I was speaking of actions as they presented themselves to us moderns: and it will hardly be said that the European mind, in our day, has much more affinity with the times of Macbeth than with those of Oedipus. As moderns, it seems to me, we have no longer any direct affinity with the circumstances and feelings of either. As individuals, we are attracted towards this or that personage, we have a capacity for imagining him, irrespective of his times, solely according to a law of personal sympathy; and those subjects for which we feel this personal attraction most strongly, we may hope to treat successfully. Prometheus or Joan of Arc,° Charlemagne or Agamemnon,—one of these is not really nearer to us now than another. Each can be made present only by an act of poetic imagination; but this man's imagination has an affinity for one of them, and that man's for another.

It has been said° that I wish to limit the poet, in his choice of subjects, to the period of Greek and Roman antiquity; but it is not so. I only counsel him to choose for his subjects great actions, without regarding to what time they belong. Nor do I deny that the poetic faculty can and does manifest itself in treating the most trifling action, the most hopeless subject. But it is a pity that power should be wasted; and that the poet should be compelled to impart interest and force to his subject, instead of receiving them from it, and thereby doubling his impressiveness. There is, it has been excellently said, an immortal strength in the stories of great actions; the most gifted poet, then, may well be glad to supplement with it that mortal weakness, which, in presence of the vast spectacle of life and the world, he must for ever feel to be his individual portion.

Again, with respect to the study of the classical writers of antiquity: it has been said that we should emulate rather than imitate them.° I make no objection; all I say is, let us study them. They can help to cure us of what is, it seems to me, the great vice of our intellect, manifesting itself in our incredible vagaries in literature, in art, in religion, in morals: namely, that it is *fantastic*, and wants *sanity*. Sanity,—that is the great virtue of the ancient literature; the want of that is the great defect of the modern, in spite of all its variety and power. It is impossible to read carefully the great ancients, without losing something of our caprice and eccentricity; and to emulate them we must at least read them.

Sohrab and Rustum

AN EPISODE

And the first grey of morning fill'd the east,
And the fog rose out of the Oxus stream.
But all the Tartar camp along the stream°
Was hush'd, and still the men were plunged in sleep;
Sohrab alone, he slept not; all night long
He had lain wakeful, tossing on his bed;
But when the grey dawn stole into his tent,
He rose, and clad himself, and girt his sword,
And took his horseman's cloak, and left his tent,
And went abroad into the cold wet fog,   10
Through the dim camp to Peran-Wisa's tent.
   Through the black Tartar tents he pass'd, which stood
Clustering like bee-hives on the low flat strand
Of Oxus, where the summer-floods o'erflow°
When the sun melts the snows in high Pamere;
Through the black tents he pass'd, o'er that low strand,
And to a hillock came, a little back
From the stream's brink—the spot where first a boat,
Crossing the stream in summer, scrapes the land.
The men of former times had crown'd the top   20
With a clay fort; but that was fall'n, and now
The Tartars built there Peran-Wisa's tent,
A dome of laths, and o'er it felts were spread.
And Sohrab came there, and went in, and stood
Upon the thick piled carpets in the tent,
And found the old man sleeping on his bed
Of rugs and felts, and near him lay his arms.
And Peran-Wisa heard him, though the step
Was dull'd; for he slept light, an old man's sleep;
And he rose quickly on one arm, and said:—   30
   'Who art thou? for it is not yet clear dawn.
Speak! is there news, or any night alarm?'
   But Sohrab came to the bedside, and said:—
'Thou know'st me, Peran-Wisa! it is I.
The sun is not yet risen, and the foe
Sleep; but I sleep not; all night long I lie
Tossing and wakeful, and I come to thee.

For so did King Afrasiab bid me seek°
Thy counsel, and to heed thee as thy son,
In Samarcand, before the army march'd;   40
And I will tell thee what my heart desires.
Thou know'st if, since from Ader-baijan first
I came among the Tartars and bore arms,
I have still served Afrasiab well, and shown,
At my boy's years, the courage of a man.
This too thou know'st, that while I still bear on
The conquering Tartar ensigns through the world,
And beat the Persians back on every field,
I seek one man, one man, and one alone—
Rustum, my father; who I hoped should greet,   50
Should one day greet, upon some well-fought field,
His not unworthy, not inglorious son.
So I long hoped, but him I never find.
Come then, hear now, and grant me what I ask.
Let the two armies rest to-day; but I
Will challenge forth the bravest Persian lords
To meet me, man to man; if I prevail,
Rustum will surely hear it; if I fall—
Old man, the dead need no one, claim no kin.
Dim is the rumour of a common fight,   60
Where host meets host, and many names are sunk;
But of a single combat fame speaks clear.'
    He spoke; and Peran-Wisa took the hand
Of the young man in his, and sigh'd, and said:—
    'O Sohrab, an unquiet heart is thine!
Canst thou not rest among the Tartar chiefs,
And share the battle's common chance with us
Who love thee, but must press for ever first,
In single fight incurring single risk,
To find a father thou hast never seen?   70
That were far best, my son, to stay with us
Unmurmuring; in our tents, while it is war,
And when 'tis truce, then in Afrasiab's towns.
But, if this one desire indeed rules all,°
To seek out Rustum—seek him not through fight!
Seek him in peace, and carry to his arms,
O Sohrab, carry an unwounded son!
But far hence seek him, for he is not here.

For now it is not as when I was young,
When Rustum was in front of every fray;   80
But now he keeps apart, and sits at home,
In Seistan, with Zal, his father old.
Whether that his own mighty strength at last
Feels the abhorr'd approaches of old age,
Or in some quarrel with the Persian King.
There go!—Thou wilt not? Yet my heart forebodes
Danger or death awaits thee on this field.
Fain would I know thee safe and well, though lost
To us; fain therefore send thee hence, in peace
To seek thy father, not seek single fights   90
In vain;—but who can keep the lion's cub
From ravening, and who govern Rustum's son?
Go, I will grant thee what thy heart desires.'
   So said he, and dropp'd Sohrab's hand, and left
His bed, and the warm rugs whereon he lay;
And o'er his chilly limbs his woollen coat
He pass'd, and tied his sandals on his feet,
And threw a white cloak round him, and he took
In his right hand a ruler's staff, no sword;
And on his head he set his sheep-skin cap,   100
Black, glossy, curl'd, the fleece of Kara-Kul;
And raised the curtain of his tent, and call'd
His herald to his side, and went abroad.
   The sun by this had risen, and clear'd the fog
From the broad Oxus and the glittering sands.
And from their tents the Tartar horsemen filed
Into the open plain; so Haman bade—
Haman, who next to Peran-Wise ruled
The host, and still was in his lusty prime.
From their black tents, long files of horse, they stream'd;°110
As when some grey November morn the files,
In marching order spread, of long-neck'd cranes
Stream over Casbin and the southern slopes
Of Elburz, from the Aralian estuaries,
Or some frore Caspian reed-bed, southward bound
For the warm Persian sea-board—so they stream'd.
The Tartars of the Oxus, the King's guard,
First, with black sheep-skin caps and with long spears;
Large men, large steeds; who from Bokhara come

And Khiva, and ferment the milk of mares.   120
Next, the more temperate Toorkmuns of the south,
The Tukas, and the lances of Salore,
And those from Attruck and the Caspian sands;
Light men and on light steeds, who only drink
The acrid milk of camels, and their wells.
And then a swarm of wandering horse, who came
From far, and a more doubtful service own'd;
The Tartars of Ferghana, from the banks
Of the Jaxartes, men with scanty beards
And close-set skull-caps; and those wilder hordes   130
Who roam o'er Kipchak and the northern waste,
Kalmucks and unkempt Kuzzaks, tribes who stray
Nearest the Pole, and wandering Kirghizzes,
Who come on shaggy ponies from Pamere;
These all filed out from camp into the plain.
And on the other side the Persians form'd;—
First a light cloud of horse, Tartars they seem'd,
The Ilyats of Khorassan; and behind,
The royal troops of Persia, horse and foot,
Marshall'd battalions bright in burnish'd steel.   140
But Peran-Wisa with his herald came,
Threading the Tartar squadrons to the front,
And with his staff kept back the foremost ranks.
And when Ferood, who led the Persians, saw
That Peran-Wisa kept the Tartars back,
He took his spear, and to the front he came,
And check'd his ranks, and fix'd them where they stood.
And the old Tartar came upon the sand
Betwixt the silent hosts, and spake, and said:—
    'Ferood, and ye, Persians and Tartars, hear!   150
Let there be truce between the hosts to-day.
But choose a champion from the Persian lords
To fight our champion Sohrab, man to man.'
    As, in the country, on a morn in June,
When the dew glistens on the pearlèd ears,
A shiver runs through the deep corn for joy—
So, when they heard what Peran-Wisa said,
A thrill through all the Tartar squadrons ran
Of pride and hope for Sohrab, whom they loved.
    But as a troop of pedlars, from Cabool,   160

Cross underneath the Indian Caucasus,
That vast sky-neighbouring mountain of milk snow;
Crossing so high, that, as they mount, they pass
Long flocks of travelling birds dead on the snow,
Choked by the air, and scarce can they themselves
Slake their parch'd throats with sugar'd mulberries—
In single file they move, and stop their breath,
For fear they should dislodge the o'erhanging snows—
So the pale Persians held their breath with fear.

    And to Ferood his brother chiefs came up   170
To counsel; Gudurz and Zoarrah came,
And Feraburz, who ruled the Persian host
Second, and was the uncle of the King;
These came and counsell'd, and then Gudurz said:—
    'Ferood, shame bids us take their challenge up,
Yet champion have we none to match this youth.
He has the wild stag's foot, the lion's heart.
But Rustum came last night; aloof he sits°
And sullen, and has pitch'd his tents apart.
Him will I seek, and carry to his ear   180
The Tartar challenge, and this young man's name.
Haply he will forget his wrath, and fight.
Stand forth the while, and take their challenge up.'
    So spake he; and Ferood stood forth and cried:—
'Old man, be it agreed as thou hast said!
Let Sohrab arm, and we will find a man.'
    He spake: and Peran-Wisa turn'd, and strode
Back through the opening squadrons to his tent.
But through the anxious Persians Gudurz ran,
And cross'd the camp which lay behind, and reach'd,   190
Out on the sands beyond it, Rustum's tents.
Of scarlet cloth they were, and glittering gay,
Just pitch'd; the high pavilion in the midst
Was Rustum's, and his men lay camp'd around.
And Gudurz enter'd Rustum's tent, and found
Rustum; his morning meal was done, but still
The table stood before him, charged with food—
A side of roasted sheep, and cakes of bread,
And dark green melons; and there Rustum sate
Listless, and held a falcon on his wrist,   200
And play'd with it; but Gudurz came and stood

Before him; and he look'd, and saw him stand,
And with a cry sprang up and dropp'd the bird,
And greeted Gudurz with both hands, and said:—
    'Welcome! these eyes could see no better sight.
What news? but sit down first, and eat and drink.'
    But Gudurz stood in the tent-door, and said:—
'Not now! a time will come to eat and drink,
But not to-day; to-day has other needs.
The armies are drawn out, and stand at gaze;   210
For from the Tartars is a challenge brought
To pick a champion from the Persian lords
To fight their champion—and thou know'st his name—
Sohrab men call him, but his birth is hid.
O Rustum, like thy might is this young man's!
He has the wild stag's foot, the lion's heart;
And he is young, and Iran's chiefs are old,
Or else too weak; and all eyes turn to thee.
Come down and help us, Rustum, or we lose!'
    He spoke; but Rustum answer'd with a smile:—   220
'Go to! if Iran's chiefs are old, then I
Am older; if the young are weak, the King
Errs strangely; for the King, for Kai Khosroo,
Himself is young, and honours younger men,
And lets the aged moulder to their graves.
Rustum he loves no more, but loves the young—
The young may rise at Sohrab's vaunts, not I.
For what care I, though all speak Sohrab's fame?
For would that I myself had such a son,
And not that one slight helpless girl I have—°   230
A son so famed, so brave, to send to war,
And I to tarry with the snow-hair'd Zal,
My father, whom the robber Afghans vex,°
And clip his borders short, and drive his herds,
And he has none to guard his weak old age.
There would I go, and hang my armour up,
And with my great name fence that weak old man,
And spend the goodly treasures I have got,
And rest my age, and hear of Sohrab's fame,
And leave to death the hosts of thankless kings,   240
And with these slaughterous hands draw sword no more.'
    He spoke, and smiled; and Gudurz made reply:—°

'What then, O Rustum, will men say to this,
When Sohrab dares our bravest forth, and seeks
Thee most of all, and thou, whom most he seeks,
Hidest thy face? Take heed lest men should say:
Like some old miser, Rustum hoards his fame,
And shuns to peril it with younger men.'
    And, greatly moved, then Rustum made reply:—
'O Gudurz, wherefore dost thou say such words?   250
Thou knowest better words than this to say.
What is one more, one less, obscure or famed,
Valiant or craven, young or old, to me?
Are not they mortal, am not I myself?
But who for men of nought would do great deeds?
Come, thou shalt see how Rustum hoards his fame!
But I will fight unknown, and in plain arms;
Let not men say of Rustum, he was match'd
In single fight with any mortal man.'
    He spoke, and frown'd; and Gudurz turn'd, and ran   260
Back quickly through the camp in fear and joy—
Fear at his wrath, but joy that Rustum came.
But Rustum strode to his tent-door, and call'd
His followers in, and bade them bring his arms,
And clad himself in steel; the arms he chose
Were plain, and on his shield was no device,
Only his helm was rich, inlaid with gold,
And, from the fluted spine atop, a plume
Of horsehair waved, a scarlet horsehair plume.
So arm'd, he issued forth; and Ruksh, his horse,   270
Follow'd him like a faithful hound at heel—
Ruksh, whose renown was noised through all the earth,
The horse, whom Rustum on a foray once
Did in Bokhara by the river find
A colt beneath its dam, and drove him home,
And rear'd him; a bright bay, with lofty crest,
Dight with a saddle-cloth of broider'd green
Crusted with gold, and on the ground were work'd
All beasts of chase, all beasts which hunters know.
So follow'd, Rustum left his tents, and cross'd   280
The camp, and to the Persian host appear'd.
And all the Persians knew him, and with shouts
Hail'd; but the Tartars knew not who he was.

And dear as the wet diver to the eyes°
Of his pale wife who waits and weeps on shore,
By sandy Bahrein, in the Persian Gulf,
Plunging all day in the blue waves, at night,
Having made up his tale of precious pearls,
Rejoins her in their hut upon the sands—
So dear to the pale Persians Rustum came.   290
    And Rustum to the Persian front advanced,
And Sohrab arm'd in Haman's tent, and came.
And as afield the reapers cut a swath
Down through the middle of a rich man's corn,
And on each side are squares of standing corn,
And in the midst a stubble, short and bare—
So on each side were squares of men, with spears
Bristling, and in the midst, the open sand.
And Rustum came upon the sand, and cast
His eyes toward the Tartar tents, and saw   300
Sohrab come forth, and eyed him as he came.
    As some rich woman, on a winter's morn,°
Eyes through her silken curtains the poor drudge
Who with numb blacken'd fingers makes her fire—
At cock-crow, on a starlit winter's morn,
When the frost flowers the whiten'd window-panes—
And wonders how she lives, and what the thoughts
Of that poor drudge may be; so Rustum eyed
The unknown adventurous youth, who from afar
Came seeking Rustum, and defying forth   310
All the most valiant chiefs; long he perused
His spirited air, and wonder'd who he was.
For very young he seem'd, tenderly rear'd;
Like some young cypress, tall, and dark, and straight,
Which in a queen's secluded garden throws
Its slight dark shadow on the moonlit turf,
By midnight, to a bubbling fountain's sound—
So slender Sohrab seem'd, so softly rear'd.
And a deep pity enter'd Rustum's soul
As he beheld him coming; and he stood,   320
And beckon'd to him with his hand, and said:—
    'O thou young man, the air of Heaven is soft,
And warm, and pleasant; but the grave is cold!
Heaven's air is better than the cold dead grave.

Behold me! I am vast, and clad in iron,
And tried; and I have stood on many a field
Of blood, and I have fought with many a foe—
Never was that field lost, or that foe saved.
O Sohrab, wherefore wilt thou rush on death?
Be govern'd! quit the Tartar host, and come   330
To Iran, and be as my son to me,
And fight beneath my banner till I die!
There are no youths in Iran brave as thou.'
     So he spake, mildly; Sohrab heard his voice,
The might voice of Rustum, and he saw
His giant figure planted on the sand,
Sole, like some single tower, which a chief
Hath builded on the waste in former years
Against the robbers; and he saw that head,
Streak'd with its first grey hairs;—hope filled his soul,   340
And he ran forward and embraced his knees,
And clasp'd his hand within his own, and said:—
     'O, by thy father's head! by thine own soul!
Art thou not Rustum? speak! art thou not he?'
     But Rustum eyed askance the kneeling youth,
And turn'd away, and spake to his own soul:—
     'Ah me, I muse what this young fox may mean!
False, wily, boastful, are these Tartar boys.
For if I now confess this thing he asks,
And hide it not, but say: *Rustum is here!*   350
He will not yield indeed, nor quit our foes,
But he will find some pretext not to fight,
And praise my fame, and proffer courteous gifts,
A belt or sword perhaps, and go his way.
And on a feast-tide, in Afrasiab's hall,
In Samarcand, he will arise and cry:
"I challenged once, when the two armies camp'd
Beside the Oxus, all the Persian lords
To cope with me in single fight; but they
Shrank, only Rustum dared; then he and I   360
Changed gifts, and went on equal terms away."
So will he speak, perhaps, while men applaud;
Then were the chiefs of Iran shamed through me.'
     And then he turn'd, and sternly spake aloud:—
'Rise! wherefore dost thou vainly question thus

Of Rustum? I am here, whom thou hast call'd
By challenge forth; make good thy vaunt, or yield!
Is it with Rustum only thou wouldst fight?
Rash boy, men look on Rustum's face and flee!
For well I know, that did great Rustum stand°   370
Before thy face this day, and were reveal'd,
There would be then no talk of fighting more.
But being what I am, I tell thee this—
Do thou record it in thine inmost soul:
Either thou shalt renounce thy vaunt and yield,
Or else thy bones shall strew this sand, till winds
Bleach them, or Oxus with his summer-floods,
Oxus in summer wash them all away.'
  He spoke; and Sohrab answer'd, on his feet:—
'Art thou so fierce? Thou wilt not fright me so!   380
I am no girl, to be made pale by words.
Yet this thou hast said well, did Rustum stand
Here on this field, there were no fighting then.
But Rustum is far hence, and we stand here.
Begin! thou art more vast, more dread than I,
And thou art proved, I know, and I am young—
But yet success sways with the breath of Heaven.
And though thou thinkest that thou knowest sure
Thy victory, yet thou canst not surely know.
For we are all, like swimmers in the sea,°   390
Poised on the top of a huge wave of fate,
Which hangs uncertain to which side to fall.
And whether it will heave us up to land,
Or whether it will roll us out to sea,
Back out to sea, to the deep waves of death,
We know not, and no search will make us know;
Only the event will teach us in its hour.'
  He spoke, and Rustum answer'd not, but hurl'd
His spear; down from the shoulder, down it came,
As on some partridge in the corn a hawk,   400
That long has tower'd in the airy clouds,
Drops like a plummet; Sohrab saw it come,
And sprang aside, quick as a flash; the spear
Hiss'd, and went quivering down into the sand,
Which it sent flying wide;—then Sohrab threw
In turn, and full struck Rustum's shield; sharp rang,

The iron plates rang sharp, but turn'd the spear.
And Rustum seized his club, which none but he
Could wield; an unlopp'd trunk it was, and huge,°
Still rough—like those which men in treeless plains   410
To build them boats fish from the flooded rivers,
Hyphasis or Hydaspes, when, high up
By their dark springs, the wind in winter-time
Hath made in Himalayan forests wrack,
And strewn the channels with torn boughs—so huge
The club which Rustum lifted now, and struck
One stroke; but again Sohrab sprang aside,
Lithe as the glancing snake, and the club came
Thundering to earth, and leapt from Rustum's hand.
And Rustum follow'd his own blow, and fell   420
To his knees, and with his fingers clutch'd the sand;
And now might Sohrab have unsheathed his sword,
And pierced the mighty Rustum while he lay
Dizzy, and on his knees, and choked with sand;
But he look'd on, and smiled, nor bared his sword,
But courteously drew back, and spoke, and said:—

   'Thou strik'st too hard! that club of thine will float
Upon the summer-floods, and not my bones.
But rise, and be not wroth! not wroth am I;
No, when I see thee, wrath forsakes my soul.   430
Thou say'st, thou art not Rustum; be it so!
Who art thou then, that canst so touch my soul?
Boy as I am, I have seen battles too—
Have waded foremost in their bloody waves,
And heard their hollow roar of dying men;
But never was my heart thus touch'd before.
Are they from Heaven, these softenings of the heart?
O thou old warrior, let us yield to Heaven!
Come, plant we here in earth our angry spears,°
And make a truce, and sit upon this sand,   440
And pledge each other in red wine, like friends,
And thou shalt talk to me of Rustum's deeds.
There are enough foes in the Persian host,
Whom I may meet, and strike, and feel no pang;
Champions enough Afrasiab has, whom thou
Mayst fight; fight *them*, when they confront thy spear!
But oh, let there be peace 'twixt thee and me!'

He ceased, but while he spake, Rustum had risen,°
And stood erect, trembling with rage; his club
He left to lie, but had regain'd his spear,   450
Whose fiery point now in his mail'd right-hand
Blazed bright and baleful, like that autumn-star,
The baleful sign of fevers; dust had soil'd
His stately crest, and dimm'd his glittering arms.
His breast heaved, his lips foam'd, and twice his voice
Was choked with rage; at last these words broke way:—
    'Girl! nimble with thy feet, not with thy hands!
Curl'd minion, dancer, coiner of sweet words!
Fight, let me hear thy hateful voice no more!
Thou art not in Afrasiab's gardens now   460
With Tartar girls, with whom thou art wont to dance;
But on the Oxus-sands, and in the dance
Of battle, and with me, who make no play
Of war; I fight it out, and hand to hand.
Speak not to me of truce, and pledge, and wine!
Remember all thy valour; try thy feints
And cunning! all the pity I had is gone;
Because thou hast shamed me before both the hosts
With thy light skipping tricks, and thy girl's wiles.'
    He spoke, and Sohrab kindled at his taunts,   470
And he too drew his sword; at once they rush'd
Together, as two eagles on one prey
Come rushing down together from the clouds,
One from the east, one from the west; their shields
Dash'd with a clang together, and a din
Rose, such as that the sinewy woodcutters
Make often in the forest's heart at morn,
Of hewing axes, crashing trees—such blows
Rustum and Sohrab on each other hail'd.
And you would say that sun and stars took part   480
In that unnatural conflict; for a cloud
Grew suddenly in Heaven, and dark'd the sun
Over the fighters' heads; and a wind rose
Under their feet, and moaning swept the plain,
And in a sandy whirlwind wrapp'd the pair.
In gloom they twain were wrapp'd, and they alone;
For both the on-looking hosts on either hand
Stood in broad daylight, and the sky was pure,

And the sun sparkled on the Oxus stream.
But in the gloom they fought, with bloodshot eyes   490
And labouring breath; first Rustum struck the shield
Which Sohrab held stiff out; the steel-spiked spear
Rent the tough plates, but fail'd to reach the skin,
And Rustum pluck'd it back with angry groan.
Then Sohrab with his sword smote Rustum's helm,
Nor clove its steel quite through; but all the crest
He shore away, and that proud horsehair plume,
Never till now defiled, sank to the dust;
And Rustum bow'd his head; but then the gloom
Grew blacker, thunder rumbled in the air,   500
And lightnings rent the cloud; and Ruksh, the horse,
Who stood at hand, utter'd a dreadful cry;—
No horse's cry was that, most like the roar
Of some pain'd desert-lion, who all day
Hath trail'd the hunter's javelin in his side,
And comes at night to die upon the sand.
The two hosts heard that cry, and quaked for fear,
And Oxus curdled as it cross'd his stream.
But Sohrab heard, and quail'd not, but rush'd on,
And struck again; and again Rustum bow'd   510
His head; but this time all the blade, like glass,
Sprang in a thousand shivers on the helm,
And in the hand the hilt remain'd alone.
Then Rustum raised his head; his dreadful eyes
Glared, and he shook on high his menacing spear,
And shouted: *Rustum!*—Sohrab heard that shout,
And shrank amazed; back he recoil'd one step,
And scann'd with blinking eyes the advancing form;
And then he stood bewilder'd; and he dropp'd
His covering shield, and the spear pierced his side.   520
He reel'd, and staggering back, sank to the ground;
And then the gloom dispersed, and the wind fell,
And the bright sun broke forth, and melted all
The cloud; and the two armies saw the pair—
Saw Rustum standing, safe upon his feet,
And Sohrab, wounded, on the bloody sand.
     Then, with a bitter smile, Rustum began:—°
'Sohrab, thou thoughtest in thy mind to kill
A Persian lord this day, and strip his corpse,

And bear thy trophies to Afrasiab's tent.   530
Or else that the great Rustum would come down
Himself to fight, and that thy wiles would move
His heart to take a gift, and let thee go.
And then that all the Tartar host would praise
Thy courage or thy craft, and spread thy fame,
To glad thy father in his weak old age.
Fool, thou art slain, and by an unknown man!
Dearer to the red jackals shalt thou be
Than to thy friends, and to thy father old.'

    And, with a fearless mien, Sohrab replied:—   540
'Unknown thou art; yet thy fierce vaunt is vain.°
Thou dost not slay me, proud and boastful man!
No! Rustum slays me, and this filial heart.
For were I match'd with ten such men as thee,
And I were that which till to-day I was,
They should be lying here, I standing there.
But that belovéd name unnerved my arm—
That name, and something, I confess, in thee,
Which troubles all my heart, and made my shield
Fall; and thy spear transfix'd an unarm'd foe.   550
And now thou boastest, and insult'st my fate.
But hear thou this, fierce man, tremble to hear:
The mighty Rustum shall avenge my death!
My father, whom I seek through all the world,
He shall avenge my death, and punish thee!'

    As when some hunter in the spring hath found°
A breeding eagle sitting on her nest,
Upon the craggy isle of a hill-lake,
And pierced her with an arrow as she rose,
And follow'd her to find her where she fell   560
Far off;—anon her mate comes winging back
From hunting, and a great way off descries
His huddling young left sole; at that, he checks
His pinion, and with short uneasy sweeps
Circles above his eyry, with loud screams
Chiding his mate back to her nest; but she
Lies dying, with the arrow in her side,
In some far stony gorge out of his ken,
A heap of fluttering feathers—never more
Shall the lake glass her, flying over it;   570

Never the black and dripping precipices
Echo her stormy scream as she sails by—
As that poor bird flies home, nor knows his loss,
So Rustum knew not his own loss, but stood
Over his dying son, and knew him not.

    But, with a cold incredulous voice, he said:—
'What prate is this of fathers and revenge?
The mighty Rustum never had a son.'

    And, with a failing voice, Sohrab replied:—
'Ah yes, he had! and that lost son am I.   580
Surely the news will one day reach his ear,
Reach Rustum, where he sits, and tarries long,
Somewhere, I know not where, but far from here;
And pierce him like a stab, and make him leap
To arms, and cry for vengeance upon thee.
Fierce man, bethink thee, for an only son!
What will that grief, what will that vengeance be?
Oh, could I live, till I that grief had seen!
Yet him I pity not so much, but her,
My mother, who in Ader-baijan dwells   590
With that old king, her father, who grows grey
With age, and rules over the valiant Koords.
Her most I pity, who no more will see
Sohrab returning from the Tartar camp,
With spoils and honour, when the war is done.
But a dark rumour will be bruited up,
From tribe to tribe, until it reach her ear;
And then will that defenceless woman learn
That Sohrab will rejoice her sight no more,
But that in battle with a nameless foe,   600
By the far-distant Oxus, he is slain.'

    He spoke; and as he ceased, he wept aloud,
Thinking of her he left, and his own death.
He spoke; but Rustum listen'd, plunged in thought.
Nor did he yet believe it was his son
Who spoke, although he call'd back names he knew;
For he had had sure tidings that the babe,
Which was in Ader-baijan born to him,
Had been a puny girl, no boy at all—
So that sad mother sent him word, for fear   610
Rustum should seek the boy, to train in arms.

And so he deem'd that either Sohrab took,
By a false boast, the style of Rustum's son;
Or that men gave it him, to swell his fame.
So deem'd he; yet he listen'd, plunged in thought
And his soul set to grief, as the vast tide
Of the bright rocking Ocean sets to shore
At the full moon; tears gather'd in his eyes;
For he remember'd his own early youth,°
And all its bounding rapture; as, at dawn,   620
The shepherd from his mountain-lodge descries
A far, bright city, smitten by the sun,
Through many rolling clouds—so Rustum saw
His youth; saw Sohrab's mother, in her bloom;
And that old king, her father, who loved well
His wandering guest, and gave him his fair child
With joy; and all the pleasant life they led,
They three, in that long-distant summer-time—
The castle, and the dewy woods, and hunt
And hound, and morn on those delightful hills   630
In Ader-baijan. And he saw that youth,
Of age and looks to be his own dear son,
Piteous and lovely, lying on the sand,
Like some rich hyacinth which by the scythe°
Of an unskilful gardener has been cut,
Mowing the garden grass-plots near its bed,
And lies, a fragrant tower of purple bloom,
On the mown, dying grass—so Sohrab lay,
Lovely in death, upon the common sand.
And Rustum gazed on him with grief, and said:—   640
    'O Sohrab, thou indeed art such a son
Whom Rustum, wert thou his, might well have loved.
Yet here thou errest, Sohrab, or else men
Have told thee false—thou art not Rustum's son.
For Rustum had no son; one child he had—
But one—a girl; who with her mother now
Plies some light female task, nor dreams of us—
Of us she dreams not, nor of wounds, nor war.'
    But Sohrab answer'd him in wrath; for now
The anguish of the deep-fix'd spear grew fierce,   650
And he desired to draw forth the steel,
And let the blood flow free, and so to die—

But first he would convince his stubborn foe;
And, rising sternly on one arm, he said:—
　'Man, who art thou who dost deny my words?
Truth sits upon the lips of dying men,
And falsehood, while I lived, was far from mine.
I tell thee, prick'd upon this arm I bear°
That seal which Rustum to my mother gave,
That she might prick it on the babe she bore.'   660
　He spoke; and all the blood left Rustum's cheeks,
And his knees totter'd, and he smote his hand
Against his breast, his heavy mailéd hand,
That the hard iron corslet clank'd aloud;
And to his heart he press'd the other hand,
And in a hollow voice he spake, and said:—
　'Sohrab, that were a proof which could not lie!
If thou show this, then art thou Rustum's son.'
　Then, with weak hasty fingers, Sohrab loosed
His belt, and near the shoulder bared his arm,   670
And show'd a sign in faint vermilion points
Prick'd; as a cunning workman, in Pekin,
Pricks with vermilion some clear porcelain vase,
An emperor's gift—at early morn he paints,
And all day long, and, when night comes, the lamp
Lights up his studious forehead and thin hands—
So delicately prick'd the sign appear'd
On Sohrab's arm, the sign of Rustum's seal.
It was that griffin, which of old rear'd Zal,°
Rustum's great father, whom they left to die,   680
A helpless babe, among the mountain-rocks;
Him that kind creature found, and rear'd, and loved—
Then Rustum took it for his glorious sign.
And Sohrab bared that image on his arm,
And himself scann'd it long with mournful eyes,
And then he touch'd it with his hand and said:—
　'How say'st thou? Is that sign the proper sign
Of Rustum's son, or of some other man's?'
　He spoke; but Rustum gazed, and gazed, and stood
Speechless; and then he utter'd one sharp cry: 690
O boy—thy father!—and his voice choked there.
And then a dark cloud pass'd before his eyes,
And his head swam, and he sank down to earth.

But Sohrab crawl'd to where he lay, and cast
His arms about his neck, and kiss'd his lips,
And with fond faltering fingers stroked his cheeks,
Trying to call him back to life; and life
Came back to Rustum, and he oped his eyes,
And they stood wide with horror; and he seized
In both his hands the dust which lay around,   700
And threw it on his head, and smirch'd his hair,—
His hair, and face, and beard, and glittering arms;
And strong convulsive groanings shook his breast,
And his sobs choked him; and he clutch'd his sword,
To draw it, and for ever let life out.
But Sohrab saw his thought, and held his hands,
And with a soothing voice he spake, and said:—
    'Father, forbear! for I but meet to-day
The doom which at my birth was written down
In Heaven, and thou art Heaven's unconscious hand.   710
Surely my heart cried out that it was thou,
When first I saw thee; and thy heart spoke too,
I know it! but fate trod those promptings down
Under its iron heel; fate, fate engaged
The strife, and hurl'd me on my father's spear.
But let us speak no more of this! I find
My father; let me feel that I have found!
Come, sit beside me on this sand, and take
My head betwixt thy hands, and kiss my cheeks,
And wash them with thy tears, and say: *My son!*   720
Quick! quick! for number'd are my sands of life,
And swift; for like the lightning to this field
I came, and like the wind I go away—
Sudden, and swift, and like a passing wind.
But it was writ in Heaven that this should be.'
    So said he, and his voice released the heart
Of Rustum, and his tears broke forth; he cast
His arms round his son's neck, and wept aloud,
And kiss'd him. And awe fell on both the hosts,
When they saw Rustum's grief; and Ruksh, the horse,°   730
With his head bowing to the ground and mane
Sweeping the dust, came near, and in mute woe
First to the one then to the other moved
His head, as if inquiring what their grief

Might mean; and from his dark, compassionate eyes,
The big warm tears roll'd down, and caked the sand.
But Rustum chid him with stern voice, and said:—
　'Ruksh, now thou grievest; but, O Ruksh, thy feet
Should first have rotted on their nimble joints,
Or ere they brought thy master to this field!'   740
　But Sohrab look'd upon the horse and said;—
'Is this, then, Ruksh! How often, in past days,
My mother told me of thee, thou brave steed,
My terrible father's terrible horse! and said,
That I should one day find thy lord and thee.
Come, let me lay my hand upon thy mane!
O Ruksh, thou art more fortunate than I;
For thou hast gone where I shall never go,
And snuff'd the breezes of my father's home.
And thou hast trod the sands of Seistan,   750
And seen the River of Helmund, and the Lake
Of Zirrah; and the aged Zal himself
Has often stroked thy neck, and given thee food,
Corn in a golden platter soak'd with wine,
And said: *O Ruksh! bear Rustum well!*—but I
Have never known my grandsire's furrow'd face,
Nor seen his lofty house in Seistan,
Nor slaked my thirst at the clear Helmund stream;
But lodged among my father's foes, and seen
Afrasiab's cities only, Samarcand,   760
Bokhara, and lone Khiva in the waste,
And the black Toorkmun tents; and only drunk
The desert rivers, Moorghab and Tejend,
Kohik, and where the Kalmuks feed their sheep,
The northern Sir; and this great Oxus stream,
The yellow Oxus, by whose brink I die.'
　Then, with a heavy groan, Rustum bewail'd:—
'Oh, that its waves were flowing over me!
Oh, that I saw its grains of yellow silt
Roll tumbling in the current o'er my head!'   770
　But, with a grave mild voice, Sohrab replied:—
'Desire not that, my father! thou must live.
For some are born to do great deeds, and live,
As some are born to be obscured, and die.
Do thou the deeds I die too young to do,

And reap a second glory in thine age;
Thou art my father, and thy gain is mine.
But come! thou seest this great host of men
Which follow me; I pray thee, slay not these!
Let me entreat for them; what have they done?   780
They follow'd me, my hope, my fame, my star.
Let them all cross the Oxus back in peace.
But me thou must bear hence, not send with them,°
But carry me with thee to Seistan,
And place me on a bed, and mourn for me,
Thou, and the snow-hair'd Zal, and all thy friends.
And thou must lay me in that lovely earth,
And heap a stately mound above my bones,
And plant a far-seen pillar over all.
That so the passing horseman on the waste   790
May see my tomb a great way off, and cry:
Sohrab, the mighty Rustum's son, lies there,
Whom his great father did in ignorance kill!
And I be not forgotten in my grave.'
     And, with a mournful voice, Rustum replied:—
'Fear not! as thou hast said, Sohrab, my son,
So shall it be; for I will burn my tents,
And quit the host, and bear thee hence with me,
And carry thee away to Seistan,
And place thee on a bed, and mourn for thee,   800
With the snow-headed Zal, and all my friends.
And I will lay thee in that lovely earth,
And heap a stately mound above thy bones,
And plant a far-seen pillar over all,
And men shall not forget thee in thy grave.
And I will spare thy host; yea, let them go!
Let them all cross the Oxus back in peace!
What should I do with slaying any more?
For would that all whom I have ever slain
Might be once more alive; my bitterest foes,   810
And they who were call'd champions in their time,
And through whose death I won that fame I have—
And I were nothing but a common man,
A poor, mean soldier, and without renown,
So thou mightest live too, my son, my son!°
Or rather would that I, even I myself,

Might now be lying on this bloody sand,
Near death, and by an ignorant stroke of thine,
Not thou of mine! and I might die, not thou;
And I, not thou, be borne to Seistan;   820
And Zal might weep above my grave, not thine;
And say: *O son, I weep thee not too sore,*
For willingly, I know, thou met'st thine end!
But now in blood and battles was my youth,
And full of blood and battles is my age,
And I shall never end this life of blood.'
   Then, at the point of death, Sohrab replied:—
'A life of blood indeed, thou dreadful man!
But thou shalt yet have peace; only not now,
Not yet! but thou shalt have it on that day,°   830
When thou shalt sail in a high-masted ship,
Thou and the other peers of Kai Khosroo,
Returning home over the salt blue sea,
From laying thy dear master in his grave.'
   And Rustum gazed in Sohrab's face, and said:—
'Soon be that day, my son, and deep that sea!
Till then, if fate so wills, let me endure.'
   He spoke; and Sohrab smiled on him, and took
The spear, and drew it from his side, and eased
His wound's imperious anguish; but the blood   840
Came welling from the open gash, and life
Flow'd with the stream;—all down his cold white side
The crimson torrent ran, dim now and soil'd,
Like the soil'd tissue of white violets
Left, freshly gather'd, on their native bank,
By children whom their nurses call with haste
Indoors from the sun's eye; his head droop'd low,
His limbs grew slack; motionless, white, he lay—
White, with eyes closed; only when heavy grasps,
Deep heavy gasps quivering through all his frame,   850
Convulsed him back to life, he open'd them,
And fix'd them feebly on his father's face;
Till now all strength was ebb'd, and from his limbs°
Unwillingly the spirit fled away,
Regretting the warm mansion which it left,
And youth, and bloom, and this delightful world.°
   So, on the bloody sand, Sohrab lay dead;

And the great Rustum drew his horseman's cloak
Down o'er his face, and sate by his dead son.
As those black granite pillars, once high-rear'd°   860
By Jemshid in Persepolis, to bear
His house, now 'mid their broken flights of steps
Lie prone, enormous, down the mountain side—
So in the sand lay Rustum by his son.
    And night came down over the solemn waste,
And the two gazing hosts, and that sole pair,
And darken'd all; and a cold fog, with night,
Crept from the Oxus. Soon a hum arose,
As of a great assembly loosed, and fires°
Began to twinkle through the fog; for now   870
Both armies moved to camp, and took their meal;
The Persians took it on the open sands
Southward, the Tartars by the river marge;
And Rustum and his son were left alone.
    But the majestic river floated on,°
Out of the mist and hum of that low land,
Into the frosty starlight, and there moved,
Rejoicing, through the hush'd Chorasmian waste,°
Under the solitary moon;—he flow'd
Right for the polar star, past Orgunjé,   880
Brimming, and bright, and large; then sands begin
To hem his watery march, and dam his streams,
And split his currents; that for many a league
The shorn and parcell'd Oxus strains along
Through beds of sand and matted rushy isles—
Oxus, forgetting the bright speed he had
In his high mountain-cradle in Pamere,
A foil'd circuitous wanderer—till at last
The long'd-for dash of waves is heard, and wide
His luminous home of waters opens, bright°   890
And tranquil, from whose floor the new-bathed stars°
Emerge, and shine upon the Aral Sea.

The Scholar-Gipsy

Go, for they call you, shepherd, from the hill;°
　Go, shepherd, and untie the wattled cotes!°
　　No longer leave thy wistful flock unfed,
　Nor let thy bawling fellows rack their throats,
　　Nor the cropp'd herbage shoot another head.
　　　But when the fields are still,
　And the tired men and dogs all gone to rest,
　　And only the white sheep are sometimes seen
　　Cross and recross the strips of moon-blanch'd green,
Come, shepherd, and again begin the quest!°　　　　10

Here, where the reaper was at work of late—
　In this high field's dark corner, where he leaves
　　His coat, his basket, and his earthen cruse,
　And in the sun all morning binds the sheaves,
　　Then here, at noon, comes back his stores to use—
　　　Here will I sit and wait,
　While to my ear from uplands far away
　　The bleating of the folded flocks is borne,
　　With distant cries of reapers in the corn—
All the live murmur of a summer's day.　　　　20

Screen'd is this nook o'er the high, half-reap'd field,
　And here till sun-down, shepherd! will I be.
　　Through the thick corn the scarlet poppies peep,
　And round green roots and yellowing stalks I see
　　Pale pink convolvulus in tendrils creep;
　　　And air-swept lindens yield
　Their scent, and rustle down their perfumed showers
　　Of bloom on the bent grass where I am laid,
　　And bower me from the August sun with shade;
And the eye travels down to Oxford's towers.　　　　30

And near me on the grass lies Glanvil's book—
　Come, let me read the oft-read tale again!
　　The story of the Oxford scholar poor,
　Of pregnant parts and quick inventive brain,°

Who, tired of knocking at preferment's door,
      One summer-morn forsook
His friends, and went to learn the gipsy-lore,
      And roam'd the world with that wild brotherhood,
      And came, as most men deem'd, to little good,
But came to Oxford and his friends no more.   40

But once, years after, in the country-lanes,
      Two scholars, whom at college erst he knew,
      Met him, and of his way of life enquired;
Whereat he answer'd, that the gipsy-crew,
      His mates, had arts to rule as they desired
            The workings of men's brains,
And they can bind them to what thoughts they will.
      'And I,' he said, 'the secret of their art,
      When fully learn'd, will to the world impart;
But it needs heaven-sent moments for this skill.'°   50

This said, he left them, and return'd no more.—
      But rumours hung about the country-side,
      That the lost Scholar long was seen to stray,
Seen by rare glimpses, pensive and tongue-tied,°
      In hat of antique shape, and cloak of grey,
            The same the gipsies wore.
Shepherds had met him on the Hurst in spring;°
      At some lone alehouse in the Berkshire moors,
      On the warm ingle-bench, the smock-frock'd boors
Had found him seated at their entering,   60

But, 'mid their drink and clatter, he would fly.
And I myself seem half to know thy looks,
      And put the shepherds, wanderer! on thy trace;
And boys who in lone wheatfields scare the rooks
      I ask if thou hast pass'd their quiet place;
            Or in my boat I lie
Moor'd to the cool bank in the summer-heats,
      'Mid wide grass meadows which the sunshine fills,
      And watch the warm, green-muffled Cumner hills,
And wonder if thou haunt'st their shy retreats.   70

For most, I know, thou lov'st retired ground!
  Thee at the ferry Oxford riders blithe,
    Returning home on summer-nights, have met
  Crossing the stripling Thames at Bab-lock-hithe,
    Trailing in the cool stream thy fingers wet,
      As the punt's rope chops round;
  And leaning backward in a pensive dream,
    And fostering in thy lap a heap of flowers
    Pluck'd in shy fields and distant Wychwood bowers,
  And thine eyes resting on the moonlit stream.   80

And then they land, and thou art seen no more!—
  Maidens, who from the distant hamlets come
    To dance around the Fyfield elm in May,
  Oft through the darkening fields have seen thee roam,
    Or cross a stile into the public way.
      Oft thou hast given them store
  Of flowers—the frail-leaf'd, white anemony,
    Dark bluebells drench'd with dews of summer eves,
    And purple orchises with spotted leaves—
  But none hath words she can report of thee.   90

And, above Godstow Bridge, when hay-time's here
  In June, and many a scythe in sunshine flames,
    Men who through those wide fields of breezy grass
  Where black-wing'd swallows haunt the glittering Thames,
    To bathe in the abandon'd lasher pass,°
      Have often pass'd thee near
  Sitting upon the river bank o'ergrown;
    Mark'd thine outlandish garb, thy figure spare,
    Thy dark vague eyes, and soft abstracted air—
  But, when they came from bathing, thou wast gone!   100

At some lone homestead in the Cumner hills,
  Where at her open door the housewife darns,
    Thou hast been seen, or hanging on a gate
  To watch the threshers in the mossy barns.
    Children, who early range these slopes and late
      For cresses from the rills,

Have known thee eying, all an April-day,
    The springing pastures and the feeding kine;
    And mark'd thee, when the stars come out and shine,
Through the long dewy grass move slow away.   110

In autumn, on the skirts of Bagley Wood—
    Where most the gipsies by the turf-edged way
        Pitch their smoked tents, and every bush you see
With scarlet patches tagg'd and shreds of grey,
        Above the forest-ground called Thessaly—
            The blackbird, picking food,
    Sees thee, nor stops his meal, nor fears at all;
        So often has he known thee past him stray,
        Rapt, twirling in thy hand a wither'd spray,
And waiting for the spark from heaven to fall.°   120

And once, in winter, on the causeway chill
    Where home through flooded fields foot-travellers go,
        Have I not pass'd thee on the wooden bridge,
Wrapt in thy cloak and battling with the snow,
        Thy face tow'rd Hinksey and its wintry ridge?
            And thou hast climb'd the hill,
    And gain'd the white brow of the Cumner range;
        Turn'd once to watch, while thick the snowflakes fall.
        The line of festal light in Christ-Church hall—
Then sought thy straw in some sequester'd grange.   130

But what—I dream! Two hundred years are flown
    Since first thy story ran through Oxford halls,
        And the grave Glanvil did the tale inscribe
That thou wert wander'd from the studious walls
        To learn strange arts, and join a gipsy-tribe;
            And thou from earth art gone
    Long since, and in some quiet churchyard laid—
        Some country-nook, where o'er thy unknown grave
        Tall grasses and white flowering nettles wave,
Under a dark, red-fruited yew-tree's shade.   140

—No, no, thou hast not felt the lapse of hours!
  For what wears out the life of mortal men?
    'Tis that from change to change their being rolls;
  'Tis that repeated shocks, again, again,
    Exhaust the energy of strongest souls
      And numb the elastic powers.
  Till having used our nerves with bliss and teen,°
    And tired upon a thousand schemes our wit,
    To the just-pausing Genius we remit°
  Our worn-out life, and are—what we have been.   150

Thou hast not lived, why should'st thou perish, so?
  Thou hadst *one* aim, *one* business, *one* desire;
    Else wert thou long since number'd with the dead!
  Else hadst thou spent, like other men, thy fire!
    The generations of thy peers are fled,
      And we ourselves shall go;
  But thou possessest an immortal lot,
    And we imagine thee exempt from age
    And living as thou liv'st on Glanvil's page,
  Because thou hadst—what we, alas! have not.   160

For early didst thou leave the world, with powers
  Fresh, undiverted to the world without,
    Firm to their mark, not spent on other things;
  Free from the sick fatigue, the languid doubt,°
    Which much to have tried, in much been baffled, brings.
      O life unlike to ours!
  Who fluctuate idly without term or scope,°
  Of whom each strives, nor knows for what he strives,
    And each half lives a hundred different lives;
  Who wait like thee, but not, like thee, in hope.   170

Thou waitest for the spark from heaven! and we,
  Light half-believers of our casual creeds,
    Who never deeply felt, nor clearly will'd,
  Whose insight never has borne fruit in deeds,
    Whose vague resolves never have been fulfill'd;
      For whom each year we see

Breeds new beginnings, disappointments new;
   Who hesitate and falter life away,
   And lose to-morrow the ground won to-day—
Ah! do not we, wanderer! await it too?   180

Yes, we await it!—but it still delays,
   And then we suffer! and amongst us one,°
     Who most has suffer'd, takes dejectedly
His seat upon the intellectual throne;
     And all his store of sad experience he
       Lays bare of wretched days;
   Tells us his misery's birth and growth and signs,
     And how the dying spark of hope was fed,
     And how the breast was soothed, and how the head,
And all his hourly varied anodynes.   190

This for our wisest! and we others pine,
   And wish the long unhappy dream would end,
     And waive all claim to bliss, and try to bear;
With close-lipp'd patience for our only friend,
     Sad patience, too near neighbour to despair—
       But none has hope like thine!
   Thou through the fields and through the woods dost stray,
     Roaming the country-side, a truant boy,
     Nursing thy project in unclouded joy,
And every doubt long blown by time away.   200

O born in days when wits were fresh and clear,°
   And life ran gaily as the sparkling Thames;
     Before this strange disease of modern life,
With its sick hurry, its divided aims,
     Its heads o'ertax'd, its palsied hearts, was rife—°
       Fly hence, our contact fear!
   Still fly, plunge deeper in the bowering wood!
     Averse, as Dido did with gesture stern
     From her false friend's approach in Hades turn,
Wave us away, and keep thy solitude!   210

Still nursing the unconquerable hope,
 Still clutching the inviolable shade,
  With a free, onward impulse brushing through,
 By night, the silver'd branches of the glade—
  Far on the forest-skirts, where none pursue.
   On some mild pastoral slope
 Emerge, and resting on the moonlit pales°
  Freshen thy flowers as in former years
  With dew, or listen with enchanted ears,
From the dark dingles, to the nightingales!°   220

But fly our paths, our feverish contact fly!
 For strong the infection of our mental strife,
  Which, though it gives no bliss, yet spoils for rest;
 And we should win thee from thy own fair life,
  Like us distracted, and like us unblest.
   Soon, soon thy cheer would die,
 Thy hopes grow timorous, and unfix'd thy powers,
  And thy clear aims be cross and shifting made;
  And then thy glad perennial youth would fade,
Fade, and grow old at last, and die like ours.   230

Then fly our greetings, fly our speech and smiles!
 —As some grave Tyrian trader, from the sea,°
  Descried at sunrise an emerging prow
 Lifting the cool-hair'd creepers stealthily,
  The fringes of a southward-facing brow
   Among the Aegean isles;
 And saw the merry Grecian coaster come,
  Freighted with amber grapes, and Chian wine,
  Green, bursting figs, and tunnies steep'd in brine—
And knew the intruders on his ancient home,   240

The young light-hearted masters of the waves—
 And snatch'd his rudder, and shook out more sail;
  And day and night held on indignantly
 O'er the blue Midland waters with the gale,
  Betwixt the Syrtes and soft Sicily,
   To where the Atlantic raves

Outside the western straits; and unbent sails
   There, where down cloudy cliffs, through sheets of foam,
   Shy traffickers, the dark Iberians come;
And on the beach undid his corded bales.   250

Requiescat

   Strew on her roses, roses,
      And never a spray of yew!
   In quiet she reposes;
      Ah, would that I did too!

   Her mirth the world required;°
      She bathed it in smiles of glee.
   But her heart was tired, tired,
      And now they let her be.

   Her life was turning, turning,
      In mazes of heart and sound.   10
   But for peace her soul was yearning,
      And now peace laps her round.

   Her cabin'd, ample spirit,°
      It flutter'd and fail'd for breath.
   To-night it doth inherit
      The vasty hall of death.

Philomela

   Hark! ah, the nightingale—
   The tawn-throated!
   Hark, from that moonlit cedar what a burst!
   What triumph! hark!—what pain!

O wanderer from a Grecian shore,
Still, after many years, in distant lands,
Still nourishing in thy bewilder'd brain
That wild, unquench'd, deep-sunken, old-world pain—
Say, will it never heal?
And can this fragrant lawn   10
With its cool trees, and night,
And the sweet, tranquil Thames,
And moonshine, and the dew,
To thy rack'd heart and brain
Afford no balm?

Dost thou to-night behold,
Here, through the moonlight on this English grass,
The unfriendly palace in the Thracian wild?
Dost thou again peruse
With hot cheeks and sear'd eyes   20
The too clear web, and thy dumb sister's shame?
Dost thou once more assay
Thy flight, and feel come over thee,
Poor fugitive, the feathery change
Once more, and once more seem to make resound
With love and hate, triumph and agony,
Lone Daulis, and the high Cephissian vale?°
Listen, Eugenia—
How thick the bursts come crowding through the leaves!
Again—thou hearest?   30
Eternal passion!
Eternal pain!

Haworth Churchyard

APRIL 1855

Where, under Loughrigg, the stream°
Of Rotha sparkles through fields
Vested for ever with green,
Four years since, in the house°
Of a gentle spirit, now dead—

Wordsworth's son-in-law, friend—
I saw the meeting of two°
Gifted women. The one,
Brilliant with recent renown,°
Young, unpractised, had told   10
With a master's accent her feign'd
Story of passionate life;
The other, maturer in fame,°
Earning, she too, her praise
First in fiction, had since
Widen'd her sweep, and survey'd
History, politics, mind.

   The two held converse; they wrote
In a book which of world-famous souls°
Kept the memorial;—bard,   20
Warrior, statesman, had sign'd
Their names; chief glory of all,
Scott had bestow'd there his last
Breathings of song, with a pen
Tottering, a death-stricken hand.

Hope at that meeting smiled fair.
Years in number, it seem'd,
Lay before both, and a fame
Heighten'd, and multiplied power.—
Behold! The elder, to-day,°   30
Lies expecting from death,
In mortal weakness, a last
Summons! the younger is dead!

First to the living we pay
Mournful homage;—the Muse
Gains not an earth-deafen'd ear.

Hail to the steadfast soul,
Which, unflinching and keen,
Wrought to erase from its depth
Mist and illusion and fear!   40
Hail to the spirit which dared
Trust its own thoughts, before yet

Echoed her back by the crowd!
Hail to the courage which gave
Voice to its creed, ere the creed
Won consecration from time!

Turn we next to the dead.
—How shall we honour the young,
The ardent, the gifted? how mourn?
Console we cannot, her ear   50
Is deaf. Far northward from here,
In a churchyard high 'mid the moors
Of Yorkshire, a little earth
Stops it for ever to praise.

Where, behind Keighley, the road
Up to the heart of the moors
Between heath-clad showery hills
Runs, and colliers' carts
Poach the deep ways coming down,
And a rough, grimed race have their homes—   60
There on its slope is built
The moorland town. But the church
Stands on the crest of the hill,
Lonely and bleak;—at its side
The parsonage-house and the graves.°

Strew with laurel the grave
Of the early-dying! Alas,
Early she goes on the path
To the silent country, and leaves
Half her laurels unwon,   70
Dying too soon!—yet green
Laurels she had, and a course
Short, but redoubled by fame.

And not friendless, and not
Only with strangers to meet,
Faces ungreeting and cold,
Thou, O mourn'd one, to-day
Enterest the house of the grave!
Those of thy blood, whom thou lov'dst,°

Have preceded thee—young,   80
Loving, a sisterly band;
Some in art, some in gift
Inferior—all in fame.
They, like friends, shall receive
This comer, greet her with joy;
Welcome the sister, the friend;
Hear with delight of thy fame!

Round thee they lie—the grass
Blows from their graves to thy own!
She, whose genius, though not   90
Puissant like thine, was yet
Sweet and graceful;—and she
(How shall I sing her?) whose soul
Knew no fellow for might,
Passion, vehemence, grief,
Daring, since Byron died,
That world-famed son of fire—she, who sank
Baffled, unknown, self-consumed;
Whose too bold dying song
Stirr'd, like a clarion-blast, my soul.   100

Of one, too, I have heard,°
A brother—sleeps he here?
Of all that gifted race
Not the least gifted; young,
Unhappy, eloquent—the child
Of many hopes, of many tears.
O boy, if here thou sleep'st, sleep well!
On thee too did the Muse
Bright in thy cradle smile;
But some dark shadow came   110
(I know not what) and interposed.

Sleep, O cluster of friends,°
Sleep!—or only when May,
Brought by the west-wind, returns
Back to your native heaths,
And the plover is heard on the moors,
Yearly awake to behold

The opening summer, the sky,
The shining moorland—to hear
The drowsy bee, as of old,   120
Hum o'er the thyme, the grouse
Call from the heather in bloom!
Sleep, or only for this
Break your united repose!

EPILOGUE

So I sang; but the Muse,
Shaking her head, took the harp—
Stern interrupted my strain,
Angrily smote on the chords.

April showers
Rush o'er the Yorkshire moors.   130
Stormy, through driving mist,
Loom the blurr'd hills; the rain
Lashes the newly-made grave.

Unquiet souls!
—In the dark fermentation of earth,
In the never idle workshop of nature,
In the eternal movement,
Ye shall find yourselves again!

Rugby Chapel

NOVEMBER 1857

Coldly, sadly descends°
The autumn-evening. The field
Strewn with its dank yellow drifts
Of wither'd leaves, and the elms,
Fade into dimness apace,
Silent;—hardly a shout
From a few boys late at their play!
The lights come out in the street,
In the school-room windows;—but cold,
Solemn, unlighted, austere,°   10

Through the gathering darkness, arise
The chapel-walls, in whose bound°
Thou, my father! art laid.

There thou dost lie, in the gloom
Of the autumn evening. But ah!
That word, *gloom*, to my mind
Brings thee back, in the light
Of thy radiant vigour, again;
In the gloom of November we pass'd
Days not dark at thy side;   20
Seasons impair'd not the ray
Of thy buoyant cheerfulness clear.
Such thou wast! and I stand
In the autumn evening, and think
Of bygone autumns with thee.

Fifteen years have gone round
Since thou arosest to tread,
In the summer-morning, the road
Of death, at a call unforeseen,
Sudden. For fifteen years,   30
We who till then in thy shade
Rested as under the boughs
Of a mighty oak, have endured
Sunshine and rain as we might,
Bare, unshaded, alone,
Lacking the shelter of thee.

O strong soul, by what shore°
Tarriest thou now? For that force,
Surely, has not been left vain!
Somewhere, surely, afar,   40
In the sounding labour-house vast°
Of being, is practised that strength,
Zealous, beneficent, firm!

Yes, in some far-shining sphere,
Conscious or not of the past,
Still thou performest the word
Of the Spirit in whom thou dost live—°

Prompt, unwearied, as here!
Still thou upraisest with zeal
The humble good from the ground,   50
Sternly repressest the bad!
Still, like a trumpet, dost rouse
Those who with half-open eyes
Tread the border-land dim
'Twixt vice and virtue; reviv'st,
Succourest!—this was thy work,
This was thy life upon earth.

What is the course of the life°
Of mortal men on the earth?—
Most men eddy about°   60
Here and there—eat and drink,
Chatter and love and hate,
Gather and squander, are raised°
Aloft, are hurl'd in the dust,
Striving blindly, achieving
Nothing; and then they die—
Perish;—and no one asks°
Who or what they have been,
More than he asks what waves,
In the moonlit solitudes mild   70
Of the midmost Ocean, have swell'd,
Foam'd for a moment, and gone.

And there are some, whom a thirst
Ardent, unquenchable, fires,
Not with the crowd to be spent,
Not without aim to go round
In an eddy of purposeless dust,
Effort unmeaning and vain.
Ah yes! some of us strive
Not without action to die   80
Fruitless, but something to snatch
From dull oblivion, nor all
Glut the devouring grave!°
We, we have chosen our path—
Path to a clear-purposed goal,
Path of advance!—but it leads

A long, steep journey, through sunk
Gorges, o'er mountains in snow.
Cheerful, with friends, we set forth—
Then, on the height, comes the storm.   90
Thunder crashes from rock
To rock, the cataracts reply,
Lightnings dazzle our eyes.
Roaring torrents have breach'd
The track, the stream-bed descends
In the place where the wayfarer once
Planted his footstep—the spray
Boils o'er its borders! aloft
The unseen snow-beds dislodge
Their hanging ruin; alas,   100
Havoc is made in our train!
Friends, who set forth at our side,
Falter, are lost in the storm.
We, we only are left!
With frowning foreheads, with lips
Sternly compress'd, we strain on,
On—and at nightfall at last
Come to the end of our way,
To the lonely inn 'mid the rocks;
Where the gaunt and taciturn host   110
Stands on the threshold, the wind
Shaking his thin white hairs—
Holds his lantern to scan
Our storm-beat figures, and asks:
Whom in our party we bring?
Whom we have left in the snow?

Sadly we answer: We bring
Only ourselves! we lost
Sight of the rest in the storm.
Hardly ourselves we fought through,   120
Stripp'd, without friends, as we are.
Friends, companions, and train,
The avalanche swept from our side.

But thou would'st not *alone*°
Be saved, my father! *alone*

Conquer and come to thy goal,
Leaving the rest in the wild.
We were weary, and we°
Fearful, and we in our march
Fain to drop down and to die.   130
Still thou turnedst, and still
Beckonedst the trembler, and still
Gavest the weary thy hand.

If, in the paths of the world,
Stones might have wounded thy feet,
Toil or dejection have tried
Thy spirit, of that we saw
Nothing—to us thou wast still
Cheerful, and helpful, and firm!
Therefore to thee it was given   140
Many to save with thyself;
And, at the end of thy day,
O faithful shepherd! to come,
Bringing thy sheep in thy hand.

And through thee I believe°
In the noble and great who are gone;
Pure souls honour'd and blest
By former ages, who else—
Such, so soulless, so poor,
Is the race of men whom I see—   150
Seem'd but a dream of the heart,
Seem'd but a cry of desire.
Yes! I believe that there lived
Others like thee in the past,
Not like the men of the crowd
Who all round me to-day
Bluster or cringe, and make life
Hideous, and arid, and vile;
But souls temper'd with fire,
Fervent, heroic, and good,   160
Helpers and friends of mankind.

Servants of God!—or sons°
Shall I not call you? because

Not as servants ye knew
Your Father's innermost mind,
His, who unwillingly sees°
One of his little ones lost—
Yours is the praise, if mankind
Hath not as yet in its march
Fainted, and fallen, and died!   170

See! In the rocks of the world
Marches the host of mankind,
A feeble, wavering line.
Where are they tending?—A God°
Marshall'd them, gave them their goal.
Ah, but the way is so long!
Years they have been in the wild!°
Sore thirst plagues them, the rocks,
Rising all round, overawe;
Factions divide them, their host   180
Threatens to break, to dissolve.
—Ah, keep, keep them combined!
Else, of the myriads who fill
That army, not one shall arrive;
Sole they shall stray; in the rocks
Stagger for ever in vain,
Die one by one in the waste.

Then, in such hour of need
Of your fainting, dispirited race,
Ye, like angels, appear,   190
Radiant with ardour divine!

Beacons of hope, ye appear!
Languor is not in your heart,
Weakness is not in your word,
Weariness not on your brow.
Ye alight in our van! at your voice,
Panic, despair, flee away.
Ye move through the ranks, recall
The stragglers, refresh the outworn,
Praise, re-inspire the brave!   200
Order, courage, return.

Eyes rekindling, and prayers,
Follow your steps as ye go.
Ye fill up the gaps in our files,
Strengthen the wavering line,
Stablish, continue our march,
On, to the bound of the waste,
On, to the City of God.°

A Southern Night

The sandy spits, the shore-lock'd lakes,
   Melt into open, moonlit sea;
The soft Mediterranean breaks
     At my feet, free.

Dotting the fields of corn and vine,
   Like ghosts the huge, gnarl'd olives stand
Behind, that lovely mountain-line!
     While, by the strand,

Cette, with its glistening houses white,
   Curves with the curving beach away   10
To where the lighthouse beacons bright
     Far in the bay.

Ah! such a night, so soft, so lone,°
   So moonlit, saw me once of yore
Wander unquiet, and my own
     Vext heart deplore.

But now that trouble is forgot;
   Thy memory, thy pain, to-night,
My brother! and thine early lot,
     Possess me quite.   20

The murmur of this Midland deep
   Is heard to-night around thy grave,
There, where Gibraltar's cannon'd steep
     O'erfrowns the wave.

For there, with bodily anguish keen,
  With Indian heats at last fordone,
With public toil and private teen—°
    Thou sank'st, alone.

Slow to a stop, at morning grey,
  I see the smoke-crown'd vessel come;   30
Slow round her paddles dies away
    The seething foam.

A boat is lower'd from her side;
  Ah, gently place him on the bench!
That spirit—if all have not yet died—
    A breath might quench.

Is this the eye, the footstep fast,
  The mien of youth we used to see,
Poor, gallant boy!—for such thou wast,
    Still art, to me.   40

The limbs their wonted tasks refuse;
  The eyes are glazed, thou canst not speak;
And whiter than thy white burnous
    That wasted cheek!

Enough! The boat, with quiet shock,
  Unto its haven coming nigh,
Touches, and on Gibraltar's rock
    Lands thee to die.

Ah me! Gibraltar's strand is far,
  But farther yet across the brine°   50
Thy dear wife's ashes buried are,
    Remote from thine.

For there, where morning's sacred fount
  Its golden rain on earth confers,
The snowy Himalayan Mount
    O'ershadows hers.

Strange irony of fate, alas,
   Which, for two jaded English, saves,
When from their dusty life they pass,
      Such peaceful graves!   60

In cities should we English lie,
   Where cries are rising ever new,
And men's incessant stream goes by—
      We who pursue

Our business with unslackening stride,
   Traverse in troops, with care-fill'd breast,
The soft Mediterranean side,
      The Nile, the East,

And see all sights from pole to pole,
   And glance, and nod, and bustle by,   70
And never once possess our soul
      Before we die.

Not by those hoary Indian hills,
   Not by this gracious Midland sea
Whose floor to-night sweet moonshine fills,
      Should our graves be.°

Some sage, to whom the world was dead,
   And men were specks, and life a play;
Who made the roots of trees his bed,
      And once a day   80

With staff and gourd his way did bend
   To villages and homes of man,
For food to keep him till he end
      His mortal span

And the pure goal of being reach;
   Hoar-headed, wrinkled, clad in white,
Without companion, without speech,
      By day and night

Pondering God's mysteries untold,
  And tranquil as the glacier-snows   90
He by those Indian mountains old
    Might well repose.

Some grey crusading knight austere,
  Who bore Saint Louis company,°
And came home hurt to death, and here
    Landed to die;

Some youthful troubadour, whose tongue°
  Fill'd Europe once with his love-pain,
Who here outworn had sunk, and sung
    His dying strain;   100

Some girl, who here from castle-bower,
  With furtive step and cheek of flame,
'Twixt myrtle-hedges all in flower
    By moonlight came

To meet her pirate-lover's ship;
  And from the wave-kiss'd marble stair
Beckon'd him on, with quivering lip
    And floating hair;

And lived some moons in happy trance,
  Then learnt his death and pined away— 110
Such by these waters of romance
    'Twas meet to lay.

But you—a grave for knight or sage,
  Romantic, solitary, still,
O spent ones of a work-day age!
    Befits you ill.

So sang I; but the midnight breeze,
  Down to the brimm'd, moon-charméd main,
Comes softly through the olive-trees,
    And checks my strain.   120

I think of her, whose gentle tongue
   All plaint in her own cause controll'd;
Of thee I think, my brother! young
     In heart, high-soul'd—

That comely face, that cluster'd brow,
   That cordial hand, that bearing free,
I see them still, I see them now,
     Shall always see!

And what but gentleness untired,
   And what but noble feeling warm,   130
Wherever shown, howe'er inspired,
     Is grace, is charm?

What else is all these waters are,
   What else is steep'd in lucid sheen,
What else is bright, what else is fair,
     What else serene?

Mild o'er her grave, ye mountains, shine!
   Gently by his, ye waters, glide!
To that in you which is divine
     They were allied.   140

Heine's Grave

'HENRI HEINE'——'tis here!
That black tombstone, the name
Carved there—no more! and the smooth,
Swarded alleys, the limes
Touch'd with yellow by hot
Summer, but under them still,
In September's bright afternoon,
Shadow, and verdure, and cool.
Trim Montmartre! the faint
Murmur of Paris outside;   10
Crisp everlasting-flowers,
Yellow and black, on the graves.

Half blind, palsied, in pain'°
Hither to come, from the streets'
Uproar, surely not loath
Was thou, Heine!—to lie
Quiet, to ask for closed
Shutters, and darken'd room,
And cool drinks, and an eased
Posture, and opium, no more;   20
Hither to come, and to sleep
Under the wings of Renown.

Ah! not little, when pain
Is most quelling, and man
Easily quell'd, and the fine
Temper of genius so soon
Thrills at each smart, is the praise,
Not to have yielded to pain!
No small boast, for a weak
Son of mankind, to the earth   30
Pinn'd by the thunder, to rear
His bolt-scathed front to the stars;
And, undaunted, retort
'Gainst thick-crashing, insane,
Tyrannous tempests of bale,
Arrowy lightnings of soul.

Hark! through the alley resounds
Mocking laughter! A film
Creeps o'er the sunshine; a breeze
Ruffles the warm afternoon,   40
Saddens my soul with its chill.
Gibing of spirits in scorn
Shakes every leaf of the grove,
Mars the benignant repose
Of this amiable home of the dead.

Bitter spirits, ye claim°
Heine?—Alas, he is yours!
Only a moment I long'd
Here in the quiet to snatch
From such mates the outworn   50

Poet, and steep him in calm.
Only a moment! I knew
Whose he was who is here
Buried—I knew he was yours!
Ah, I knew that I saw
Here no sepulchre built
In the laurell'd rock, o'er the blue
Naples bay, for a sweet
Tender Virgil! no tomb
On Ravenna sands, in the shade   60
Of Ravenna pines, for a high
Austere Dante! no grave°
By the Avon side, in the bright
Stratford meadows, for thee,
Shakespeare! loveliest of souls,
Peerless in radiance, in joy.

What, then, so harsh and malign,
Heine! distils from thy life?
Poisons the peace of thy grave?

I chide with thee not, that thy sharp   70
Upbraidings often assail'd
England, my country—for we,
Heavy and sad, for her sons,
Long since, deep in our hearts,
Echo the blame of her foes.
We, too, sigh that she flags;
We, too, say that she now—°
Scarce comprehending the voice
Of her greatest, golden-mouth'd sons
Of a former age any more—   80
Stupidly travels her round
Of mechanic business, and lets
Slow die out of her life
Glory, and genius, and joy.

So thou arraign'st her, her foe;
So we arraign her, her sons.

     Yes, we arraign her! but she,

The weary Titan, with deaf
Ears, and labour-dimm'd eyes,
Regarding neither to right   90
Nor left, goes passively by,
Staggering on to her goal;
Bearing on shoulders immense,
Atlanteän, the load,
Wellnigh not to be borne,
Of the too vast orb of her fate.

But was it thou—I think°
Surely it was!—that bard
Unnamed, who, Goethe said,
Had every other gift, but wanted love;   100
Love, without which the tongue
Even of angels sounds amiss?

Charm is the glory which makes
Song of the poet divine,
Love is the fountain of charm.
How without charm wilt thou draw,
Poet! the world to thy way?
Not by the lightnings of wit—
Not by the thunder of scorn!
These to the world, too, are given;   110
Wit it possesses, and scorn—
Charm is the poet's alone.
Hollow and dull are the great,
And artists envious, and the mob profane.
We know all this, we know!
Cam'st thou from heaven, O child°
Of light! but this to declare?
Alas, to help us forget
Such barren knowledge awhile,
God gave the poet his song!   120

Therefore a secret unrest
Tortured thee, brilliant and bold!
Therefore triumph itself
Tasted amiss to thy soul.
Therefore, with blood of thy foes,

Trickled in silence thine own.
Therefore the victor's heart
Broke on the field of his fame.

Ah! as of old, from the pomp°
Of Italian Milan, the fair   130
Flower of marble of white
Southern palaces—steps
Border'd by statues, and walks
Terraced, and orange-bowers
Heavy with fragrance—the blond
German Kaiser full oft
Long'd himself back to the fields,
Rivers, and high-roof'd towns
Of his native Germany; so,
So, how often! from hot   140
Paris drawing-rooms, and lamps
Blazing, and brilliant crowds,
Starr'd and jewell'd, of men
Famous, of women the queens
Of dazzling converse—from fumes
Of praise, hot, heady fumes, to the poor brain
That mount, that madden—how oft
Heine's spirit outworn
Long'd itself out of the din,
Back to the tranquil, the cool   150
Far German home of his youth!

See! in the May-afternoon,°
O'er the fresh, short turf of the Hartz,
A youth, with the foot of youth,
Heine! thou climbest again!
Up, through the tall dark firs
Warming their heads in the sun,
Chequering the grass with their shade—
Up, by the stream, with its huge
Moss-hung boulders, and thin   160
Musical water half-hid—
Up, o'er the rock-strewn slope,
With the sinking sun, and the air
Chill, and the shadows now

Long on the grey hill-side—
To the stone-roof'd hut at the top!

Or, yet later, in watch
On the roof of the Brocken-tower
Thou standest, gazing!—to see
The broad red sun, over field,   170
Forest, and city, and spire,
And mist-track'd stream of the wide,
Wide, German land, going down
In a bank of vapours——again
Standest, at nightfall, alone!

Or, next morning, with limbs
Rested by slumber, and heart
Freshen'd and light with the May,
O'er the gracious spurs coming down
Of the Lower Hartz, among oaks,   180
And beechen coverts, and copse
Of hazels green in whose depth
Isle, the fairy transform'd,
In a thousand water-breaks light
Pours her petulant youth—
Climbing the rock which juts
O'er the valley, the dizzily perch'd
Rock—to its iron cross
Once more thou cling'st; to the Cross
Clingest! with smiles, with a sigh!   190

Goethe, too, had been there.
In the long-past winter he came
To the frozen Hartz, with his soul
Passionate, eager—his youth
All in ferment!—but he
Destined to work and to live
Left it, and thou, alas!
Only to laugh and to die.

But something prompts me: Not thus
Take leave of Heine! not thus   200
Speak the last word at his grave!

Not in pity, and not
With half censure—with awe
Hail, as it passes from earth
Scattering lightnings, that soul!

The Spirit of the world,°
Beholding the absurdity of men—
Their vaunts, their feats—let a sardonic smile,
For one short moment, wander o'er his lips.
That smile was Heine!—for its earthly hour   210
The strange guest sparkled; now 'tis pass'd away.

That was Heine! and we,
Myriads who live, who have lived,
What are we all, but a mood,
A single mood, of the life
Of the Spirit in whom we exist,
Who alone is all things in one?

Spirit, who fillest us all!
Spirit, who utterest in each
New-coming son of mankind   220
Such of thy thoughts as thou wilt!
O thou, one of whose moods,
Bitter and strange, was the life
Of Heine—his strange, alas,
His bitter life!—may a life
Other and milder be mine!
May'st thou a mood more serene,°
Happier, have utter'd in mine!
May'st thou the rapture of peace
Deep have embreathed at its core;   230
Made it a ray of thy thought,
Made it a beat of thy joy!

The Terrace at Berne

from 'SWITZERLAND'

Ten years!—and to my waking eye
Once more the roofs of Berne appear;
The rocky banks, the terrace high,
The stream!—and do I linger here?

The clouds are on the Oberland,
The Jungfrau snows look faint and far;
But bright are those green fields at hand,
And through those fields comes down the Aar,

And from the blue twin-lakes it comes,°
Flows by the town, the churchyard fair;   10
And 'neath the garden-walk it hums,
The house!—and is my Marguerite there?

Ah, shall I see thee, while a flush
Of startled pleasure floods thy brow,
Quick through the oleanders brush,
And clap thy hands, and cry: 'Tis thou!

Or hast thou long since wander'd back,
Daughter of France! to France, thy home;
And flitted down the flowery track
Where feet like thine too lightly come?   20

Doth riotous laughter now replace
Thy smile; and rouge, with stony glare,
Thy cheek's soft hue; and fluttering lace
The kerchief that enwound thy hair?

Or is it over?—art thou dead?—
Dead!—and no warning shiver ran
Across my heart, to say thy thread
Of life was cut, and closed thy span!

Could from earth's ways that figure slight
Be lost, and I not feel 'twas so?   30
Of that fresh voice the gay delight°
Fail from earth's air, and I not know?

Or shall I find thee still, but changed,
But not the Marguerite of thy prime?
With all thy being re-arranged,
Pass'd through the crucible of time;

With spirit vanish'd, beauty waned,
And hardly yet a glance, a tone,
A gesture—anything—retain'd
Of all that was my Marguerite's own?   40

I will not know! For wherefore try,
To things by mortal course that live,
A shadowy durability,
For which they were not meant, to give?

Like driftwood spars, which meet and pass°
Upon the boundless ocean-plain,
So on the sea of life, alas!
Man meets man—meets, and quits again.

I knew it when my life was young;
I feel it still, now youth is o'er.   50
—The mists are on the mountain hung,
And Marguerite I shall see no more.

from *Rachel*

In Paris all look'd hot and like to fade.
Sere, in the garden of the Tuileries,
Sere with September, droop'd the chestnut-trees.
'Twas dawn; a brougham roll'd through the streets and made

Halt at the white and silent colonnade
Of the French Theatre. Worn with disease,
Rachel, with eyes no gazing can appease,
Sate in the brougham and those blank walls survey'd.

She follows the gay world, whose swarms have fled
To Switzerland, to Baden, to the Rhine;   10
Why stops she by this empty play-house drear?

Ah, where the spirit its highest life hath led,
All spots, match'd with that spot, are less divine;
And Rachel's Switzerland, her Rhine, is here!

Austerity of Poetry

That son of Italy who tried to blow,°
Ere Dante came, the trump of sacred song,
In his light youth amid a festal throng
Sate with his bride to see a public show.

Fair was the bride, and on her front did glow
Youth like a star; and what to youth belong—°
Gay raiment, sparkling gauds, elation strong.
A prop gave way! crash fell a platform! lo,

'Mid struggling sufferers, hurt to death, she lay!
Shuddering, they drew her garments off—and found   10
A robe of sackcloth next the smooth, white skin.

Such, poets, is your bride, the Muse! young, gay,°
Radiant, adorn'd outside; a hidden ground
Of thought and of austerity within.

Palladium

Set where the upper streams of Simois flow
Was the Palladium, high 'mid rock and wood;
And Hector was in Ilium, far below,
And fought, and saw it not—but there it stood!

It stood, and sun and moonshine rain'd their light
On the pure columns of its glen-built hall.
Backward and forward roll'd the waves of fight
Round Troy—but while this stood, Troy could not fall.

So, in its lovely moonlight, lives the soul.°
Mountains surround it, and sweet virgin air;   10
Cold plashing, past it, crystal waters roll;
We visit it by moments, ah, too rare!°

We shall renew the battle in the plain
To-morrow;—red with blood will Xanthus be;
Hector and Ajax will be there again,
Helen will come upon the wall to see.°

Then we shall rust in shade, or shine in strife,°
And fluctuate 'twixt blind hopes and blind despairs,°
And fancy that we put forth all our life,
And never know how with the soul it fares.   20

Still doth the soul, from its lone fastness high,
Upon our life a ruling effluence send.
And when it fails, fight as we will, we die;
And while it lasts, we cannot wholly end.

Thyrsis

A MONODY, *to commemorate the author's friend,*
ARTHUR HUGH CLOUGH, *who died at Florence,* 1861

How changed is here each spot man makes or fills!
    In the two Hinkseys nothing keeps the same;
        The village street its haunted mansion lacks,
    And from the sign is gone Sibylla's name,°
            And from the roofs the twisted chimney-stacks—
            Are ye too changed, ye hills?
    See, 'tis no foot of unfamiliar men
        To-night from Oxford up your pathway strays!
        Here came I often, often, in old days—
    Thyrsis and I; we still had Thyrsis then.   10

Runs it not here, the track by Childsworth Farm,°
    Past the high wood, to where the elm-tree crowns
        The hill behind whose ridge the sunset flames?
    The signal-elm, that looks on Ilsley Downs,

The Vale, the three lone weirs, the youthful Thames?—
   This winter-eve is warm,°
Humid the air! leafless, yet soft as spring,
   The tender purple spray on copse and briers!
    And that sweet city with her dreaming spires,°
She needs not June for beauty's heightening,   20

Lovely all times she lies, lovely to-night!—
   Only, methinks, some loss of habit's power
    Befalls me wandering through this upland dim.
Once pass'd I blindfold here, at any hour;
   Now seldom come I, since I came with him.
    That single elm-tree bright°
Against the west—I miss it! is it gone?
   We prized it dearly; while it stood, we said,
   Our friend, the Gipsy-Scholar, was not dead;
While the tree lived, he in these fields lived on.   30

Too rare, too rare, grow now my visits here,°
   But once I knew each field, each flower, each stick;
    And with the country-folk acquaintance made
By barn in threshing-time, by new-built rick.
    Here, too, our shepherd-pipes we first assay'd.°
     Ah me! this many a year
My pipe is lost, my shepherd's holiday!
   Needs must I lose them, needs with heavy heart
   Into the world and wave of men depart;
But Thyrsis of his own will went away.   40

It irk'd him to be here, he could not rest.°
   He loved each simple joy the country yields,
    He loved his mates; but yet he could not keep,
For that a shadow lour'd on the fields,
    Here with the shepherds and the silly sheep.
     Some life of men unblest
He knew, which made him droop, and fill'd his head.
   He went; his piping took a troubled sound
   Of storms that rage outside our happy ground;
He could not wait their passing, he is dead.   50

So, some tempestuous morn in early June,°
  When the year's primal burst of bloom is o'er,
    Before the roses and the longest day—
  When garden-walks and all the grassy floor
    With blossoms red and white of fallen May
      And chestnut-flowers are strewn—
  So have I heard the cuckoo's parting cry,°
    From the wet field, through the vext garden-trees,
    Come with the volleying rain and tossing breeze:
  The bloom is gone, and with the bloom go I!   60

Too quick despairer, wherefore wilt thou go?°
  Soon will the high Midsummer pomps come on,
    Soon will the musk carnations break and swell,
  Soon shall we have gold-dusted snapdragon,
    Sweet-William with his homely cottage-smell,
      And stocks in fragrant blow;
  Roses that down the alleys shine afar,
    And open, jasmine-muffled lattices,
    And groups under the dreaming garden-trees,
  And the full moon, and the white evening-star.   70

He hearkens not! light comer, he is flown!
  What matters it? next year he will return,
    And we shall have him in the sweet spring-days,
  With whitening hedges, and uncrumpling fern,
    And blue-bells trembling by the forest-ways,
      And scent of hay new-mown.
  But Thyrsis never more we swains shall see;
    See him come back, and cut a smoother reed,°
    And blow a strain the world at last shall heed—
  For Time, not Corydon, hath conquer'd thee!°   80

Alack, for Corydon no rival now!—
  But when Sicilian shepherds lost a mate,°
    Some good survivor with his flute would go,
  Piping a ditty sad for Bion's fate;
    And cross the unpermitted ferry's flow,
      And relax Pluto's brow,

And make leap up with joy the beauteous head
  Of Proserpine, among whose crownéd hair
  Are flowers first open'd on Sicilian air,
And flute his friend, like Orpheus, from the dead.   90

O easy access to the hearer's grace
  When Dorian shepherds sang to Proserpine!
    For she herself had trod Sicilian fields,
  She knew the Dorian water's gush divine,
    She knew each lily white which Enna yields,
      Each rose with blushing face;
  She loved the Dorian pipe, the Dorian strain.
    But ah, of our poor Thames she never heard!
    Her foot the Cumner cowslips never stirr'd;
And we should tease her with our plaint in vain!   100

Well! wind-dispersed and vain the words will be,
  Yet, Thyrsis, let me give my grief its hour
    In the old haunt, and find our tree-topp'd hill!
  Who, if not I, for questing here hath power?
    I know the wood which hides the daffodil,
      I know the Fyfield tree,°
  I know what white, what purple fritillaries
    The grassy harvest of the river-fields,
    Above by Ensham, down by Sandford, yields,
And what sedged brooks are Thames's tributaries;   110

I know these slopes; who knows them if not I?—
  But many a dingle on the loved hill-side,
    With thorns once studded, old, white-blossom'd trees,
  Where thick the cowslips grew, and far descried
    High tower'd the spikes of purple orchises,
      Hath since our day put by
  The coronals of that forgotten time;
    Down each green bank hath gone the ploughboy's team,
    And only in the hidden brookside gleam
Primroses, orphans of the flowery prime.   120

Where is the girl, who by the boatman's door,
  Above the locks, above the boating throng,
    Unmoor'd our skiff when through the Wytham flats,
Red loosestrife and blond meadow-sweet among
  And darting swallows and light water-gnats,
    We track'd the shy Thames shore?
Where are the mowers, who, as the tiny swell
  Of our boat passing heaved the river-grass,
    Stood with suspended scythe to see us pass?—
They all are gone, and thou art gone as well!   130

Yes, thou art gone! and round me too the night
  In ever-nearing circle weaves her shade.
    I see her veil draw soft across the day,
  I feel her slowly chilling breath invade
    The cheek grown thin, the brown hair sprent with grey;
      I feel her finger light
Laid pausefully upon life's headlong train;—°
    The foot less prompt to meet the morning dew,
    The heart less bounding at emotion new,
And hope, once crush'd, less quick to spring again.   140

And long the way appears, which seem'd so short
  To the less practised eye of sanguine youth;
    And high the mountain-tops, in cloudy air,
The mountain-tops where is the throne of Truth,
    Tops in life's morning-sun so bright and bare!
      Unbreachable the fort
Of the long-batter'd world uplifts its wall;
    And strange and vain the earthly turmoil grows,
    And near and real the charm of thy repose,
And night as welcome as a friend would fall.   150

But hush! the upland hath a sudden loss
  Of quiet!—Look, adown the dusk hill-side,
    A troop of Oxford hunters going home,°
As in old days, jovial and talking, ride!
    From hunting with the Berkshire hounds they come.
      Quick! let me fly, and cross

Into yon farther field!—'Tis done; and see,
  Back'd by the sunset, which doth glorify
  The orange and pale violet evening-sky,
Bare on its lonely ridge, the Tree! the Tree!   160

I take the omen! Eve lets down her veil,
  The white fog creeps from bush to bush about,
    The west unflushes, the high stars grow bright
And in the scatter'd farms the lights come out.
    I cannot reach the signal-tree to-night,
      Yet, happy omen, hail!
Hear it from thy broad lucent Arno-vale°
  (For there thine earth-forgetting eyelids keep
  The morningless and unawakening sleep
Under the flowery oleanders pale),   170

Hear it, O Thyrsis, still our tree is there!—
  Ah, vain! These English fields, this upland dim,
    These brambles pale with mist engarlanded,
That lone, sky-pointing tree, are not for him;
    To a boon southern country he is fled,°
      And now in happier air,
Wandering with the great Mother's train divine°
  (And purer or more subtle soul than thee,
  I trow, the mighty Mother doth not see)
Within a folding of the Apennine,   180

Thou hearest the immortal chants of old!—
  Putting his sickle to the perilous grain°
    In the hot cornfield of the Phrygian king,
For thee the Lityerses-song again
    Young Daphnis with his silver voice doth sing;
      Sings his Sicilian fold,
His sheep, his hapless love, his blinded eyes—
  And how a call celestial round him rang,
  And heavenward from the fountain-brink he sprang,
And all the marvel of the golden skies.   190

There thou art gone, and me thou leavest here
  Sole in these fields! yet will I not despair.
    Despair I will not, while I yet descry
'Neath the mild canopy of English air
    That lonely tree against the western sky.
      Still, still these slopes, 'tis clear,
  Our Gipsy-Scholar haunts, outliving thee!
    Fields where soft sheep from cages pull the hay,
    Woods with anemonies in flower till May,
Know him a wanderer still; then why not me?   200

A fugitive and gracious light he seeks,°
  Shy to illumine; and I seek it too.
    This does not come with houses or with gold,°
With place, with honour, and a flattering crew;
      'Tis not in the world's market bought and sold—
      But the smooth-slipping weeks
Drop by, and leave its seeker still untired;
    Out of the heed of mortals he is gone,
    He wends unfollow'd, he must house alone;
Yet on he fares, by his own heart inspired.   210

Thou too, O Thyrsis, on like quest wast bound;
  Thou wanderedst with me for a little hour!
    Men gave thee nothing; but this happy quest,
If men esteem'd thee feeble, gave thee power,
    If men procured thee trouble, gave thee rest.
      And this rude Cumner ground,
Its fir-topped Hurst, its farms, its quiet fields,
    Here cam'st thou in thy jocund youthful time,
    Here was thine height of strength, thy golden prime!
And still the haunt beloved a virtue yields.   220

What though the music of thy rustic flute
  Kept not for long its happy, country tone;
    Lost it too soon, and learnt a stormy note
Of men contention-tost, of men who groan,°
      Which task'd thy pipe too sore, and tired thy throat—
        It fail'd, and thou was mute!°

Yet hadst thou alway visions of our light,
  And long with men of care thou couldst not stay,
  And soon thy foot resumed its wandering way,
Left human haunt, and on alone till night.   230

Too rare, too rare, grow now my visits here!
  'Mid city-noise, not, as with thee of yore,
    Thyrsis! in reach of sheep-bells is my home.
  —Then through the great town's harsh, heart-wearying roar,
    Let in thy voice a whisper often come,
      To chase fatigue and fear:
Why faintest thou? I wander'd till I died.
Roam on! The light we sought is shining still.
Dost thou ask proof? Our tree yet crowns the hill,
Our Scholar travels yet the loved hill-side.   240

Epilogue to Lessing's 'Laocoön'

One morn as through Hyde Park we walk'd,
My friend and I, by chance we talk'd
Of Lessing's famed Laocoön;
And after we awhile had gone
In Lessing's track, and tried to see
What painting is, what poetry—
Diverging to another thought,
'Ah,' cries my friend, 'but who hath taught
Why music and the other arts
Oftener perform aright their parts   10
Than poetry? why she, than they,
Fewer fine successes can display?

'For 'tis so, surely! Even in Greece,
Where best the poet framed his piece,
Even in that Phoebus-guarded ground
Pausanias on his travels found°
Good poems, if he look'd, more rare
(Though many) than good statues were—
For these, in truth, were everywhere.
Of bards full many a stroke divine   20
In Dante's, Petrarch's, Tasso's line,
The land of Ariosto show'd;
And yet, e'en there, the canvas glow'd

With triumphs, a yet ampler brood,
Of Raphael and his brotherhood.
And nobly perfect, in our day
Of haste, half-work, and disarray,
Profound yet touching, sweet yet strong,
Hath risen Goethe's, Wordsworth's song;°
Yet even I (and none will bow   30
Deeper to these) must needs allow,
They yield us not, to soothe our pains,
Such multitude of heavenly strains
As from the kings of sound are blown,
Mozart, Beethoven, Mendelssohn.'

While thus my friend discoursed, we pass
Out of the path, and take the grass.
The grass had still the green of May,
And still the unblacken'd elms were gay;
The kine were resting in the shade,   40
The flies a summer-murmur made.
Bright was the morn and south the air;
The soft-couch'd cattle were as fair°
As those which pastured by the sea,
That old-world morn, in Sicily,
When on the beach the Cyclops lay,
And Galatea from the bay
Mock'd her poor lovelorn giant's lay.
'Behold', I said, 'the painter's sphere!
The limits of his art appear.   50
The passing group, the summer-morn,
The grass, the elms, that blossom'd thorn—
Those cattle couch'd, or, as they rise,
Their shining flanks, their liquid eyes—
These, or much greater things, but caught
Like these, and in one aspect brought!
In outward semblance he must give°
A moment's life of things that live;
Then let him choose his moment well,
With power divine its story tell.'   60

Still we walk'd on, in thoughtful mood,
And now upon the bridge we stood.°

Full of sweet breathings was the air,
Of sudden stirs and pauses fair.
Down o'er the stately bridge the breeze
Came rustling from the garden-trees
And on the sparkling waters play'd;
Light-plashing waves an answer made,
And mimic boats their haven near'd.
Beyond, the Abbey-towers appear'd,°   70
By mist and chimneys unconfined,
Free to the sweep of light and wind;
While through their earth-moor'd nave below
Another breath of wind doth blow,
Sound as of wandering breeze—but sound
In laws by human artists bound.
'The world of music!' I exclaim'd:—
'This breeze that rustles by, that famed
Abbey recall it! what a sphere
Large and profound, hath genius here!   80
The inspired musician what a range,
What power of passion, wealth of change!
Some source of feeling he must choose
And its lock'd fount of beauty use,
And through the stream of music tell
Its else unutterable spell;
To choose it rightly is his part,
And press into its inmost heart.

'Miserere, Domine!
The words are utter'd, and they flee.   90
Deep is their penitential moan,
Mighty their pathos, but 'tis gone.
They have declared the spirit's sore
Sore load, and words can do no more.
Beethoven takes them then—those two°
Poor, bounded words—and makes them new;
Infinite makes them, makes them young;
Transplants them to another tongue,
Where they can now, without constraint,
Pour all the soul of their complaint,   100
And roll adown a channel large
The wealth divine they have in charge.

Page after page of music turn,
And still they live and still they burn,
Eternal, passion-fraught, and free—
Miserere, Domine!'

Onward we moved, and reach'd the Ride°
Where gaily flows the human tide.
Afar, in rest the cattle lay;
We head, afar, faint music play;   110
But agitated, brisk, and near,
Men, with their stream of life, were here.
Some hang upon the rails, and some
On foot behind them go and come.
This through the ride upon his steed
Goes slowly by, and this at speed.
The young, the happy, and the fair,
The old, the sad, the worn, were there;
Some vacant, and some musing went,
And some in talk and merriment.   120
Nods, smiles, and greetings, and farewells!
And now and then, perhaps, there swells
A sigh, a tear—but in the throng
All changes fast, and hies along.
Hies, ah, from whence, what native ground?
And to what goal, what ending, bound?
'Behold, at last the poet's sphere!
But who', I said, 'suffices here?

'For, ah! so much he has to do;°
Be painter and musician too!   130
The aspect of the moment show,
The feeling of the moment know!
The aspect not, I grant, express
Clear as the painter's art can dress;
The feeling not, I grant, explore
So deep as the musician's lore—
But clear as words can make revealing,
And deep as words can follow feeling.
But, ah! then comes his sorest spell
Of toil—he must life's *movement* tell!   140
The thread which binds it all in one,°
And not its separate parts alone.

The *movement* he must tell of life,
Its pain and pleasure, rest and strife;
His eye must travel down, at full,
The long, unpausing spectacle;
With faithful unrelaxing force
Attend it from its primal source,
From change to change and year to year
Attend it of its mid career,   150
Attend it to the last repose
And solemn silence of its close.

'The cattle rising from the grass
His thought must follow where they pass;
The penitent with anguish bow'd
His thought must follow through the crowd.
Yes! all this eddying, motley throng
That sparkles in the sun along,
Girl, statesman, merchant, soldier bold,
Master and servant, young and old,   160
Grave, gay, child, parent, husband, wife,
He follows home, and lives their life.

'And many, many are the souls
Life's movement fascinates, controls;
It draws them on, they cannot save
Their feet from its alluring wave;
They cannot leave it, they must go
With its unconquerable flow.
But ah! how few, of all that try
This mighty march, do aught but die!   170
For ill-endow'd for such a way,
Ill-stored in strength, in wits, are they.
They faint, they stagger to and fro,
And wandering from the stream they go;
In pain, in terror, in distress,
They see, all round, a wilderness.
Sometimes a momentary gleam°
They catch of the mysterious stream;
Sometimes, a second's space, their ear
The murmur of its waves doth hear.   180
That transient glimpse in song they say,
But not as painter can portray—

That transient sound in song they tell,
But not, as the musician, well.
And when at last their snatches cease,
And they are silent and at peace,
The stream of life's majestic whole
Hath ne'er been mirror'd on their soul.

'Only a few the life-stream's shore
With safe unwandering feet explore;   190
Untired its movement bright attend,
Follow its windings to the end.
Then from its brimming waves their eye°
Drinks up delighted ecstasy,
And its deep-toned, melodious voice
For ever makes their ear rejoice.
They speak! the happiness divine
They feel, runs o'er in every line;
Its spell is round them like a shower—
It gives them pathos, gives them power.   200
No painter yet hath such a way,
Nor no musician made, as they,
And gather'd on immortal knolls
Such lovely flowers for cheering souls.
Beethoven, Raphael, cannot reach
The charm which Homer, Shakespeare, teach.
To these, to these, their thankful race
Gives, then, the first, the fairest place;
And brightest is their glory's sheen,
For greatest hath their labour been.'   210

Obermann Once More

Savez-vous quelque bien qui console du regret d'un monde?

OBERMANN.

Glion?——Ah, twenty years, it cuts°
All meaning from a name!
White houses prank where once were huts.
Glion, but not the same!

And yet I know not! All unchanged
The turf, the pines, the sky!
The hills in their old order ranged;
The lake, with Chillon by!

And, 'neath those chestnut-trees, where stiff
And stony mounts the way,   10
The crackling husk-heaps burn, as if
I left them yesterday!

Across the valley, on that slope,
The huts of Avant shine!
Its pines, under their branches, ope
Ways for the pasturing kine.

Full-foaming milk-pails, Alpine fare,
Sweet heaps of fresh-cut grass,
Invite to rest the traveller there
Before he climb the pass—   20

The gentian-flower'd pass, its crown°
With yellow spires aflame;
Whence drops the path to Allière down,
And walls where Byron came,°

By their green river, who doth change
His birth-name just below;
Orchard, and croft, and full-stored grange
Nursed by his pastoral flow.

But stop!—to fetch back thoughts that stray
Beyond this gracious bound,   30
The cone of Jaman, pale and grey,°
See, in the blue profound!

Ah, Jaman! delicately tall
Above his sun-warm'd firs—
What thoughts to me his rocks recall,
What memories he stirs!

And who but thou must be, in truth,
Obermann! with me here?
Thou master of my wandering youth,
But left this many a year!   40

Yes, I forget the world's work wrought,
Its warfare waged with pain;
An eremite with thee, in thought
Once more I slip my chain,

And to thy mountain-chalet come,
And lie beside its door,
And hear the wild bee's Alpine hum,
And thy sad, tranquil lore!

Again I feel the words inspire
Their mournful calm; serene,   50
Yet tinged with infinite desire
For all that *might* have been—

The harmony from which man swerved°
Made his life's rule once more!
The universal order served,
Earth happier than before!

—While thus I mused, night gently ran
Down over hill and wood.
Then, still and sudden, Obermann
On the grass near me stood.   60

Those pensive features well I knew,
On my mind, years before,
Imaged so oft! imaged so true!
—A shepherd's garb he wore,

A mountain-flower was in his hand,°
A book was in his breast.
Bent on my face, with gaze which scann'd
My soul, his eyes did rest.

'And is it thou,' he cried, 'so long
Held by the world which we   70
Loved not, who turnest from the throng
Back to thy youth and me?

'And from thy world, with heart opprest,
Choosest thou *now* to turn!—
Ah me! we anchorites read things best,
Clearest their course discern!

'Thou fledst me when the ungenial earth,
Man's work-place, lay in gloom.
Return'st thou in her hour of birth,°
Of hopes and hearts in bloom?   80

'Perceiv'st thou not the change of day?
Ah! Carry back thy ken,
What, some two thousand years! Survey
The world as it was then!

'Like ours it look'd in outward air.°
Its head was clear and true,
Sumptuous its clothing, rich its fare,
No pause its action knew;

'Stout was its arm, each thew and bone
Seem'd puissant and alive—   90
But, ah! its heart, its heart was stone,
And so it could not thrive!

'On that hard Pagan world disgust
And secret loathing fell.
Deep weariness and sated lust
Made human life a hell.

'In his cool hall, with haggard eyes,°
The Roman noble lay;
He drove abroad, in furious guise,
Along the Appian way.   100

'He made a feast, drank fierce and fast,
And crown'd his hair with flowers—
No easier nor no quicker pass'd
The impracticable hours.

'The brooding East with awe beheld
Her impious younger world.
The Roman tempest swell'd and swell'd,
And on her head was hurl'd.

'The East bow'd low before the blast
In patient, deep disdain;　　　　　　　　　　　　　　110
She let the legions thunder past,
And plunged in thought again.

'So well she mused, a morning broke
Across her spirit grey;
A conquering, new-born joy awoke,°
And fill'd her life with day.

'"Poor world," she cried, "so deep accurst,
That runn'st from pole to pole
To seek a draught to slake thy thirst—
Go, seek it in thy soul!"　　　　　　　　　　　　120

'She heard it, the victorious West,°
In crown and sword array'd!
She felt the void which mined her breast,
She shiver'd and obey'd.

'She veil'd her eagles, snapp'd her sword,
And laid her sceptre down;
Her stately purple she abhorr'd,
And her imperial crown.

'She broke her flutes, she stopp'd her sports,
Her artists could not please;　　　　　　　　　　130
She tore her books, she shut her courts,
She fled her palaces;

'Lust of the eye and pride of life
She left it all behind,
And hurried, torn with inward strife,
The wilderness to find.°

'Tears wash'd the trouble from her face!
She changed into a child!
'Mid weeds and wrecks she stood—a place°
Of ruin—but she smiled!   140

'Oh, had I lived in that great day,
How had its glory new
Fill'd earth and heaven, and caught away
My ravish'd spirit too!

'No thoughts that to the world belong
Had stood against the wave°
Of love which set so deep and strong
From Christ's then open grave.

'No cloister-floor of humid stone
Had been too cold for me.   150
For me no Eastern desert lone
Had been too far to flee.

'No lonely life had pass'd too slow,
When I could hourly scan
Upon his Cross, with head sunk low,
That nail'd, thorn-crownéd Man!

'Could see the Mother with her Child
Whose tender winning arts
Have to his little arms beguiled
So many wounded hearts!   160

'And centuries came and ran their course,
And unspent all that time
Still, still went forth that Child's dear force,
And still was at its prime.

'Ay, ages long endured his span
Of life—'tis true received—
That gracious Child, that thorn-crown'd Man!
—He lived while we believed.

'While we believed, on earth he went,
And open stood his grave.   170
Men call'd from chamber, church, and tent;
And Christ was by to save.

'Now he is dead! Far hence he lies°
In the lorn Syrian town;
And on his grave, with shining eyes,
The Syrian stars look down.

'In vain men still, with hoping new,
Regard his death-place dumb,
And say the stone is not yet to,
And wait for words to come.   180

'Ah, o'er that silent sacred land,
Of sun, and arid stone,
And crumbling wall, and sultry sand,
Sounds now one word alone!

'*Unduped of fancy, henceforth man*
Must labour!—must resign
His all too human creeds, and scan
Simply the way divine!

'But slow that tide of common thought,°
Which bathed our life, retired;   190
Slow, slow the old world wore to nought,
And pulse by pulse expired.

'Its frame yet stood without a breach
When blood and warmth were fled;
And still it spake its wonted speech—
But every word was dead.

'And oh, we cried, that on this corse
Might fall a freshening storm!
Rive its dry bones, and with new force
A new-sprung world inform!   200

'—Down came the storm! O'er France it pass'd°
In sheets of scathing fire;
All Europe felt that fiery blast,
And shook as it rush'd by her.

'Down came the storm! In ruins fell
The worn-out world we knew.
It pass'd, that elemental swell!
Again appear'd the blue;

'The sun shone in the new-wash'd sky,
And what from heaven saw he?   210
Blocks of the past, like icebergs high,
Float on a rolling sea!

'Upon them plies the race of man
All it before endeavour'd;
"Ye live," I cried, "ye work and plan,
And know not ye are sever'd!

'"Poor fragments of a broken world
Whereon men pitch their tent!
Why were ye too to death not hurl'd
When your world's day was spent?   220

' "That glow of central fire is done°
Which with its fusing flame
Knit all your parts, and kept you one—
But ye, ye are the same!

' "The past, its mask of union on,
Had ceased to live and thrive.
The past, its mask of union gone,
Say, is it more alive?

' "Your creeds are dead, your rites are dead,
Your social order too!   230
Where tarries he, the Power who said:
See, I make all things new?°

' "The millions suffer still, and grieve,
And what can helpers heal
With old-world cures men half believe
For woes they wholly feel?

' "And yet men have such need of joy!°
But joy whose grounds are true;
And joy that should all hearts employ
As when the past was new.   240

' "Ah, not the emotion of that past,
Its common hope, were vain!
Some new such hope must dawn at last,
Or man must toss in pain.

' "But now the old is out of date,°
The new is not yet born,
And who can be *alone* elate,
While the world lies forlorn?"

'Then to the wilderness I fled.—
There among Alpine snows   250
And pastoral huts I hid my head,
And sought and found repose.

'It was not yet the appointed hour.
Sad, patient, and resign'd,
I watch'd the crocus fade and flower,
I felt the sun and wind.

'The day I lived in was not mine,
Man gets no second day.
In dreams I saw the future shine—
But ah! I could not stay!   260

'Action I had not, followers, fame;
I pass'd obscure, alone.
The after-world forgets my name,
Nor do I wish it known.

'Composed to bear, I lived and died,
And knew my life was vain,
With fate I murmur not, nor chide,
At Sèvres by the Seine°

'(If Paris that brief flight allow)
My humble tomb explore!   270
It bears: *Eternity, be thou
My refuge!* and no more.

'But thou, whom fellowship of mood
Did make from haunts of strife
Come to my mountain-solitude,
And learn my frustrate life;

'O thou, who, ere thy flying span
Was past of cheerful youth,
Didst find the solitary man
And love his cheerless truth—   280

'Despair not thou as I despair'd,
Nor be cold gloom thy prison!
Forward the gracious hours have fared,°
And see! the sun is risen!

'He breaks the winter of the past;
A green, new earth appears.°
Millions, whose life in ice lay fast,
Have thoughts, and smiles, and tears.

'What though there still need effort, strife?
Though much be still unwon?   290
Yet warm it mounts, the hour of life!
Death's frozen hour is done!

'The world's great order dawns in sheen,°
After long darkness rude,
Divinelier imaged, clearer seen,
With happier zeal pursued.

'With hope extinct and brow composed
I mark'd the present die;
Its term of life was nearly closed,
Yet it had more than I.   300

'But thou, though to the world's new hour
Thou come with aspect marr'd,
Shorn of the joy, the bloom, the power
Which best befits its bard—

'Though more than half thy years be past,
And spent thy youthful prime;
Though, round thy firmer manhood cast,°
Hang weeds of our sad time

'Whereof thy youth felt all the spell,
And traversed all the shade—   310
Though late, though dimm'd, though weak, yet tell
Hope to a world new-made!

'Help it to fill that deep desire,
The want which rack'd our brain,
Consumed our heart with thirst like fire,
Immedicable pain;°

'Which to the wilderness drove out
Our life, to Alpine snow,
And palsied all our word with doubt,
And all our work with woe—   320

'What still of strength is left, employ
That end to help attain:
One common wave of thought and joy°
Lifting mankind again!'

—The vision ended. I awoke
As out of sleep, and no
Voice moved;—only the torrent broke
The silence, far below.

Soft darkness on the turf did lie.
Solemn, o'er hut and wood,   330
In the yet star-sown nightly sky,
The peak of Jaman stood.

Still in my soul the voice I heard
Of Obermann!——away
I turned; by some vague impulse stirr'd,
Along the rocks of Naye

Past Sonchaud's piny flanks I gaze
And the blanch'd summit bare
Of Malatrait, to where in haze
The Valais opens fair,   340

And the domed Velan, with his snows,
Behind the upcrowding hills,
Doth all the heavenly opening close
Which the Rhone's murmur fills;—

And glorious there, without a sound,
Across the glimmering lake,
High in the Valais-depth profound,
I saw the morning break.°

Persistency of Poetry

Though the Muse be gone away,
Though she move not earth to-day,
Souls, erewhile who caught her word,
Ah! still harp on what they heard.

Bacchanalia; or, The New Age

I

The evening comes, the fields are still.
The tinkle of the thirsty rill,
Unheard all day, ascends again;
Deserted is the half-mown plain,
Silent the swaths! the ringing wain,
The mower's cry, the dog's alarms,
All housed within the sleeping farms!
The business of the day is done,
The last-left haymaker is gone.
And from the thyme upon the height,   10
And from the elder-blossom white
And pale dog-roses in the hedge,
And from the mint-plant in the sedge,
In puffs of balm the night-air blows
The perfume which the day forgoes.
And on the pure horizon far,
See, pulsing with the first-born star,
The liquid sky above the hill!
The evening comes, the fields are still.

     Loitering and leaping,   20
     With saunter, with bounds—
     Flickering and circling
     In files and in rounds—
     Gaily their pine-staff green
     Tossing in air,
     Loose o'er their shoulders white
     Showering their hair—
     See! the wild Maenads
     Break from the wood,
     Youth and Iacchus°   30
     Maddening their blood.
     See! through the quiet land
     Rioting they pass—
     Fling the fresh heaps about,
     Trample the grass.

Tear from the rifled hedge
Garlands, their prize;
Fill with their sports the field,
Fill with their cries.

    Shepherd, what ails thee, then?   40
Shepherd, why mute?
Forth with thy joyous song!
Forth with thy flute!
Tempts not the revel blithe?
Lure not their cries?
Glow not their shoulders smooth?
Melt not their eyes?
Is not, on cheeks like those,
Lovely the flush?
—*Ah, so the quiet was!*   50
So was the hush!

II

The epoch ends, the world is still.
The age has talk'd and work'd its fill—
The famous orators have shone,
The famous poets sung and gone,
The famous men of war have fought,
The famous speculators thought,
The famous players, sculptors, wrought,
The famous painters fill'd their wall,
The famous critics judged it all.
The combatants are parted now—   10
Uphung the spear, unbent the bow,
The puissant crown'd, the weak laid low.
And in the after-silence sweet,
Now strifes are hush'd, our ears doth meet,
Ascending pure, the bell-like fame
Of this or that down-trodden name,
Delicate spirits, push'd away
In the hot press of the noon-day.
And o'er the plain, where the dead age
Did its now silent warfare wage—   20
O'er that wide plain, now wrapt in gloom,

Where many a splendour finds its tomb,
Many spent fames and fallen mights—
The one or two immortal lights
Rise slowly up into the sky
To shine there everlastingly,
Like stars over the bounding hill.
The epoch ends, the world is still.

Thundering and bursting
In torrents, in waves—   30
Carolling and shouting
Over tombs, amid graves—
See! on the cumber'd plain
Clearing a stage,
Scattering the past about,
Comes the new age.
Bards make new poems,
Thinkers new schools,
Statesmen new systems,
Critics new rules.   40
All things begin again;
Life is their prize;
Earth with their deeds they fill,
Fill with their cries.

Poet, what ails thee, then?
Say, why so mute?
Forth with thy praising voice!
Forth with thy flute!
Loiterer! why sittest thou
Sunk in thy dream?   50
Tempts not the bright new age?
Shines not its stream?
Look, ah, what genius,
Art, science, wit!
Soldiers like Caesar,
Statesmen like Pitt!
Sculptors like Phidias,
Raphaels in shoals,
Poets like Shakespeare—
Beautiful souls!°   60

See, on their glowing cheeks
Heavenly the flush!
—*Ah, so the silence was!*
So was the hush!

The world but feels the present's spell,
The poet feels the past as well;
Whatever men have done, might do,°
Whatever thought, might think it too.

Growing Old

What is it to grow old?
Is it to lose the glory of the form,
The lustre of the eye?
Is it for beauty to forego her wreath?
—Yes, but not this alone.

Is it to feel our strength—
Not our bloom only, but our strength—decay?
Is it to feel each limb
Grow stiffer, every function less exact,
Each nerve more loosely strung?   10

Yes, this, and more; but not
Ah, 'tis not what in youth we dream'd 'twould be!
'Tis not to have our life
Mellow'd and soften'd as with sunset-glow,
A golden day's decline.

'Tis not to see the world°
As from a height, with rapt prophetic eyes,
And heart profoundly stirr'd;
And weep, and feel the fulness of the past,°
The years that are no more.   20

It is to spend long days
And not once feel that we were ever young;
It is to add, immured°
In the hot prison of the present, month
To month with weary pain.

It is to suffer this,
And feel but half, and feebly, what we feel.
Deep in our hidden heart
Festers the dull remembrance of a change,
But no emotion—none.   30

It is—last stage of all—°
When we are frozen up within, and quite
The phantom of ourselves,
To hear the world applaud the hollow ghost
Which blamed the living man.

The Progress of Poesy

A VARIATION

Youth rambles on life's arid mount,
And strikes the rock, and finds the vein,
And brings the water from the fount,
The fount which shall not flow again.

The man mature with labour chops
For the bright stream a channel grand,
And sees not that the sacred drops
Ran off and vanish'd out of hand.

And then the old man totters nigh,
And feebly rakes among the stones.   10
The mount is mute, the channel dry;
And down he lays his weary bones.

The Last Word

Creep into thy narrow bed,
Creep, and let no more be said!
Vain thy onset! all stands fast.
Thou thyself must break at last.

Let the long contention cease!
Geese are swans, and swans are geese.
Let them have it how they will!
Thou art tired; best be still.

They out-talk'd thee, hiss'd thee, tore thee?
Better men fared thus before thee;   10
Fired their ringing shot and pass'd,
Hotly charged—and sank at last.

Charge once more, then, and be dumb!°
Let the victors, when they come,
When the forts of folly fall,
Find thy body by the wall!

'Below the Surface-Stream'

Below the surface-stream, shallow and light,
Of what we *say* we feel—below the stream,
As light, of what we *think* we feel—there flows
With noiseless current strong, obscure and deep,
The central stream of what we feel indeed.

Poor Matthias

Poor Matthias!—Found him lying
Fall'n beneath his perch and dying?
Found him stiff, you say, though warm—
All convulsed his little form?
Poor canary! many a year
Well he knew his mistress dear;°

Now in vain you call his name,
Vainly raise his rigid frame,
Vainly warm him in your breast,
Vainly kiss his golden crest,   10
Smooth his ruffled plumage fine,
Touch his trembling beak with wine.
One more gasp—it is the end!
Dead and mute our tiny friend!
—Songster thou of many a year,
Now thy mistress bring thee here,
Says, it fits that I rehearse,
Tribute due to thee, a verse,
Meed for daily song of yore
Silent now for evermore.   20

Poor Matthias! Wouldst thou have
More than pity? claim'st a stave?
—Friends more near us than a bird
We dismiss'd without a word.
Rover, with the good brown head,°
Great Atossa, they are dead;
Dead, and neither prose nor rhyme
Tells the praises of their prime.
Thou didst know them old and grey,
Know them in their sad decay.   30
Thou hast seen Atossa sage
Sit for hours beside thy cage;
Thou wouldst chirp, thou foolish bird,
Flutter, chirp—she never stirr'd!
What were now these toys to her?
Down she sank amid her fur;
Eyed thee with a soul resign'd—
And thou deemedst cats were kind!
—Cruel, but composed and bland,
Dumb, inscrutable and grand,   40
So Tiberius might have sat,
Had Tiberius been a cat.

Rover died—Atossa too.
Less than they to us are you!

Nearer human were their powers,
Closer knit their life with ours.
Hands had stroked them, which are cold,°
Now for years, in churchyard mould;
Comrades of our past were they,
Of that unreturning day.   50
Changed and aging, they and we
Dwelt, it seem'd, in sympathy.
Alway from their presence broke
Somewhat which remembrance woke
Of the loved, the lost, the young—
Yet they died, and died unsung.

    Geist came next, our little friend;
Geist had verse to mourn his end.
Yes, but that enforcement strong
Which compell'd for Geist a song—   60
All that gay courageous cheer,
All that human pathos dear;
Soul-fed eyes with suffering worn,
Pain heroically borne,
Faithful love in depth divine—
Poor Matthias, were they thine?

    Max and Kaiser we to-day
Greet upon the lawn at play;
Max a dachshound without blot—
Kaiser should be, but is not.°   70
Max, with shining yellow coat,
Prinking ears and dewlap throat—
Kaiser, with his collie face,
Penitent for want of race.
—Which may be the first to die,
Vain to augur, they or I?
But, as age comes on, I know,
Poet's fire gets faint and low;
If so be that travel they
First the inevitable way,   80
Much I doubt if they shall have
Dirge from me to crown their grave.

Yet, poor bird, thy tiny corse
Moves me, somehow, to remorse;
Something haunts my conscience, brings
Sad, compunctious visitings.°
Other favourites, dwelling here,
Open lived to us, and near;
Well we knew when they were glad,
Plain we saw if they were sad,   90
Joy'd with them when they were gay,
Soothed them in their last decay;
Sympathy could feel and show
Both in weal of theirs and woe.

Birds, companions more unknown,
Live beside us, but alone;
Finding not, do all they can,
Passage from their souls to man.
Kindness we bestow, and praise,
Laud their plumage, greet their lays;   100
Still, beneath their feather'd breast,
Stirs a history unexpress'd.
Wishes there, and feelings strong,
Incommunicably throng;
What they want, we cannot guess,
Fail to track their deep distress—
Dull look on when death is nigh,
Note no change, and let them die.
Poor Matthias! couldst thou speak,
What a tale of thy last week!   110
Every morning did we pay
Stupid salutations gay,
Suited well to health, but how
Mocking, how incongruous now!
Cake we offer'd, sugar, seed,
Never doubtful of thy need;
Praised, perhaps, thy courteous eye,
Praised thy golden livery.
Gravely thou the while, poor dear!
Sat'st upon thy perch to hear,   120
Fixing with a mute regard
Us, thy human keepers hard,

Troubling, with our chatter vain,
Ebb of life, and mortal pain—
Us, unable to divine
Our companion's dying sign,
Or o'erpass the severing sea
Set betwixt ourselves and thee,
Till the sand thy feathers smirch
Fallen dying off thy perch!   130

    Was it, as the Grecian sings,°
Birds were born the first of things,
Before the sun, before the wind,
Before the gods, before mankind,
Airy, ante-mundane throng—
Witness their unworldly song!
Proof they give, too, primal powers,
Of a prescience more than ours—
Teach us, while they come and go,
When to sail, and when to sow.   140
Cuckoo calling from the hill,
Swallow skimming by the mill,
Swallows trooping in the sedge,
Starlings swirling from the hedge,
Mark the seasons, map our year,
As they show and disappear.
But, with all this travail sage
Brought from that anterior age,
Goes an unreversed decree
Whereby strange are they and we,   150
Making want of theirs, and plan,
Indiscernible by man.

    No, away with tales like these
Stol'n from Aristophanes!
Does it, if we miss your mind,
Prove us so remote in kind?
Birds! we but repeat on you
What amongst ourselves we do.
Somewhat more or somewhat less.
'Tis the same unskilfulness.   160

What you feel, escapes our ken—
Know we more our fellow men?
Human suffering at our side,
Ah, like yours is undescried!
Human longings, human fears,
Miss our eyes and miss our ears.
Little helping, wounding much,
Dull of heart, and hard of touch,
Brother man's despairing sign
Who may trust us to divine?   170
Who assure us, sundering powers
Stand not 'twixt his soul and ours?

  Poor Matthias! See, thy end
What a lesson doth it lend!
For that lesson thou shalt have,
Dead canary bird, a stave!
Telling how, one stormy day,
Stress of gale and showers of spray
Drove my daughter small and me
Inland from the rocks and sea.   180
Driv'n inshore, we follow down
Ancient streets of Hastings town—
Slowly thread them—when behold,
French canary-merchant old
Shepherding his flock of gold
In a low dim-lighted pen
Scann'd of tramps and fishermen!
There a bird, high-coloured, fat,
Proud of port, though something squat—
Pursy, play'd-out Philistine—   190
Dazzled Nelly's youthful eyne.
But, far in, obscure, there stirr'd
On his perch a sprightlier bird,
Courteous-eyed, erect and slim;
And I whisper'd: 'Fix on *him*!'
Home we brought him, young and fair,
Songs to trill in Surrey air.
Here Matthias sang his fill,
Saw the cedars of Pains Hill;

Here he pour'd his little soul,   200
Heard the murmur of the Mole.°
Eight in number now the years
He hath pleased our eyes and ears;
Other favourites he hath known
Go, and now himself is gone.
—Fare thee well, companion dear!
Fare for ever well, nor fear,
Tiny though thou art, to stray
Down the uncompanion'd way!°
We without thee, little friend,   210
Many years have not to spend;
What are left, will hardly be
Better than we spent with thee.

Kaiser Dead

APRIL 6, 1887

What, Kaiser dead? The heavy news
Post-haste to Cobham calls the Muse,
From where in Farringford she brews°
          The ode sublime,
Or with Pen-bryn's bold bard pursues°
          A rival rhyme.

Kai's bracelet tail, Kai's busy feet,
Were known to all the village-street.
'What, poor Kai dead?' say all I meet;
          'A loss indeed!'   10
O for the croon pathetic, sweet,°
          Of Robin's reed!

Six years ago I brought him down,
A baby dog, from London town;
Round his small throat of black and brown
          A ribbon blue,
And vouch'd by glorious renown
          A dachshound true.

His mother, most majestic dame,
Of blood-unmix'd, from Potsdam came;   20
And Kaiser's race we deem'd the same—
          No lineage higher.
And so he bore the imperial name.
          But ah, his sire!

Soon, soon the days conviction bring.
The collie hair, the collie swing,
The tail's indomitable ring,
          The eye's unrest—
The case was clear; a mongrel thing
          Kai stood confest.   30

But all those virtues, which commend
The humbler sort who serve and tend,
Were thine in store, thou faithful friend.
          What sense, what cheer!
To us, declining tow'rds our end,
          A mate how dear!

For Max, thy brother-dog, began
To flag, and feel his narrowing span.
And cold, besides, his blue blood ran,
          Since, 'gainst the classes,°   40
He heard, of late, the Grand Old Man
          Incite the masses.

Yes, Max and we grew slow and sad;
But Kai, a tireless shepherd-lad,
Teeming with plans, alert, and glad
          In work or play,
Like sunshine went and came, and bade
          Live out the day!

Still, still I see the figure smart—
Trophy in mouth, agog to start,   50
Then, home return'd, once more depart;
          Or prest together
Against thy mistress, loving heart,
          In winter weather.

I see the tail, like bracelet twirl'd,
In moments of disgrace uncurl'd,
Then at a pardoning word re-furl'd,
      A conquering sign;
Crying, 'Come on, and range the world,
      And never pine.'   60

Thine eye was bright, thy coat it shone;
Thou hadst thine errands, off and on;
In joy thy last morn flew; anon,
      A fit! All's over;
And thou art gone where Geist hath gone,°
      And Toss, and Rover.

Poor Max, with downcast, reverent head,
Regards his brother's form outspread;
Full well Max knows the friend is dead
      Whose cordial talk,   70
And jokes in doggish language said,
      Beguiled his walk.

And Glory, stretch'd at Burwood gate,°
Thy passing by doth vainly wait;
And jealous Jock, thy only hate,
      The chiel from Skye,
Lets from his shaggy Highland pate
      Thy memory die.

Well, fetch his graven collar fine,
And rub the steel, and make it shine,   80
And leave it round thy neck to twine,
      Kai, in thy grave.
There of thy master keep that sign,
      And this plain stave.

Letters to Arthur Hugh Clough

[London]
[shortly after 6 December 1847]

My dear Clough

I sent you a beastly vile note the other day: but I was all rasped by influenza and a thousand other bodily discomforts. Upon this came all the exacerbation produced by your apostrophes to duty: and put me quite wrong: so that I did not at all do justice to the great precision and force you have attained in those inward ways. I do think however that rare as individuality is you have to be on your guard against it—you particularly:—tho: indeed I do not really know that I think so. Shakspeare says° that if imagination would apprehend some joy it comprehends some bringer of that joy: and this latter operation which makes palatable the bitterest or most arbitrary original apprehension you seem to me to despise. Yet to *solve* the Universe as you try to do is as irritating as Tennyson's dawdling° with its painted shell is fatiguing to me to witness: and yet I own that to *re-construct* the Universe is not a satisfactory attempt either—I keep saying, Shakspeare, Shakspeare, you are as obscure as life is: yet this unsatisfactoriness goes against the poetic office in general: for this must I think certainly be its end. But have I been inside you, or Shakspeare? Never. Therefore heed me not, but come to what you can. Still my first note was cynical and beastly-vile. To compensate it, I have got you the Paris diamond edition of Béranger,° like mine. Tell me when you are coming up hither. I think it possible Tom may have trotted into Arthur's Bosom° in some of the late storms; which would have been a pity as he meant to enjoy himself in New Zealand.° It is like your noble abstemiousness not to have shown him the Calf Poem:° he would have worshipped like the children of Israel. Farewell.

yours most truly
M. ARNOLD

London, Tuesday
[December 1847 or early 1848]

My dearest Clough

My heart warms to the kindness of your letter: it is necessity not inclination indeed that ever repels me from you.

I forget what I said to provoke your explosion about Burbidge:° au reste,° I have formed my opinion of him, as Nelson said of Mack.° One does not always remember that one of the signs of the Decadence of a

literature, one of the factors of its decadent condition indeed, is this—
that new authors attach themselves to the poetic expression the founders
of a literature have flowered into, which may be *learned* by a sensitive
person, to the neglect of an inward poetic life. The strength of the
German literature consists in this—that having no national models from
whence to get an idea of *style* as half the work, they were thrown upon
themselves, and driven to make the fulness of the content of a work atone
for deficiencies of form. Even Goethe at the end of his life has not the
inversions, the taking tourmenté° style we admire in the Latins, in some
of the Greeks, & in the great French & English authors. And had
Shakspeare & Milton lived in the atmosphere of modern feeling, had
they had the multitude of new thoughts & feelings to deal with a modern
has, I think it likely the style of each would have been far less *curious* &
exquisite. For in a *man* style is the saying in the best way *what you have to
say*. The *what you have to say* depends on your age. In the 17th century it
was *a smaller harvest than now*, & sooner to be reaped: & therefore to its
reaper was left time to stow it more finely & curiously. Still more was this
the case in the ancient world. The poet's matter being *the hitherto
experience of the world, & his own*, increases with every century. Burbidge
lives quite beside the true poetical life, under a little gourd.° So much for
him. For me you may often hear my sinews cracking under the effort to
unite matter°

London. Monday.
[1848 (after September) or 1849]

My dearest Clough
    What a brute you were to tell me to read Keats' Letters.° However it is
over now: and reflexion resumes her power over agitation.
    What harm he has done in English Poetry. As Browning° is a man with
a moderate gift passionately desiring movement & fulness, and obtaining
but a confused multitudinousness, so Keats with a very high gift, is yet
also consumed by this desire: & cannot produce the truly living &
moving, as his conscience keeps telling him. They will not be patient
neither understand that they must begin with an Idea of the world in
order not to be prevailed over by the world's multitudinousness: or if they
cannot get that, at least with isolated ideas: & all other things shall
(perhaps) be added unto them.°
    —I recommend you to follow up these letters with the Laocoon of
Lessing:° it is not quite satisfactory, & a little mare's nesty—but very
searching.
    —I have had that desire of fulness without respect of the means, which

may become almost maniacal: but nature had placed a bar thereto not only in the conscience (as with all men) but in a great numbness in that direction. But what perplexity Keats Tennyson et id genus omne° must occasion to young writers of the ὁπλίτης° sort: yes & those d–d Elizabethan poets generaliy. Those who cannot read G[ree]k sh[ou]ld read nothing but Milton & parts of Wordsworth: the state should see to it: for the failures of the στάσιμοι° may leave them good citizens enough, as Trench:° but the others go to the dogs failing or succeeding.

So much for this inspired 'cheeper'° as they are saying on the moors. My own good man farewell.

M.A.

L[ansdowne] H[ouse]
Friday [early February 1849]

My dear Clough—

If I were to say the real truth as to your poems in general, as they impress me—it would be this—that they are not *natural*.

Many persons with far lower gifts than yours yet seem to find their natural mode of expression in poetry, and tho: the contents may not be very valuable they appeal with justice from the judgment of the mere thinker to the world's general appreciation of naturalness—i.e. an absolute propriety—of form, as the sole *necessary* of Poetry as such: whereas the greatest wealth & depth of matter is merely a superfluity in the Poet *as such*.

—Form of conception comes by nature certainly, but is generally developed late: but this lower form, of expression, is found from the beginning amongst all born poets, even feeble thinkers, and in an unpoetical age: as Collins, Green° and fifty more, in England only.

The question is not of congruity between conception & expression: which when both are poetical, is the poet's highest result:—you say what you mean to say: but in such a way as to leave it doubtful whether your mode of expression is not quite arbitrarily adopted.

I often think that even a slight gift of poetical expression which in a common person might have developed itself easily & naturally, is overlaid and crushed in a profound thinker so as to be of no use to him to help him to express himself.—The trying to go into and to the bottom of an object instead of grouping *objects* is as fatal to the sensuousness of poetry as the mere painting, (for, *in Poetry*, this is not *grouping*) is to its airy & rapidly moving life.

'Not deep the Poet sees, but wide':°—think of this as you gaze from the Cumner Hill° towards Cirencester & Cheltenham.

—You succeed best you see, in fact, in the hymn, where man, his deepest personal feelings being in play, finds poetical expression as *man* only, not as artist:—but consider whether you attain the *beautiful*, and whether your product gives PLEASURE, not excites curiosity & reflexion. Forgive me all this: but I am always prepared myself to give up the attempt, on conviction: & so, I know, are you: & I only urge you to reflect whether you are advancing. Reflect too, as I cannot but do here more & more, in spite of all the nonsense some people talk, how deeply *unpoetical* the age and all one's surroundings are. Not unprofound, not ungrand, not unmoving:—but un*poetical*.

Ever yrs
M.A.

[About 1 March 1849]

Dear Clough

The Iliad translation° is better, but not anglicised enough I think. I am told that Germans who are ignorant of the original complain that they cannot understand Voss.° Carlyle's Dante° seemed to me clearer.

—It is true about form: something of the same sort is in my letter which crossed yours on the road. On the other hand, there are two offices of Poetry—one to add to one's store of thoughts & feelings—another to compose & elevate the mind by a sustained tone, numerous allusions, and a grand style. What other process is Milton's than this last, in Comus for instance. There is no fruitful analysis of character: but a great effect is produced. What is Keats? A style & form seeker, & this with an impetuosity that heightens the effect of his style almost painfully. Nay in Sophocles what is valuable is not so much his contributions to psychology & the anatomy of sentiment, as the grand moral effects produced by *style*. For the style is the expression of the nobility of the poet's character, as the matter is the expression of the richness of his mind: but on men character produces as great an effect as mind.

This however does not save Burbidge who planes & polishes to the forgetting of matter without ever arriving at style. But my Antigone° supports me & in some degree subjugates destiny.

—I have had a very enthusiastic letter from Brodie and from John Coleridge,° to my surprize. These are all I have heard from. Shairp is δεξιά.°

Yours,
M.A.

Thun.° Sunday. Sept^{ber} 23 [1849]

My dear Clough

I wrote to you from this place last year. It is long since I have communicated with you and I often think of you among the untoward generation with whom I live and of whom all I read testifies. With me it is curious at present: I am getting to feel more independent & unaffectible as to all intellectual & poetical performance the impatience at being faussé° in which drove me some time since so strongly into myself, and more snuffing after a moral atmosphere to respire in than ever before in my life. Marvel not that I say unto you, ye must be born again.° While I will not much talk of these things, yet the considering of them has led me constantly to you the only living one almost that I know of of

> The children of the second birth
> Whom the world could not tame—°

for my dear Tom° has not sufficient besonnenheit° for it to be any *rest* to think of him any more than it is a *rest* to think of mystics & such cattle—not that Tom is in any sense cattle or even a mystic but he has not a 'still, considerate mind'.°

What I must tell you is that I have never yet succeeded in any one great occasion in consciously mastering myself: I can go thro: the imaginary process of mastering myself & see the whole affair as it would then stand, but at the critical point I am too apt to hoist up the mainsail to the wind & let her drive. However as I get more awake to this it will I hope mend for I find that with me a clear almost palpable intuition (damn the logical senses of the word)° is necessary before I get into prayer: unlike many people who set to work at their duty self-denial &c like furies in the dark hoping to be gradually illuminated as they persist in this course. Who also perhaps may be sheep but not of my fold,° whose one natural craving is not for profound thoughts, mighty spiritual workings &c &c but a distinct seeing of my way as far as my own nature is concerned: which I believe to be the reason why the mathematics were ever foolishness to me.°

—I am here in a curious & not altogether comfortable state: however tomorrow I carry my aching head to the mountains & to my cousin the Blümlis Alp.°

> Fast, fast by my window
> The rushing winds go
> Towards the ice-cumber'd gorges,
> The vast fields of snow.

> There the torrents drive upward
> Their rock strangled hum,
> And the avalanche thunders
> The hoarse torrent dumb.
> I come, O ye mountains—
> Ye torrents, I come,°

Yes, I come, but in three or four days I shall be back here, & then I must try how soon I can ferociously turn towards England.

My dearest Clough these are damned times—everything is against one—the height to which knowledge is come, the spread of luxury, our physical enervation, the absence of great *natures*, the unavoidable contact with millions of small ones, newspapers, cities, light profligate friends, moral desperadoes° like Carlyle, our own selves, and the sickening consciousness of our difficulties: but for God's sake let us neither be fanatics nor yet chalf blown by the wind but let us be ὡς ὁ φρόνιμος διορίσειεν° and not as any one else διορίσειεν. When I come to town I tell you beforehand I will have a real effort at managing myself as to newspapers & the talk of the day. Why the devil do I read about L^d. Grey's sending convicts to the Cape,° & excite myself thereby, when I can thereby produce no possible good. But public opinion consists in a multitude of such excitements. Thou fool°—that which is morally worthless remains so, & undesired by Heaven, whatever results flow from it: & which of the units which has felt the excitement caused by reading of Lord Grey's conduct has been made one iota a better man thereby, or can honestly call his excitement a *moral* feeling.

You will not I know forget me. You cannot answer this letter for I know not how I come home.

Yours faithfully,
M.A.

[London] Tuesday. [May 1850]

Dear Clough

Or my memory bewrayeth me or thou promisedst to come & breakfast with me tomorrow morning. Maskelyne° has offered himself for that day: & two is to my mind naught at breakfast—besides thou lovest not that young man. Forster° wants to see you: come therefore on Friday at 9¼ & you shall meet him. I am engaged the evenings of this week: still I would fain see thee as I have at Quillinan's sollicitation° dirged W. W. *in the grand style*° & need thy rapture therewith.

F. Newman's book° I saw yestern at our ouse. He seems to have written himself down an hass. It is a display of the theological mind, which I am accustomed to regard as a suffetation, existing in a man from the beginning, colouring his whole being, and being him in short. One would think to read him that enquiries into articles, biblical inspiration, &c &c were as much the natural functions of a man as to eat & copulate. This sort of man is only possible in Great Britain & North Germany, thanks be to God for it. Ireland even spews him out.

The world in general has always stood towards religions & their doctors in the attitude of a half-astonished clown acquiescingly ducking at their grand words & thinking it must be very fine, but for its soul not being able to make out what it is all about. This beast talks of such matters as if they were meat & drink. What a miserable place Oxford & the society of the serious middle classes must have been 20 years ago. He bepaws the religious sentiment so much that he quite effaces it to me. This sentiment now, I think, is best not regarded alone, but considered in conjunction with the grandeur of the world, love of kindred, love, gratitude &c &c.

Il faut feuilleter seulement cet ouvrage: wenn man es durchlesen sollte, so wäre es gar zu ekelhaft.°

Yours dear
M.A.

Milford B.[oys] S.[chool] Oct^{ber} 28/52

My dear Clough

I have got your note: Shairp I hope will come to me for a day, and then he can bring the money.

As to that article.° I am anxious to say that so long as I am prosperous, nothing would please me more than for you to make use of me, at any time, as if I were your brother.

And now what shall I say? First as to the poems.° Write me from America° concerning them, but do not read them in the hurry of this week. Keep them, as the Solitary° did his Bible, for the silent deep.

More and more I feel that the difference between a mature and a youthful age of the world compels the poetry of the former to use great plainness of speech as compared with that of the latter: and that Keats and Shelley were on a false track when they set themselves to reproduce the exuberance of expression, the charm, the richness of images, and the felicity, of the Elizabethan poets. Yet critics cannot get to learn this, because the Elizabethan poets are our greatest, and our canons of poetry are founded on their works. They still think that the object of poetry is to produce exquisite bits and images—such as Shelley's *clouds shepherded by the slow unwilling wind*,° and Keats passim: whereas modern poetry can only subsist by its *contents*: by becoming a complete magister vitae° as the poetry of the ancients did: by including, as theirs did, religion with poetry, instead of existing as poetry only, and leaving religious wants to be supplied by the Christian religion, as a power existing independent of the poetical power. But the language, style and general proceedings of a poetry which has such an immense task to perform, must be very plain direct and severe: and it must not lose itself in parts and episodes and ornamental work, but must press forwards to the whole.

A new sheet will cut short my discourse: however, let us, as far as we can, continue to exchange our thoughts, as with all our differences we agree more with one another than with the rest of the world, I think. What do you say to a bi-monthly mail?

It was perhaps as well that the Rugby meeting was a Bacchic rout, for after all on those occasions there is nothing to be said.—God bless you wherever you go—with all my scepticism I can still say that. I shall go over and see Miss Smith from Hampton in December, and perhaps take Fanny Lucy° with me. I am not very well or in very good spirits, but I subsist:—what a difference there is between reading in poetry & morals of the loss of youth, and experiencing it! And after all there is so much to be done, if one could but do it.—

Goodbye again and again, my dear Clough—

your ever affectionate
M. ARNOLD

Battersea. Dec^{ber} 14th 1852

My dear Clough

—I write to you from an evening sitting of the candidates for certificates at the Training School here. It is a Church of England place° but such is my respectability that I am admitted to their mysteries.

I have no doubt that you will do well *socially* in the U[nited] States: you are English, you are well introduced—and you have personal merit—the object for you is to do well *commercially*. Value the first only so far as it helps the second. It would be a poor consolation for having not established oneself at the end of a year and a half to be able to say—I have got into the best American Society. If you are to succeed there, you will begin at once, and will be the fashion as a tutor.

What sort of beings are the Yankees really? Better or worse in masses than they are individually? They me font l'effet° of a nation not having on a wedding garment.° It is true that the well born, the well mannered, the highly cultivated—are called no longer, because they have shown such incapacity for administering the world: but it is too bad that when our

Heavenly Father has whipped in these long ugly yellow rascallions from the highways & hedges they should not clean and polish themselves a little before taking the places of honour.

As for my poems they have weight, I think, but little or no charm

> What Poets feel not, when they make,
> A pleasure in creating,
> The world, in *its* turn, will not take
> Pleasure in contemplating.°

There is an oracular quatrain for you, terribly true. I feel now where my poems (this set) are all wrong, which I did not a year ago: but I doubt whether I shall ever have heat & radiance enough to pierce the clouds that are massed round me. Not in my little social sphere indeed, with you & Walrond:° there I could crackle to my grave—but vis à vis of the world.—This volume is going off though: a nice notice of it was in the Guardian°—and Froude will review it in the April Westminster,° calling me by my name. He is much pleased.—You must tell me what Emerson says—make him look at it. *You* in your heart are saying *mollis et exspes*° over again. But woe was upon me if I analysed not my situation: & Werter René° and such like none of them analyse the modern situation in its true *blankness* and *barrenness*, and un*poetrylessness*.

Now my dear Clough you were a good boy to write when you did, but you are not to write me scraps across the ὑγρὰ κέλευθα° of the Atlantic, or I shall dry up as a correspondent. But write me a nice long letter as if I was an ἀνάλογον° (at least) of Miss Blanche Smith—& then we will establish a regular bimonthly mail. God bless you.

Flu & I shall go & see Miss Smith from Hampton.

> ever yours affectionately
> M.A.

Edgbaston. February 12th 1853

My dear Clough

I received your letter ten days since—just as I was leaving London— but I have since that time had too much to do to attempt answering it, or indeed to attempt anything else that needed any thing of 'recueille-ment'.° I do not like to put off writing any longer, but to say the truth I do not feel in the vein to write even now, nor do I feel certain that I can write as I should wish. I am past thirty, and three parts iced over—and my pen, it seems to me is even stiffer and more cramped than my feeling.

But I will write historically, as I can write naturally in no other way. I did not really think you had been hurt at anything I did or left undone while we were together in town: that is, I did not think any impression of hurt you might have had for a moment, had lasted. I remember your being annoyed once or twice, and that I was vexed with myself: but at that time I was absorbed in my speculations and plans and agitations respecting Fanny Lucy, and was as egoistic and anti-social as possible. People in the condition in which I then was always are. I thought I had said this and explained one or two pieces of apparent carelessness in this way: and that you had quite understood it. So entirely indeed am I convinced that being in love generally unfits a man for the society of his friends, that I remember often smiling to myself at my own selfishness in half compelling you several times to meet me in the last few months before you left England, and thinking that it was only I who could make such unreasonable demands or find pleasure in meeting and being with a person, for the mere sake of meeting and being with them, without regarding whether they would be absent and preoccupied or not. I never, while we were both in London, had any feeling towards you but one of attachment and affection: if I did not enter into much explanation when you expressed annoyance, it was really because I thought the mention of my circumstances accounted for all and more than all that had annoyed you. I remember Walrond telling me you were vexed one day that on a return to town after a longish absence I let him stop in Gordon Square° without me: I was then expecting to find a letter—or something of that sort—it all seems trivial now, but it was enough at the time to be the cause of heedlessnesses selfishnesses and heartlessnesses—in all directions but one—without number. It ought not to have been so perhaps—but it was so—and I quite thought you had understood that it was so.

There was one time indeed—shortly after you had published the Bothie°—that I felt a strong disposition to intellectual seclusion, and to the barring out all influences that I felt troubled without advancing me: but I soon found that it was needless to secure myself against a danger from which my own weakness even more than my strength—my coldness and want of intellectual robustness—sufficiently exempted me—and besides your company and mode of being always had a charm and a salutary effect for me, and I could not have foregone these on a mere theory of intellectual dietetics.

In short, my dear Clough, I cannot say more than that I really have clung to you in spirit more than to any other man—and have never been seriously estranged from you at any time—for the estrangement I have just spoken of was merely a contemplated one and it never took place: I remember saying something about it to you at the time—and your answer, which struck me for the genuineness and faith it exhibited as compared with my own—not want of faith exactly—but invincible languor of spirit, and fickleness and insincerity even in the gravest matters. All this is dreary work—and I cannot go on with it now: but tomorrow night I will try again—for I have one or two things more to say. Goodnight now.—

Sunday. 6 P.M.

I will not look at what I wrote last night—one endeavours to write deliberately out what is in one's mind, without any veils of flippancy levity metaphor or demi-mot,° and one succeeds only in putting upon the paper a string of dreary dead sentences that correspond to nothing in one's inmost heart or mind, and only represent themselves. It was your own fault partly for forcing me to it. I will not go on with it: only remember, *pray* remember that I am and always shall be, whatever I do or say, powerfully attracted towards you, and vitally connected with you: this I am sure of: the period of my developement (God forgive me the d–d expression!) coincides with that of my friendship with you so exactly that I am for ever linked with you by intellectual bonds—the strongest of all: more than you are with me: for your developement was really over before you knew me, and you had properly speaking come to your *assiette*° for life. You ask me in what I think or have thought you going wrong: in this: that you would never take your assiette as something determined final and unchangeable for you and proceed to work away on the basis of that: but were always poking and patching and cobbling at the assiette itself— could never finally, as it seemed—'resolve to be thyself'°—but were looking for this and that experience, and doubting whether you ought not to adopt this or that mode of being of persons qui ne vous valaient pas° because it might possibly be nearer the truth than your own: you had no reason for thinking it *was*, but it *might* be—and so you would try to adapt yourself to it. You have I am convinced lost infinite time in this way: it is what I call your morbid conscientiousness—you are the most conscientious man I ever knew: but on some lines morbidly so, and it spoils your action.

There—but now we will have done with this: we are each very near to the other—write and tell me that you feel this: as to my behaviour in London I have told you the simple truth: it is I fear too simple than that (excuse the idiom) you with your raffinements° should believe and appreciate it.

There is a power of truth in your letter and in what you say about America and this country: yes—*congestion of the brain* is what we suffer from—I always feel it and say it—and cry for air like my own

Empedocles. But this letter shall be what it is. I have a number of things I want to talk to you about—they shall wait till I have heard again from you. Pardon me, but we *will* exchange intellectual aperçus°—we shall both be the better for it. Only let us pray all the time—God keep us both from aridity! *Arid*—that is what the times are.—Write soon and tell me you are well—I was sure you were not well. God bless you. Flu sends her kindest remembrances. ever yours

M.A.

We called the other day at Combe Hurst° but found vacuas sedes et inania arcana.° But we shall meet in town. What does Emerson say to my poems—or is he gone crazy as Miss Martineau° says. But this is probably one of her d–d lies. Once more fare*well*, in every sense.

23. Grosvenor St. West
Grosvenor Place [London]
March 21/53

My dear Clough

I got your letter at Halstead in Essex° on Friday evening last. This is the thinnest paper° I can lay my hand upon: would that I could but write upon it. We will not discuss what is past any more: as to the Italian poem,° if I forbore to comment it was that I had nothing special to say—what is to be said when a thing does not suit you—suiting and not suiting is a subjective affair and only time determines, by the colour a thing takes with years, whether it *ought* to have suited or no.

I am glad to hear a good account of Emerson's health—I thought his insanity was one of Miss Martineau's terrific lies: sane he certainly is, though somewhat incolore° as the French say—very thin and ineffectual, and self-defensive only. Tell me when you can something about his life and manner of going on—and his standing in the Boston world.

Margaret Fuller°—what do you think of her? I have given, after some hesitation, half a guinea for the three volumes concerning her—partly moved by the low price partly by interest about that partly brazen female. I incline to think that the meeting with her would have made me return all the contents of my spiritual stomach but through the screen of a book I willingly look at her and allow her her exquisite intelligence and fineness of aperçus. But my G–d what rot did she and the other female dogs of Boston talk about the Greek mythology! The absence of men of any culture in America, where everybody knows that the Earth is an oblate spheroid° and nobody knows anything worth knowing, must have made her run riot so wildly, and for many years made her insufferable.

Miss Bronte° has written a hideous undelightful convulsed constricted novel—what does Thackeray° say to it. It is one of the most utterly disagreeable books I ever read—and having seen her makes it more so. She is so entirely—what Margaret Fuller was partially—a fire without aliment—one of the most distressing barren sights one can witness. Religion or devotion or whatever it is to be called may be impossible for such people now: but they have at any rate not found a substitute for it and it was better for the world when they comforted themselves with it.

Thackeray's Esmond you know everyone here calls a failure—but I do not think so—it is one of the most readable books I ever met —and Thackeray is certainly a first rate journeyman though not a great artist:—It gives you an insight into the *heaven born* character of Waverley and Indiana° and such like when you read the undeniably powerful but most un-heaven-born productions of the present people —Thackeray—the woman Stowe° &c. The woman Stowe by her picture must be a Gorgon—I can quite believe all you tell me of her—a strong Dissenter-religious middle-class person—she will never go far, I think.

Look at Alexander Smith's poems° which some people speak of and let me know what you think of them. The article on Wordsworth,° I hear, is Lockhart's, very just though cold. Perhaps it does not sufficiently praise his *diction*: his *manner* was often bad, but his diction scarcely ever—and Byron's Moore's° &c—constantly.

Goodnight—no more paper.

Read some of the articles of Ampère's° in the Revue des 2 Mondes on America: what he says is so cool clear désabusé° and true that it will do you good in the atmosphere of inflation exaggeration and intoxication in which you live.

We will yet see the young lady—though not soon, I fear. I am frightfully worked at present. I read Homer and toujours Homer.°

ever yours
M.A.

Susy° is going to be married to John Cropper—second son of the principal Liverpool Cropper.

London. May 1st 1853.

My dear Clough

I do not know that the tone of your letters exactly facilitates correspondence—however, let it be as you will. I for my part think that what Curran said° of the constitution of the state holds true of individual moral constitutions: it does not do to lay bare their foundations too constantly. It is very true I am not myself in writing—but it is of no use reproaching me with it, since so it must be.

I do not think we did each other harm at Oxford. I look back to that time with pleasure. All activity to which the conscience does not give its consent is mere *philisterey*,° and it is always a good thing to have been preserved from this. I catch myself desiring now at times political life, and this and that; and I say to myself—you do not desire these things because you are really adapted to them, and therefore the desire for them is merely contemptible—and it is so. I am nothing and very probably never shall be anything—but there are characters which are truest to themselves by never being anything, when circumstances do not suit. If you had never met me, I do not think you would have been the happier or the wiser on that account: though I do not think I have increased your stock of happiness. You have, however, on the whole, added to mine.

You do not tell what you are doing: Mrs. Lingen° told me last night you had six pupils: she is a great friend you know of *your* friend's:° we talked a great deal about you—not that I like her (Mrs. Lingen) much. Your friend has been to see Fanny Lucy—but I was from home: I shall manage to see her some day. You will come all right, I think, when you are once married.

If you have been looking over North's Plutarch° lately you are probably right about it: but I cannot help thinking (I am going on with this Tuesday night May 3rd) that there is a freshness in his style and language which is like a new world to one—it produces the same effect on me as Cotton's Montaigne° does:—if North could be read *safely*, without one's continually suspecting an error, and in a handy volume, I think he would be delightful reading. You are quite right to incorporate Long.° I should much like to see what you have done. Stick to literature—it is the great comforter after all.

I should like to read an article of yours on me°—I should read it with a curious feeling—my version of Tristram and Iseult° comes from an article in the Revue de Paris, on Fauriel, I think: the story of Merlin is imported from the Morte d'Arthur. If I republish that poem I shall try to make it more intelligible: I wish I had you with me to put marks against the places where something is wanted. The whole affair is by no means thoroughly successful.

I have just got through a thing° which pleases me better than anything I have yet done—but it is pain and grief composing with such interruptions as I have: however in this case the material was a thoroughly good one, and what a thing is this! and how little do young writers feel what a thing it is—how it is *everything*.

I feel immensely—more and more clearly—what I *want*—what I have (I believe) lost and choked by my treatment of myself and the studies to which I have addicted myself. But what ought I to have done in preference to what I have done? there is the question.

As to Alexander Smith I have not read him—I shrink from what is so intensely immature—but I think the extracts I have seen most remarkable—and I think at the same time that he will not go far. I have not room or time for my reasons—but I think so. This kind does not go far: it dies like Keats or loses itself like Browning.

You know (or you do not know) that Froude, who is one of the very few people who much liked my last vein, or [thought me]° to be other than the black villain my Maker made me [, will review my *Poems* for the *Westminster*]. Tell me about yourself—and above all do not dream of my using you as food for speculation: that is simply a morbid suspicion: I like to hear all about you because I am fond of you.

Good-bye again—

Your incorrigible and affectionate
M.A.

My father's journals° are out—they are a mere bookseller's catchpenny, in my judgment: but they are a convenient size—but there is nothing new in them.

Derby. Nov^{ber} 25/53.

My dear Clough

Just read through Tennyson's Morte d'Arthur° and Sohrab & Rustum one after the other, and you will see the difference in the *tissue* of the style of the two poems, and in its *movement*. I think the likeness, where there is likeness, (except in the two last lines° which I own are a regular slip) proceeds from our both having imitated Homer. But never mind—you are a dear soul. I am in great hopes you will one day like the poem—really like it. There is no one to whose aperçus I attach the value I do to yours—but I think you are sometimes—with regard to *me* especially—a little cross and wilful.

I send you two letters—not that you may see the praise of me in them (and I can sincerely say that praise of *myself*—talking about imagination—genius and so on—does not give me, at heart, the slightest flutter of pleasure—seeing people interested in what I have made, does—) but that you may see how heartily two very different people seem to have taken to Sohrab and Rustum. This is something, at any rate. Hill's criticism° is always delicate and good—and his style in prose has something of the beauty of his father in law's. How well all the third page is written.

Return me the letters—write a line to P. O. Lincoln. I am worked to death. God bless you. ever yours

M.A.

Democracy

I know that, since the Revolution, along with many dangerous, many useful powers of Government have been weakened.

BURKE (1770)°

IN giving an account of education in certain countries of the Continent, I have often spoken of the State and its action in such a way as to offend, I fear, some of my readers, and to surprise others. With many Englishmen, perhaps with the majority, it is a maxim that the State, the executive power, ought to be entrusted with no more means of action than those which it is impossible to withhold from it; that the State neither would nor could make a safe use of any more extended liberty; would not, because it has in itself a natural instinct of despotism, which, if not jealously checked, would become outrageous; could not, because it is, in truth, not at all more enlightened, or fit to assume a lead, than the mass of this enlightened community.

No sensible man will lightly go counter to an opinion firmly held by a great body of his countrymen. He will take for granted, that for any opinion which has struck deep root among a people so powerful, so successful, and so well worthy of respect as the people of this country, there certainly either are, or have been, good and sound reasons. He will venture to impugn such an opinion with real hesitation, and only when he thinks he perceives that the reasons which once supported it exist no longer, or at any rate seem about to disappear very soon. For undoubtedly there arrive periods, when, the circumstances and conditions of government having changed, the guiding maxims of government ought to change also. *J'ai dit souvent*, says Mirabeau,[1] admonishing the Court of France in 1790, *qu'on devait changer de manière de gouverner, lorsque le gouvernement n'est plus le même.*° And these decisive changes in the political situation of a people happen gradually as well as violently. 'In the silent lapse of events,' says Burke,[2] writing in England twenty years before the French Revolution, 'as material alterations have been insensibly brought about in the policy and character of governments and nations, as those which have been marked by the tumult of public revolutions.'

---

[1] *Correspondance entre le Comte de Mirabeau et le Comte de la Marck*, publiée par M. [Adolphe] de Bacourt, Paris, 1851, vol. ii. p. 143.

[2] Burke's *Works*° (edit. of 1852), vol. iii. p. 115.

I propose to submit to those who have been accustomed to regard all State-action with jealousy, some reasons for thinking that the circumstances which once made that jealousy prudent and natural have undergone an essential change. I desire to lead them to consider with me, whether, in the present altered conjuncture, that State-action, which was once dangerous, may not become, not only without danger in itself, but the means of helping us against dangers from another quarter. To combine and present the considerations upon which these two propositions are based, is a task of some difficulty and delicacy. My aim is to invite impartial reflection upon the subject, not to make a hostile attack against old opinions, still less to set on foot and fully equip a new theory. In offering, therefore, the thoughts which have suggested themselves to me, I shall studiously avoid all particular applications of them likely to give offence, and shall use no more illustration and development than may be indispensable to enable the reader to seize and appreciate them.

The dissolution of the old political parties which have governed this country since the Revolution of 1688 has long been remarked. It was repeatedly declared to be happening long before it actually took place, while the vital energy of these parties still subsisted in full vigour, and was threatened only by some temporary obstruction. It has been eagerly deprecated long after it had actually begun to take place, when it was in full progress, and inevitable. These parties, differing in so much else, were yet alike in this, that they were both, in a certain broad sense, *aristocratical* parties. They were combinations of persons considerable, either by great family and estate, or by Court favour, or lastly, by eminent abilities and popularity; this last body, however, attaining participation in public affairs only through a conjunction with one or other of the former. These connections, though they contained men of very various degrees of birth and property, were still wholly leavened with the feelings and habits of the upper class of the nation. They had the bond of a common culture; and, however their political opinions and acts might differ, what they said and did had the stamp and style imparted by this culture, and by a common and elevated social condition.

Aristocratical bodies have no taste for a very imposing executive, or for a very active and penetrating domestic administration. They have a sense of equality among themselves, and of constituting in themselves what is greatest and most dignified in the realm, which makes their pride revolt against the overshadowing greatness and dignity of a commanding executive. They have a temper of independence, and a habit of uncontrolled action, which makes them impatient of encountering, in the management of the interior concerns of the country, the machinery and regulations of a superior and peremptory power. The different parties amongst them, as they successively get possession of the government, respect this jealous disposition in their opponents, because they share it themselves. It is a disposition proper to them as great personages, not as ministers; and as they are great personages for their whole life, while they may probably be ministers but for a very short time, the instinct of their social condition avails more with them than the instinct of their official function. To administer as little as possible, to make its weight felt in foreign affairs rather than in domestic, to see in ministerial station rather the means of power and dignity than a means of searching and useful administrative activity, is the natural tendency of an aristocratic executive. It is a tendency which is creditable to the good sense of aristocracies, honourable to their moderation, and at the same time fortunate for their country, of whose internal development they are not fitted to have the full direction.

One strong and beneficial influence, however, the administration of a vigorous and high-minded aristocracy is calculated to exert upon a robust and sound people. I have had occasion, in speaking of Homer, to say very often, and with much emphasis, that he is *in the grand style.*° It is the chief virtue of a healthy and uncorrupted aristocracy, that it is, in general, in this grand style. That elevation of character, that noble way of thinking and behaving, which is an eminent gift of nature to some individuals, is also often generated in whole classes of men (at least when these come of a strong and good race) by the possession of power, by the importance and responsibility of high station, by habitual dealing with great things, by being placed above the necessity of constantly struggling for little things. And it is the source of great virtues. It may go along with a not very quick or open intelligence; but it cannot well go along with a conduct vulgar and ignoble. A governing class imbued with it may not be capable of intelligently leading the masses of a people to the highest pitch of welfare for them; but it sets them an invaluable example of qualities without which no really high welfare can exist. This has been done for their nation by the best aristocracies. The Roman aristocracy did it; the English aristocracy has done it. They each fostered in the mass of the peoples they governed,—peoples of sturdy moral constitution and apt to learn such lessons,—a greatness of spirit, the natural growth of the condition of magnates and rulers, but not the natural growth of the condition of the common people. They made, the one of the Roman, the other of the English people, in spite of all the shortcomings of each, great peoples, peoples *in the grand style*. And this they did, while wielding the people according to their own notions, and in the direction which seemed good to them; not as servants and instruments of the people, but as its commanders and heads; solicitous for the good of their country, indeed, but taking for granted that of that good they themselves were the supreme judges, and were to fix the conditions.

The time has arrived, however, when it is becoming impossible for the aristocracy of England to conduct and wield the English nation any longer. It still, indeed, administers public affairs; and it is a great error to suppose, as many persons in England suppose, that it administers but does not govern. He who administers, governs,[1] because he infixes his own mark and stamps his own character on all public affairs as they pass through his hands; and, therefore, so long as the English aristocracy administers the commonwealth, it still governs it. But signs not to be mistaken show that its headship and leadership of the nation, by virtue of the substantial acquiescence of the body of the nation in its predominance and right to lead, is nearly over. That acquiescence was the tenure by which it held its power; and it is fast giving way. The superiority of the upper class over all others is no longer so great; the willingness of the others to recognise that superiority is no longer so ready.

This change has been brought about by natural and inevitable causes, and neither the great nor the multitude are to be blamed for it. The growing demands and audaciousness of the latter, the encroaching spirit of democracy, are, indeed, matters of loud complaint with some persons. But these persons are complaining of human nature itself, when they thus complain of a manifestation of its native and ineradicable impulse. Life itself consists, say the philosophers, in the effort *to affirm one's own essence*;° meaning by this, to develop one's own existence fully and freely, to have ample light and air, to be neither cramped nor overshadowed. Democracy is trying *to affirm its own essence*; to live, to enjoy, to possess the world, as aristocracy has tried, and successfully tried, before it. Ever since Europe emerged from barbarism, ever since the condition of the common people began a little to improve, ever since their minds began to stir, this effort of democracy has been gaining strength; and the more their condition improves, the more strength this effort gains. So potent is the charm of life and expansion upon the living; the moment men are aware of them, they begin to desire them, and the more they have of them, the more they crave.

This movement of democracy, like other operations of nature, merits properly neither blame nor praise. Its partisans are apt to give it credit

---

[1] *Administrer, c'est gouverner*, says Mirabeau; *gouverner, c'est régner; tout se réduit là.*°

which it does not deserve, while its enemies are apt to upbraid it unjustly. Its friends celebrate it as the author of all freedom. But political freedom may very well be established by aristocratic founders; and, certainly, the political freedom of England owes more to the grasping English barons than to democracy. Social freedom,—equality,—that is rather the field of the conquests of democracy. And here what I must call the injustice of its enemies comes in. For its seeking after equality, democracy is often, in this country above all, vehemently and scornfully blamed; its temper contrasted with that worthier temper which can magnanimously endure social distinctions; its operations all referred, as of course, to the stirrings of a base and malignant envy. No doubt there is a gross and vulgar spirit of envy, prompting the hearts of many of those who cry for equality. No doubt there are ignoble natures which prefer equality to liberty. But what we have to ask is, when the life of democracy is admitted as something natural and inevitable, whether this or that product of democracy is a necessary growth from its parent stock, or merely an excrescence upon it. If it be the latter, certainly it may be due to the meanest and most culpable passions. But if it be the former, then this product, however base and blameworthy the passions which it may sometimes be made to serve, can in itself be no more reprehensible than the vital impulse of democracy is in itself reprehensible; and this impulse is, as has been shown, identical with the ceaseless vital effort of human nature itself.

Now, can it be denied, that a certain approach to equality, at any rate a certain reduction of signal inequalities, is a natural, instinctive demand of that impulse which drives society as a whole,—no longer individuals and limited classes only, but the mass of a community,—to develop itself with the utmost possible fulness and freedom? Can it be denied, that to live in a society of equals tends in general to make a man's spirits expand, and his faculties work easily and actively; while, to live in a society of superiors, although it may occasionally be a very good discipline, yet in general tends to tame the spirits and to make the play of the faculties less secure and active? Can it be denied, that to be heavily overshadowed, to be profoundly insignificant, has, on the whole, a depressing and benumb-ing effect on the character? I know that some individuals react against the strongest impediments, and owe success and greatness to the efforts which they are thus forced to make. But the question is not about individuals. The question is about the common bulk of mankind, persons without extraordinary gifts or exceptional energy, and who will ever require, in order to make the best of themselves, encouragement and directly favouring circumstances. Can any one deny, that for these the spectacle, when they would rise, of a condition of splendour,

grandeur, and culture, which they cannot possibly reach, has the effect of
making them flag in spirit, and of disposing them to sink despondingly
back into their own condition? Can any one deny, that the knowledge
how poor and insignificant the best condition of improvement and
culture attainable by them must be esteemed by a class incomparably
richer-endowed, tends to cheapen this modest possible amelioration in
the account of those classes also for whom it would be relatively a real
progress, and to disenchant their imaginations with it? It seems to me
impossible to deny this. And therefore a philosophic observer,[1] with no
love for democracy, but rather with a terror of it, has been constrained to
remark, that 'the common people is more uncivilised in aristocratic
countries than in any others'; because there 'the lowly and the poor feel
themselves, as it were, overwhelmed with the weight of their own
inferiority'. He has been constrained to remark,[2] that 'there is such a
thing as a manly and legitimate passion for equality, prompting men to
desire to be, *all* of them, in the enjoyment of power and consideration'.
And, in France, that very equality, which is by us so impetuously decried,
while it has by no means improved (it is said) the upper classes of French
society, has undoubtedly given to the lower classes, to the body of the
common people, a self-respect, an enlargement of spirit, a consciousness
of counting for something in their country's action, which has raised
them in the scale of humanity. The common people, in France, seems to
me the soundest part of the French nation. They seem to me more free
from the two opposite degradations of multitudes, brutality and servility,
to have a more developed human life, more of what distinguishes
elsewhere the cultured classes from the vulgar, than the common people
in any other country with which I am acquainted.

I do not say that grandeur and prosperity may not be attained by a
nation divided into the most widely distinct classes, and presenting the
most signal inequalities of rank and fortune. I do not say that great
national virtues may not be developed in it. I do not even say that a
popular order, accepting this demarcation of classes as an eternal
providential arrangement, not questioning the natural right of a superior
order to lead it, content within its own sphere, admiring the grandeur

---

[1] M. [Alexis] de Tocqueville. See his *Démocratie en Amérique* (edit. of 1835), vol. i.
p. 11. 'Le peuple est plus grossier dans les pays aristocratiques que partout ailleurs.
Dans ces lieux, où se rencontrent des hommes si forts et si riches, les faibles et les
pauvres se sentent comme accablés de leur bassesse; ne découvrant aucun point par
lequel ils puissent regagner l'égalité, ils désespèrent entièrement d'eux-mêmes, et se
laissent tomber au-dessous de la dignité humaine.'°

[2] *Démocratie en Amérique*, vol. i. p. 60.

and high-mindedness of its ruling class, and catching on its own spirit some reflex of what it thus admires, may not be a happier body, as to the eye of the imagination it is certainly a more beautiful body, than a popular order, pushing, excited, and presumptuous; a popular order, jealous of recognising fixed superiorities, petulantly claiming to be as good as its betters, and tastelessly attiring itself with the fashions and designations which have become unalterably associated with a wealthy and refined class, and which, tricking out those who have neither wealth nor refinement, are ridiculous. But a popular order of that old-fashioned stamp exists now only for the imagination. It is not the force with which modern society has to reckon. Such a body may be a sturdy, honest, and sound-hearted lower class; but it is not a democratic people. It is not that power, which at the present day in all nations is to be found existing; in some, has obtained the mastery; in others, is yet in a state of expectation and preparation.

The power of France in Europe is at this day mainly owing to the completeness with which she has organised democratic institutions. The action of the French State is excessive; but it is too little understood in England that the French people has adopted this action for its own purposes, has in great measure attained those purposes by it, and owes to its having done so the chief part of its influence in Europe. The growing power in Europe is democracy; and France has organised democracy with a certain indisputable grandeur and success. The ideas of 1789 were working everywhere in the eighteenth century; but it was because in France the State adopted them that the French Revolution became an historic epoch for the world, and France the lode-star of Continental democracy. Her airs of superiority and her overweening pretensions come from her sense of the power which she derives from this cause. Every one knows how Frenchmen proclaim France to be at the head of civilisation, the French army to be the soldier of God, Paris to be the brain of Europe, and so on. All this is, no doubt, in a vein of sufficient fatuity and bad taste; but it means, at bottom, that France believes she has so organised herself as to facilitate for all members of her society full and free expansion; that she believes herself to have remodelled her institutions with an eye to reason rather than custom, and to right rather than fact; it means, that she believes the other peoples of Europe to be preparing themselves, more or less rapidly, for a like achievement, and that she is conscious of her power and influence upon them as an initiatress and example. In this belief there is a part of truth and a part of delusion. I think it is more profitable for a Frenchman to consider the part of delusion contained in it; for an Englishman, the part of truth.

It is because aristocracies almost inevitably fail to appreciate justly, or even to take into their mind, the instinct pushing the masses towards expansion and fuller life, that they lose their hold over them. It is the old story of the incapacity of aristocracies for ideas,—the secret of their want of success in modern epochs. The people treats them with flagrant injustice, when it denies all obligation to them. They can, and often do, impart a high spirit, a fine ideal of grandeur, to the people; thus they lay the foundations of a great nation. But they leave the people still the multitude, the crowd; they have small belief in the power of the ideas which are its life. Themselves a power reposing on all which is most solid, material, and visible, they are slow to attach any great importance to influences impalpable, spiritual, and viewless. Although, therefore, a disinterested looker-on might often be disposed, seeing what has actually been achieved by aristocracies, to wish to retain or replace them in their preponderance, rather than commit a nation to the hazards of a new and untried future; yet the masses instinctively feel that they can never consent to this without renouncing the inmost impulse of their being; and that they should make such a renunciation cannot seriously be expected of them. Except on conditions which make its expansion, in the sense understood by itself, fully possible, democracy will never frankly ally itself with aristocracy; and on these conditions perhaps no aristocracy will ever frankly ally itself with it. Even the English aristocracy, so politic, so capable of compromises, has shown no signs of being able so to transform itself as to render such an alliance possible. The reception given by the Peers to the bill for establishing life-peerages° was, in this respect, of ill omen. The separation between aristocracy and democracy will probably, therefore, go on still widening.

And it must in fairness be added, that as in one most important part of general human culture,—openness to ideas and ardour for them,—aristocracy is less advanced than democracy, to replace or keep the latter under the tutelage of the former would in some respects be actually unfavourable to the progress of the world. At epochs when new ideas are powerfully fermenting in a society, and profoundly changing its spirit, aristocracies, as they are in general not long suffered to guide it without question, so are they by nature not well fitted to guide it intelligently.

In England, democracy has been slow in developing itself, having met with much to withstand it, not only in the worth of the aristocracy, but also in the fine qualities of the common people. The aristocracy has been more in sympathy with the common people than perhaps any other aristocracy. It has rarely given them great umbrage; it has neither been frivolous, so as to provoke their contempt, nor impertinent, so as to provoke their irritation. Above all, it has in general meant to act with justice, according to its own notions of justice. Therefore the feeling of admiring deference to such a class was more deep-rooted in the people of this country, more cordial, and more persistent, than in any people of the Continent. But, besides this, the vigour and high spirit of the English common people bred in them a self-reliance which disposed each man to act individually and independently; and so long as this disposition prevails through a nation divided into classes, the predominance of an aristocracy, of the class containing the greatest and strongest individuals of the nation, is secure. Democracy is a force in which the concert of a great number of men makes up for the weakness of each man taken by himself; democracy accepts a certain relative rise in their condition, obtainable by this concert for a great number, as something desirable in itself, because though this is undoubtedly far below grandeur, it is yet a good deal above insignificance. A very strong, self-reliant people neither easily learns to act in concert, nor easily brings itself to regard any middling good, any good short of the best, as an object ardently to be coveted and striven for. It keeps its eye on the grand prizes, and these are to be won only by distancing competitors, by getting before one's comrades, by succeeding all by one's self; and so long as a people works thus individually, it does not work democratically. The English people has all the qualities which dispose a people to work individually; may it never lose them! A people without the salt of these qualities, relying wholly on mutual co-operation, and proposing to itself second-rate ideals, would arrive at the pettiness and stationariness of China. But the English people is no longer so entirely ruled by them as not to show visible beginnings of democratic action; it becomes more and more sensible to the irresistible seduction of democratic ideas, promising to each individual of the multitude increased self-respect and expansion with the increased importance and authority of the multitude to which he belongs, with the diminished preponderance of the aristocratic class above him.

While the habit and disposition of deference are thus dying out among the lower classes of the English nation, it seems to me indisputable that the advantages which command deference, that eminent superiority in high feeling, dignity, and culture, tend to diminish among the highest class. I shall not be suspected of any inclination to underrate the aristocracy of this country. I regard it as the worthiest, as it certainly has been the most successful, aristocracy of which history makes record. If it has not been able to develop excellences which do not belong to the nature of an aristocracy, yet it has been able to avoid defects to which the nature of an aristocracy is peculiarly prone. But I cannot read the history of the flowering time of the English aristocracy, the eighteenth century, and then look at this aristocracy in our own century, without feeling that there has been a change. I am not now thinking of private and domestic virtues, of morality, of decorum. Perhaps with respect to these there has in this class, as in society at large, been a change for the better. I am thinking of those public and conspicuous virtues by which the multitude is captivated and led,—lofty spirit, commanding character, exquisite culture. It is true that the advance of all classes in culture and refinement may make the culture of one class, which, isolated, appeared remarkable, appear so no longer; but exquisite culture and great dignity are always something rare and striking, and it is the distinction of the English aristocracy, in the eighteenth century, that not only was their culture something rare by comparison with the rawness of the masses, it was something rare and admirable in itself. It is rather that this rare culture of the highest class has actually somewhat declined,[1] than that it has come to look less by juxtaposition with the augmented culture of other classes.

Probably democracy has something to answer for in this falling off of her rival. To feel itself raised on high, venerated, followed, no doubt stimulates a fine nature to keep itself worthy to be followed, venerated, raised on high; hence that lofty maxim, *noblesse oblige*. To feel its culture something precious and singular, makes such a nature zealous to retain and extend it. The elation and energy thus fostered by the sense of its advantages, certainly enhances the worth, strengthens the behaviour, and quickens all the active powers of the class enjoying it. *Possunt quia posse videntur.°* The removal of the stimulus a little relaxes their energy. It is not so much that they sink to be somewhat less than themselves, as that they cease to be somewhat more than themselves. But, however this may be, whencesoever the change may proceed, I cannot doubt that in the aristocratic virtue, in the intrinsic commanding force of the English upper class, there is a diminution. Relics of a great generation are still, perhaps, to be seen amongst them, surviving exemplars of noble manners and consummate culture; but they disappear one after the other, and no one of their kind takes their place. At the very moment when democracy becomes less and less disposed to follow and to admire, aristocracy becomes less and less qualified to command and to captivate.

---

[1] This will appear doubtful to no one well acquainted with the literature and memoirs of the last century. To give but two illustrations out of a thousand. Let the reader refer to the anecdote told by Robert Wood in his *Essay on the Genius of Homer* (London, 1775), p. vii, and to Lord Chesterfield's *Letters°* (edit. of 1845), vol. i. pp. 115, 143; vol. ii. p. 54; and then say, whether the culture there indicated as the culture of a *class* has maintained itself at that level.

On the one hand, then, the masses of the people in this country are preparing to take a much more active part than formerly in controlling its destinies; on the other hand, the aristocracy (using this word in the widest sense, to include not only the nobility and landed gentry, but also those reinforcements from the classes bordering upon itself, which this class constantly attracts and assimilates), while it is threatened with losing its hold on the rudder of government, its power to give to public affairs its own bias and direction, is losing also that influence on the spirit and character of the people which it long exercised.

I know that this will be warmly denied by some persons. Those who have grown up amidst a certain state of things, those whose habits, and interests, and affections, are closely concerned with its continuance, are slow to believe that it is not a part of the order of nature, or that it can ever come to an end. But I think that what I have here laid down will not appear doubtful either to the most competent and friendly foreign observers of this country, or to those Englishmen who, clear of all influences of class or party, have applied themselves steadily to see the tendencies of their nation as they really are. Assuming it to be true, a great number of considerations are suggested by it; but it is my purpose here to insist upon one only.

That one consideration is: On what action may we rely to replace, for some time at any rate, that action of the aristocracy upon the people of this country, which we have seen exercise an influence in many respects elevating and beneficial, but which is rapidly, and from inevitable causes, ceasing? In other words, and to use a short and significant modern expression which every one understands, what influence may help us to prevent the English people from becoming, with the growth of democracy, *Americanised*? I confess I am disposed to answer: On the action of the State.

I know what a chorus of objectors will be ready. One will say: Rather repair and restore the influence of aristocracy. Another will say: It is not a bad thing, but a good thing, that the English people should be Americanised. But the most formidable and the most widely entertained objection, by far, will be that which founds itself upon the present actual state of things in another country; which says: Look at France!° there you have a signal example of the alliance of democracy with a powerful State-action, and see how it works.

This last and principal objection I will notice at once. I have had occasion to touch upon the first already, and upon the second I shall touch presently. It seems to me, then, that one may save one's self from much idle terror at names and shadows if one will be at the pains to remember what different conditions the different character of two nations must necessarily impose on the operation of any principle. That which operates noxiously in one, may operate wholesomely in the other; because the unsound part of the one's character may be yet further inflamed and enlarged by it, the unsound part of the other's may find in it a corrective and an abatement. This is the great use which two unlike characters may find in observing each other. Neither is likely to have the other's faults, so each may safely adopt as much as suits him of the other's qualities. If I were a Frenchman I should never be weary of admiring the independent, individual, local habits of action in England, of directing attention to the evils occasioned in France by the excessive action of the State; for I should be very sure that, say what I might, the part of the State would never be too small in France, nor that of the individual too large. Being an Englishman, I see nothing but good in freely recognising the coherence, rationality, and efficaciousness which characterise the strong State-action of France, of acknowledging the want of method, reason, and result which attend the feeble State-action of England; because I am very sure that, strengthen in England the action of the State as one may, it will always find itself sufficiently controlled. But when either the *Constitutionnel* sneers at the do-little talkativeness of parliamentary government, or when the *Morning Star* inveighs against the despotism of a centralised administration, it seems to me that they lose their labour, because they are hardening themselves against dangers to which they are neither of them liable. Both the one and the other, in plain truth,

> 'Compound for sins they are inclined to,
> By damning those they have no mind to.'⁰

They should rather exchange doctrines one with the other, and each might thus, perhaps, be profited.

So that the exaggeration of the action of the State, in France, furnishes no reason for absolutely refusing to enlarge the action of the State in England; because the genius and temper of the people of this country are such as to render impossible that exaggeration which the genius and temper of the French rendered easy. There is no danger at all that the native independence and individualism of the English character will ever belie itself, and become either weakly prone to lean on others, or blindly confiding in them.

English democracy runs no risk of being overmastered by the State; it is almost certain that it will throw off the tutelage of aristocracy. Its real danger is, that it will have far too much its own way, and be left far too much to itself. 'What harm will there be in that?' say some; 'are we not a self-governing people?' I answer: 'We have never yet been a *self-governing democracy*, or anything like it.' The difficulty for democracy is, how to find and keep high ideals. The individuals who compose it are, the bulk of them, persons who need to follow an ideal, not to set one; and one ideal of greatness, high feeling, and fine culture, which an aristocracy once supplied to them, they lose by the very fact of ceasing to be a lower order and becoming a democracy. Nations are not truly great solely because the individuals composing them are numerous, free, and active; but they are great when these numbers, this freedom, and this activity are employed in the service of an ideal higher than that of an ordinary man, taken by himself. Our society is probably destined to become much more democratic; who or what will give a high tone to the nation then? That is the grave question.

The greatest men of America, her Washingtons, Hamiltons, Madisons, well understanding that aristocratical institutions are not in all times and places possible; well perceiving that in their Republic there was no place for these; comprehending, therefore, that from these that security for national dignity and greatness, an ideal commanding popular reverence, was not to be obtained, but knowing that this ideal was indispensable, would have been rejoiced to found a substitute for it in the dignity and authority of the State. They deplored the weakness and insignificance of the executive power as a calamity. When the inevitable course of events has made our self-government something really like that of America, when it has removed or weakened that security for national dignity, which we possessed in *aristocracy*, will the substitute of the *State* be equally wanting to us? If it is, then the dangers of America will really be ours; the dangers which come from the multitude being in power, with no adequate ideal to elevate or guide the multitude.

It would really be wasting time to contend at length, that to give more prominence to the idea of the State is now possible in this country, without endangering liberty. In other countries the habits and dispositions of the people may be such that the State, if once it acts, may be easily suffered to usurp exorbitantly; here they certainly are not. Here the people will always sufficiently keep in mind that any public authority is a trust delegated by themselves, for certain purposes, and with certain limits; and if that authority pretends to an absolute, independent character, they will soon enough (and very rightly) remind it of its error. Here there can be no question of a paternal government, of an irresponsible executive power, professing to act for the people's good, but without the people's consent, and, if necessary, against the people's wishes; here no one dreams of removing a single constitutional control, of abolishing a single safeguard for securing a correspondence between the acts of government and the will of the nation. The question is, whether, retaining all its power of control over a government which should abuse its trust, the nation may not now find advantage in voluntarily allowing to it purposes somewhat ampler, and limits somewhat wider within which to execute them, than formerly; whether the nation may not thus acquire in the State an ideal of high reason and right feeling, representing its best self, commanding general respect, and forming a rallying-point for the intelligence and for the worthiest instincts of the community, which will herein find a true bond of union.

I am convinced that if the worst mischiefs of democracy ever happen in England, it will be, not because a new condition of things has come upon us unforeseen, but because, though we all foresaw it, our efforts to deal with it were in the wrong direction. At the present time, almost every one believes in the growth of democracy, almost every one talks of it, almost every one laments it; but the last thing people can be brought to do is to make timely preparation for it. Many of those who, if they would, could do most to forward this work of preparation, are made slack and hesitating by the belief that, after all, in England, things may probably never go very far; that it will be possible to keep much more of the past than speculators say. Others, with a more robust faith, think that all democracy wants is vigorous putting-down; and that, with a good will and strong hand, it is perfectly possible to retain or restore the whole system of the Middle Ages. Others, free from the prejudices of class and position which warp the judgment of these, and who would, I believe, be the first and greatest gainers by strengthening the hands of the State, are averse from doing so by reason of suspicions and fears, once perfectly well-grounded, but, in this age and in the present circumstances, well-grounded no longer.

I speak of the middle classes. I have already shown how it is the natural disposition of an aristocratical class to view with jealousy the development of a considerable State-power. But this disposition has in England found extraordinary favour and support in regions not aristocratical,— from the middle classes; and, above all, from the kernel of these classes, the Protestant Dissenters. And for a very good reason. In times when passions ran high, even an aristocratical executive was easily stimulated into using, for the gratification of its friends and the abasement of its enemies, those administrative engines which, the moment it chose to stretch its hand forth, stood ready for its grasp. Matters of domestic concern, matters of religious profession and religious exercise, offered a peculiar field for an intervention gainful and agreeable to friends, injurious and irritating to enemies. Such an intervention was attempted and practised. Government lent its machinery and authority to the aristocratical and ecclesiastical party, which it regarded as its best support. The party which suffered comprised the flower and strength of that middle class of society, always very flourishing and robust in this country. That powerful class, from this specimen of the administrative activity of government, conceived a strong antipathy against all intervention of the State in certain spheres. An active, stringent administration in those spheres, meant at that time a High Church and Prelatic administration in them, an administration galling to the Puritan party and to the middle class; and this aggrieved class had naturally no proneness to draw nice philosophical distinctions between State-action in these spheres, as a thing for abstract consideration, and State-action in them as they practically felt it and supposed themselves likely long to feel it, guided by their adversaries. In the minds of the English middle class, therefore, State-action in social and domestic concerns became inextricably associated with the idea of a Conventicle Act, a Five-Mile Act, an Act of Uniformity.° Their abhorrence of such a State-action as this they extended to State-action in general; and, having never known a beneficent and just State-power, they enlarged their hatred of a cruel and partial State-power, the only one they had ever known, into a maxim that no State-power was to be trusted, that the least action, in certain provinces, was rigorously to be denied to the State, whenever this denial was possible.

Thus that jealousy of an important, sedulous, energetic executive, natural to grandees unwilling to suffer their personal authority to be circumscribed, their individual grandeur to be eclipsed, by the authority and grandeur of the State, became reinforced in this country by a like sentiment among the middle classes, who had no such authority or grandeur to lose, but who, by a hasty reasoning, had theoretically condemned for ever an agency which they had practically found at times oppressive. *Leave us to ourselves!* magnates and middle classes alike cried to the State. Not only from those who were full and abounded° went up this prayer, but also from those whose condition admitted of great amelioration. Not only did the whole repudiate the physician, but also those who were sick.°

For it is evident, that the action of a diligent, an impartial, and a national government, while it can do little to better the condition, already fortunate enough, of the highest and richest class of its people, can really do much, by institution and regulation, to better that of the middle and lower classes. The State can bestow certain broad collective benefits, which are indeed not much if compared with the advantages already possessed by individual grandeur, but which are rich and valuable if compared with the make-shifts of mediocrity and poverty. A good thing meant for the many cannot well be so exquisite as the good things of the few; but it can easily, if it comes from a donor of great resources and wide power, be incomparably better than what the many could, unaided, provide for themselves.

In all the remarks which I have been making, I have hitherto abstained from any attempt to suggest a positive application of them. I have limited myself to simply pointing out in how changed a world of ideas we are living; I have not sought to go further, and to discuss in what particular manner the world of facts is to adapt itself to this changed world of ideas. This has been my rule so far; but from this rule I shall here venture to depart, in order to dwell for a moment on a matter of practical institution, designed to meet new social exigencies: on the intervention of the State in public education.

The public secondary schools of France, decreed by the Revolution and established under the Consulate, are said by many good judges to be inferior to the old colleges.° By means of the old colleges and of private tutors, the French aristocracy could procure for its children (so it is said, and very likely with truth) a better training than that which is now given in the lyceums. Yes; but the boon conferred by the State, when it founded the lyceums, was not for the aristocracy; it was for the vast middle class of Frenchmen. This class, certainly, had not already the means of a better training for its children, before the State interfered. This class, certainly, would not have succeeded in procuring by its own efforts a better training for its children, if the State had not interfered. Through the intervention of the State this class enjoys better schools for its children, not than the great and rich enjoy (that is not the question), but than the same class enjoys in any country where the State has not interfered to found them. The lyceums may not be so good as Eton or Harrow; but they are a great deal better than a *Classical and Commercial Academy*.°

The aristocratic classes in England may, perhaps, be well content to rest satisfied with their Eton and Harrow. The State is not likely to do better for them. Nay, the superior confidence, spirit, and style, engendered by a training in the great public schools, constitute for these classes a real privilege, a real engine of command, which they might, if they were selfish, be sorry to lose by the establishment of schools great enough to beget a like spirit in the classes below them. But the middle classes in England have every reason not to rest content with their private schools; the State can do a great deal better for them. By giving to schools for these classes a public character, it can bring the instruction in them under a criticism which the stock of knowledge and judgment in our middle classes is not of itself at present able to supply. By giving to them a national character, it can confer on them a greatness and a noble spirit, which the tone of these classes is not of itself at present adequate to impart. Such schools would soon prove notable competitors with the existing public schools; they would do these a great service by stimulating them, and making them look into their own weak points more closely. Economical, because with charges uniform and under severe revision, they would do a great service to that large body of persons who, at present, seeing that on the whole the best secondary instruction to be found is that of the existing public schools, obtain it for their children from a sense of duty, although they can ill afford it, and although its cost is certainly exorbitant. Thus the middle classes might, by the aid of the State, better their instruction, while still keeping its cost moderate. This in itself would be a gain; but this gain would be slight in comparison with that of acquiring the sense of belonging to great and honourable seats of learning, and of breathing in their youth the air of the best culture of their nation. This sense would be an educational influence for them of the highest value. It would really augment their self-respect and moral force; it would truly fuse them with the class above, and tend to bring about for them the equality which they are entitled to desire.

So it is not State-action in itself which the middle and lower classes of a nation ought to deprecate; it is State-action exercised by a hostile class, and for their oppression. From a State-action reasonably, equitably, and nationally exercised, they may derive great benefit; greater, by the very nature and necessity of things, than can be derived from this source by the class above them. For the middle or lower classes to obstruct such a State-action, to repel its benefits, is to play the game of their enemies, and to prolong for themselves a condition of real inferiority.

This, I know, is rather dangerous ground to tread upon. The great middle classes of this country are conscious of no weakness, no inferiority; they do not want any one to provide anything for them. Such as they are, they believe that the freedom and prosperity of England are their work, and that the future belongs to them. No one esteems them more than I do; but those who esteem them most, and who most believe in their capabilities, can render them no better service than by pointing out in what they underrate their deficiencies, and how their deficiencies, if unremedied, may impair their future. They want culture and dignity; they want ideas. Aristocracy has culture and dignity; democracy has readiness for new ideas, and ardour for what ideas it possesses. Of these, our middle class has the last only: ardour for the ideas it already possesses. It believes ardently in liberty, it believes ardently in industry; and, by its zealous belief in these two ideas, it has accomplished great things. What it has accomplished by its belief in industry is patent to all the world. The liberties of England are less its exclusive work than it supposes; for these, aristocracy has achieved nearly as much. Still, of one inestimable part of liberty, liberty of thought, the middle class has been (without precisely intending it) the principal champion. The intellectual action of the Church of England upon the nation has been insignificant; its social action has been great. The social action of Protestant Dissent, that genuine product of the English middle class, has not been civilising; its positive intellectual action has been insignificant; its negative intellectual action,—in so far as by strenuously maintaining for itself, against persecution, liberty of conscience and the right of free opinion, it at the same time maintained and established this right as a universal principle,—has been invaluable. But the actual results of this negative intellectual service rendered by Protestant Dissent,—by the middle class,—to the whole community, great as they undoubtedly are, must not be taken for something which they are not. It is a very great thing to be able to think as you like; but, after all, an important question remains: *what* you think. It is a fine thing to secure a free stage and no favour; but, after all, the part which you play on that stage will have to be criticised. Now, all the liberty and industry in the world will not ensure these two things: a high reason and a fine culture. They may favour them, but they will not of themselves produce them; they may exist without them. But it is by the appearance of these two things, in some shape or other, in the life of a nation, that it becomes something more than an independent, an energetic, a successful nation,—that it becomes a *great* nation.

In modern epochs the part of a high reason, of ideas, acquires constantly increasing importance in the conduct of the world's affairs. A fine culture is the complement of a high reason, and it is in the conjunction of both with character, with energy, that the ideal for men and nations is to be placed. It is common to hear remarks on the frequent divorce between culture and character, and to infer from this that culture is a mere varnish, and that character only deserves any serious attention. No error can be more fatal. Culture without character is, no doubt, something frivolous, vain, and weak; but character without culture is, on the other hand, something raw, blind, and dangerous. The most interesting, the most truly glorious peoples, are those in which the alliance of the two has been effected most successfully, and its result spread most widely. This is why the spectacle of ancient Athens has such profound interest for a rational man; that it is the spectacle of the culture of a *people*. It is not an aristocracy, leavening with its own high spirit the multitude which it wields, but leaving it the unformed multitude still; it is not a democracy, acute and energetic, but tasteless, narrow-minded, and ignoble; it is the middle and lower classes in the highest development of their humanity that these classes have yet reached. It was the *many* who relished those arts, who were not satisfied with less than those monuments. In the conversations recorded by Plato, or even by the matter-of-fact Xenophon, which for the free yet refined discussion of ideas have set the tone for the whole cultivated world, shopkeepers and tradesmen of Athens mingle as speakers. For any one but a pedant, this is why a handful of Athenians of two thousand years ago are more interesting than the millions of most nations our contemporaries. Surely, if they knew this, those friends of progress, who have confidently pronounced the remains of the ancient world to be so much lumber, and a classical education an aristocratic impertin-ence, might be inclined to reconsider their sentence.

The course taken in the next fifty years by the middle classes of this nation will probably give a decisive turn to its history. If they will not seek the alliance of the State for their own elevation, if they go on exaggerating their spirit of individualism, if they persist in their jealousy of all governmental action, if they cannot learn that the antipathies and the shibboleths of a past age are now an anachronism for them—that will not prevent them, probably, from getting the rule of their country for a season, but they will certainly *Americanise* it. They will rule it by their energy, but they will deteriorate it by their low ideals and want of culture. In the decline of the aristocratical element, which in some sort supplied an ideal to ennoble the spirit of the nation and to keep it together, there will be no other element present to perform this service. It is of itself a serious calamity for a nation that its tone of feeling and grandeur of spirit should be lowered or dulled. But the calamity appears far more serious still when we consider that the middle classes, remaining as they are now, with their narrow, harsh, unintelligent, and unattractive spirit and culture, will almost certainly fail to mould or assimilate the masses below them, whose sympathies are at the present moment actually wider and more liberal than theirs. They arrive, these masses, eager to enter into possession of the world, to gain a more vivid sense of their own life and activity. In this their irrepressible development, their natural educators and initiators are those immediately above them, the middle classes. If these classes cannot win their sympathy or give them their direction, society is in danger of falling into anarchy.

Therefore, with all the force I can, I wish to urge upon the middle classes of this country, both that they might be very greatly profited by the action of the State, and also that they are continuing their opposition to such action out of an unfounded fear. But at the same time I say that the middle classes have the right, in admitting the action of government, to make the condition that this government shall be one of their own adoption, one that they can trust. To ensure this is now in their own power. If they do not as yet ensure this, they ought to do so, they have the means of doing so. Two centuries ago they had not; now they have. Having this security, let them now show themselves jealous to keep the action of the State equitable and rational, rather than to exclude the action of the State altogether. If the State acts amiss, let them check it, but let them no longer take it for granted that the State cannot possibly act usefully.

The State—but what is *the State?* cry many. Speculations on the idea of a State abound, but these do not satisfy them; of that which is to have practical effect and power they require a plain account. The full force of the term, *the State*, as the full force of any other important term, no one will master without going a little deeply, without resolutely entering the world of ideas; but it is possible to give in very plain language an account of it sufficient for all practical purposes. The State is properly just what Burke called it—*the nation in its collective and corporate character.*° The State is the representative acting-power of the nation; the action of the State is the representative action of the nation. Nominally emanating from the Crown, as the ideal unity in which the nation concentrates itself, this action, by the constitution of our country, really emanates from the ministers of the Crown. It is common to hear the depreciators of State-action run through a string of ministers' names, and then say: 'Here is really your *State*; would you accept the action of these men as your own representative action? In what respect is their judgment on national affairs likely to be any better than that of the rest of the world?' In the first place I answer: Even supposing them to be originally no better or wiser than the rest of the world, they have two great advantages from their position: access to almost boundless means of information, and the enlargement of mind which the habit of dealing with great affairs tends to produce. Their position itself, therefore, if they are men of only average honesty and capacity, tends to give them a fitness for acting on behalf of the nation superior to that of other men of equal honesty and capacity who are not in the same position. This fitness may be yet further increased by treating them as persons on whom, indeed, a very grave responsibility has fallen, and from whom very much will be expected;—nothing less than the representing, each of them in his own department, under the control of Parliament, and aided by the suggestions of public opinion, the collective energy and intelligence of his nation. By treating them as men on whom all this devolves to do, to their honour if they do it well, to their shame if they do it ill, one probably augments their faculty of well-doing; as it is excellently said: 'To treat men as if they were better than they are, is the surest way to *make* them better than they are.' But to treat them as if they had been shuffled into their places by a lucky accident, were most likely soon to be shuffled out of them again, and meanwhile ought to magnify themselves and their office as little as possible; to treat them as if they and their functions could without much inconvenience be quite dispensed with, and they ought perpetually to be admiring their own inconceivable good fortune in being permitted to discharge them;—this is the way to paralyse all high effort in the executive government, to extinguish all lofty sense of responsibility; to make its members either merely solicitous for the gross advantages, the emolument and self-importance, which they derive from their offices, or else timid, apologetic, and self-mistrustful in filling them; in either case, formal and inefficient.

But in the second place I answer: If the executive government is really in the hands of men no wiser than the bulk of mankind, of men whose action an intelligent man would be unwilling to accept as representative of his own action, whose fault is that? It is the fault of the nation itself, which, not being in the hands of a despot or an oligarchy, being free to control the choice of those who are to sum up and concentrate its action, controls it in such a manner that it allows to be chosen agents so little in its confidence, or so mediocre, or so incompetent, that it thinks the best thing to be done with them is to reduce their action as near as possible to a nullity. Hesitating, blundering, unintelligent, inefficacious, the action of the State may be; but, such as it is, it is the collective action of the nation itself, and the nation is responsible for it. It is our own action which we suffer to be thus unsatisfactory. Nothing can free us from this responsibility. The conduct of our affairs is in our own power. To carry on into its executive proceedings the indecision, conflict, and discordance of its parliamentary debates, may be a natural defect of a free nation, but it is certainly a defect; it is a dangerous error to call it, as some do, a perfection. The want of concert, reason, and organisation in the State, is the want of concert, reason, and organisation in the collective nation.

Inasmuch, therefore, as collective action is more efficacious than isolated individual efforts, a nation having great and complicated matters to deal with must greatly gain by employing the action of the State. Only, the State-power which it employs should be a power which really represents its best self, and whose action its intelligence and justice can heartily avow and adopt; not a power which reflects its inferior self, and of whose action, as of its own second-rate action, it has perpetually to be ashamed. To offer a worthy initiative, and to set a standard of rational and equitable action,—this is what the nation should expect of the State; and the more the State fulfils this expectation, the more will it be accepted in practice for what in idea it must always be. People will not then ask the State, what title it has to commend or reward genius and merit, since commendation and reward imply an attitude of superiority, for it will then be felt that the State truly acts for the English nation; and the genius of the English nation is greater than the genius of any individual, greater even than Shakspeare's genius, for it includes the genius of Newton also.

I will not deny that to give a more prominent part to the State would be a considerable change in this country; that maxims once very sound, and habits once very salutary, may be appealed to against it. The sole question is, whether those maxims and habits are sound and salutary at this moment. A yet graver and more difficult change,—to reduce the all-effacing prominence of the State, to give a more prominent part to the individual,—is imperiously presenting itself to other countries. Both are the suggestions of one irresistible force, which is gradually making its way everywhere, removing old conditions and imposing new, altering long-fixed habits, undermining venerable institutions, even modifying national character: *the modern spirit.*

Undoubtedly we are drawing on towards great changes; and for every nation the thing most needful is to discern clearly its own condition, in order to know in what particular way it may best meet them. Openness and flexibility of mind are at such a time the first of virtues. *Be ye perfect,*° said the Founder of Christianity; *I count not myself to have apprehended,*° said its greatest Apostle. Perfection will never be reached; but to recognise a period of transformation when it comes, and to adapt themselves honestly and rationally to its laws, is perhaps the nearest approach to perfection of which men and nations are capable. No habits or attachments should prevent their trying to do this; nor indeed, in the long run, can they. Human thought, which made all institutions, inevitably saps them, resting only in that which is absolute and eternal.

The Function of Criticism at the Present Time

MANY objections have been made to a proposition which, in some remarks of mine on translating Homer, I ventured to put forth; a proposition about criticism, and its importance at the present day. I said: 'Of the literature of France and Germany, as of the intellect of Europe in general, the main effort, for now many years, has been a critical effort; the endeavour, in all branches of knowledge, theology, philosophy, history, art, science, to see the object as in itself it really is.' I added, that owing to the operation in English literature of certain causes, 'almost the last thing for which one would come to English literature is just that very thing which now Europe most desires,—criticism';° and that the power and value of English literature was thereby impaired. More than one rejoinder declared that the importance I here assigned to criticism was excessive, and asserted the inherent superiority of the creative effort of the human spirit over its critical effort. And the other day, having been led by Mr Shairp's excellent notice° of Wordsworth[1] to turn again to his biography, I found, in the words of this great man, whom I, for one, must always listen to with the profoundest respect, a sentence passed on the critic's business, which seems to justify every possible disparagement of it. Wordsworth says in one of his letters:—

'The writers in these publications' (the Reviews), 'while they prosecute their inglorious employment, can not be supposed to be in a state of mind very favourable for being affected by the finer influences of a thing so pure as genuine poetry.'°

And a trustworthy reporter of his conversation quotes a more elaborate judgment to the same effect:—

'Wordsworth holds the critical power very low, infinitely lower than the inventive; and he said to-day that if the quantity of time consumed in writing critiques on the works of others were given to original composition, of whatever kind it might be, it would be much better employed; it would make a man find out sooner his own level, and it would do

---

[1] I cannot help thinking that a practice,° common in England during the last century, and still followed in France, of printing a notice of this kind,—a notice by a competent critic,—to serve as an introduction to an eminent author's works, might be revived among us with advantage. To introduce all succeeding editions of Wordsworth, Mr Shairp's notice might, it seems to me, excellently serve; it is written from the point of view of an admirer, nay, of a disciple, and that is right; but then the disciple must be also, as in this case he is, a critic, a man of letters, not, as too often happens, some relation or friend with no qualification for his task except affection for his author.°

infinitely less mischief. A false or malicious criticism may do much injury to the minds of others; a stupid invention, either in prose or verse, is quite harmless.'°

It is almost too much to expect of poor human nature, that a man capable of producing some effect in one line of literature, should, for the greater good of society, voluntarily doom himself to impotence and obscurity in another. Still less is this to be expected from men addicted to the composition of the 'false or malicious criticism' of which Words- worth speaks. However, everybody would admit that a false or malicious criticism had better never have been written. Everybody, too, would be willing to admit, as a general proposition, that the critical faculty is lower than the inventive. But is it true that criticism is really, in itself, a baneful and injurious employment; is it true that all time given to writing critiques on the works of others would be much better employed if it were given to original composition, of whatever kind this may be? Is it true that Johnson had better have gone on producing more *Irenes*° instead of writing his *Lives of the Poets*; nay, is it certain that Wordsworth himself was better employed in making his Ecclesiastical Sonnets than when he made his celebrated Preface,° so full of criticism, and criticism of the works of others? Wordsworth was himself a great critic, and it is to be sincerely regretted that he has not left us more criticism; Goethe was one of the greatest of critics, and we may sincerely congratulate ourselves that he has left us so much criticism. Without wasting time over the exaggeration which Wordsworth's judgment on criticism clearly con- tains, or over an attempt to trace the causes,—not difficult, I think, to be traced,—which may have led Wordsworth to this exaggeration, a critic may with advantage seize an occasion for trying his own conscience, and for asking himself of what real service at any given moment the practice of criticism either is or may be made to his own mind and spirit, and to the minds and spirits of others.

The critical power is of lower rank than the creative. True; but in assenting to this proposition, one or two things are to be kept in mind. It is undeniable that the exercise of a creative power, that a free creative activity, is the highest function of man; it is proved to be so by man's finding in it his true happiness. But it is undeniable, also, that men may have the sense of exercising this free creative activity in other ways than in producing great works of literature or art; if it were not so, all but a very few men would be shut out from the true happiness of all men. They may have it in well-doing, they may have it in learning, they may have it even in criticising. This is one thing to be kept in mind. Another is, that the exercise of the creative power in the production of great works of literature or art, however high this exercise of it may rank, is not at all epochs and under all conditions possible; and that therefore labour may be vainly spent in attempting it, which might with more fruit be used in preparing for it, in rendering it possible. This creative power works with elements, with materials; what if it has not those materials, those elements, ready for its use? In that case it must surely wait till they are ready. Now, in literature,—I will limit myself to literature, for it is about literature that the question arises,—the elements with which the creative power works are ideas; the best ideas, on every matter which literature touches, current at the time.° At any rate we may lay it down as certain that in modern literature no manifestation of the creative power not working with these can be very important or fruitful. And I say *current* at the time, not merely accessible at the time; for creative literary genius does not principally show itself in discovering new ideas, that is rather the business of the philosopher. The grand work of literary genius is a work of synthesis and exposition, not of analysis and discovery; its gift lies in the faculty of being happily inspired by a certain intellectual and spiritual atmosphere, by a certain order of ideas, when it finds itself in them; of dealing divinely with these ideas, presenting them in the most effective and attractive combinations,—making beautiful works with them, in short. But it must have the atmosphere, it must find itself amidst the order of ideas, in order to work freely; and these it is not so easy to command. This is why great creative epochs in literature are so rare, this is why there is so much that is unsatisfactory in the productions of many men of real genius; because for the creation of a master-work of literature two powers must concur, the power of the man and the power of the moment, and the man is not enough without the moment; the creative power has, for its happy exercise, appointed elements, and those elements are not in its own control.

Nay, they are more within the control of the critical power. It is the business of the critical power, as I said in the words already quoted, 'in all branches of knowledge, theology, philosophy, history, art, science, to see the object as in itself it really is'. Thus it tends, at last, to make an intellectual situation of which the creative power can profitably avail itself. It tends to establish an order of ideas, if not absolutely true, yet true by comparison with that which it displaces; to make the best ideas prevail. Presently these new ideas reach society, the touch of truth is the touch of life, and there is a stir and growth everywhere; out of this stir and growth come the creative epochs of literature.

Or, to narrow our range, and quit these considerations of the general march of genius and of society,—considerations which are apt to become too abstract and impalpable,—every one can see that a poet, for instance, ought to know life and the world before dealing with them in poetry; and life and the world being in modern times very complex things, the creation of a modern poet, to be worth much, implies a great critical effort behind it; else it must be a comparatively poor, barren, and short-lived affair. This is why Byron's poetry had so little endurance in it, and Goethe's so much; both Byron and Goethe had a great productive power, but Goethe's was nourished by a great critical effort providing the true materials for it, and Byron's was not; Goethe knew life and the world, the poet's necessary subjects, much more comprehensively and thoroughly than Byron. He knew a great deal more of them, and he knew them much more as they really are.

It has long seemed to me that the burst of creative activity° in our literature, through the first quarter of this century, had about it in fact something premature; and that from this cause its productions are doomed, most of them, in spite of the sanguine hopes which accompanied and do still accompany them, to prove hardly more lasting than the productions of far less splendid epochs. And this prematureness comes from its having proceeded without having its proper data, without sufficient materials to work with. In other words, the English poetry of the first quarter of this century, with plenty of energy, plenty of creative force, did not know enough. This makes Byron so empty of matter, Shelley so incoherent, Wordsworth even, profound as he is, yet so wanting in completeness and variety. Wordsworth cared little for books, and disparaged Goethe. I admire Wordsworth, as he is, so much that I cannot wish him different; and it is vain, no doubt, to imagine such a man different from what he is, to suppose that he *could* have been different. But surely the one thing wanting to make Wordsworth an even greater poet than he is,—his thought richer, and his influence of wider application,—was that he should have read more books, among them, no doubt, those of that Goethe whom he disparaged without reading him.

But to speak of books and reading may easily lead to a misunderstanding here. It was not really books and reading that lacked to our poetry at this epoch; Shelley had plenty of reading, Coleridge had immense reading. Pindar and Sophocles—as we all say so glibly, and often with so little discernment of the real import of what we are saying—had not many books; Shakspeare was no deep reader. True; but in the Greece of Pindar and Sophocles, in the England of Shakspeare, the poet lived in a current of ideas in the highest degree animating and nourishing to the creative power; society was, in the fullest measure, permeated by fresh thought, intelligent and alive. And this state of things is the true basis for the creative power's exercise, in this it finds its data, its materials, truly ready for its hand; all the books and reading in the world are only valuable as they are helps to this. Even when this does not actually exist, books and reading may enable a man to construct a kind of semblance of it in his own mind, a world of knowledge and intelligence in which he may live and work. This is by no means an equivalent to the artist for the nationally diffused life and thought of the epochs of Sophocles or Shakspeare; but, besides that it may be a means of preparation for such epochs, it does really constitute, if many share in it, a quickening and sustaining atmosphere of great value. Such an atmosphere the many-sided learning and the long and widely-combined critical effort of Germany formed for Goethe, when he lived and worked. There was no national glow of life and thought there as in the Athens of Pericles or the England of Elizabeth. That was the poet's weakness. But there was a sort of equivalent for it in the complete culture and unfettered thinking of a large body of Germans. That was his strength. In the England of the first quarter of this century there was neither a national glow of life and thought, such as we had in the age of Elizabeth, nor yet a culture and a force of learning and criticism such as were to be found in Germany. Therefore the creative power of poetry wanted, for success in the highest sense, materials and a basis; a thorough interpretation of the world was necessarily denied to it.

At first sight it seems strange that out of the immense stir of the French Revolution and its age should not have come a crop of works of genius equal to that which came out of the stir of the great productive time of Greece, or out of that of the Renascence, with its powerful episode the Reformation. But the truth is that the stir of the French Revolution took a character which essentially distinguished it from such movements as these. These were, in the main, disinterestedly intellectual and spiritual movements; movements in which the human spirit looked for its satisfaction in itself and in the increased play of its own activity. The French Revolution took a political, practical character. The movement which went on in France under the old *régime*, from 1700 to 1789, was far more really akin than that of the Revolution itself to the movement of the Renascence; the France of Voltaire and Rousseau told far more powerfully upon the mind of Europe than the France of the Revolution. Goethe reproached this last expressly with having 'thrown quiet culture back'.° Nay, and the true key to how much in our Byron, even in our Wordsworth, is this!—that they had their source in a great movement of feeling, not in a great movement of mind. The French Revolution, however,—that object of so much blind love and so much blind hatred,—found undoubtedly its motive-power in the intelligence of men, and not in their practical sense; this is what distinguishes it from the English Revolution of Charles the First's time. This is what makes it a more spiritual event than our Revolution, an event of much more powerful and world-wide interest, though practically less successful; it appeals to an order of ideas which are universal, certain, permanent. 1789 asked of a thing, Is it rational? 1642 asked of a thing, Is it legal? or, when it went furthest, Is it according to conscience? This is the English fashion, a fashion to be treated, within its own sphere, with the highest respect; for its success, within its own sphere, has been prodigious. But what is law in one place is not law in another; what is law here to-day is not law even here tomorrow; and as for conscience, what is binding on one man's conscience is not binding on another's. The old woman° who threw her stool at the head of the surpliced minister in St. Giles's Church at Edinburgh obeyed an impulse to which millions of the human race may be permitted to remain strangers. But the prescriptions of reason are absolute, unchanging, of universal validity; *to count by tens is the easiest way of counting*—that is a proposition of which every one, from here to the Antipodes, feels the force; at least I should say so if we did not live in a country where it is not impossible that any morning we may find a letter in the *Times* declaring that a decimal coinage° is an absurdity. That a whole nation should have been penetrated with an enthusiasm for pure reason, and with an ardent zeal for making its prescriptions triumph, is a very remarkable thing, when we consider how little of mind, or anything so worthy and quickening as mind, comes into the motives which alone, in general, impel great masses of men. In spite of the extravagant direction given to this enthusiasm, in spite of the crimes and follies in which it lost itself, the French Revolution derives from the force, truth, and universality of the ideas which it took for its law, and from the passion with which it could inspire a multitude for these ideas, a unique and still living power; it is—it will probably long remain—the greatest, the most animating event in history. And as no sincere passion for the things of the mind, even though it turn out in many respects an unfortunate passion, is ever quite thrown away and quite barren of good, France has reaped from hers one fruit—the natural and legitimate fruit, though not precisely the grand fruit she expected: she is the country in Europe where *the people* is most alive.

But the mania for giving an immediate political and practical application to all these fine ideas of the reason was fatal. Here an Englishman is in his element: on this theme we can all go on for hours. And all we are in the habit of saying on it has undoubtedly a great deal of truth. Ideas cannot be too much prized in and for themselves, cannot be too much lived with; but to transport them abruptly into the world of politics and practice, violently to revolutionise this world to their bidding,—that is quite another thing. There is the world of ideas and there is the world of practice; the French are often for suppressing the one and the English the other; but neither is to be suppressed. A member of the House of Commons said to me the other day: 'That a thing is an anomaly, I consider to be no objection to it whatever.' I venture to think he was wrong; that a thing is an anomaly *is* an objection to it, but absolutely and in the sphere of ideas: it is not necessarily, under such and such circumstances, or at such and such a moment, an objection to it in the sphere of politics and practice. Joubert has said beautifully: 'C'est la force et le droit qui règlent toutes choses dans le monde; la force en attendant le droit.'° (Force and right are the governors of this world; force till right is ready.) *Force till right is ready*; and till right is ready, force, the existing order of things, is justified, is the legitimate ruler. But right is something moral, and implies inward recognition, free assent of the will; we are not ready for right,—*right*, so far as we are concerned, *is not ready*,—until we have attained this sense of seeing it and willing it. The way in which for us it may change and transform force, the existing order of things, and become, in its turn, the legitimate ruler of the world, should depend on the way in which, when our time comes, we see it and will it. Therefore for other people enamoured of their own newly discerned right, to attempt to impose it upon us as ours, and violently to substitute their right for our force, is an act of tyranny, and to be resisted. It sets at nought the second great half of our maxim, *force till right is ready*. This was the grand error of the French Revolution; and its movement of ideas, by quitting the intellectual sphere and rushing furiously into the political sphere, ran, indeed, a prodigious and memorable course, but produced no such intellectual fruit as the movement of ideas of the Renascence, and created, in opposition to itself, what I may call an *epoch of concentration*. The great force of that epoch of concentration was England; and the great voice of that epoch of concentration was Burke. It is the fashion to treat Burke's writings on the French Revolution as superannuated and conquered by the event; as the eloquent but unphilosophical tirades of bigotry and prejudice. I will not deny that they are often disfigured by the violence and passion of the moment, and that in some directions Burke's view was bounded, and his observation therefore at fault. But on the whole, and for those who can make the needful corrections, what distinguishes these writings is their profound, permanent, fruitful, philosophical truth. They contain the true philosophy of an epoch of concentration, dissipate the heavy atmosphere which its own nature is apt to engender round it, and make its resistance rational instead of mechanical.

But Burke is so great because, almost alone in England, he brings thought to bear upon politics, he saturates politics with thought. It is his accident that his ideas were at the service of an epoch of concentration, not of an epoch of expansion; it is his characteristic that he so lived by ideas, and had such a source of them welling up within him, that he could float even an epoch of concentration and English Tory politics with them. It does not hurt him that Dr Price° and the Liberals were enraged with him; it does not even hurt him that George the Third and the Tories were enchanted with him. His greatness is that he lived in a world which neither English Liberalism nor English Toryism is apt to enter;—the world of ideas, not the world of catchwords and party habits. So far is it from being really true of him that he 'to party gave up what was meant for mankind',° that at the very end of his fierce struggle with the French Revolution, after all his invectives against its false pretensions, hollowness, and madness, with his sincere conviction of its mischievousness, he can close a memorandum on the best means of combating it, some of the last pages he ever wrote,°—the *Thoughts on French Affairs*, in December 1791,—with these striking words:—

'The evil is stated, in my opinion, as it exists. The remedy must be where power, wisdom, and information, I hope, are more united with good intentions than they can be with me. I have done with this subject, I believe, for ever. It has given me many anxious moments for the last two years. *If a great change is to be made in human affairs, the minds of men will be fitted to it; the general opinions and feelings will draw that way. Every fear, every hope will forward it; and then they who persist in opposing this mighty current in human affairs, will appear rather to resist the decrees of Providence itself, than the mere designs of men. They will not be resolute and firm, but perverse and obstinate.'*

That return of Burke upon himself has always seemed to me one of the finest things in English literature, or indeed in any literature. That is what I call living by ideas: when one side of a question has long had your earnest support, when all your feelings are engaged, when you hear all round you no language but one, when your party talks this language like a steam-engine and can imagine no other,—still to be able to think, still to be irresistibly carried, if so it be, by the current of thought to the opposite side of the question, and, like Balaam, to be unable to speak anything *but what the Lord has put in your mouth.*° I know nothing more striking, and I must add that I know nothing more un-English.

For the Englishman in general is like my friend the Member of Parliament, and believes, point-blank, that for a thing to be an anomaly is absolutely no objection to it whatever. He is like the Lord Auckland° of Burke's day, who, in a memorandum on the French Revolution, talks of 'certain miscreants, assuming the name of philosophers, who have presumed themselves capable of establishing a new system of society'. The Englishman has been called a political animal, and he values what is political and practical so much that ideas easily become objects of dislike in his eyes, and thinkers 'miscreants', because ideas and thinkers have rashly meddled with politics and practice. This would be all very well if the dislike and neglect confined themselves to ideas transported out of their own sphere, and meddling rashly with practice; but they are inevitably extended to ideas as such, and to the whole life of intelligence; practice is everything, a free play of the mind is nothing. The notion of the free play of the mind upon all subjects being a pleasure in itself, being an object of desire, being an essential provider of elements without which a nation's spirit, whatever compensations it may have for them, must, in the long run, die of inanition, hardly enters into an Englishman's thoughts. It is noticeable that the word *curiosity*, which in other languages is used in a good sense, to mean, as a high and fine quality of man's nature, just this disinterested love of a free play of the mind on all subjects, for its own sake,—it is noticeable, I say, that this word has in our language no sense of the kind, no sense but a rather bad and disparaging one. But criticism, real criticism, is essentially the exercise of this very quality. It obeys an instinct prompting it to try to know the best that is known and thought in the world, irrespectively of practice, politics, and everything of the kind; and to value knowledge and thought as they approach this best, without the intrusion of any other considerations whatever. This is an instinct for which there is, I think, little original sympathy in the practical English nature, and what there was of it has undergone a long benumbing period of blight and suppression in the epoch of concentration which followed the French Revolution.

But epochs of concentration cannot well endure for ever; epochs of expansion, in the due course of things, follow them. Such an epoch of expansion seems to be opening in this country. In the first place all danger of a hostile forcible pressure of foreign ideas upon our practice has long disappeared; like the traveller in the fable,° therefore, we begin to wear our cloak a little more loosely. Then, with a long peace,° the ideas of Europe steal gradually and amicably in, and mingle, though in infinitesimally small quantities at a time, with our own notions. Then, too, in spite of all that is said about the absorbing and brutalising

influence of our passionate material progress, it seems to me indisputable that this progress is likely, though not certain, to lead in the end to an
apparition of intellectual life; and that man, after he has made himself
perfectly comfortable and has now to determine what to do with himself
next, may begin to remember that he has a mind, and that the mind may
be made the source of great pleasure. I grant it is mainly the privilege of
faith, at present, to discern this end to our railways, our business, and our
fortune-making; but we shall see if, here as elsewhere, faith is not in the
end the true prophet. Our ease, our travelling, and our unbounded
liberty to hold just as hard and securely as we please to the practice to
which our notions have given birth, all tend to beget an inclination to deal
a little more freely with these notions themselves, to canvass them a little,
to penetrate a little into their real nature. Flutterings of curiosity, in the
foreign sense of the word, appear amongst us, and it is in these that
criticism must look to find its account. Criticism first; a time of true
creative activity, perhaps,—which, as I have said, must inevitably be
preceded amongst us by a time of criticism,—hereafter, when criticism
has done its work.

It is of the last importance that English criticism should clearly discern
what rule for its course, in order to avail itself of the field now opening to
it, and to produce fruit for the future, it ought to take. The rule may be
summed up in one word,—*disinterestedness*.° And how is criticism to show
disinterestedness? By keeping aloof from what is called 'the practical
view of things'; by resolutely following the law of its own nature, which is
to be a free play of the mind on all subjects which it touches. By steadily
refusing to lend itself to any of those ulterior, political, practical
considerations about ideas, which plenty of people will be sure to attach
to them, which perhaps ought often to be attached to them, which in this
country at any rate are certain to be attached to them quite sufficiently,
but which criticism has really nothing to do with. Its business is, as I have
said, simply to know the best that is known and thought in the world, and
by in its turn making this known, to create a current of true and fresh
ideas. Its business is to do this with inflexible honesty, with due ability;
but its business is to do no more, and to leave alone all questions of
practical consequences and applications, questions which will never fail
to have due prominence given to them. Else criticism, besides being
really false to its own nature, merely continues in the old rut which it has
hitherto followed in this country, and will certainly miss the chance now
given to it. For what is at present the bane of criticism in this country? It is
that practical considerations cling to it and stifle it. It subserves interests
not its own. Our organs of criticism are organs of men and parties having

practical ends to serve, and with them those practical ends are the first thing and the play of mind the second; so much play of mind as is compatible with the prosecution of those practical ends is all that is wanted. An organ like the *Revue des Deux Mondes*, having for its main function to understand and utter the best that is known and thought in the world, existing, it may be said, as just an organ for a free play of the mind, we have not. But we have the *Edinburgh Review*, existing as an organ of the old Whigs, and for as much play of the mind as may suit its being that; we have the *Quarterly Review*, existing as an organ of the Tories, and for as much play of mind as may suit its being that; we have the *British Quarterly Review*, existing as an organ of the political Dissenters, and for as much play of mind as may suit its being that; we have the *Times*, existing as an organ of the common, satisfied, well-to-do Englishman, and for as much play of mind as may suit its being that. And so on through all the various fractions, political and religious, of our society; every fraction has, as such, its organ of criticism, but the notion of combining all fractions in the common pleasure of a free disinterested play of mind meets with no favour. Directly this play of mind wants to have more scope, and to forget the pressure of practical considerations a little, it is checked, it is made to feel the chain. We saw this the other day in the extinction, so much to be regretted, of the *Home and Foreign Review*.° Perhaps in no organ of criticism in this country was there so much knowledge, so much play of mind; but these could not save it. The *Dublin Review* subordinates play of mind to the practical business of English and Irish Catholicism, and lives. It must needs be that men should act in sects and parties, that each of these sects and parties should have its organ, and should make this organ subserve the interests of its action; but it would be well, too, that there should be a criticism, not the minister of these interests, not their enemy, but absolutely and entirely independent of them. No other criticism will ever attain any real authority or make any real way towards its end,—the creating a current of true and fresh ideas.

It is because criticism has so little kept in the pure intellectual sphere, has so little detached itself from practice, has been so directly polemical and controversial, that it has so ill accomplished, in this country, its best spiritual work; which is to keep man from a self-satisfaction which is retarding and vulgarising, to lead him towards perfection, by making his mind dwell upon what is excellent in itself, and the absolute beauty and fitness of things. A polemical practical criticism makes men blind even to the ideal imperfection of their practice, makes them willingly assert its ideal perfection, in order the better to secure it against attack; and clearly this is narrowing and baneful for them. If they were reassured on the practical side, speculative considerations of ideal perfection they might be brought to entertain, and their spiritual horizon would thus gradually widen. Sir Charles Adderley° says to the Warwickshire farmers:—

'Talk of the improvement of breed! Why, the race we ourselves represent, the men and women, the old Anglo-Saxon race, are the best breed in the whole world. . . . The absence of a too enervating climate, too unclouded skies, and a too luxurious nature, has produced so vigorous a race of people, and has rendered us so superior to all the world.'

Mr. Roebuck° says to the Sheffield cutlers:—

'I look around me and ask what is the state of England? Is not property safe? Is not every man able to say what he likes? Can you not walk from one end of England to the other in perfect security? I ask you whether, the world over or in past history, there is anything like it? Nothing. I pray that our unrivalled happiness may last.'

Now obviously there is a peril for poor human nature in words and thoughts of such exuberant self-satisfaction, until we find ourselves safe in the streets of the Celestial City.

'Das wenige verschwindet leicht dem Blicke
Der vorwärts sieht, wie viel noch übrig bleibt—'°

says Goethe; 'the little that is done seems nothing when we look forward and see how much we have yet to do'. Clearly this is a better line of reflection for weak humanity, so long as it remains on this earthly field of labour and trial.

But neither Sir Charles Adderley nor Mr. Roebuck is by nature inaccessible to considerations of this sort. They only lose sight of them owing to the controversial life we all lead, and the practical form which all speculation takes with us. They have in view opponents whose aim is not ideal, but practical; and in their zeal to uphold their own practice against these innovators, they go so far as even to attribute to this practice an ideal perfection. Somebody has been wanting to introduce a six-pound franchise, or to abolish church-rates, or to collect agricultural statistics by force, or to diminish local self-government.° How natural, in reply to such proposals, very likely improper or ill-timed, to go a little beyond the mark, and to say stoutly, 'Such a race of people as we stand, so superior to all the world! The old Anglo-Saxon race, the best breed in the whole world! I pray that our unrivalled happiness may last! I ask you whether, the world over or in past history, there is anything like it?' And so long as criticism answers this dithyramb by insisting that the old Anglo-Saxon race would be still more superior to all others if it had no church-rates, or that our unrivalled happiness would last yet longer with a six-pound franchise, so long will the strain, 'The best breed in the whole world!' swell louder and louder, everything ideal and refining will be lost out of sight, and both the assailed and their critics will remain in a sphere, to say the truth, perfectly unvital, a sphere in which spiritual progression is impossible. But let criticism leave church-rates and the franchise alone, and in the most candid spirit, without a single lurking thought of practical innovation, confront with our dithyramb this paragraph on which I stumbled in a newspaper immediately after reading Mr Roebuck:—

'A shocking child murder has just been committed at Nottingham. A girl named Wragg left the workhouse there on Saturday morning with her young illegitimate child. The child was soon afterwards found dead on Mapperly Hills, having been strangled. Wragg is in custody.'°

Nothing but that; but, in juxtaposition with the absolute eulogies of Sir Charles Adderley and Mr Roebuck, how eloquent, how suggestive are those few lines! 'Our old Anglo-Saxon breed, the best in the whole world!'—how much that is harsh and ill-favoured there is in this best! *Wragg!* If we are to talk of ideal perfection, of 'the best in the whole world', has any one reflected what a touch of grossness in our race, what an original shortcoming in the more delicate spiritual perceptions, is shown by the natural growth amongst us of such hideous names,— Higginbottom, Stiggins, Bugg! In Ionia and Attica they were luckier in this respect than 'the best race in the world'; by the Ilissus° there was no Wragg, poor thing! And 'our unrivalled happiness';—what an element of grimness, bareness, and hideousness mixes with it and blurs it; the workhouse, the dismal Mapperly Hills,—how dismal those who have seen them will remember;—the gloom, the smoke, the cold, the strangled illegitimate child! 'I ask you whether, the world over or in past history, there is anything like it?' Perhaps not, one is inclined to answer; but at any rate, in that case, the world is [not] very much to be pitied. And the final touch,—short, bleak, and inhuman: *Wragg is in custody*. The sex lost in the confusion of our unrivalled happiness; or (shall I say?) the superfluous Christian name lopped off° by the straightforward vigour of our old Anglo-Saxon breed! There is profit for the spirit in such contrasts as this; criticism serves the cause of perfection by establishing them. By eluding sterile conflict, by refusing to remain in the sphere where alone narrow and relative conceptions have any worth and validity, criticism may diminish its momentary importance, but only in this way has it a chance of gaining admittance for those wider and more perfect conceptions to which all its duty is really owed. Mr Roebuck will have a poor opinion of an adversary who replies to his defiant songs of triumph only by murmuring under his breath, *Wragg is in custody*; but in no other way will these songs of triumph be induced gradually to moderate themselves, to get rid of what in them is excessive and offensive, and to fall into a softer and truer key.

It will be said that it is a very subtle and indirect action which I am thus prescribing for criticism, and that, by embracing in this manner the Indian virtue of detachment° and abandoning the sphere of practical life, it condemns itself to a slow and obscure work. Slow and obscure it may be, but it is the only proper work of criticism. The mass of mankind will never have any ardent zeal for seeing things as they are; very inadequate° ideas will always satisfy them. On these inadequate ideas reposes, and must repose, the general practice of the world. That is as much as saying that whoever sets himself to see things as they are will find himself one of a very small circle; but it is only by this small circle resolutely doing its own work that adequate ideas will ever get current at all. The rush and roar of practical life will always have a dizzying and attracting effect upon the most collected spectator, and tend to draw him into its vortex; most of all will this be the case where that life is so powerful as it is in England. But it is only by remaining collected, and refusing to lend himself to the point of view of the practical man, that the critic can do the practical man any service; and it is only by the greatest sincerity in pursuing his own course, and by at last convincing even the practical man of his sincerity, that he can escape misunderstandings which perpetually threaten him.

For the practical man is not apt for fine distinctions, and yet in these distinctions truth and the highest culture greatly find their account. But it is not easy to lead a practical man,—unless you reassure him as to your practical intentions, you have no chance of leading him,—to see that a thing which he has always been used to look at from one side only, which he greatly values, and which, looked at from that side, quite deserves, perhaps, all the prizing and admiring which he bestows upon it,—that this thing, looked at from another side, may appear much less beneficent and beautiful, and yet retain all its claims to our practical allegiance. Where shall we find language innocent enough, how shall we make the spotless purity of our intentions evident enough, to enable us to say to the political Englishman that the British Constitution itself, which, seen from the practical side, looks such a magnificent organ of progress and virtue, seen from the speculative side,—with its compromises, its love of facts, its horror of theory, its studied avoidance of clear thoughts,—that, seen from this side, our august Constitution sometimes looks,—forgive me, shade of Lord Somers!°—a colossal machine for the manufacture of

Philistines?° How is Cobbett° to say this and not be misunderstood, blackened as he is with the smoke of a lifelong conflict in the field of political practice? how is Mr Carlyle to say it and not be misunderstood, after his furious raid into this field with his *Latter-day Pamphlets*?° how is Mr Ruskin, after his pugnacious political economy? I say, the critic must keep out of the region of immediate practice in the political, social, humanitarian sphere, if he wants to make a beginning for that more free speculative treatment of things, which may perhaps one day make its benefits felt even in this sphere, but in a natural and thence irresistible manner.

Do what he will, however, the critic will still remain exposed to frequent misunderstandings, and nowhere so much as in this country. For here people are particularly indisposed even to comprehend that without this free disinterested treatment of things, truth and the highest culture are out of the question. So immersed are they in practical life, so accustomed to take all their notions from this life and its processes, that they are apt to think that truth and culture themselves can be reached by the processes of this life, and that it is an impertinent singularity to think of reaching them in any other. 'We are all *terrae filii*',° cries their eloquent advocate; 'all Philistines together. Away with the notion of proceeding by any other course than the course dear to the Philistines; let us have a social movement, let us organise and combine a party to pursue truth and new thought, let us call it *the liberal party*, and let us all stick to each other, and back each other up. Let us have no nonsense about independent criticism, and intellectual delicacy, and the few and the many. Don't let us trouble ourselves about foreign thought; we shall invent the whole thing for ourselves as we go along. If one of us speaks well, applaud him; if one of us speaks ill, applaud him too; we are all in the same movement, we are all liberals, we are all in pursuit of truth.' In this way the pursuit of truth becomes really a social, practical, pleasurable affair, almost requiring a chairman, a secretary, and advertisements; with the excitement of an occasional scandal, with a little resistance to give the happy sense of difficulty overcome; but, in general, plenty of bustle and very little thought. To act is so easy, as Goethe says;° to think is so hard! It is true that the critic has many temptations to go with the stream, to make one of the party of movement, one of these *terrae filii*; it seems ungracious to refuse to be a *terrae filius*, when so many excellent people are; but the critic's duty is to refuse, or, if resistance is vain, at least to cry with Obermann: *Périssons en résistant.*°

How serious a matter it is to try and resist, I had ample opportunity of experiencing when I ventured some time ago to criticise the celebrated first volume of Bishop Colenso.[1] The echoes of the storm which was then raised I still, from time to time, hear grumbling round me. That storm arose out of a misunderstanding almost inevitable. It is a result of no little culture to attain to a clear perception that science and religion are two wholly different things. The multitude will for ever confuse them; but happily that is of no great real importance, for while the multitude imagines itself to live by its false science, it does really live by its true religion. Dr Colenso, however, in his first volume did all he could to strengthen the confusion,[2] and to make it dangerous. He did this with the best intentions, I freely admit, and with the most candid ignorance that this was the natural effect of what he was doing; but, says Joubert,° 'Ignorance, which in matters of morals extenuates the crime, is itself, in intellectual matters, a crime of the first order.' I criticised Bishop Colenso's speculative confusion. Immediately there was a cry raised: 'What is this? here is a liberal attacking a liberal. Do not you belong to the movement? are not you a friend of truth? Is not Bishop Colenso in pursuit of truth? then speak with proper respect of his book. Dr Stanley° is another friend of truth, and you speak with proper respect of his book; why make these invidious differences? both books are excellent, admirable, liberal; Bishop Colenso's perhaps the most so, because it is the boldest, and will have the best practical consequences for the liberal cause. Do you want to encourage to the attack of a brother liberal his, and your, and our implacable enemies, the *Church and State Review* or the *Record*,—the High Church rhinoceros and the Evangelical hyaena? Be silent, therefore; or rather speak, speak as loud as ever you can! and go into ecstasies over the eighty and odd pigeons.'°

But criticism cannot follow this coarse and indiscriminate method. It is unfortunately possible for a man in pursuit of truth to write a book which reposes upon a false conception. Even the practical consequences of a book are to genuine criticism no recommendation of it, if the book is,

[1] So sincere is my dislike to all personal attack and controversy, that I abstain from reprinting, at this distance of time from the occasion which called them forth, the essays in which I criticised Dr Colenso's book;° I feel bound, however, after all that has passed, to make here a final declaration of my sincere impenitence for having published them. Nay, I cannot forbear repeating yet once more, for his benefit and that of his readers, this sentence from my original remarks upon him: *There is truth of science and truth of religion; truth of science does not become truth of religion till it is made religious.* And I will add: Let us have all the science there is from the men of science; from the men of religion let us have religion.

[2] It has been said° I make it 'a crime against literary criticism and the higher culture to attempt to inform the ignorant'. Need I point out that the ignorant are not informed by being confirmed in a confusion?

in the highest sense, blundering. I see that a lady° who herself, too, is in pursuit of truth, and who writes with great ability, but a little too much, perhaps, under the influence of the practical spirit of the English liberal movement, classes Bishop Colenso's book and M. Renan's° together, in her survey of the religious state of Europe, as facts of the same order, works, both of them, of 'great importance'; 'great ability, power, and skill'; Bishop Colenso's, perhaps, the most powerful; at least, Miss Cobbe gives special expression to her gratitude that to Bishop Colenso 'has been given the strength to grasp, and the courage to teach, truths of such deep import'. In the same way, more than one popular writer has compared him to Luther. Now it is just this kind of false estimate which the critical spirit is, it seems to me, bound to resist. It is really the strongest possible proof of the low ebb at which, in England, the critical spirit is, that while the critical hit in the religious literature of Germany is Dr Strauss's book,° in that of France M. Renan's book, the book of Bishop Colenso is the critical hit in the religious literature of England. Bishop Colenso's book reposes on a total misconception of the essential elements of the religious problem, as that problem is now presented for solution. To criticism, therefore, which seeks to have the best that is known and thought on this problem, it is, however well meant, of no importance whatever. M. Renan's book attempts a new synthesis of the elements furnished to us by the Four Gospels. It attempts, in my opinion, a synthesis, perhaps premature, perhaps impossible, certainly not successful. Up to the present time, at any rate, we must acquiesce in Fleury's sentence on such recastings of the Gospel-story: *Quiconque s'imagine la pouvoir mieux écrire, ne l'entend pas.*° M. Renan had himself passed by anticipation a like sentence on his own work, when he said: 'If a new presentation of the character of Jesus were offered to me, I would not have it; its very clearness would be, in my opinion, the best proof of its insufficiency.'° His friends may with perfect justice rejoin that at the sight of the Holy Land, and of the actual scene of the Gospel-story, all the current of M. Renan's thoughts may have naturally changed, and a new casting of that story irresistibly suggested itself to him; and that this is just a case for applying Cicero's maxim: Change of mind is not inconsistency—*nemo doctus unquam mutationem consilii inconstantiam dixit esse.*° Nevertheless, for criticism, M. Renan's first thought must still be the truer one, as long as his new casting so fails more fully to commend itself, more fully (to use Coleridge's happy phrase° about the Bible) to *find* us. Still M. Renan's attempt is, for criticism, of the most real interest and importance, since, with all its difficulty, a fresh synthesis of the New Testament *data*,—not a making war on them, in Voltaire's fashion, not a leaving them out of mind, in the world's fashion, but the putting a new construction upon them, the taking them from under the old, traditional, conventional point of view and placing them under a new one,—is the very essence of the religious problem, as now presented; and only by efforts in this direction can it receive a solution.

Again, in the same spirit in which she judges Bishop Colenso, Miss Cobbe, like so many earnest liberals of our practical race, both here and in America, herself sets vigorously about a positive reconstruction of religion, about making a religion of the future out of hand, or at least setting about making it. We must not rest, she and they are always thinking and saying, in negative criticism, we must be creative and constructive; hence we have such works as her recent *Religious Duty*,° and works still more considerable, perhaps, by others, which will be in every one's mind. These works often have much ability; they often spring out of sincere convictions, and a sincere wish to do good; and they sometimes, perhaps, do good. Their fault is (if I may be permitted to say so) one which they have in common with the British College of Health,° in the New Road. Every one knows the British College of Health; it is that building with the lion and the statue of the Goddess Hygeia before it; at least I am sure about the lion, though I am not absolutely certain about the Goddess Hygeia. This building does credit, perhaps, to the resources of Dr Morison and his disciples; but it falls a good deal short of one's idea of what a British College of Health ought to be. In England, where we hate public interference and love individual enterprise, we have a whole crop of places like the British College of Health; the grand name without the grand thing. Unluckily, creditable to individual enterprise as they are, they tend to impair our taste by making us forget what more grandiose, noble, or beautiful character properly belongs to a public institution. The same may be said of the religions of the future of Miss Cobbe and others. Creditable, like the British College of Health, to the resources of their authors, they yet tend to make us forget what more grandiose, noble, or beautiful character properly belongs to religious constructions. The historic religions, with all their faults, have had this; it certainly belongs to the religious sentiment, when it truly flowers, to have this; and we impoverish our spirit if we allow a religion of the future without it. What then is the duty of criticism here? To take the practical point of view, to applaud the liberal movement and all its works,—its New Road religions of the future into the bargain,—for their general utility's sake? By no means; but to be perpetually dissatisfied with these works, while they perpetually fall short of a high and perfect ideal.

For criticism, these are elementary laws; but they never can be popular, and in this country they have been very little followed, and one meets with immense obstacles in following them. That is a reason for asserting them again and again. Criticism must maintain its independence of the practical spirit and its aims. Even with well-meant efforts of the practical spirit it must express dissatisfaction, if in the sphere of the ideal they seem impoverishing and limiting. It must not hurry on to the goal because of its practical importance. It must be patient, and know how to wait; and flexible, and know how to attach itself to things and how to withdraw from them. It must be apt to study and praise elements that for the fulness of spiritual perfection are wanted, even though they belong to a power which in the practical sphere may be maleficent. It must be apt to discern the spiritual shortcomings or illusions of powers that in the practical sphere may be beneficent. And this without any notion of favouring or injuring, in the practical sphere, one power or the other; without any notion of playing off, in this sphere, one power against the other. When one looks, for instance, at the English Divorce Court,—an institution which perhaps has its practical conveniences, but which in the ideal sphere is so hideous; an institution which neither makes divorce impossible nor makes it decent, which allows a man to get rid of his wife, or a wife of her husband, but makes them drag one another first, for the public edification, through a mire of unutterable infamy,—when one looks at this charming institution, I say, with its crowded trials, its newspaper reports, and its money compensations, this institution in which the gross unregenerate British Philistine has indeed stamped an image of himself,—one may be permitted to find the marriage theory of Catholicism refreshing and elevating. Or when Protestantism, in virtue of its supposed rational and intellectual origin, gives the law to criticism too magisterially, criticism may and must remind it that its pretensions, in this respect, are illusive and do it harm; that the Reformation was a moral rather than an intellectual event; that Luther's theory of grace no more exactly reflects the mind of the spirit than Bossuet's° philosophy of history reflects it; and that there is no more antecedent probability of the Bishop of Durham's° stock of ideas being agreeable to perfect reason than of Pope Pius the Ninth's. But criticism will not on that account forget the achievements of Protestantism in the practical and moral sphere; nor that, even in the intellectual sphere, Protestantism, though in a blind and stumbling manner, carried forward the Renascence, while Catholicism threw itself violently across its path.

I lately heard a man of thought and energy contrasting the want of ardour and movement which he now found amongst young men in this country with what he remembered in his own youth, twenty years ago.

'What reformers we were then!' he exclaimed; 'what a zeal we had! how we canvassed every institution in Church and State, and were prepared to remodel them all on first principles!' He was inclined to regret, as a spiritual flagging, the lull which he saw. I am disposed rather to regard it as a pause in which the turn to a new mode of spiritual progress is being accomplished. Everything was long seen, by the young and ardent amongst us, in inseparable connection with politics and practical life. We have pretty well exhausted the benefits of seeing things in this connection, we have got all that can be got by so seeing them. Let us try a more disinterested mode of seeing them; let us betake ourselves more to the serener life of the mind and spirit. This life, too, may have its excesses and dangers; but they are not for us at present. Let us think of quietly enlarging our stock of true and fresh ideas, and not, as soon as we get an idea or half an idea, be running out with it into the street, and trying to make it rule there. Our ideas will, in the end, shape the world all the better for maturing a little. Perhaps in fifty years' time it will in the English House of Commons be an objection to an institution that it is an anomaly, and my friend the Member of Parliament will shudder in his grave. But let us in the meanwhile rather endeavour that in twenty years' time it may, in English literature, be an objection to a proposition that it is absurd. That will be a change so vast, that the imagination almost fails to grasp it. *Ab integro saeclorum nascitur ordo.*°

If I have insisted so much on the course which criticism must take where politics and religion are concerned, it is because, where these burning matters are in question, it is most likely to go astray. I have wished, above all, to insist on the attitude which criticism should adopt towards things in general; on its right tone and temper of mind. But then comes another question as to the subject-matter which literary criticism should most seek. Here, in general, its course is determined for it by the idea which is the law of its being; the idea of a disinterested endeavour to learn and propagate the best that is known and thought in the world, and thus to establish a current of fresh and true ideas. By the very nature of things, as England is not all the world, much of the best that is known and thought in the world cannot be of English growth, must be foreign; by the nature of things, again, it is just this that we are least likely to know, while English thought is streaming in upon us from all sides, and takes excellent care that we shall not be ignorant of its existence. The English critic of literature, therefore, must dwell much on foreign thought, and with particular heed on any part of it, which, while significant and fruitful in itself, is for any reason specially likely to escape him. Again, judging is often spoken of as the critic's one business, and so in some sense it is; but the judgment which almost insensibly forms itself in a fair and clear mind, along with fresh knowledge, is the valuable one; and thus knowledge, and ever fresh knowledge, must be the critic's great concern for himself. And it is by communicating fresh knowledge, and letting his own judgment pass along with it,—but insensibly, and in the second place, not the first, as a sort of companion and clue, not as an abstract lawgiver,—that the critic will generally do most good to his readers. Sometimes, no doubt, for the sake of establishing an author's place in literature, and his relation to a central standard (and if this is not done, how are we to get at our *best in the world*?) criticism may have to deal with a subject-matter so familiar that fresh knowledge is out of the question, and then it must be all judgment; an enunciation and detailed application of principles. Here the great safeguard is never to let oneself become abstract, always to retain an intimate and lively consciousness of the truth of what one is saying, and, the moment this fails us, to be sure that something is wrong. Still, under all circumstances, this mere judgment and application of principles is, in itself, not the most satisfactory work to the critic; like mathematics, it is tautological, and cannot well give us, like fresh learning, the sense of creative activity.

But stop, some one will say; all this talk is of no practical use to us whatever; this criticism of yours is not what we have in our minds when we speak of criticism; when we speak of critics and criticism, we mean critics and criticism of the current English literature of the day; when you offer to tell criticism its function, it is to this criticism that we expect you to address yourself. I am sorry for it, for I am afraid I must disappoint these expectations. I am bound by my own definition of criticism: *a disinterested endeavour to learn and propagate the best that is known and thought in the world*. How much of current English literature comes into this 'best that is known and thought in the world'? Not very much, I fear; certainly less, at this moment, than of the current literature of France or Germany. Well, then, am I to alter my definition of criticism, in order to meet the requirements of a number of practising English critics, who, after all, are free in their choice of a business? That would be making criticism lend itself just to one of those alien practical considerations, which, I have said, are so fatal to it. One may say, indeed, to those who have to deal with the mass—so much better disregarded—of current English literature, that they may at all events endeavour, in dealing with this, to try it, so far as they can, by the standard of the best that is known and thought in the world; one may say, that to get anywhere near this standard, every critic should try and possess one great literature, at least, besides his own; and the more unlike his own, the better. But, after all, the criticism I am really concerned with,—the criticism which alone can much help us for the future, the criticism which, throughout Europe, is at the present day meant, when so much stress is laid on the importance of criticism and the critical spirit,—is a criticism which regards Europe as being, for intellectual and spiritual purposes, one great confederation, bound to a joint action and working to a common result; and whose members have, for their proper outfit, a knowledge of Greek, Roman, and Eastern antiquity, and of one another. Special, local, and temporary advantages being put out of account, that modern nation will in the intellectual and spiritual sphere make most progress, which most thoroughly carries out this programme. And what is that but saying that we too, all of us, as individuals, the more thoroughly we carry it out, shall make the more progress?

There is so much inviting us!—what are we to take? what will nourish us in growth towards perfection? That is the question which, with the immense field of life of literature lying before him, the critic has to answer; for himself first, and afterwards for others. In this idea of the critic's business the essays brought together in the following pages have had their origin; in this idea, widely different as are their subjects, they have, perhaps, their unity.

I conclude with what I said at the beginning: to have the sense of creative activity is the great happiness and the great proof of being alive, and it is not denied to criticism to have it; but then criticism must be sincere, simple, flexible, ardent, ever widening its knowledge. Then it may have, in no contemptible measure, a joyful sense of creative activity; a sense which a man of insight and conscience will prefer to what he might derive from a poor, starved, fragmentary, inadequate creation. And at some epochs no other creation is possible.

Still, in full measure, the sense of creative activity belongs only to genuine creation; in literature we must never forget that. But what true man of letters ever can forget it? It is no such common matter for a gifted nature to come into possession of a current of true and living ideas, and to produce amidst the inspiration of them, that we are likely to underrate it. The epochs of Aeschylus and Shakspeare make us feel their pre-eminence. In an epoch like those is, no doubt, the true life of literature; there is the promised land, towards which criticism can only beckon. That promised land it will not be ours to enter, and we shall die in the wilderness:° but to have desired to enter it, to have saluted it from afar, is already, perhaps, the best distinction among contemporaries; it will certainly be the best title to esteem with posterity.

Preface to Essays in Criticism (*1865/1869*)

SEVERAL of the Essays which are here collected and reprinted had the good or the bad fortune to be much criticised at the time of their first appearance. I am not now going to inflict upon the reader a reply to those criticisms; for one or two explanations which are desirable, I shall elsewhere, perhaps, be able some day to find an opportunity; but, indeed, it is not in my nature,—some of my critics would rather say, not in my power,°—to dispute on behalf of any opinion, even my own, very obstinately. To try and approach truth on one side after another, not to strive or cry,° nor to persist in pressing forward, on any one side, with violence and self-will,—it is only thus, it seems to me, that mortals may hope to gain any vision of the mysterious Goddess, whom we shall never see except in outline, but only thus even in outline. He who will do nothing but fight impetuously towards her on his own, one, favourite, particular line, is inevitably destined to run his head into the folds of the black robe in which she is wrapped.

So it is not to reply to my critics that I write this preface, but to prevent a misunderstanding, of which certain phrases that some of them use make me apprehensive. Mr Wright,° one of the many translators of Homer, has published a letter to the Dean of Canterbury, complaining of some remarks of mine, uttered now a long while ago, on his version of the *Iliad*. One cannot be always studying one's own works, and I was really under the impression, till I saw Mr Wright's complaint, that I had spoken of him with all respect. The reader may judge of my astonishment, therefore, at finding, from Mr Wright's pamphlet, that I had 'declared with much solemnity that there is not any proper reason for his existing'. That I never said; but, on looking back at my Lectures on translating Homer, I find that I did say, not that Mr Wright, but that Mr Wright's version of the *Iliad*, repeating in the main the merits and defects of Cowper's version, as Mr Sotheby's repeated those of Pope's version, had, if I might be pardoned for saying so, no proper reason for existing.° Elsewhere I expressly spoke of the merit of his version; but I confess that the phrase, qualified as I have shown, about its want of a proper reason for existing, I used. Well, the phrase had, perhaps, too much vivacity; we have all of us a right to exist, we and our works; an unpopular author should be the last person to call in question this right. So I gladly withdraw the offending phrase, and I am sorry for having used it; Mr Wright, however, would perhaps be more indulgent to my vivacity, if he considered that we are none of us likely to be lively much longer. My vivacity is but the last sparkle of flame before we are all in the dark, the last glimpse of colour before we all go into drab,—the drab of the earnest, prosaic, practical, austerely literal future. Yes, the world will soon be the Philistines'! and then, with every voice, not of thunder, silenced, and the whole earth filled and ennobled every morning by the magnificent roaring of the young lions of the *Daily Telegraph*, we shall all yawn in one another's faces with the dismallest, the most unimpeachable gravity.

But I return to my design in writing this Preface. That design was, after apologising to Mr Wright for my vivacity of five years ago, to beg him and others to let me bear my own burdens, without saddling the great and famous University to which I have the honour to belong with any portion of them. What I mean to deprecate is such phrases as, 'his professorial assault', 'his assertions issued *ex cathedrâ*', 'the sanction of his name as the representative of poetry', and so on. Proud as I am of my connection with the University of Oxford,[1] I can truly say, that knowing how unpopular a task one is undertaking when one tries to pull out a few more stops in that powerful but at present somewhat narrow-toned organ, the modern Englishman, I have always sought to stand by myself, and to compromise others as little as possible. Besides this, my native modesty is such, that I have always been shy of assuming the honourable style of Professor, because this is a title I share with so many distinguished men,—Professor Pepper, Professor Anderson, Professor Frickel, and others,°—who adorn it, I feel, much more than I do.

However, it is not merely out of modesty that I prefer to stand alone, and to concentrate on myself, as a plain citizen of the republic of letters, and not as an office-bearer in a hierarchy, the whole responsibility for all I write; it is much more out of genuine devotion to the University of Oxford, for which I feel, and always must feel, the fondest, the most reverential attachment. In an epoch of dissolution and transformation, such as that on which we are now entered, habits, ties, and associations are inevitably broken up, the action of individuals becomes more distinct, the shortcomings, errors, heats, disputes, which necessarily attend individual action, are brought into greater prominence. Who would not gladly keep clear, from all these passing clouds, an august institution which was there before they arose, and which will be there when they have blown over?

It is true, the *Saturday Review* maintains that our epoch of transforma-

[1] When the above was written the author had still the Chair of Poetry at Oxford, which he has since vacated.

tion is finished; that we have found our philosophy; that the British nation has searched all anchorages for the spirit, and has finally anchored itself, in the fulness of perfected knowledge, on Benthamism. This idea at first made a great impression on me; not only because it is so consoling in itself, but also because it explained a phenomenon which in the summer of last year had, I confess, a good deal troubled me. At that time my avocations led me to travel almost daily on one of the Great Eastern Lines,—the Woodford Branch. Every one knows that the murderer, Müller,° perpetrated his detestable act on the North London Railway, close by. The English middle class, of which I am myself a feeble unit, travel on the Woodford Branch in large numbers. Well, the demoralisa- tion of our class,—the class which (the newspapers are constantly saying it, so I may repeat it without vanity) has done all the great things which have ever been done in England,—the demoralisation, I say, of our class, caused by the Bow tragedy, was something bewildering. Myself a transcendentalist (as the *Saturday Review* knows), I escaped the infec- tion; and, day after day, I used to ply my agitated fellow-travellers with all the consolations which my transcendentalism would naturally suggest to me. I reminded them how Caesar refused to take precautions against assassination, because life was not worth having at the price of an ignoble solicitude for it. I reminded them what insignificant atoms we all are in the life of the world. 'Suppose the worst to happen,' I said, addressing a portly jeweller from Cheapside;° 'suppose even yourself to be the victim; *il n'y a pas d'homme nécessaire.*° We should miss you for a day or two upon the Woodford Branch; but the great mundane movement would still go on, the gravel walks of your villa would still be rolled, dividends would still be paid at the Bank, omnibuses would still run, there would still be the old crush at the corner of Fenchurch Street.' All was of no avail. Nothing could moderate, in the bosom of the great English middle-class, their passionate, absorbing, almost blood-thirsty clinging to life. At the moment I thought this over-concern a little unworthy; but the *Saturday Review* suggests a touching explanation of it. What I took for the ignoble clinging to life of a comfortable worldling, was, perhaps, only the ardent longing of a faithful Benthamite, traversing an age still dimmed by the last mists of transcendentalism, to be spared long enough to see his religion in the full and final blaze of its triumph. This respectable man, whom I imagined to be going up to London to serve his shop, or to buy shares, or to attend an Exeter Hall° meeting, or to assist at the deliberations of the Marylebone Vestry,° was even, perhaps, in real truth, on a pious pilgrimage, to obtain from Mr Bentham's executors a sacred bone of his great, dissected master.°

And yet, after all, I cannot but think that the *Saturday Review* has here, for once, fallen a victim to an idea,—a beautiful but deluding idea,—and that the British nation has not yet, so entirely as the reviewer seems to imagine, found the last word of its philosophy. No, we are all seekers still! seekers often make mistakes, and I wish mine to redound to my own discredit only, and not to touch Oxford. Beautiful city! so venerable, so lovely, so unravaged by the fierce intellectual life of our century, so serene!

'There are our young barbarians, all at play!'°

And yet, steeped in sentiment as she lies, spreading her gardens to the moonlight, and whispering from her towers the last enchantments of the Middle Age, who will deny that Oxford, by her ineffable charm, keeps ever calling us nearer to the true goal of all of us, to the ideal, to perfection,—to beauty, in a word, which is only truth seen from another side?—nearer, perhaps, than all the science of Tübingen.° Adorable dreamer, whose heart has been so romantic! who hast given thyself so prodigally, given thyself to sides and to heroes not mine, only never to the Philistines! home of lost causes, and forsaken beliefs, and unpopular names, and impossible loyalties! what example could ever so inspire us to keep down the Philistine in ourselves, what teacher could ever so save us from that bondage to which we are all prone, that bondage which Goethe, in his incomparable lines on the death of Schiller, makes it his friend's highest praise (and nobly did Schiller deserve the praise) to have left miles out of sight behind him;—the bondage of '*was uns alle bändigt*, DAS GEMEINE'!° She will forgive me, even if I have unwittingly drawn upon her a shot or two aimed at her unworthy son; for she is generous, and the cause in which I fight is, after all, hers. Apparitions of a day, what is our puny warfare against the Philistines, compared with the warfare which this queen of romance has been waging against them for centuries, and will wage after we are gone?

Culture and Its Enemies

In one of his speeches a short time ago, that fine speaker and famous Liberal, Mr Bright,° took occasion to have a fling at the friends and preachers of culture. 'People who talk about what they call *culture*!' said he, contemptuously; 'by which they mean a smattering of the two dead languages of Greek and Latin.' And he went on to remark, in a strain with which modern speakers and writers have made us very familiar, how poor a thing this culture is, how little good it can do to the world, and how absurd it is for its possessors to set much store by it. And the other day a younger Liberal than Mr Bright, one of a school whose mission it is to bring into order and system that body of truth with which the earlier Liberals merely fumbled, a member of the University of Oxford, and a very clever writer, Mr Frederic Harrison, developed, in the systematic and stringent manner of his school, the thesis which Mr Bright had propounded in only general terms. 'Perhaps the very silliest cant of the day',° said Mr Frederic Harrison, 'is the cant about culture. Culture is a desirable quality in a critic of new books, and sits well on a professor of *belles-lettres*; but as applied to politics, it means simply a turn for small fault-finding, love of selfish ease, and indecision in action. The man of culture is in politics one of the poorest mortals alive. For simple pedantry and want of good sense no man is his equal. No assumption is too unreal, no end is too unpractical for him. But the active exercise of politics requires common sense, sympathy, trust, resolution, and enthusiasm, qualities which your man of culture has carefully rooted up, lest they damage the delicacy of his critical olfactories. Perhaps they are the only class of responsible beings in the community who cannot with safety be entrusted with power.'

Now for my part I do not wish to see men of culture asking to be entrusted with power; and, indeed, I have freely said, that in my opinion the speech most proper, at present, for a man of culture to make to a body of his fellow-countrymen who get him into a committee-room, is Socrates's: *Know thyself!*° and this is not a speech to be made by men wanting to be entrusted with power. For this very indifference to direct political action I have been taken to task by the *Daily Telegraph*,° coupled, by a strange perversity of fate, with just that very one of the Hebrew prophets whose style I admire the least, and called 'an elegant Jeremiah'. It is because I say (to use the words which the *Daily Telegraph* puts in my mouth):—'You mustn't make a fuss because you have no vote,—that is vulgarity; you mustn't hold big meetings to agitate for reform bills and to repeal corn laws,—that is the very height of vulgarity',—it is for this reason that I am called, sometimes an elegant Jeremiah, sometimes a spurious Jeremiah, a Jeremiah about the reality of whose mission the writer in the *Daily Telegraph* has his doubts. It is evident, therefore, that I have so taken my line as not to be exposed to the whole brunt of Mr Frederic Harrison's censure. Still, I have often spoken in praise of culture, I have striven to make all my works and ways serve the interests of culture. I take culture to be something a great deal more than what Mr Frederic Harrison and others call it: 'a desirable quality in a critic of new books'. Nay, even though to a certain extent I am disposed to agree with Mr Frederic Harrison, that men of culture are just the class of responsible beings in this community of ours who cannot properly, at present, be entrusted with power, I am not sure that I do not think this the fault of our community rather than of the men of culture. In short, although, like Mr Bright and Mr Frederic Harrison, and the editor of the *Daily Telegraph*, and a large body of valued friends of mine, I am a Liberal, yet I am a Liberal tempered by experience, reflection, and renouncement, and I am, above all, a believer in culture. Therefore I propose now to try and inquire, in the simple unsystematic way which best suits both my taste and my powers, what culture really is, what good it can do, what is our own special need of it; and I shall seek to find some plain grounds on which a faith in culture,—both my own faith in it and the faith of others,—may rest securely.

The disparagers of culture make its motive curiosity; sometimes, indeed, they make its motive mere exclusiveness and vanity. The culture which is supposed to plume itself on a smattering of Greek and Latin is a culture which is begotten by nothing so intellectual as curiosity; it is valued either out of sheer vanity and ignorance or else as an engine of social and class distinction, separating its holder, like a badge or title, from other people who have not got it. No serious man would call this *culture*, or attach any value to it, as culture, at all. To find the real ground for the very different estimate which serious people will set upon culture, we must find some motive for culture in the terms of which may lie a real ambiguity; and such a motive the word *curiosity* gives us.

I have before now pointed out° that we English do not, like the foreigners, use this word in a good sense as well as in a bad sense. With us the word is always used in a somewhat disapproving sense. A liberal and intelligent eagerness about the things of the mind may be meant by a foreigner when he speaks of curiosity, but with us the word always conveys a certain notion of frivolous and unedifying activity. In the

Quarterly Review,° some little time ago, was an estimate of the celebrated French critic, M. Sainte-Beuve, and a very inadequate estimate it in my judgment was. And its inadequacy consisted chiefly in this: that in our English way it left out of sight the double sense really involved in the word *curiosity*, thinking enough was said to stamp M. Sainte-Beuve with blame if it was said that he was impelled in his operations as a critic by curiosity, and omitting either to perceive that M. Sainte-Beuve himself, and many other people with him, would consider that this was praise-worthy and not blameworthy, or to point out why it ought really to be accounted worthy of blame and not of praise. For as there is a curiosity about intellectual matters which is futile, and merely a disease, so there is certainly a curiosity,—a desire after the things of the mind simply for their own sakes and for the pleasure of seeing them as they are,—which is, in an intelligent being, natural and laudable. Nay, and the very desire to see things as they are implies a balance and regulation of mind which is not often attained without fruitful effort, and which is the very opposite of the blind and diseased impulse of mind which is what we mean to blame when we blame curiosity. Montesquieu says:° 'The first motive which ought to impel us to study is the desire to augment the excellence of our nature, and to render an intelligent being yet more intelligent.' This is the true ground to assign for the genuine scientific passion, however manifested, and for culture, viewed simply as a fruit of this passion; and it is a worthy ground, even though we let the term *curiosity* stand to describe it.

But there is of culture another view, in which not solely the scientific passion, the sheer desire to see things as they are, natural and proper in an intelligent being, appears as the ground of it. There is a view in which all the love of our neighbour, the impulses towards action, help, and beneficence, the desire for removing human error, clearing human confusion, and diminishing human misery, the noble aspiration to leave the world better and happier than we found it,—motives eminently such as are called social,—come in as part of the grounds of culture, and the main and pre-eminent part. Culture is then properly described not as having its origin in curiosity, but as having its origin in the love of perfection; it is *a study of perfection*. It moves by the force, not merely or primarily of the scientific passion for pure knowledge, but also of the moral and social passion for doing good. As, in the first view of it, we took for its worthy motto Montesquieu's words: 'To render an intelligent being yet more intelligent!' so, in the second view of it, there is no better motto which it can have than these words of Bishop Wilson:° 'To make reason and the will of God prevail!'

Only, whereas the passion for doing good is apt to be over-hasty in determining what reason and the will of God say, because its turn is for acting rather than thinking and it wants to be beginning to act; and whereas it is apt to take its own conceptions, which proceed from its own state of development and share in all the imperfections and immaturities of this, for a basis of action; what distinguishes culture is, that it is possessed by the scientific passion as well as by the passion of doing good; that it demands worthy notions of reason and the will of God, and does not readily suffer its own crude conceptions to substitute themselves for them. And knowing that no action or institution can be salutary and stable which is not based on reason and the will of God, it is not so bent on acting and instituting, even with the great aim of diminishing human error and misery ever before its thoughts, but that it can remember that acting and instituting are of little use, unless we know how and what we ought to act and to institute.

This culture is more interesting and more far-reaching than that other, which is founded solely on the scientific passion for knowing. But it needs times of faith and ardour, times when the intellectual horizon is opening and widening all round us, to flourish in. And is not the close and bounded intellectual horizon° within which we have long lived and moved now lifting up, and are not new lights finding free passage to shine in upon us? For a long time there was no passage for them to make their way in upon us, and then it was of no use to think of adapting the world's action to them. Where was the hope of making reason and the will of God prevail among people who had a routine which they had christened reason and the will of God, in which they were inextricably bound, and beyond which they had no power of looking? But now the iron force of adhesion to the old routine,—social, political, religious,—has wonderfully yielded; the iron force of exclusion of all which is new has wonderfully yielded. The danger now is, not that people should obstinately refuse to allow anything but their old routine to pass for reason and the will of God, but either that they should allow some novelty or other to pass for these too easily, or else that they should underrate the importance of them altogether, and think it enough to follow action for its own sake, without troubling themselves to make reason and the will of God prevail therein. Now, then, is the moment for culture to be of service, culture which believes in making reason and the will of God prevail, believes in perfection, is the study and pursuit of perfection, and is no longer debarred, by a rigid invincible exclusion of whatever is new, from getting acceptance for its ideas, simply because they are new.

The moment this view of culture is seized, the moment it is regarded not solely as the endeavour to see things as they are, to draw towards a knowledge of the universal order which seems to be intended and aimed at in the world, and which it is a man's happiness to go along with or his misery to go counter to,—to learn, in short, the will of God,—the moment, I say, culture is considered not merely as the endeavour to *see* and *learn* this, but as the endeavour, also, to make it *prevail*, the moral, social, and beneficent character of culture becomes manifest. The mere endeavour to see and learn the truth for our own personal satisfaction is indeed a commencement for making it prevail, a preparing the way for this, which always serves this, and is wrongly, therefore, stamped with blame absolutely in itself and not only in its caricature and degeneration. But perhaps it has got stamped with blame, and disparaged with the dubious title of curiosity, because in comparison with this wider endeavour of such great and plain utility it looks selfish, petty, and unprofitable.°

And religion, the greatest and most important of the efforts by which the human race has manifested its impulse to perfect itself,—religion, that voice of the deepest human experience,—does not only enjoin and sanction the aim which is the great aim of culture, the aim of setting ourselves to ascertain what perfection is and to make it prevail; but also, in determining generally in what human perfection consists, religion comes to a conclusion identical with that which culture,—culture seeking the determination of this question through *all* the voices of human experience which have been heard upon it, of art, science, poetry, philosophy, history, as well as of religion, in order to give a greater fulness and certainty to its solution,—likewise reaches. Religion says: *The kingdom of God is within you*;° and culture, in like manner, places human perfection in an *internal* condition, in the growth and predominance of our humanity proper, as distinguished from our animality. It places it in the ever-increasing efficacy and in the general harmonious expansion of those gifts of thought and feeling, which make the peculiar dignity, wealth, and happiness of human nature. As I have said on a former occasion:° 'It is in making endless additions to itself, in the endless expansion of its powers, in endless growth in wisdom and beauty, that the spirit of the human race finds its ideal. To reach this ideal, culture is an indispensable aid, and that is the true value of culture.' Not a having and a resting, but a growing and a becoming, is the character of perfection as culture conceives it; and here, too, it coincides with religion.

And because men are all members of one great whole, and the sympathy which is in human nature will not allow one member to be indifferent to the rest or to have a perfect welfare independent of the rest, the expansion of our humanity, to suit the idea of perfection which culture forms, must be a *general* expansion. Perfection, as culture conceives it, is not possible while the individual remains isolated. The individual is required, under pain of being stunted and enfeebled in his own development if he disobeys, to carry others along with him in his march towards perfection, to be continually doing all he can to enlarge and increase the volume of the human stream sweeping thitherward. And here, once more, culture lays on us the same obligation as religion, which says, as Bishop Wilson has admirably put it, that 'to promote the kingdom of God is to increase and hasten one's own happiness'.°

But, finally, perfection,—as culture from a thorough disinterested study of human nature and human experience learns to conceive it,—is a harmonious expansion° of *all* the powers which make the beauty and worth of human nature, and is not consistent with the over-development of any one power at the expense of the rest. Here culture goes beyond religion, as religion is generally conceived by us.

If culture, then, is a study of perfection, and of harmonious perfection, general perfection, and perfection which consists in becoming something rather than in having something, in an inward condition of the mind and spirit, not in an outward set of circumstances,—it is clear that culture, instead of being the frivolous and useless thing which Mr Bright, and Mr Frederic Harrison, and many other Liberals are apt to call it, has a very important function to fulfil for mankind. And this function is particularly important in our modern world, of which the whole civilisation is, to a much greater degree than the civilisation of Greece and Rome, mechanical and external, and tends constantly to become more so. But above all in our own country has culture a weighty part to perform, because here that mechanical character, which civilisation tends to take everywhere, is shown in the most eminent degree. Indeed nearly all the characters of perfection, as culture teaches us to fix them, meet in this country with some powerful tendency which thwarts them and sets them at defiance. The idea of perfection as an *inward* condition of the mind and spirit is at variance with the mechanical and material civilisation in esteem with us, and nowhere, as I have said, so much in esteem as with us. The idea of perfection as a *general* expansion of the human family is at variance with our strong individualism, our hatred of all limits to the unrestrained swing of the individual's personality, our maxim of 'every man for himself'. Above all, the idea of perfection as a *harmonious* expansion of human nature is at variance with our want of flexibility, with our inaptitude for seeing more than one side of a thing, with our intense energetic absorption in the particular pursuit we happen to be following.

So culture has a rough task to achieve in this country. Its preachers have, and are likely long to have, a hard time of it, and they will much oftener be regarded, for a great while to come, as elegant or spurious Jeremiahs than as friends and benefactors. That, however, will not prevent their doing in the end good service if they persevere. And, meanwhile, the mode of action they have to pursue, and the sort of habits they must fight against, ought to be made quite clear for every one to see, who may be willing to look at the matter attentively and dispassionately.

Faith in machinery° is, I said, our besetting danger; often in machinery most absurdly disproportioned to the end which this machinery, if it is to do any good at all, is to serve; but always in machinery, as if it had a value in and for itself. What is freedom but machinery? what is population but machinery? what is coal but machinery? what are railroads but machinery? what is wealth but machinery? what are, even, religious organisations but machinery? Now almost every voice in England is accustomed to speak of these things as if they were precious ends in themselves, and therefore had some of the characters of perfection indissolubly joined to them. I have before now noticed Mr Roebuck's stock argument° for proving the greatness and happiness of England as she is, and for quite stopping the mouths of all gainsayers. Mr Roebuck is never weary of reiterating this argument of his, so I do not know why I should be weary of noticing it. 'May not every man in England say what he likes?'—Mr Roebuck perpetually asks; and that, he thinks, is quite sufficient, and when every man may say what he likes, our aspirations ought to be satisfied. But the aspirations of culture, which is the study of perfection, are not satisfied, unless what men say, when they may say what they like, is worth saying,—has good in it, and more good than bad. In the same way the *Times*, replying to some foreign strictures on the dress, looks, and behaviour of the English abroad, urges that the English ideal is that every one should be free to do and to look just as he likes. But culture indefatigably tries, not to make what each raw person may like the rule by which he fashions himself; but to draw ever nearer to a sense of what is indeed beautiful, graceful, and becoming, and to get the raw person to like that.

And in the same way with respect to railroads and coal. Every one must have observed the strange language current during the late discussions as to the possible failure of our supplies of coal. Our coal, thousands of people were saying, is the real basis of our national greatness; if our coal runs short, there is an end of the greatness of England. But what *is* greatness?—culture makes us ask. Greatness is a spiritual condition worthy to excite love, interest, and admiration; and the outward proof of possessing greatness is that we excite love, interest, and admiration. If England were swallowed up by the sea tomorrow, which of the two, a hundred years hence, would most excite the love, interest, and admiration of mankind,—would most, therefore, show the evidences of having possessed greatness,—the England of the last twenty years, or the England of Elizabeth, of a time of splendid spiritual effort, but when our coal, and our industrial operations depending on coal, were very little developed? Well, then, what an unsound habit of mind it must be which makes us talk of things like coal or iron as constituting the greatness of England, and how salutary a friend is culture, bent on seeing things as they are, and thus dissipating delusions of this kind and fixing standards of perfection that are real!

Wealth, again, that end to which our prodigious works for material advantage are directed,—the commonest of commonplaces tells us how men are always apt to regard wealth as a precious end in itself; and certainly they have never been so apt thus to regard it as they are in England at the present time. Never did people believe anything more firmly than nine Englishmen out of ten at the present day believe that our greatness and welfare are proved by our being so very rich. Now, the use of culture is that it helps us, by means of its spiritual standard of perfection, to regard wealth as but machinery, and not only to say as a matter of words that we regard wealth as but machinery, but really to perceive and feel that it is so. If it were not for this purging effect wrought upon our minds by culture, the whole world, the future as well as the present, would inevitably belong to the Philistines. The people who believe most that our greatness and welfare are proved by our being very rich, and who most give their lives and thoughts to becoming rich, are just the very people whom we call Philistines.° Culture says: 'Consider these people, then, their way of life, their habits, their manners, the very tones of their voice; look at them attentively; observe the literature they read, the things which give them pleasure, the words which come forth out of their mouths,° the thoughts which make the furniture of their minds; would any amount of wealth be worth having with the condition that one was to become just like these people by having it?' And thus culture begets a dissatisfaction which is of the highest possible value in stemming the common tide of men's thoughts in a wealthy and industrial community, and which saves the future, as one may hope, from being vulgarised, even if it cannot save the present.

Population, again, and bodily health and vigour, are things which are nowhere treated in such an unintelligent, misleading, exaggerated way as in England. Both are really machinery; yet how many people all around

us do we see rest in them and fail to look beyond them! Why, one has
heard people, fresh from reading certain articles of the *Times*° on the
Registrar-General's returns of marriages and births in this country, who
would talk of our large English families in quite a solemn strain, as if they
had something in itself beautiful, elevating, and meritorious in them; as if
the British Philistine would have only to present himself before the Great
Judge with his twelve children, in order to be received among the sheep°
as a matter of right!

But bodily health and vigour, it may be said, are not to be classed with
wealth and population as mere machinery; they have a more real and
essential value. True; but only as they are more intimately connected
with a perfect spiritual condition than wealth or population are. The
moment we disjoin them from the idea of a perfect spiritual condition,
and pursue them, as we do pursue them, for their own sake and as ends in
themselves, our worship of them becomes as mere worship of
machinery, as our worship of wealth or population, and as unintelligent
and vulgarising a worship as that is. Every one with anything like an
adequate idea of human perfection has distinctly marked this subordina-
tion to higher and spiritual ends of the cultivation of bodily vigour and
activity. 'Bodily exercise profiteth little; but godliness is profitable unto
all things', says the author of the Epistle to Timothy.° And the utilitarian
Franklin says° just as explicitly:—'Eat and drink such an exact quantity
as suits the constitution of thy body, *in reference to the services of the mind.*'
But the point of view of culture, keeping the mark of human perfection
simply and broadly in view, and not assigning to this perfection, as
religion or utilitarianism assigns to it, a special and limited character, this
point of view, I say, of culture is best given by these words of Epictetus:°—'It is a sign of ἀφυΐα', says he,—that is, of a nature not finely
tempered,—'to give yourselves up to things which relate to the body; to
make, for instance, a great fuss about exercise, a great fuss about eating, a
great fuss about drinking, a great fuss about walking, a great fuss about
riding. All these things ought to be done merely by the way: the formation
of the spirit and character must be our real concern.' This is admirable;
and, indeed, the Greek word εὐφυΐα, a finely tempered nature, gives
exactly the notion of perfection as culture brings us to conceive it: a
harmonious perfection, a perfection in which the characters of beauty
and intelligence are both present, which unites 'the two noblest of
things',—as Swift, who of one of the two, at any rate, had himself all too
little, most happily calls them° in his *Battle of the Books*,—'the two noblest
of things, *sweetness and light*'. The εὐφυής is the man who tends towards
sweetness and light; the ἀφυής, on the other hand, is our Philistine. The

immense spiritual significance of the Greeks is due to their having been inspired with this central and happy idea of the essential character of human perfection; and Mr Bright's misconception of culture, as a smattering of Greek and Latin, comes itself, after all, from this wonderful significance of the Greeks having affected the very machinery of our education, and is in itself a kind of homage to it.

In thus making sweetness and light to be characters of perfection, culture is of like spirit with poetry, follows one law with poetry. Far more than on our freedom, our population, and our industrialism, many amongst us rely upon our religious organisations to save us.° I have called religion a yet more important manifestation of human nature than poetry, because it has worked on a broader scale for perfection, and with greater masses of men. But the idea of beauty and of a human nature perfect on all its sides, which is the dominant idea of poetry, is a true and invaluable idea, though it has not yet had the success that the idea of conquering the obvious faults of our animality, and of a human nature perfect on the moral side,—which is the dominant idea of religion,—has been enabled to have; and it is destined, adding to itself the religious idea of a devout energy, to transform and govern the other.

The best art and poetry of the Greeks, in which religion and poetry are one, in which the idea of beauty and of a human nature perfect on all sides adds to itself a religious and devout energy, and works in the strength of that, is on this account of such surpassing interest and instructiveness for us, though it was,—as, having regard to the human race in general, and, indeed, having regard to the Greeks themselves, we must own,—a premature attempt, an attempt which for success needed the moral and religious fibre in humanity to be more braced and developed than it had yet been. But Greece did not err in having the idea of beauty, harmony, and complete human perfection, so present and paramount. It is impossible to have this idea too present and paramount; only, the moral fibre must be braced too. And we, because we have braced the moral fibre, are not on that account in the right way, if at the same time the idea of beauty, harmony, and complete human perfection, is wanting or misapprehended amongst us; and evidently it *is* wanting or misapprehended at present. And when we rely as we do on our religious organisations, which in themselves do not and cannot give us this idea, and think we have done enough if we make them spread and prevail, then, I say, we fall into our common fault of overvaluing machinery.

Nothing is more common than for people to confound the inward peace and satisfaction which follows the subduing of the obvious faults of our animality with what I may call absolute inward peace and satisfaction,—the peace and satisfaction which are reached as we draw near to complete spiritual perfection, and not merely to moral perfection, or rather to relative moral perfection. No people in the world have done more and struggled more to attain this relative moral perfection than our English race has. For no people in the world has the command to *resist the devil*, to *overcome the wicked one*,° in the nearest and most obvious sense of those words, had such a pressing force and reality. And we have had our reward, not only in the great worldly prosperity which our obedience to this command has brought us, but also, and far more, in great inward peace and satisfaction. But to me few things are more pathetic than to see people, on the strength of the inward peace and satisfaction which their rudimentary efforts towards perfection have brought them, employ, concerning their incomplete perfection and the religious organisations within which they have found it, language which properly applies only to complete perfection, and is a far-off echo of the human soul's prophecy of it. Religion itself, I need hardly say, supplies them in abundance with this grand language. And very freely do they use it; yet it is really the severest possible criticism of such an incomplete perfection as alone we have yet reached through our religious organisations.

The impulse of the English race towards moral development and self-conquest has nowhere so powerfully manifested itself as in Puritanism. Nowhere has Puritanism found so adequate an expression as in the religious organisation of the Independents. The modern Independents° have a newspaper, the *Nonconformist*, written with great sincerity and ability. The motto, the standard, the profession of faith which this organ of theirs carries aloft, is: 'The Dissidence of Dissent and the Protestantism of the Protestant religion.' There is sweetness and light, and an ideal of complete harmonious human perfection! One need not go to culture and poetry to find language to judge it. Religion, with its instinct for perfection, supplies language to judge it, language, too, which is in our mouths every day. 'Finally, be of one mind, united in feeling', says St Peter.° There is an ideal which judges the Puritan ideal: 'The Dissidence of Dissent and the Protestantism of the Protestant religion'! And religious organisations like this are what people believe in, rest in, would give their lives for! Such, I say, is the wonderful virtue of even the beginnings of perfection, of having conquered even the plain faults of our animality, that the religious organisation which has helped us to do it can seem to us something precious, salutary, and to be propagated, even when it wears such a brand of imperfection on its forehead as this. And men have got such a habit of giving to the language of religion a special application, of making it a mere jargon, that for the condemnation which religion itself passes on the shortcomings of their religious organisations they have no ear; they are sure to cheat themselves and to explain this condemnation away. They can only be reached by the criticism which culture, like poetry, speaking a language not to be sophisticated, and resolutely testing these organisations by the ideal of a human perfection complete on all sides, applies to them.

But men of culture and poetry, it will be said, are again and again failing, and failing conspicuously, in the necessary first stage to a harmonious perfection, in the subduing of the great obvious faults of our animality, which it is the glory of these religious organisations to have helped us to subdue. True, they do often so fail. They have often been without the virtues as well as the faults of the Puritan; it has been one of their dangers that they so felt the Puritan's faults that they too much neglected the practice of his virtues. I will not, however, exculpate them at the Puritan's expense. They have often failed in morality, and morality is indispensable. And they have been punished for their failure, as the Puritan has been rewarded for his performance. They have been punished wherein they erred; but their ideal of beauty, of sweetness and light, and a human nature complete on all its sides, remains the true ideal of perfection still; just as the Puritan's ideal of perfection remains narrow and inadequate, although for what he did well he has been richly rewarded. Notwithstanding the mighty results of the Pilgrim Fathers' voyage, they and their standard of perfection are rightly judged when we figure to ourselves Shakspeare or Virgil,—souls in whom sweetness and light, and all that in human nature is most humane, were eminent,— accompanying them on their voyage, and think what intolerable company Shakspeare and Virgil would have found them! In the same way let us judge the religious organisations which we see all around us. Do not let us deny the good and the happiness which they have accomplished; but do not let us fail to see clearly that their idea of human perfection is narrow and inadequate, and that the Dissidence of Dissent and the Protestantism of the Protestant religion will never bring humanity to its true goal. As I said with regard to wealth: Let us look at the life of those who live in and for it,—so I say with regard to the religious organisations. Look at the life imaged in such a newspaper as the *Nonconformist*,—a life of jealousy of the Establishment, disputes, tea-meetings, openings of chapels, sermons; and then think of it as an ideal of a human life completing itself on all sides, and aspiring with all its organs after sweetness, light, and perfection!

Another newspaper, representing, like the *Nonconformist*, one of the religious organisations of this country, was a short time ago giving an account of the crowd at Epsom on the Derby day, and of all the vice and hideousness which was to be seen in that crowd; and then the writer turned suddenly round upon Professor Huxley,° and asked him how he proposed to cure all this vice and hideousness without religion. I confess I felt disposed to ask the asker this question: And how do you propose to cure it with such a religion as yours? How is the ideal of a life so unlovely, so unattractive, so incomplete, so narrow, so far removed from a true and satisfying ideal of human perfection, as is the life of your religious organisation as you yourself reflect it, to conquer and transform all this vice and hideousness? Indeed, the strongest plea for the study of perfection as pursued by culture, the clearest proof of the actual inadequacy of the idea of perfection held by the religious organisa- tions,—expressing, as I have said, the most widespread effort which the human race has yet made after perfection,—is to be found in the state of our life and society with these in possession of it, and having been in possession of it I know not how many hundred years. We are all of us included in some religious organisation or other; we all call ourselves, in the sublime and aspiring language of religion which I have before noticed, *children of God.*° Children of God;—it is an immense preten- sion!—and how are we to justify it? By the works which we do, and the words which we speak. And the work which we collective children of God do, our grand centre of life, our *city* which we have builded for us to dwell in,° is London! London, with its unutterable external hideousness, and with its internal canker of *publicè egestas, privatim opulentia,*°—to use the words which Sallust puts into Cato's mouth about Rome,— unequalled in the world! The word, again, which we children of God speak, the voice which most hits our collective thought, the newspaper with the largest circulation in England, nay, with the largest circulation in the whole world, is the *Daily Telegraph*! I say that when our religious organisations,—which I admit to express the most considerable effort after perfection that our race has yet made,—land us in no better result than this, it is high time to examine carefully their idea of perfection, to see whether it does not leave out of account sides and forces of human nature which we might turn to great use; whether it would not be more operative if it were more complete. And I say that the English reliance on our religious organisations and on their ideas of human perfection just as they stand, is like our reliance on freedom, on muscular Christianity,° on population, on coal, on wealth,—mere belief in machinery, and unfruit- ful; and that it is wholesomely counteracted by culture, bent on seeing things as they are, and on drawing the human race onwards to a more complete, a harmonious perfection.

Culture, however, shows its single-minded love of perfection, its desire simply to make reason and the will of God prevail, its freedom from fanaticism, by its attitude towards all this machinery, even while it insists that it *is* machinery. Fanatics, seeing the mischief men do themselves by their blind belief in some machinery or other,—whether it is wealth and industrialism, or whether it is the cultivation of bodily strength and activity, or whether it is a political organisation, or whether it is a religious organisation,—oppose with might and main the tendency to this or that political and religious organisation, or to games and athletic exercises, or to wealth and industrialism, and try violently to stop it. But the flexibility which sweetness and light give, and which is one of the rewards of culture pursued in good faith, enables a man to see that a tendency may be necessary, and even, as a preparation for something in the future, salutary, and yet that the generations or individuals who obey this tendency are sacrificed to it, that they fall short of the hope of perfection by following it; and that its mischiefs are to be criticised, lest it should take too firm a hold and last after it has served its purpose.

Mr Gladstone well pointed out, in a speech at Paris,°—and others have pointed out the same thing,—how necessary is the present great movement towards wealth and industrialism, in order to lay broad foundations of material well-being for the society of the future. The worst of these justifications is, that they are generally addressed to the very people engaged, body and soul, in the movement in question; at all events, that they are always seized with the greatest avidity by these people, and taken by them as quite justifying their life; and that thus they tend to harden them in their sins. Now, culture admits the necessity of the movement towards fortune-making and exaggerated industrialism, readily allows that the future may derive benefit from it; but insists, at the same time, that the passing generations of industrialists,—forming, for the most part, the stout main body of Philistinism,—are sacrificed to it. In the same way, the result of all the games and sports which occupy the passing generation of boys and young men may be the establishment of a better and sounder physical type for the future to work with. Culture does not set itself against the games and sports; it congratulates the future, and hopes it will make a good use of its improved physical basis; but it points out that our passing generation of boys and young men is, meantime, sacrificed. Puritanism was perhaps necessary to develop the moral fibre of the English race, Nonconformity to break the yoke of ecclesiastical domination over men's minds and to prepare the way for freedom of thought in the distant future; still, culture points out that the harmonious perfection of generations of Puritans and Nonconformists have been, in consequence, sacrificed. Freedom of speech may be necessary for the society of the future, but the young lions of the *Daily Telegraph* in the meanwhile are sacrificed. A voice for every man in his country's government may be necessary for the society of the future, but meanwhile Mr Beales and Mr Bradlaugh° are sacrificed.

Oxford, the Oxford of the past, has many faults; and she has heavily paid for them in defeat, in isolation, in want of hold upon the modern world. Yet we in Oxford, brought up amidst the beauty and sweetness of that beautiful place, have not failed to seize one truth,—the truth that beauty and sweetness are essential characters of a complete human perfection. When I insist on this, I am all in the faith and tradition of Oxford. I say boldly that this our sentiment for beauty and sweetness, our sentiment against hideousness and rawness, has been at the bottom of our attachment to so many beaten causes, of our opposition to so many triumphant movements. And the sentiment is true, and has never been wholly defeated, and has shown its power even in its defeat. We have not won our political battles, we have not carried our main points, we have not stopped our adversaries' advance, we have not marched victoriously with the modern world; but we have told silently upon the mind of the country; we have prepared currents of feeling which sap our adversaries' position when it seems gained, we have kept up our own communications with the future. Look at the course of the great movement which shook Oxford° to its centre some thirty years ago! It was directed, as any one who reads Dr Newman's *Apology* may see, against what in one word may be called 'Liberalism'. Liberalism prevailed; it was the appointed force to do the work of the hour; it was necessary, it was inevitable that it should prevail. The Oxford movement was broken, it failed; our wrecks are scattered on every shore:—

> Quae regio in terris nostri non plena laboris?°

But what was it, this liberalism, as Dr Newman saw it, and as it really broke the Oxford movement? It was the great middle-class liberalism, which had for the cardinal points of its belief the Reform Bill of 1832, and local self-government, in politics; in the social sphere, free-trade, unrestricted competition, and the making of large industrial fortunes; in the religious sphere, the Dissidence of Dissent and the Protestantism of the Protestant religion. I do not say that other and more intelligent forces than this were not opposed to the Oxford movement: but this was the force which really beat it; this was the force which Dr Newman felt himself fighting with; this was the force which till only the other day seemed to be the paramount force in this country, and to be in possession of the future; this was the force whose achievements fill Mr Lowe° with such inexpressible admiration, and whose rule he was so horror-struck to see threatened. And where is this great force of Philistinism now? It is thrust into the second rank, it is become a power of yesterday, it has lost the future. A new power° has suddenly appeared, a power which it is impossible yet to judge fully, but which is certainly a wholly different force from middle-class liberalism; different in its cardinal points of belief, different in its tendencies in every sphere. It loves and admires neither the legislation of middle-class Parliaments, nor the local self-government of middle-class vestries, nor the unrestricted competition of middle-class industrialists, nor the dissidence of middle-class Dissent and the Protestantism of middle-class Protestant religion. I am not now praising this new force, or saying that its own ideals are better; all I say is, that they are wholly different. And who will estimate how much the currents of feeling created by Dr Newman's movement, the keen desire for beauty and sweetness which it nourished, the deep aversion it manifested to the hardness and vulgarity of middle-class liberalism, the strong light it turned on the hideous and grotesque illusions of middle-class Protestantism,—who will estimate how much all these contributed to swell the tide of secret dissatisfaction which has mined the ground under the self-confident liberalism of the last thirty years, and has prepared the way for its sudden collapse and supersession? It is in this manner that the sentiment of Oxford for beauty and sweetness conquers, and in this manner long may it continue to conquer!

In this manner it works to the same end as culture, and there is plenty of work for it yet to do. I have said that the new and more democratic force which is now superseding our old middle-class liberalism cannot yet be rightly judged. It has its main tendencies still to form. We hear promises of its giving us administrative reform, law reform, reform of education, and I know not what; but those promises come rather from its advocates, wishing to make a good plea for it and to justify it for superseding middle-class liberalism, than from clear tendencies which it has itself yet developed. But meanwhile it has plenty of well-intentioned friends against whom culture may with advantage continue to uphold steadily its ideal of human perfection; that this is *an inward spiritual activity, having for its characters increased sweetness, increased light, increased life, increased sympathy*. Mr Bright, who has a foot in both worlds, the world of middle-class liberalism and the world of democracy, but who brings most of his ideas from the world of middle-class liberalism in which he was bred, always inclines to inculcate that faith in machinery to which, as we have seen, Englishmen are so prone, and which has been the bane of middle-class liberalism. He complains with a sorrowful indignation of people who 'appear to have no proper estimate of the value of the franchise'; he leads his disciples to believe,—what the Englishman is always too ready to believe,—that the having a vote, like the having a large family, or a large business, or large muscles, has in itself some edifying and perfecting effect upon human nature. Or else he cries out to the democracy,—'the men', as he calls them, 'upon whose shoulders the greatness of England rests',—he cries out to them: 'See what you have done! I look over this country and see the cities you have built, the railroads you have made, the manufactures you have produced, the cargoes which freight the ships of the greatest mercantile navy the world has ever seen! I see that you have converted by your labours what was once a wilderness, these islands, into a fruitful garden; I know that you have created this wealth, and are a nation whose name is a word of power throughout all the world.'° Why, this is just the very style of laudation with which Mr Roebuck or Mr Lowe debauches the minds of the middle classes, and makes such Philistines of them. It is the same fashion of teaching a man to value himself not on what he *is*, not on his progress in sweetness and light, but on the number of the railroads he has constructed, or the bigness of the tabernacle° he has built. Only the middle classes are told they have done it all with their energy, self-reliance, and capital, and the democracy are told they have done it all with their hands and sinews. But teaching the democracy to put its trust in achievements of this kind is merely training them to be Philistines to take the place of the Philistines whom they are superseding; and they too, like the middle class, will be encouraged to sit down at the banquet of the future without having on a wedding garment,° and nothing excellent can then come from them. Those who know their besetting faults, those who have watched them and listened to them, or those who will read the instructive account recently given of them by one of themselves, the *Journeyman Engineer*,° will agree that the idea which culture sets before us of perfection,—an increased spiritual activity, having for its characters increased sweetness, increased light, increased life, increased sympathy,—is an idea which the new democracy needs far more than the idea of the blessedness of the franchise, or the wonderfulness of its own industrial performances.

Other well-meaning friends of this new power are for leading it, not in the old ruts of middle-class Philistinism, but in ways which are naturally alluring to the feet of democracy, though in this country they are novel and untried ways. I may call them the ways of Jacobinism.° Violent indignation with the past, abstract systems of renovation applied wholesale, a new doctrine drawn up in black and white for elaborating down to the very smallest details a rational society for the future,—these are the ways of Jacobinism. Mr Frederic Harrison and other disciples of Comte,—one of them, Mr Congreve,° is an old friend of mine, and I am glad to have an opportunity of publicly expressing my respect for his talents and character,—are among the friends of democracy who are for leading it in paths of this kind. Mr Frederic Harrison is very hostile to culture, and from a natural enough motive; for culture is the eternal opponent of the two things which are the signal marks of Jacobinism,— its fierceness, and its addiction to an abstract system. Culture is always assigning to system-makers and systems a smaller share in the bent of human destiny than their friends like. A current in people's minds° sets towards new ideas; people are dissatisfied with their old narrow stock of Philistine ideas, Anglo-Saxon ideas, or any other; and some man, some Bentham or Comte, who has the real merit of having early and strongly felt and helped the new current, but who brings plenty of narrowness and mistakes of his own into his feeling and help of it, is credited with being the author of the whole current, the fit person to be entrusted with its regulation and to guide the human race.

The excellent German historian of the mythology of Rome, Preller,° relating the introduction at Rome under the Tarquins of the worship of Apollo, the god of light, healing, and reconciliation, will have us observe that it was not so much the Tarquins who brought to Rome the new worship of Apollo, as a current in the mind of the Roman people which set powerfully at that time towards a new worship of this kind, and away from the old run of Latin and Sabine religious ideas. In a similar way, culture directs our attention to the natural current there is in human affairs, and to its continual working, and will not let us rivet our faith upon any one man and his doings. It makes us see not only his good side, but also how much in him was of necessity limited and transient; nay, it even feels a pleasure, a sense of an increased freedom and of an ampler future, in so doing.

I remember, when I was under the influence of a mind to which I feel the greatest obligations, the mind of a man who was the very incarnation of sanity and clear sense, a man the most considerable, it seems to me, whom America has yet produced,—Benjamin Franklin,—I remember the relief with which, after long feeling the sway of Franklin's imperturbable common-sense, I came upon a project of his for a new version of the Book of Job,° to replace the old version, the style of which, says Franklin, has become obsolete, and thence less agreeable. 'I give', he continues, 'a few verses, which may serve as a sample of the kind of version I would recommend.' We all recollect the famous verse in our translation: 'Then Satan answered the Lord and said: "Doth Job fear God for nought?" ' Franklin makes this: 'Does your Majesty imagine that Job's good conduct is the effect of mere personal attachment and affection?' I well remember how, when first I read that, I drew a deep breath of relief, and said to myself: 'After all, there is a stretch of humanity beyond Franklin's victorious good sense!' So, after hearing Bentham cried loudly up as the renovator of modern society, and Bentham's mind and ideas proposed as the rulers of our future, I open the *Deontology*.° There I read: 'While Xenophon was writing his history and Euclid teaching geometry, Socrates and Plato were talking nonsense under pretence of teaching wisdom and morality. This morality of theirs consisted in words; this wisdom of theirs was the denial of matters known to every man's experience.' From the moment of reading that, I am delivered from the bondage of Bentham! the fanaticism of his adherents can touch me no longer. I feel the inadequacy of his mind and ideas for supplying the rule of human society, for perfection.

Culture tends always thus to deal with the men of a system, of disciples, of a school; with men like Comte, or the late Mr Buckle,° or Mr Mill. However much it may find to admire in these personages, or in some of them, it nevertheless remembers the text: 'Be not ye called Rabbi!'° and it soon passes on from any Rabbi. But Jacobinism loves a Rabbi; it does not want to pass on from its Rabbi in pursuit of a future and still unreached perfection; it wants its Rabbi and his ideas to stand for perfection, that they may with the more authority recast the world; and for Jacobinism, therefore, culture,—eternally passing onwards and seeking,—is an impertinence and an offence. But culture, just because it resists this tendency of Jacobinism to impose on us a man with limitations and errors of his own along with the true ideas of which he is the organ, really does the world and Jacobinism itself a service.

So, too, Jacobinism, in its fierce hatred of the past and of those whom it makes liable for the sins of the past, cannot away with the inexhaustible indulgence proper to culture, the consideration of circumstances, the severe judgment of actions joined to the merciful judgment of persons. 'The man of culture is in politics', cries Mr Frederic Harrison, 'one of the poorest mortals alive!' Mr Frederic Harrison wants to be doing business, and he complains that the man of culture stops him with a 'turn for small fault-finding, love of selfish ease, and indecision in action'. Of what use is culture, he asks, except for 'a critic of new books or a professor of *belles-lettres*?'° Why, it is of use because, in presence of the fierce exasperation which breathes, or rather, I may say, hisses through the whole production in which Mr Frederic Harrison asks that question, it reminds us that the perfection of human nature is sweetness and light. It is of use because, like religion,—that other effort after perfection,—it testifies that, where bitter envying and strife° are, there is confusion and every evil work.

The pursuit of perfection,° then, is the pursuit of sweetness and light. He who works for sweetness and light, works to make reason and the will of God prevail. He who works for machinery, he who works for hatred, works only for confusion. Culture looks beyond machinery, culture hates hatred; culture has one great passion, the passion for sweetness and light. It has one even yet greater!—the passion for making them *prevail*. It is not satisfied till we *all* come to a perfect man; it knows that the sweetness and light of the few must be imperfect until the raw and unkindled masses of humanity are touched with sweetness and light. If I have not shrunk from saying that we must work for sweetness and light, so neither have I shrunk from saying that we must have a broad basis, must have sweetness and light for as many as possible. Again and again I have insisted how those are the happy moments of humanity, how those are the marking epochs of a people's life, how those are the flowering times for literature and art and all the creative power of genius, when there is a *national* glow of life and thought, when the whole of society is in the fullest measure permeated by thought, sensible to beauty, intelligent and alive. Only it must be *real* thought and *real* beauty; *real* sweetness and *real* light. Plenty of people will try to give the masses, as they call them, an intellectual food prepared and adapted in the way they think proper for the actual condition of the masses. The ordinary popular literature is an example of this way of working on the masses. Plenty of people will try to indoctrinate the masses with the set of ideas and judgments constituting the creed of their own profession or party. Our religious and political organisations give an example of this way of working on the masses. I condemn neither way; but culture works differently. It does not try to teach down to the level of inferior classes; it does not try to win them for this or that sect of its own, with ready-made judgments and watchwords. It seeks to do away with classes; to make the best that has been thought and known in the world current everywhere; to make all men live in an atmosphere of sweetness and light, where they may use ideas, as it uses them itself, freely,—nourished, and not bound by them.

This is the *social idea*; and the men of culture are the true apostles of equality. The great men of culture are those who have had a passion for diffusing, for making prevail, for carrying from one end of society to the other, the best knowledge, the best ideas of their time; who have laboured to divest knowledge of all that was harsh, uncouth, difficult, abstract, professional, exclusive; to humanise it, to make it efficient outside the clique of the cultivated and learned, yet still remaining the *best* knowledge and thought of the time, and a true source, therefore, of sweetness and light. Such a man was Abelard in the Middle Ages, in spite of all his imperfections; and thence the boundless emotion and enthusiasm which Abelard excited. Such were Lessing and Herder in Germany, at the end of the last century; and their services to Germany were in this way inestimably precious. Generations will pass, and literary monuments will accumulate, and works far more perfect than the works of Lessing and Herder will be produced in Germany; and yet the names of these two men will fill a German with a reverence and enthusiasm such as the names of the most gifted masters will hardly awaken. And why? Because they *humanised* knowledge; because they broadened the basis of life and intelligence; because they worked powerfully to diffuse sweetness and light, to make reason and the will of God prevail. With Saint Augustine° they said: 'Let us not leave thee alone to make in the secret of thy knowledge, as thou didst before the creation of the firmament, the division of light from darkness; let the children of thy spirit, placed in their firmament, make their light shine upon the earth, mark the division of night and day, and announce the revolution of the times; for the old order is passed, and the new arises; the night is spent, the day is come forth; and thou shalt crown the year with thy blessing, when thou shalt send forth labourers into thy harvest sown by other hands than theirs; when thou shalt send forth new labourers to new seed-times, whereof the harvest shall be not yet.'

A Psychological Parallel

WHOEVER has to impugn the soundness of popular theology will most certainly find parts in his task which are unwelcome and painful. Other parts in it, however, are full of reward. And none more so than those, in which the work to be done is positive, not negative, and uniting, not dividing; in which what survives in Christianity is dwelt upon, not what perishes; and what offers us points of contact with the religion of the community, rather than motives for breaking with it. Popular religion is too forward to employ arguments which may well be called arguments of despair. 'Take me in the lump,' it cries, 'or give up Christianity altogether. Construe the Bible as I do, or renounce my public worship and solemnities; renounce all communion with me, as an imposture and falsehood on your part. Quit, as weak-minded, deluded blunderers, all those doctors and lights of the Church who have long served you, aided you, been dear to you. Those teachers set forth what are, in your opinion, errors, and go on grounds which you believe to be hollow. Whoever thinks as you do, ought, if he is courageous and consistent, to trust such blind guides no more, but to remain staunch by his new lights and himself.'

It happens, I suppose, to most people who treat an interesting subject, and it happens to me, to receive from those whom the subject interests, and who may have in general followed one's treatment of it with sympathy, avowals of difficulty upon certain points, requests for explanation. But the discussion of a subject, more especially of a religious subject, may easily be pursued longer than is advisable. On the immense difference which there seems to me to be between the popular conception of Christianity and the true conception of it, I have said what I wished to say. I wished to say it, partly in order to aid those whom the popular conception embarrassed; partly because, having frequently occasion to assert the truth and importance of Christianity against those who disparaged them, I was bound in honesty to make clear what sort of Christianity I meant. But having said, however imperfectly, what I wished, I leave, and am glad to leave, a discussion where the hope to do good must always be mixed with an apprehension of doing harm. Only, in leaving it, I will conclude with what cannot, one may hope, do harm: an endeavour to dispel some difficulties raised by the *arguments of despair*, as I have called them, of popular religion.

I have formerly spoken° at much length of the writings of St Paul, pointing out what a clue he gives us to the right understanding of the word *resurrection*, the great word of Christianity; and how he deserves, on this account, our special interest and study. It is the *spiritual* resurrection of which he is thus the instructive expounder to us. But undoubtedly he believed also in the miracle of the physical resurrection, both of Jesus himself and for mankind at large. This belief those who do not admit the miraculous will not share with him. And one who does not admit the miraculous, but who yet had continued to think St Paul worthy of all honour and his teaching full of instruction, brings forward to me a sentence from an eloquent and most popular author, wherein it is said that 'St Paul—surely no imbecile or credulous enthusiast—vouches for the reality of the (physical) resurrection, of the appearances of Jesus after it, and of his own vision.' Must then St Paul, he asks, if he was mistaken in thus vouching,—which whoever does not admit the miraculous cannot but suppose,—of necessity be an 'imbecile and credulous enthusiast', and his words and character of no more value to us than those of that slight sort of people? And again, my questioner finds the same author saying, that to suppose St Paul and the Evangelists mistaken about the miracles which they allege, is to 'insinuate that the faith of Christendom was founded on most facile and reprehensible credulity, and this in men who have taught the spirit of truthfulness as a primary duty of the religion which they preached'. And he inquires whether St Paul and the Evangelists, in admitting the miraculous, were really founding the faith of Christendom on most facile and reprehensible credulity, and were false to the spirit of truthfulness taught by themselves as the primary duty of the religion which they preached.

Let me answer by putting a parallel case. The argument is that St Paul, by believing and asserting the reality of the physical resurrection and subsequent appearances of Jesus, proves himself, supposing those alleged facts not to have happened, an imbecile or credulous enthusiast, and an unprofitable guide. St Paul's vision° we need not take into account, because even those who do not admit the miraculous will readily admit that he had his vision, only they say it is to be explained naturally. But they do not admit the reality of the physical resurrection of Jesus and of his appearances afterwards, while yet they must own that St Paul did. The question is, does either the belief of these things by a man of signal truthfulness, judgment, and mental power in St Paul's circumstances, prove them to have really happened; or does his believing them, in spite of their not having really happened, prove that he cannot have been a man of great truthfulness, judgment, and mental power?

Undeniably St Paul was mistaken about the imminence of the end of the world.° But this was a matter of expectation, not experience. If he was mistaken about a grave fact alleged to have already positively happened, such as the bodily resurrection of Jesus, he must, it is argued, have been a credulous and imbecile enthusiast.

II

I have already mentioned elsewhere[1] Sir Matthew Hale's belief in the reality of witchcraft. The contemporary records of this belief in our own country and among our own people, in a century of great intellectual force and achievement, and when the printing press fixed and preserved the accounts of public proceedings to which the charge of witchcraft gave rise, are of extraordinary interest. They throw an invaluable light for us on the history of the human spirit. I think it is not an illusion of national self-esteem to flatter ourselves that something of the English 'good nature and good humour'° is not absent even from these repulsive records; that from the traits of infuriated, infernal cruelty which charac- terise similar records elsewhere, particularly among the Latin nations, they are in a great measure free. They reveal, too, beginnings of that revolt of good sense, gleams of that reason, that criticism, which was presently to disperse the long-prevailing belief in witchcraft. At the beginning of the eighteenth century Addison,° though he himself looks with disfavour on a man who wholly disbelieves in ghosts and appar- itions, yet smiles at Sir Roger de Coverley's belief in witches, as a belief which intelligent men had outgrown, a survival from times of ignorance. Nevertheless, in 1716, two women were hanged at Huntingdon° for witchcraft. But they were the last victims, and in 1736 the penal statutes against witchcraft were repealed. And by the end of the eighteenth century, the majority of rational people had come to disbelieve, not in witches only, but in ghosts also. Incredulity had become the rule, credulity the exception.

But through the greater part of the seventeenth century things were just the other way. Credulity about witchcraft was the rule, incredulity the exception. It is by its all-pervadingness, its seemingly inevitable and natural character, that this credulity of the seventeenth century is distinguished from modern growths which are sometimes compared with it. In the addiction to what is called spiritualism, there is something factitious and artificial. It is quite easy to pay no attention to spiritualists

---

[1] *God and the Bible* [CPW vii, 369].

and their exhibitions; and a man of serious temper, a man even of matured sense, will in general pay none. He will instinctively apply Goethe's excellent caution: that we have all of us a nervous system which can easily be worked upon, that we are most of us very easily puzzled, and that it is foolish, by idly perplexing our understanding and playing with our nervous system, to titillate in ourselves the fibre of superstition. Whoever runs after our modern sorcerers may indeed find them. He may make acquaintance with their new spiritual visitants who have succeeded to the old-fashioned imps of the seventeenth century,—to the Jarmara, Elemauzer, Sack and Sugar, Vinegar Tom, and Grizzel Greedigut,° of our trials for witchcraft. But he may also pass his life without troubling his head about them and their masters. In the seventeenth century, on the other hand, the belief in witches and their works met a man at every turn, and created an atmosphere for his thoughts which they could not help feeling. A man who scouted the belief, who even disparaged it, was called Sadducee, atheist, and infidel.° Relations of the conviction of witches had their sharp word of 'condemnation for the particular opinion of some men who suppose there be none at all'. They had their caution to him 'to take heed how he either despised the power of God in his creatures, or vilipended the subtlety and fury of the Devil as God's minister of vengeance'.° The ministers of religion took a leading part in the proceedings against witches; the Puritan ministers were here particularly busy. Scripture had said: *Thou shalt not suffer a witch to live.*° And, strange to say, the poor creatures tried and executed for witchcraft appear to have usually been themselves firm believers in their own magic. They confess their compact with the Devil, and specify the imps, or familiars, whom they have at their disposal. All this, I say, created for the mind an atmosphere from which it was hard to escape. Again and again we hear of the 'sufficient justices of the peace and discreet magistrates', of the 'persons of great knowledge',° who were satisfied with the proofs of witchcraft offered to them. It is abundantly clear that to take as solid and convincing, where a witch was in question, evidence which would now be accepted by no reasonable man, was in the seventeenth century quite compatible with truthfulness of disposition, vigour of intelligence, and penetrating judgment on other matters.

Certainly these three advantages,—truthfulness of disposition, vigour of intelligence, and penetrating judgment,—were possessed in a signal degree by the famous Chief Justice of Charles the Second's reign, Sir Matthew Hale. Burnet° notices the remarkable mixture in him of sweetness with gravity, so to the three fore-named advantages we may add gentleness of temper. There is extant the report of a famous trial for witchcraft before Sir Matthew Hale.[1] Two widows of Lowestoft in Suffolk, named Rose Cullender and Amy Duny, were tried before him at Bury St Edmunds, at the Spring Assizes in 1664, as witches. The report was taken in Court during the trial, but was not published till eighteen years afterwards, in 1682. Every decade, at that time, saw a progressive decline in the belief in witchcraft. The person who published the report was, however, a believer; and he considered, he tells us, that 'so exact a relation of this trial would probably give more satisfaction to a great many persons, by reason that it is pure matter of fact, and that evidently demonstrated, than the arguments and reasons of other very learned men that probably may not be so intelligible to all readers; especially, this being held before a judge whom for his integrity, learning, and law, hardly any age either before or since could parallel; who not only took a great deal of pains and spent much time in this trial himself, but had the assistance and opinion of several other very eminent and learned persons'.° One of these persons was Sir Thomas Browne of Norwich, the author of the *Religio Medici* and of the book on *Vulgar Errors*.

The relation of the trial of Rose Cullender and Amy Duny° is indeed most interesting and most instructive, because it shows us so clearly how to live in a certain atmosphere of belief will govern men's conclusions from what they see and hear. To us who do not believe in witches, the evidence on which Rose Cullender and Amy Duny were convicted carries its own natural explanation with it, and itself dispels the charge against them. They were accused of having bewitched a number of children, causing them to have fits, and to bring up pins and nails. Several of the witnesses were poor ignorant people. The weighty evidence in the case was that of Samuel Pacy, a merchant of Lowestoft, two of whose children, Elizabeth and Deborah, of the ages of eleven and nine, were said to have been bewitched. The younger child was too ill to be brought to the Assizes, but the elder was produced in Court. Samuel Pacy, their father, is described as 'a man who carried himself with much soberness during the trial, from whom proceeded no words either of passion or malice, though his children were so greatly afflicted'. He deposed that his younger daughter, being lame and without power in her limbs, had on a sunshiny day in October 'desired to be carried on the east part of the house to be set upon the bank which looketh upon the sea'. While she sat there, Amy Duny, who as well as the other prisoner is shown by the evidence to have been by her neighbours commonly reputed a witch, came to the house to get some herrings. She was

[1] Reprinted in *A Collection of Rare and Curious Tracts relating to Witchcraft*. London, 1838.

refused, and went away grumbling. At the same moment the child was seized with violent fits. The doctor who attended her could not explain them. So ten days afterwards her father, according to his own deposition, 'by reason of the circumstances aforesaid, and in regard Amy Duny is a woman of an ill fame and commonly reported to be a witch and a sorceress, and for that the said child in her fits would cry out of Amy Duny as the cause of her malady, and that she did affright her with apparitions of her person, as the child in the intervals of her fits related, did suspect the said Amy Duny for a witch, and charged her with the injury and wrong to his child, and caused her to be set in the stocks'. While she was there, two women asked her the reason of the illness of Mr Pacy's child. She answered: 'Mr Pacy keeps a great stir about his child, but let him stay until he hath done as much by his children as I have done by mine.' Being asked what she had done to hers, she replied that 'she had been fain to open her child's mouth with a tap to give it victuals'. Two days afterwards Pacy's elder daughter, Elizabeth, was seized with fits like her sister's; 'insomuch that they could not open her mouth to preserve her life without the help of a tap which they were obliged to use'. The children in their fits would cry out: 'There stands Amy Duny' or 'Rose Cullender' (another reputed witch of Lowestoft); and, when the fits were over, would relate how they had seen Amy Duny and Rose Cullender shaking their fists at them and threatening them. They said that bees or flies carried into their mouths the pins and nails which they brought up in their fits. During their illness their father sometimes made them read aloud from the New Testament. He 'observed that they would read till they came to the name of *Lord*, or *Jesus*, or *Christ*, and then before they could pronounce either of the said words they would suddenly fall into their fits. But when they came to the name of *Satan* or *Devil* they would clap their fingers upon the book, crying out: "This bites, but makes me speak right well." ' And when their father asked them why they could not pronounce the words *Lord*, or *Jesus*, or *Christ*, they answered: 'Amy Duny saith, I must not use that name.'

It seems almost an impertinence nowadays to suppose, that any one can require telling how self-explanatory all this is, without recourse to witchcraft and magic. These poor rickety children, full of disease and with morbid tricks, have their imagination possessed by the two famed and dreaded witches of their native place, of whose prowess they have heard tale after tale, whom they have often seen with their own eyes, whose presence has startled one of them in her hour of suffering, and round whom all those ideas of diabolical agency, in which they have been nursed, converge and cluster. The speech of the accused witch in the stocks is the most natural speech possible, and the fulfilment which her words received in the course of Elizabeth Pacy's fits is perfectly natural also. However, Sir Thomas Browne (who appears in the report of the trial as 'Dr Brown, of Norwich, a person of great knowledge'), being desired to give his opinion on Elizabeth Pacy's case and that of two other children who on similar evidence were said to have been bewitched by the accused,—Sir Thomas Browne

'was clearly of opinion that the persons were bewitched; and said that in Denmark there had been lately a great discovery of witches, who used the very same way of afflicting persons by conveying pins into them, and crooked, as these pins were, with needles and nails. And his opinion was that the Devil in such cases did work upon the bodies of men and women upon a natural foundation, . . . for he conceived that these swooning fits were natural, and nothing else but what they call *the mother*, but only heightened to a great excess by the subtlety of the Devil, co-operating with the malice of these which we term witches, at whose instance he doth these villainies.'

That was all the light to be got from the celebrated writer on *Vulgar Errors*. Yet reason, in this trial, was not left quite without witness:—

'At the hearing the evidence, there were divers known persons, as Mr Serjeant Keeling, Mr Serjeant Earl, and Mr Serjeant Bernard, present. Mr Serjeant Keeling seemed much unsatisfied with it, and thought it not sufficient to convict the prisoners; for admitting that the children were in truth bewitched, yet, said he, it can never be applied to the prisoners upon the imagination only of the parties afflicted. For if that might be allowed, no person whatsoever can be in safety; for perhaps they might fancy another person, who might altogether be innocent in such matters.'

In order, therefore, the better to establish the guilt of the prisoners, they were made to touch the children whom they were said to have bewitched. The children screamed out at their touch. The children were 'blinded with their own aprons', and in this condition were again touched by Rose Cullender; and again they screamed out. It was objected, not that the children's heads were full of Rose Cullender and Amy Duny and of their infernal dealings with them, but that the children might be counterfeiting their malady and pretending to start at the witch's touch though it had no real power on them:—

'Wherefore, to avoid this scruple, it was privately desired by the judge, that the Lord Cornwallis, Sir Edward Bacon, Mr Serjeant Keeling, and some other gentlemen then in Court, would attend one of the distempered persons in the further part of the hall, whilst she was in her fits, and then to send for one of the witches to try what would then happen, which they did accordingly. And Amy

Duny was conveyed from the bar and brought to the maid; they put an apron before her eyes, and then one other person touched her hand, which produced the same effect as the touch of the witch did in the Court. Whereupon the gentlemen returned, openly protesting that they did believe the whole transaction of this business was a mere imposture.'

This, we are told, 'put the Court and all persons into a stand. But at length Mr Pacy did declare that possibly the maid might be deceived by a suspicion that the witch touched her when she did not.' And nothing more likely; but what does this prove? That the child's terrors were sincere; not that the so-called witch had done the acts alleged against her. However, Mr Pacy's solution of the difficulty was readily accepted. If the children were not shamming out of malice or from a love of imposture, then 'it is very evident that the parties were bewitched, and that when they apprehend that the persons who have done them this wrong are near, or touch them, then, their spirits being more than ordinarily moved with rage and anger, they do use more violent gestures of their bodies'.

Such was the evidence. The accused did not confess themselves guilty. When asked what they had to say for themselves, they replied, as well they might: 'Nothing material to anything that had been proved.' Hale then charged the jury. He did not even go over the evidence to them:—

'Only this he acquainted them: that they had two things to inquire after. First, whether or no these children were bewitched; secondly, whether the prisoners at the bar were guilty of it. That there were such creatures as witches he made no doubt at all. For, first, the Scriptures had affirmed so much; secondly, the wisdom of all nations had provided laws against such persons, which is an argument of their confidence of such a crime. And such hath been the judgment of this kingdom, as appears by that Act of Parliament which hath provided punishments proportionable to the quality of the offence. And he desired them strictly to observe their evidence, and desired the great God of Heaven to direct their hearts in this weighty thing they had in hand. For to condemn the innocent, and to let the guilty go free, were both an abomination to the Lord.'

The jury retired. In half an hour they came back with a verdict of *guilty* against both prisoners. Next morning the children who had been produced in court were brought to Hale's lodgings, perfectly restored:—

'Mr Pacy did affirm, that within less than half an hour after the witches were convicted, they were all of them restored, and slept well that night; only Susan Chandler felt a pain like pricking of pins in her stomach.'

And this seems to have removed all shadow of doubt or misgiving:—

'In conclusion, the judge and all the court were fully satisfied with the verdict, and thereupon gave judgment against the witches that they should be hanged. They were much urged to confess, but would not. That morning we departed for Cambridge; but no reprieve was granted, and they were executed on Monday, the 17th of March (1664) following, but they confessed nothing.'

Now, the inference to be drawn from this trial is not by any means that Hale was 'an imbecile or credulous enthusiast'. The whole history of his life and doings disproves it. But the belief in witchcraft was in the very atmosphere which Hale breathed, as the belief in miracle was in the very atmosphere which St Paul breathed. What the trial shows us is, that a man of veracity, judgment, and mental power, may have his mind thoroughly governed, on certain subjects, by a foregone conclusion as to what is likely and credible. But I will not further enlarge on the illustration which Hale furnishes to us of this truth. An illustration of it, with a yet closer applicability to St Paul, is supplied by another worthy of the seventeenth century.

III

The worthy in question is very little known, and I rejoice to have an opportunity of mentioning him. *John Smith*!—the name does not sound promising. He died at the age of thirty-four, having risen to no higher post in the world than a college fellowship. 'He proceeded leisurely by orderly steps,' says Simon Patrick, afterwards Bishop of Ely, who preached his funeral-sermon, 'not to what he could get, but to what he was fit to undertake.'° John Smith, born in 1618 near Oundle in Northamptonshire, was admitted a scholar of Emanuel College at Cambridge in 1636, a fellow of Queen's° College in 1644. He became a tutor and preacher in his college; died there, 'after a tedious sickness', on the 7th of August 1652, and was buried in his college-chapel. He was one of that band of Cambridge Platonists, or *latitude men*, as in their own day they were called, whom Burnet has well described° as those 'who, at Cambridge, studied to propagate better thoughts, to take men off from being in parties, or from narrow notions, from superstitious conceits and fierceness about opinions'. Principal Tulloch° has done an excellent work in seeking to reawaken our interest in this noble but neglected group. His book[1] is delightful, and it has, at the same time, the most serious value. But in his account of his worthies, Principal Tulloch has given, I cannot but think, somewhat too much space to their Platonic

---

[1] *Rational Theology and Christian Philosophy in England in the Seventeenth Century*; 2d edition, Edinburgh and London, 1874.

philosophy, to their disquisitions on spirit and incorporeal essence. It is not by these that they merited to live, or that, having passed away from men's minds, they will be brought back to them. It is by their extraordinarily simple, profound, and just conception of religion. Placed between the sacerdotal religion of the Laudian clergy° on the one side, and the notional religion of the Puritans on the other, they saw the sterility, the certain doom of both;—saw that stand permanently such developments of religion could not, inasmuch as Christianity was not what either of them supposed, but was a *temper*, a *behaviour*.

Their immediate recompense was a religious isolation of two centuries. The religious world was not then ripe for more than the High Church conception of Christianity on the one hand, or the Puritan conception on the other. The Cambridge band ceased to acquire recruits, and disappeared with the century. Individuals knew and used their writings; Bishop Wilson° of Sodor and Man, in particular, had profited by them. But they made no broad and clear mark. And this was in part for the reason already assigned, in part because what passed for their great work was that revival of a spiritualist and Platonic philosophy, to which Principal Tulloch, as I have said, seems to me to have given too much prominence. By this attempted revival they could not and cannot live. The theology and writings of Owen° are not more extinct than the *Intellectual System* of Cudworth.° But in a history of the Cambridge Platonists, works of the magnitude of Cudworth's *Intellectual System of the Universe* must necessarily, perhaps, fill a large space. Therefore it is not so much a history of this group which is wanted, as a republication of such of their utterances as show us their real spirit and power. Their spiritual brother, 'the ever memorable° Mr John Hales', must certainly, notwithstanding that he was at Oxford, not Cambridge, be classed along with them. The remains of Hales of Eton, the sermons and aphorisms of Whichcote,° the sermon preached by Cudworth before the House of Commons with the second sermon printed as a companion to it, single sayings and maxims of Henry More, and the *Select Discourses* of John Smith,—there are our documents! In them lies enshrined what the *latitude men* have of value for us. It were well if Principal Tulloch would lay us under fresh obligations by himself extracting this and giving it to us; but given some day, and by some hand,° it will surely be.

For Hales and the Cambridge Platonists here offer, formulated with sufficient distinctness, a conception of religion true, long obscured, and for which the hour of light has at last come. Their productions will not, indeed, take rank as great works of literature and style. It is not to the history of literature that Whichcote and Smith belong, but to the history of religion. Their contemporaries were Bossuet, Pascal, Taylor, Barrow.° It is in the history of literature that these men are mainly eminent, although they may also be classed, of course, among religious writers. What counts highest in the history of religion as such, is, however, to give what at critical moments the religious life of mankind needs and can use. And it will be found that the Cambridge Platonists, although neither epoch-making philosophers nor epoch-making men of letters, have in their conception of religion a boon for the religious wants of our own time, such as we shall demand in vain from the soul and poetry of Taylor, from the sense and vigour of Barrow, from the superb exercitations of Bossuet, or the passion-filled reasoning and rhetoric of Pascal.

The *Select Discourses* of John Smith, collected and published from his papers after his death, are, in my opinion, by much the most considerable work left to us by this Cambridge school. They have a right to a place in English literary history. Yet the main value of the *Select Discourses* is, I repeat, religious, not literary. Their grand merit is that they insist on the profound *natural truth* of Christianity, and thus base it upon a ground which will not crumble under our feet. Signal and rare indeed is the merit, in a theological instructor, of presenting Christianity to us in this fashion. Christianity is true; but in general the whole plan for grounding and buttressing it chosen by our theological instructors is false, and, since it is false, it must fail us sooner or later. I have often thought that if candidates for orders were simply, in preparing for their examination, to read and digest Smith's great discourse,° *On the Excellency and Nobleness of True Religion*, together with M. Reuss's *History*° *of Christian Theology at the time of the Apostles*, and nothing further except the Bible itself, we might have, perhaps, a hope of at last getting, as our national guides in religion, a clergy which could tell its bearings and steer its way, instead of being, as we now see it, too often conspicuously at a loss to do either.

Singularly enough, about fifteen years before the trial at Bury St Edmunds of the Lowestoft witches, John Smith, the author of the *Select Discourses*, had in those very eastern counties to deliver his mind on the matter of witchcraft. On Lady-day° every year, a Fellow of Queen's College, Cambridge, was required to preach at Huntingdon a sermon against witchcraft and diabolical contracts. Smith, as one of the Fellows of Queen's, had to preach this sermon.° It is printed tenth and last of his *Select Discourses*, with the title: *A Christian's Conflicts and Conquests; or, a Discourse concerning the Devil's Active Enmity and Continual Hostility against Man, the Warfare of a Christian Life, the Certainty of Success and Victory in this Spiritual Warfare, the Evil and Horridness of Magical Arts and*

Rites, Diabolical Contracts, &c. The discourse has for its text the words: 'Resist the devil, and he will flee from you.'°

The preacher sets out with the traditional account of 'the prince of darkness, who, having once stained the original beauty and glory of the divine workmanship, is continually striving to mould and shape it more and more into his own likeness'. He says:—

'It were perhaps a vain curiosity to inquire whether the number of evil spirits exceeds the number of men; but this is too, too certain, that we never want the secret and latent attendance of them. . . . Those evil spirits are not yet cast out of the world into outer darkness, though it be prepared for them; the bottomless pit hath not yet shut its mouth upon them.'

And he concludes his sermon with a reflection and a caution, called for, he says, by the particular occasion. The reflection is that—

'Did we not live in a world of professed wickedness, wherein so many men's sins go in open view before them to judgment, it might be thought needless to persuade men to resist the devil when he appears in his own colours to make merchandise of them, and comes in a formal way to bargain with them for their souls; that which human nature, however enthralled to sin and Satan in a more mysterious way, abhors, and none admit but those who are quite degenerated from human kind.'

And he adds the caution, that—

'The use of any arts, rites or ceremonies not understood, of which we can give no rational or divine account, this indeed is nothing else but a kind of magic which the devil himself owns and gives life to, though he may not be corporeally present, or require presently any further covenant from the users of them. The devil, no question, is present to all his own rites and ceremonies, though men discern him not, and may upon the use of them secretly produce those effects which may gain credit to them. Among these rites we may reckon insignificant forms of words, with their several modes and manners of pronunciation, astrological arts, and whatsoever else pretends to any strange effects which we cannot with good reason either ascribe to God or nature. As God will only be conversed withal in a way of light and understanding, so the devil loves to be conversed with in a way of darkness and obscurity.'

But between his exordium and his conclusion the real man appears. Like Hale, Smith seems to have accepted the belief in witchcraft and in diabolical contracts which was regnant in his day. But when he came to deal with the belief as an idea influencing thought and conduct, he could not take it as the people around him took it. It was his nature to seek a firm ground for the ideas admitted by him; above all, when these ideas had bearings upon religion. And for witchcraft and diabolical operation, in the common conception of them as external things, he could find no solid ground, for there was none; and therefore he could not so use them. See, therefore, how profoundly they are transformed by him! After his exordium he makes an entirely fresh departure:—'When we say the devil is continually busy with us, I mean not only some apostate spirit as one particular being, but that spirit of apostasy which is lodged in all men's natures.' Here, in this *spirit of apostasy which is lodged in all men's natures*, Smith had what was at bottom experimental and real.° And the whole effort of the sermon is to substitute this for what men call the devil, hell, fiends, and witches, as an object for their serious thought and strenuous resistance:—

'As the kingdom of heaven is not so much without men as within, as our Saviour tells us; so the tyranny of the devil and hell is not so much in some external things as in the qualities and dispositions of men's minds. And as the enjoying of God, and conversing with him, consists not so much in a change of place as in the participation of the divine nature and in our assimilation unto God; so our conversing with the devil is not so much by a mutual local presence as by an imitation of a wicked and sinful nature derived upon men's own souls. . . . He that allows himself in any sin, or useth an unnatural dalliance with any vice, does nothing else in reality than entertain an *incubus demon*.'

This, however, was by no means a view of diabolical possession acceptable to the religious world and to its Puritan ministers:—

'I know these expressions will seem to some very harsh and unwelcome; but I would beseech them to consider what they will call that spirit of malice and envy, that spirit of pride, ambition, vain-glory, covetousness, injustice, uncleanness, etc., that commonly reigns so much and acts so violently in the minds and lives of men. Let us speak the truth, and call things by their own names; so much as there is of sin in any man, so much there is of the diabolical nature. Why do we defy the devil so much with our tongues, while we entertain him in our hearts? As men's love to God is ordinarily nothing else but the mere tendency of their natures to something that hath the name of God put upon it, without any clear or distinct apprehension of him, so their hatred of the devil is commonly nothing else but an inward displacency of nature against something entitled by the devil's name. And as they commonly make a God like to themselves, such a one as they can best comply with and love, so they make a devil most unlike to themselves, which may be anything but what they themselves are, that so they may most freely spend their anger and hatred upon him; just as they say of some of the Ethiopians who used to paint the devil white because they themselves are black. This is a strange, merry kind of madness, whereby men sportingly bereave themselves of the supremest good, and insure themselves, as much as may be, to hell and misery; they may thus cheat themselves for a while, but the eternal foundation of the Divine Being is immutable and unchangeable. And where we find wisdom, justice, loveliness, goodness, love, and glory in their highest elevations and most unbounded dimensions, that is He; and where we find any true participations of these, there is a true communication of God; and a defection from these is the essence of sin and the foundation of hell.'

Finally (and I quote the more freely because the author whom I quote is so little known),—finally our preacher goes on to even confute his own exordium:—

'It was the fond error of the Manichees that there was some solid *principium mali*, which, having an eternal existence of its own, had also a mighty and uncontrollable power from within itself whereby it could forcibly enter into the souls of men, and, seating itself there, by some hidden influences irresistibly incline and inforce them to evil. But we ourselves uphold that kingdom of darkness, which else would tumble down and slide into that nothing from whence it came. *All sin and vice is our own creature*; we only give life to them which indeed are our death, and would soon wither and fade away did we substract our concurrence from them.'

O fortunate Huntingdon Church, which admitted for even one day such a counterblast to the doctrines then sounding from every pulpit, and still enjoined by Sir Robert Phillimore!°

That a man shares an error of the minds around him and of the times in which he lives, proves nothing against his being a man of veracity, judgment, and mental power. This we saw by the case of Hale. But here, in our Cambridge Platonist, we have a man who accepts the erroneous belief in witchcraft, professes it publicly, preaches on it; and yet is not only a man of veracity and intelligence, but actually manages to give to the error adopted by him a turn, an aspect, which indicates its erroneousness. Not only is he of help to us generally, in spite of his error; he is of help to us in respect of that very error itself.

Now, herein is really a most striking analogy between our little-known divine of the seventeenth century and the great Apostle of the Gentiles. St Paul's writings are in every one's hands. I have myself discussed his doctrine at length. And for our present purpose there is no need of elaborate exposition and quotation. Every one knows how St Paul declares his belief that 'Christ rose again the third day, and was seen of Cephas, then of the twelve; after that, he was seen of above five hundred brethren at once.'[1] Those who do not admit the miraculous can yet well conceive how such a belief arose, and was entertained by St Paul. *The resurrection of the just* was at that time a ruling idea of a Jew's mind. Herod at once, and without difficulty, supposed that John the Baptist° was *risen*

---

[1] 1 Cor. xv. 4, 5, 6.

from the dead. The Jewish people without difficulty supposed that Jesus might be one of the old prophets,° *risen from the dead.* In telling the story of the crucifixion men added, quite naturally, that when it was consummated, 'many bodies of the saints which slept *arose and appeared unto many*'.° Jesus himself, moreover, had in his lifetime spoken frequently of his own coming resurrection.° Such beliefs as the belief in bodily resurrection were thus a part of the mental atmosphere in which the first Christians lived. It was inevitable that they should believe their Master to have risen again in the body, and that St Paul, in becoming a Christian, should receive the belief and build upon it.

But Paul, like our Cambridge Platonist, instinctively sought in an idea used for religion a side by which the idea could enter into his religious experience and become real to him. No such side could be afforded by the mere external fact and miracle of Christ's bodily resurrection. Paul, therefore, as is well known, by a prodigy of religious insight seized another aspect for the resurrection than the aspect of physical miracle. He presented resurrection as a spiritual rising which could be appropriated and enacted in our own living experience. 'If One died° in the name of all, then all died; and he died in the name of all, that they who live should no more live unto themselves, but unto him who died and rose again in their name.'[1] Dying became thus no longer a bodily dying, but a dying to sin; rising to life no longer a bodily resurrection, but a living to God. St Paul here comes, therefore, upon that very idea of death and resurrection which was the central idea of Jesus himself. At the very same moment that he shares and professes the popular belief in Christ's miraculous bodily resurrection,—the idea by which our Saviour's own idea of resurrection has been overlaid and effaced,—St Paul seizes also this other truer idea or is seized by it, and bears unconscious witness to its unique legitimacy.

Where, then, is the force of that *argument of despair*, as we called it, that if St Paul vouches for the bodily resurrection of Jesus and for his appearance after it, and is mistaken in so vouching, then he must be 'an imbecile and credulous enthusiast', untruthful, unprofitable? We see that for a man to believe in preternatural incidents, of a kind admitted by the common belief of his time, proves nothing at all against his general truthfulness and sagacity. Nay, we see that even while affirming such preternatural incidents, he may with profound insight seize the true and natural aspect of them, the aspect which will survive and profit when the miraculous aspect has faded. He may give us, in the very same work,

---

[1] 2 Cor. v. 14, 15.

current error and also fruitful and profound new truth, the error's future corrective.

IV

But I am treating of these matters for the last time. And those who no longer admit, in religion, the old basis of the preternatural, I see them encountered by scruples of their own, as well as by scruples raised by their opponents. Their opponents, the partisans of miracle, require them if they refuse to admit miracle to throw aside as imbecile or untruthful all their instructors and inspirers who have ever admitted it. But they themselves, too, are sometimes afraid, not only of being called inconsistent and insincere, but of really meriting to be called so, if they do not break decidedly with the religion in which they have been brought up, if they at all try still to conform to it and to use it. I have now before me a remarkable letter, in which the writer says:—

'There is nothing I and many others should like better than to take service as ministers in the Church as *a national society for the promotion of goodness;*° but how can we do so, when we have first to declare our belief in a quantity of things which every intelligent man rejects?'

Now, as I have examined the question whether a man who rejects miracles must break with St Paul because Paul asserted them, so let me, before I end, examine the question whether such a man must break with the Church of his country and childhood.

Certainly it is a strong thing to suppose, as the writer of the above-quoted letter supposes, a man taking orders in the Church of England who accepts, say, the view of Christianity offered in *Literature and Dogma.* For the Church of England presents as science, and as necessary to salvation, what it is the very object of that book to show to be *not* science and *not* necessary to salvation. And at his ordination a man is required to declare that he, too, accepts this for science, as the Church does. Formerly° a deacon subscribed to the Thirty-nine Articles, and to a declaration that he acknowledged 'all and every the articles therein contained to be agreeable to the word of God'. A clerk, admitted to a benefice with cure, declared 'his unfeigned assent and consent to all the matters contained in the Articles'. At present, I think, all that is required is a general consent to whatever is contained in the Book of Common Prayer. But the Book of Common Prayer contains the Thirty-nine Articles. And the Eighth Article declares the Three Creeds° to be science, science 'thoroughly to be received and believed'. Now, whether one professes an 'unfeigned assent and consent' to this Article, as contained among the Thirty-nine Articles, or merely 'a general consent' to it, as contained in the Prayer Book, one certainly, by consenting to it at all, professes to receive the Three Creeds as science, and as true science. And this is the very point where it is important to be explicit and firm. Whatever else the Three Creeds may be, they are not science, truly formulating the Christian religion. And no one who feels convinced that they are not, can sincerely say that he gives even a general consent to whatever is contained in the Prayer Book, or can at present, therefore, be ordained a minister of the Church of England.

The obstacle, it will be observed, is in a test which lies outside of the Ordination Service itself. The test is a remnant of the system of subscriptions and tests formerly employed so vigorously. It was meant as a reduction and alleviation of that old yoke. To obtain such a reduction seemed once to generous and ardent minds, and indeed once was, a very considerable conquest. But the times move rapidly, and even the reduced test has now a great power of exclusion. If it were possible for Liberal politicians ever to deal seriously with religion, they would turn their minds to the removal of a test of this sort, instead of playing with political dissent or marriage with a deceased wife's sister.° The Ordination Service itself, on a man's entrance into orders, and the use of the Church services afterwards, are a sufficient engagement. Things were put into the Ordination Service which one might have wished otherwise. Some of them are gone. The introduction of the Oath of Supremacy was a part, no doubt, of all that *lion and unicorn* business which is too plentiful in our Prayer Book, on which Dr Newman° has showered such exquisite raillery, and of which only the Philistine element in our race prevents our seeing the ridiculousness. But the Oath of Supremacy° has now no longer a place in the Ordination Service. Apart, however, from such mere matters of taste, there was and still is the requirement, in the Ordering of Deacons, of a declaration of unfeigned belief in all the canonical Scriptures of the Old and New Testament. Perhaps this declaration can have a construction put upon it which makes it admissible. But by its form of expression it recalls, and appears to adopt, the narrow and letter-bound views of Biblical inspiration formerly prevalent,—prevalent with the Fathers as well as with the Reformers,—but which are now, I suppose, generally abandoned. I imagine the clergy themselves would be glad to substitute for this declaration the words in the Ordering of Priests, where the candidate declares himself 'persuaded that the Holy Scriptures contain sufficiently all doctrine required for eternal salvation through faith in Jesus Christ'. These words present no difficulty, nor is there any other serious difficulty, that I can see, raised by the Ordination Service for either priests or deacons. The declaration of a general consent to the Articles is another matter; although perhaps, in the present temper of men's minds, it could not easily be got rid of.

The last of Butler's jottings in his memorandum-book is a prayer to be delivered 'from *offendiculum* of scrupulousness'.° He was quite right. Religion is a matter where scrupulousness has been far too active, producing most serious mischief; and where it is singularly out of place. I am the very last person to wish to deny it. Those, therefore, who declared their consent to the Articles long ago, and who are usefully engaged in the ministry of the Church, would in my opinion do exceedingly ill to disquiet themselves about having given a consent to the Articles formerly, when things had not moved to the point where they are now, and did not appear to men's minds as they now appear. 'Forgetting those things which are behind and reaching forth unto those things which are before',° should in these cases be a man's motto. The Church is properly a national society for the promotion of goodness. For him it is such; he ministers in it as such. He has never to use the Articles, never to rehearse them. He has to rehearse the prayers and services of the Church. Much of these he may rehearse as the literal, beautiful rendering of what he himself feels and believes. The rest he may rehearse as an approximative rendering° of it;—as language *thrown out* by other men, in other times, at immense objects which deeply engaged their affections and awe, and which deeply engage his also; objects concerning which, moreover, adequate statement is impossible. To him, therefore, this approximative part of the prayers and services which he rehearses will be poetry. It is a great error to think that whatever is thus perceived to be poetry ceases to be available in religion. The noblest races are those which know how to make the most serious use of poetry.

But the Articles are plain prose. They aim at the exactitude of a legal document. They are a precise profession of belief, formulated by men of our own nation three hundred years ago, in regard, amongst other things, to parts of those services of the Church of which we have been speaking. At all points the Articles are, and must be, inadequate; but into the question of their general inadequacy we need not now enter. One point is sufficient. They present the Creeds as science, exact science; and this, at the present time of day, very many a man cannot accept. He cannot rightly, then, profess in any way to accept it; cannot, in consequence, take orders.

But it is easy for such a man to exaggerate to himself the barrier between himself and popular religion. The barrier is not so great as he may suppose; and it is expedient for him rather to think it less great than it is, than more great. It will insensibly dwindle, the more that he, and other serious men who think as he does, strive so far as they can to act as if it did not exist. It will stand stiff and bristling the more they act as if it were insurmountable. The Church of our country is to be considered as a national Christian society for the promotion of goodness, to which a man cannot but wish well, and in which he might rejoice to minister. To a right-judging mind, the cardinal points of belief for either the member or the minister of such a society are but two: *Salvation by Righteousness* and *Righteousness by Jesus Christ.* Salvation by Righteousness,—there is the sum of the Old Testament: Righteousness by Jesus Christ,—there is the sum of the New. For popular religion, the cardinal points of belief are of course a good deal more numerous. Not without adding many others could popular religion manage to benefit by the first-named two. But the first-named two have its adherence. It is from the very effort to benefit by them that it has added all the rest. The services of the Church are full of direct recognitions of the two really essential points of Christian belief: *Salvation by Righteousness* and *Righteousness by Jesus Christ.* They are full, too, of what may be called approximate recognitions of them;—efforts of the human mind, in its gradual growth, to develop them, to fix them, to buttress them, to make them clearer to itself, to bring them nearer, by the addition of miracle and metaphysic. This is poetry. The Articles say that this poetry is exact prose. But the Articles are no more a real element of the Prayer Book than Brady and Tate's metrical version° of the Psalms, which has now happily been expelled. And even while the Articles continue to stand in the Prayer Book, yet a layman can use the Prayer Book as if they and their definitions did not exist. To be ordained, however, one must adhere to their definitions. But, putting the Articles aside, will a layman, since he is free, would a clergyman, if he were free, desire to abandon the use of all those parts of the Prayer Book which are to be regarded as merely approximative recognitions of its two central truths, and as poetry? Must all such parts one day, as our experience widens and this view of their character comes to prevail, be eliminated from our public worship? The question is a most important one.

For although the Comtists, by the mouth of their most eloquent spokesman,° tell us that ''tis the pedantry of sect alone which can dare to monopolize to a special creed those precious heirlooms of a common race', the ideas and power of religion, and propose to remake religion for us with new and improved personages, and rites, and words; yet it is certain that here as elsewhere the wonderful force of habit tells, and that the power of religious ideas over us does not spring up at call, but is intimately dependent upon particular names and practices and forms of expression which have gone along with it ever since we can remember, and which have created special sentiments in us. I believe, indeed, that the eloquent spokesman of the Comtists errs at the very outset. I believe that the power of religion does of nature belong, in a unique way, to the Bible and to Christianity, and that it is no pedantry of sect which affirms this, but experience. Yet even were it as he supposes, and Christianity were not the one proper bringer-in of righteousness° and of the reign of the Spirit and of eternal life, and these were to be got as well elsewhere, but still we ourselves had learnt all we know about them from Christianity,—then for us to be taught them in some other guise, by some other instructor, would be almost impossible. Habits and associations are not formed in a day. Even if the very young have time enough before them to learn to associate religion with new personages and precepts, the middle-aged and the old have not, and must shrink from such an endeavour. *Mane nobiscum, Domine, nam advesperascit.*°

Nay, but so prodigious a revolution does the changing the whole form and feature of religion turn out to be, that it even unsettles all other things too, and brings back chaos. When it happens, the civilisation and the society to which it happens are disintegrated, and men have to begin again. This is what took place when Christianity superseded the old religion of the Pagan world. People may say that there is a fund of ideas common to all religions, at least to all religions of superior and civilised races; and that the personages and precepts, the form and feature, of one such religion may be exchanged for those of another, or for those of some new religion devised by an enlightened eclecticism, and the world may go on all the while without much disturbance. There were philosophers who thought so when Paganism was going out and Christianity coming in. But they were mistaken. The whole civilisation of the Roman world was disintegrated by the change, and men had, I say, to begin again. So immense is the sentiment created by the things to which we have been used in religion, so profound is the wrench at parting with them, so incalculable is the trouble and distraction caused by it. Now, we can hardly conceive modern civilisation breaking up as the Roman did, and men beginning again as they did in the fifth century. But the improbability of this implies the improbability, too, of our seeing all the form and feature of Christianity disappear,—of the religion of Christendom. For so vast a revolution would this be, that it would involve the other.

These considerations are of force, I think, in regard to all radical change in the language of the Prayer Book. It has created sentiments deeper than we can see or measure. Our feeling does not connect itself with *any* language about righteousness and religion, but with *that* language. Very much of it we can all use in its literal acceptation. But the question is as to those parts which we cannot. Of course, those who can take them literally will still continue to use them. But for us also, who can no longer put the literal meaning on them which others do, and which we ourselves once did, they retain a power, and something in us vibrates to them. And not unjustly. For these old forms of expression were men's sincere attempt to set forth with due honour what we honour also; and the sense of the attempt gives a beauty and an emotion to the words, and makes them poetry. The Creeds are in this way an attempt to exalt to the utmost, by assigning to him all the characters which to mankind seemed to confer exaltation, Jesus Christ. I have elsewhere called° the Apostles' Creed the popular science of Christianity, and the Nicene Creed its learned science; and in one view of them they are so. But in another and a better view of them, they are, the one its popular poetry, the other its learned or,—to borrow the word which Schopenhauer applied to Hegel's philosophy,°—its *scholastic poetry*. The one Creed exalts Jesus by concrete images, the other by an imaginative play of abstract ideas. These two Creeds are the august amplifications, or the high elucidations, which came naturally to the human spirit working in love and awe upon that inexhaustible theme of profound truth: *Salvation through Jesus Christ.*° As such, they are poetry for us; and poetry consecrated, moreover, by having been on the tongue of all our forefathers for two thousand years, and on our own tongue ever since we were born. As such, then, we can *feel* them, even when we no longer take them literally; while, as approximations to a profound truth, we can *use* them. We cannot call them science, as the Articles would have us; but we can still feel them and still use them. And if we can do this with the Creeds, still more can we do it with the rest of the services in the Prayer Book.

As to the very and true foundations, therefore, of the Christian religion,—the belief that salvation is by righteousness, and that righteousness is by Jesus Christ,—we are, in fact, at one with the religious world in general. As to the true object of the Church, that it is the promotion of goodness, we are at one with them also. And as to the form and wording of religion,—a form and wording consecrated by so many years and memories,—even as to this we need not break with them either. They and we can remain in sympathy. Some changes will no doubt befall the Prayer Book as time goes on. Certain things will drop away from its services, other things will replace them. But such change will happen, not in a sweeping way;—it will come very gradually, and by the general wish. It will be brought about, not by a spirit of scrupulosity, innovation, and negation, but by a prevalent impulse to express in our church-services somewhat which is felt to need expression, and to be not sufficiently expressed there already.

After all, the great confirmation to a man in believing that the cardinal points of our religion are far fewer and simpler than is commonly supposed, is that such was surely the belief of Jesus himself. And in like manner, the great reason for continuing to use the familiar language of the religion around us as approximative language, and as poetry, although we cannot take it literally, is that such was also the practice of Jesus. For evidently it was so. And evidently, again, the immense misapprehension of Jesus and of his meaning, by popular religion, comes in part from such having been his practice. But if Jesus used this way of speaking in spite of its plainly leading to such misapprehension, it must have been because it was the best way and the only one. For it was not by introducing a brand-new religious language, and by parting with all the old and cherished images, that popular religion could be transformed; but by keeping the old language and images, and as far as possible conveying into them the soul of the new Christian ideal.

When Jesus talked° of the Son of Man coming in his glory with the holy angels, setting the good on his right hand and the bad on his left, and sending away the bad into everlasting fire prepared for the devil and his angels, was he speaking literally? Did Jesus mean that all this would actually happen? Popular religion supposes so. Yet very many religious people, even now, suppose that Jesus was but using the figures of Messianic judgment familiar to his hearers, in order to impress upon them his main point:—what sort of spirit and of practice did really tend to salvation, and what did not. And surely almost every one must perceive, that when Jesus spoke° to his disciples of their sitting on thrones judging the twelve tribes of Israel, or of their drinking new wine with him in the kingdom of God, he was adopting their material images and beliefs, and was not speaking literally. Yet their Master's thus adopting their material images and beliefs could not but confirm the disciples in them. And so it did, and Christendom, too, after them; yet in this way, apparently, Jesus chose to proceed. But some one may say, that Jesus used this language because he himself shared the materialistic notions of his disciples about the kingdom of God, and thought that coming upon the clouds, and sitting upon thrones,° and drinking wine, would really occur in it, and was mistaken in thinking so. And yet there are plain signs that this cannot be the right account of the matter, and that Jesus did not really share the beliefs of his disciples or conceive the kingdom of God as they did. For they manifestly thought,—even the wisest of them, and after their Master's death as well as before it,—that this kingdom was to be a sudden, miraculous, outward transformation of things, which was to come about very soon and in their own lifetime.° Nevertheless they themselves report Jesus saying what is in direct contradiction to all this. They report him describing the kingdom of God as an inward change° requiring to be spread over an immense time, and coming about by natural means and gradual growth, not suddenly, miraculously. Jesus compares the kingdom of God to a grain of mustard seed and to a handful of leaven.° He says: 'So is the kingdom° of God, as a man may cast seed in the ground, and may go to bed and get up night and day, and the seed shoots and extends he knoweth not how.'[1] Jesus told his disciples, moreover, that the good news of the kingdom had to be preached *to the whole world*.° The whole world must first be evangelised, no work of one generation, but of centuries and centuries; and then, but not till then, should *the end*, the last day, the new world, the grand transformation of which Jewish heads were so full, finally come. True, the disciples also make Jesus speak as if he fancied this end to be as near as they did. But it is quite manifest that Jesus spoke to them, at different times, of two *ends*: one, the end of the Jewish state and nation, which any one who could 'discern the signs of that time'° might foresee; the other, the end of the world, the instatement of God's kingdom;—and that they confused the two ends together. Undeniably, therefore, Jesus saw things in a way very different from theirs, and much truer. And if he uses their materialising language and imagery, then, it cannot have been because he shared their illusions. Nevertheless, he uses it.

And the more we examine the whole language of the Gospels, the more we shall find it to be not language all of the speaker's own, and invented by him for the first time, but to be full of reminiscence and quotation. How deeply all the speakers' minds are governed by the contents of one or two chapters in Daniel, everybody knows. It is impossible to understand anything of the New Testament, without bearing in mind that the main pivot, on which all that is said turns, is supplied by half a dozen verses of Daniel. 'The God of heaven shall set up a kingdom which shall never be destroyed, and shall stand for ever. There shall be a time of trouble, such as never was since there was a nation even to that same time. I beheld, till the thrones were cast down, and the Ancient of days did sit; and, behold, one like the Son of man came with the clouds of heaven, and came to the Ancient of days; the judgment was set and the books were opened. And many of them that sleep in the dust of the earth shall awake, some to everlasting life, and

[1] Mark iv. 26, 27.

some to shame and everlasting contempt.'[1] The language of this group of texts, I say, governs the whole language of the New Testament speakers. The disciples use it literally, Jesus uses it as poetry. But all use it.

Those texts from Daniel almost every reader of the Bible knows. But unless a man has an exceedingly close acquaintance with the prophets, he can have no notion, I think, how very much in the speeches of Jesus is not original language of his own, but is language of the Old Testament,—the religious language on which both he and his hearers had been nourished,—adopted by Jesus, and with a sense of his own communicated to it. There is hardly a trait in the great apocalyptic speech of the twenty-fourth chapter of St Matthew, which has not its original in some prophet. Even where the scope of Jesus is most profoundly new and his own, his phrase is still, as far as may be, old. In the institution of the Lord's Supper his *new covenant*° is a phrase from the admirable and forward-pointing prophecy in the thirty-first chapter of Jeremiah.[2] The *covenant in my blood* points to Exodus,[3] and probably, also, to an expression in that strange but then popular medley, the book of Zechariah.[4] These phrases, familiar to himself and to his hearers, Jesus willingly adopted.

But if we confine to the Old Testament alone our search for parallel passages, we shall have a quite insufficient notion of the extent to which the language of Jesus is not his own original language, but language and images adopted from what was current at the time. It is this which gives such pre-eminent value to the Book of Enoch.° That book,—quoted, as every one will remember, in the Epistle of Jude,[5]—explains what would certainly appear, if we had not this explanation, to be an enlargement and heightening by Jesus, in speaking about the end of the world, of the materialistic data furnished by the Old Testament. For if he thus added to them, it may be said, he must surely have taken them literally. But the Book of Enoch exhibits just the farther stage reached by these data, between the earlier decades of the second century before Christ when the Book of Daniel was written, and the later decades to which belongs the Book of Enoch. And just this farther growth of Messianic language and imagery it was, with which the minds of the contemporaries of Jesus were familiar. And in speaking to them Jesus had to deal with this familiarity. Uncanonical, therefore, though the Book of Enoch be,—for it came too late, and perhaps contains things too strange, for admission into the Canon,—it is full of interest, and every one should read it. The

---

[1] Dan. ii. 44; xii. 1, 2; vii. 9, 10, 13.   [2] Verses 31–34.   [3] Exod. xxiv. 8.
[4] Zech. ix. 11.   [5] Verse 14.

Hebrew original and the Greek version, as is well known, are lost; but the book passed into the Æthiopic Bible, and an Æthiopic manuscript of it was brought to this country from Abyssinia by Bruce, the traveller. The first translator and editor of it, Archbishop Laurence, did his work, Orientalists say, imperfectly, and the English version cannot be trusted. There is an excellent German version; but I wish that the Bishop of Gloucester and Bristol,° who is, I believe an Æthiopic scholar, would give us the book correctly in English.

The Book of Enoch has the names and terms which are already familiar to us from the Old Testament: Head or Ancient of days, Son of man, Son of God, Messiah. It has in frequent use a designation for God, *the Lord of Spirits,* and designations for the Messiah, *the Chosen One, the Just One,* which we come upon in the New Testament,[1] but which the New Testament did not, apparently, get from the Old. It has the angels accompanying the Son of Man to judgment, and the Son of Man 'sitting on the throne of his glory'. It has, again and again, the well-known phrase of the New Testament: *the day of judgment*; it has its outer darkness and its hell-fire. It has its beautiful expression, *children of light.°* These additions to the Old Testament language had passed, when Jesus Christ came, into the religion of the time. He did not create them, but he found them and used them. He employed, as sanctions of his doctrine, his contemporaries' ready-made notions of hell and judgment, just as Socrates did. He talked of the outer darkness and the unquenchable fire, as Socrates talked of° the rivers of Tartarus. And often, when Jesus used phrases which now seem to us to be his own, he was adopting phrases made current by the Book of Enoch. When he said: 'It were better for that man he had never been born'; when he said: 'Rejoice because your names are written in heaven'; when he said: 'Their angels do always behold the face of my Father which is in heaven'; when he said: 'The brother shall deliver up the brother to death and the father the child'; when he said: 'Then shall the righteous shine forth as the sun in the kingdom of their Father', he was remembering the Book of Enoch.° When he said: 'Tell it to *the church*';° when he said to Peter:° 'Thou art Peter, and upon this rock will I build *my church*, and the gates of hell shall not prevail against it',—expressions which, because of the word *church*, some reject, and others make the foundation for the most illusory pretensions,—Jesus was but recalling the Book of Enoch. For in that book the expression, *the company* or *congregation* (in Greek *ecclesia*) *of the*

---

[1] *The Father of Spirits* in Hebrews xii. 9; *the Chosen One* in Luke xxiii. 35; *the Just One* in Acts xxii. 14.

just° or *righteous*,—of the destined rulers of the coming kingdom of the saints,—has become a consecrated phrase. The Messiah, the founder of that kingdom, is the Just One; 'the congregation of the just' are those who follow the Just One, the Just One's company or *ecclesia*. When Peter, therefore, made his ardent declaration of faith, Jesus answered: 'Rock is thy name, and on this rock will I build my company, and the power of death shall not prevail against it.' Behold at its source the colossal inscription round the dome of St Peter's: *Tu es Petrus, et super hanc petram aedificabo ecclesiam meam!*°

The practical lesson to be drawn from all this is, that we should avoid violent revolution in the words and externals of religion. Profound sentiments are connected with them; they are aimed at the highest good, however imperfectly apprehended. Their form often gives them beauty, the associations which cluster around them give them always pathos and solemnity. They are to be used as poetry; while at the same time to purge and raise our view of that ideal at which they are aimed, should be our incessant endeavour. Else the use of them is mere dilettanteism. We should seek, therefore, to use them as Jesus did. How freely Jesus himself used them, we see. And yet what a difference between the meaning he put upon them and the meaning put upon them by the Jews! In how general a sense alone can it with truth be said, that he and even his disciples had the same aspirations, the same final aim! How imperfectly did his disciples apprehend him; how imperfectly must they have reported him! But the result has justified his way of proceeding. For while he carried with him, so far as was possible, his disciples, and the world after them, and all who even now see him through the eyes of those first generations, he yet also marked his own real meaning so indelibly, that it shows and shines clearly out, to satisfy all whom,—as time goes on, and experience widens, and more things are known,—the old imperfect apprehension dissatisfies. And it is not to be supposed that a rejection of all the poetry of popular religion is necessary or advisable now, any more than when Jesus came. But it is an aim which may well indeed be pursued with enthusiasm, to make the true meaning of Jesus, in using that poetry, emerge and prevail. For the immense pathos, so perpetually enlarged upon, of his life and death, does really culminate here: that Christians have so profoundly misunderstood him.

And perhaps I may seem to have said in this essay a great deal about what was merely poetry to Jesus, but too little about what was his real meaning. What this was, however, I have tried to bring out elsewhere. Yet for fear, from my silence about it here, this essay should seem to want due balance, let me end with what a man who writes it down for himself, and meditates on it, and entitles it *Christ's religion*, will not, perhaps, go far wrong [in]. It is but a series of well-known sayings of Jesus himself, as the Gospels deliver them to us. But by putting them together in the following way, and by connecting them, we enable ourselves, I think, to understand better both what Jesus himself meant, and how his disciples came with ease,—taking the sayings singly and interpreting them by the light of their preconceptions,—to mistake them. We must begin, surely, with that wherewith both he and they began;—with that wherewith Christianity itself begins, and wherein it ends: 'the kingdom of God'.

The time is fulfilled and the kingdom of God is at hand! change the inner man and believe the good news!

He that believeth hath eternal life. He that heareth my word, and believeth him that sent me, hath eternal life, and cometh not into judgment, but hath passed from death to life. Verily, verily, I say unto you, The hour cometh and now is, when the dead shall hear the voice of the Son of God, and they that hear shall live.°

I am come forth from God and am here, for I have not come of myself, but he sent me. No man can come unto me except the Father that sent me draw him; and I will raise him up in the last day. He that is of God heareth the words of God; my doctrine is not mine but his that sent me. He that receiveth me receiveth him that sent me.°

And why call ye me Lord, Lord, and do not what I say? If ye know these things, happy are ye if ye do them. Cleanse that which is within; the evil thoughts from within, from the heart, they defile the man. And why seest thou the mote that is in thy brother's eye, but perceivest not the beam that is in thine own eye? Take heed to yourselves against insincerity; God knoweth your hearts; blessed are the pure in heart, for they shall see God!°

Come unto me, all that labour and are heavy-burdened, and I will give you rest. Take my yoke upon you, and learn of me that I am mild and lowly in heart, and ye shall find rest unto your souls. For my yoke is kindly, and my burden light.°

I am the bread of life; he that cometh to me shall never hunger, and he that believeth on me shall never thirst. I am the living bread; as the living Father sent me, and I live by the Father, so he that eateth me, even he shall live by me. It is the spirit that maketh live, the flesh profiteth nothing; the words which I have said unto you, they are spirit and they are life. If a man keep my word, he shall never see death. My sheep hear my voice, and I know them, and they follow me, and I give unto them eternal life, and they shall never perish.°

If a man serve me, let him follow me; and where I am, there shall also my servant be. Whosoever doth not carry his cross and come after me, cannot be my

*disciple. If any man will come after me, let him renounce himself, and take up his
cross daily, and follow me. For whosoever will save his life shall lose it; but
whosoever shall lose his life for my sake and the sake of the good news, the same
shall save it. For what is a man profited, if he gain the whole world, but lose
himself, be mulcted of himself? Therefore doth my Father love me, because I
lay down my life that I may take it again. A new commandment give I unto you,
that ye love one another. The Son of man came not to be served but to serve, and
to give his life a ransom for many.°*

*I am the resurrection and the life; he that believeth on me, though he die, shall
live; and he that liveth and believeth on me shall never die. I am come that ye
might have life, and that ye might have it more abundantly. I cast out devils and
I do cures to-day and to-morrow; and the third day I shall be perfected. Yet a
little while, and the world seeth me no more; but ye see me, because I live and ye
shall live. If ye keep my commandments ye shall abide in my love, like as I have
kept my Father's commandments and abide in his love. He that loveth me shall
be loved of my Father, and I will love him, and will manifest myself to him. If a
man love me, he will keep my word, and my Father will love him, and we will
come unto him, and make our abode with him.°*

*I am the good shepherd; the good shepherd lays down his life for the sheep.
And other sheep I have, which are not of this fold; them also must I bring, and
they shall be one flock, one shepherd. Fear not, little flock, for it is your Father's
good pleasure to give you the kingdom.°*

*My kingdom is not of this world; the kingdom of God cometh not with
observation; behold, the kingdom of God is within you! Whereunto shall I liken
the kingdom of God? It is like a grain of mustard seed, which a man took and
cast into his garden, and it grew, and waxed a great tree, and the fowls of the air
lodged in the branches of it. It is like leaven, which a woman took, and hid in
three measures of meal, till the whole was leavened. So is the kingdom of God, as
a man may cast seed in the ground, and may go to bed and get up night and day,
and the seed shoots and extends he knoweth not how.*

*And this good news of the kingdom shall be preached in the whole world, for a
witness to all nations; and then shall the end come.°*

With such a construction in his thoughts to govern his use of it, Jesus
loved and freely adopted the common wording and imagery of the
popular Jewish religion. In dealing with the popular religion in which we
have been ourselves bred, we may the more readily follow his example,
inasmuch as, though all error has its side of moral danger, yet, evidently,
the misconception of their religion by Christians has produced no such
grave moral perversion as we see to have been produced in the Scribes
and Pharisees° by their misconception of the religion of the Old

Testament. The fault of popular Christianity as an endeavour after *righteousness by Jesus Christ* is not, like the fault of popular Judaism as an endeavour after *salvation by righteousness*, first and foremost a moral fault. It is, much more, an intellectual one. But it is not on that account insignificant. Dr Mozley° urges, that 'no inquiry is obligatory upon religious minds in matters of the supernatural and miraculous', because, says he, though 'the human mind must refuse to submit to anything contrary to moral sense in Scripture', yet, 'there is no moral question raised by the fact of a miracle, nor does a supernatural doctrine challenge any moral resistance'. As if there were no possible resistance to religious doctrines, but a resistance on the ground of their immorality! As if intellectual resistance to them counted for nothing! The objections to popular Christianity are not moral objections, but intellectual revolt against its demonstrations by miracle and metaphysics.° To be intellectually convinced of a thing's want of conformity to truth and fact is surely an insuperable obstacle to receiving it, even though there be no moral obstacle added. And no moral advantages of a doctrine can avail to save it, in presence of the intellectual conviction of its want of conformity with truth and fact. And if the want of conformity exists, it is sure to be one day found out. 'Things are what they are, and the consequences of them will be what they will be';° and one inevitable consequence of a thing's want of conformity with truth and fact is, that sooner or later the human mind perceives it. And whoever thinks that the ground-belief of Christians is true and indispensable, but that in the account they give of it, and of the reasons for holding it, there is a want of conformity with truth and fact, may well desire to find a better account and better reasons, and to prepare the way for their admission and for their acquiring some strength and consistency in men's minds, against the day when the old means of reliance fail.

But, meanwhile, the ground-belief of all Christians, whatever account they may give to themselves of its source and sanctions, is in itself an indestructible basis of fellowship. Whoever believes the final triumph of Christianity, the Christianisation of the world, to have all the necessity and grandeur of a natural law, will never lack a bond of profound sympathy with popular religion. Compared with agreement and difference on this point, agreement and difference on other points seem trifling. To believe that, whoever are ignorant that righteousness is salvation, 'the Eternal shall have them in derision';° to believe that, whatever may be the substitute offered for the righteousness of Jesus, a substitute however sparkling, yet 'whosoever drinketh of *this* water shall thirst again'; to desire truly 'to have strength to escape all the things which shall come to pass and to stand before the Son of Man',°—is the one authentic mark and seal of the household of faith. Those who share in this belief and in this desire are fellow-citizens of the 'city which hath foundations'.° Whosoever shares in them not, is, or is in danger of any day becoming, a wanderer, as St Augustine says, through 'the waste places fertile in sorrow'; a wanderer 'seeking rest and finding none'. *In all things I sought rest; then the Creator of all things gave me commandment and said: Let thy dwelling be in Jacob, and thine inheritance in Israel! And so was I established in Sion; likewise in the beloved city he gave me rest, and in Jerusalem was my power.*

A French Critic on Milton

MR TREVELYAN'S Life of his uncle must have induced many people to read again Lord Macaulay's *Essay on Milton*. With the *Essay on Milton* began Macaulay's literary career, and, brilliant as the career was, it had few points more brilliant than its beginning. Mr Trevelyan describes with animation that decisive first success. The essay appeared in the *Edinburgh Review* in 1825. Mr Trevelyan° says, and quite truly:—

'The effect on the author's reputation was instantaneous. Like Lord Byron, he awoke one morning and found himself famous. The beauties of the work were such as all men could recognise, and its very faults pleased. . . . The family breakfast-table in Bloomsbury was covered with cards of invitation to dinner from every quarter of London. . . . A warm admirer of Robert Hall, Macaulay heard with pride how the great preacher, then well-nigh worn out with that long disease, his life, was discovered lying on the floor, employed in learning by aid of grammar and dictionary enough Italian to enable him to verify the parallel between Milton and Dante. But the compliment that, of all others, came most nearly home,—the only commendation of his literary talent which even in the innermost domestic circle he was ever known to repeat,—was the sentence with which Jeffrey° acknowledged the receipt of his manuscript: "The more I think, the less I can conceive where you picked up that style." '

And already, in the *Essay on Milton*, the style of Macaulay is, indeed, that which we know so well. A style to dazzle, to gain admirers everywhere, to attract imitators in multitude! A style brilliant, metallic, exterior; making strong points, alternating invective with eulogy, wrapping in a robe of rhetoric the thing it represents; not, with the soft play of life, following and rendering the thing's very form and pressure. For, indeed, in rendering things in this fashion, Macaulay's gift did not lie. Mr Trevelyan reminds us that in the preface to his collected Essays, Lord Macaulay himself 'unsparingly condemns the redundance of youthful enthusiasm' of the *Essay on Milton*. But the unsoundness of the essay does not spring from its 'redundance of youthful enthusiasm'. It springs from this: that the writer has not for his aim to see and to utter the real truth about his object.° Whoever comes to the *Essay on Milton* with the desire to get at the real truth about Milton, whether as a man or as a poet, will feel that the essay in nowise helps him. A reader who only wants rhetoric, a reader who wants a panegyric on Milton, a panegyric on the Puritans, will find what he wants. A reader who wants criticism will be disappointed.

This would be palpable to all the world, and every one would feel, not pleased, but disappointed, by the *Essay on Milton*, were it not that the readers who seek for criticism are extremely few; while the readers who seek for rhetoric, or who seek for praise and blame to suit their own already established likes and dislikes, are extremely many. A man who is fond of rhetoric may find pleasure in hearing that in *Paradise Lost* 'Milton's conception of love unites all the voluptuousness of the Oriental haram, and all the gallantry of the chivalric tournament, with all the pure and quiet affection of an English fireside.' He may glow at being told that 'Milton's thoughts resemble those celestial fruits and flowers which the Virgin Martyr of Massinger sent down from the gardens of Paradise to the earth, and which were distinguished from the productions of other soils not only by superior bloom and sweetness, but by miraculous efficacy to invigorate and to heal.' He may imagine that he has got something profound when he reads that, if we compare Milton and Dante in their management of the agency of supernatural beings,—'the exact details of Dante with the dim intimations of Milton',—the right conclusion of the whole matter is this:—

'Milton wrote in an age of philosophers and theologians. It was necessary, therefore, for him to abstain from giving such a shock to their understandings as might break the charm which it was his object to throw over their imaginations. It was impossible for him to adopt altogether the material or the immaterial system. He therefore took his stand on the debateable ground. He left the whole in ambiguity. He has doubtless, by so doing, laid himself open to the charge of inconsistency. But though philosophically in the wrong he was poetically in the right.'

Poor Robert Hall, 'well-nigh worn out with that long disease, his life', and, in the last precious days of it, 'discovered lying on the floor, employed in learning, by aid of grammar and dictionary, enough Italian to enable him to verify' this ingenious criticism! Alas! even had his life been prolonged like Hezekiah's,° he could not have verified it, for it is unverifiable. A poet who, writing 'in an age of philosophers and theologians', finds it 'impossible for him to adopt altogether the material or the immaterial system', who, therefore, 'takes his stand on the debateable ground', who 'leaves the whole in ambiguity', and who, in doing so, 'though philosophically in the wrong, was poetically in the right'! Substantial meaning such lucubrations have none. And in like manner, a distinct and substantial meaning can never be got out of the fine phrases about 'Milton's conception of love uniting all the voluptuousness of the Oriental haram, and all the gallantry of the chivalric

tournament, with all the pure and quiet affection of an English fireside';
or about 'Milton's thoughts resembling those celestial fruits and flowers
which the Virgin Martyr of Massinger sent down from the gardens of
Paradise to the earth'; the phrases are mere rhetoric. Macaulay's writing
passes for being admirably clear, and so externally it is; but often it is
really obscure, if one takes his deliverances seriously, and seeks to find in
them a definite meaning. However, there is a multitude of readers,
doubtless, for whom it is sufficient to have their ears tickled with fine
rhetoric; but the tickling makes a serious reader impatient.

Many readers there are, again, who come to an Essay on Milton with
their minds full of zeal for the Puritan cause, and for Milton as one of the
glories of Puritanism. Of such readers the great desire is to have the
cause and the man, who are already established objects of enthusiasm for
them, strongly praised. Certainly Macaulay will satisfy their desire. They
will hear that the Civil War was 'the great conflict between Oromasdes
and Arimanes,° liberty and despotism, reason and prejudice'; the
Puritans being Oromasdes, and the Royalists Arimanes. They will be
told that the great Puritan poet was worthy of the august cause which he
served. 'His radiant and beneficent career resembled that of the god of
light and fertility.' 'There are a few characters which have stood the
closest scrutiny and the severest tests, which have been tried in the
furnace and have proved pure, which have been declared sterling by the
general consent of mankind, and which are visibly stamped with the
image and superscription of the Most High. Of these was Milton.' To
descend a little to particulars. Milton's temper was especially admirable.
'The gloom of Dante's character discolours all the passions of men and
all the face of nature, and tinges with its own livid hue the flowers of
Paradise and the glories of the eternal throne.' But in our countryman,
although 'if ever despondency and asperity could be excused in any man,
they might have been excused in Milton', nothing 'had power to disturb
his sedate and majestic patience'. All this is just what an ardent admirer
of the Puritan cause and of Milton would most wish to hear, and when he
hears it he is in ecstasies.

But a disinterested reader, whose object is not to hear Puritanism and
Milton glorified, but to get at the truth about them, will surely be
dissatisfied. With what a heavy brush, he will say to himself, does this
man lay on his colours! The Puritans Oromasdes, and the Royalists
Arimanes? What a different strain from Chillingworth's, in his sermon at
Oxford at the beginning of the Civil War! 'Publicans and sinners on the
one side,' said Chillingworth,° 'Scribes and Pharisees on the other.' Not
at all a conflict between Oromasdes and Arimanes, but a good deal of

Arimanes on both sides. And as human affairs go, Chillingworth's version of the matter is likely to be nearer the truth than Macaulay's. Indeed, for any one who reads thoughtfully and without bias, Macaulay himself, with the inconsistency of a born rhetorician, presently confutes his own thesis. He says of the Royalists: 'They had far more both of profound and of polite learning than the Puritans. Their manners were more engaging, their tempers more amiable, their tastes more elegant, and their households more cheerful.' Is being more kindly affectioned such an insignificant superiority? The Royalists too, then, in spite of their being insufficiently jealous for civil and ecclesiastical liberty, had in them something of Oromasdes, the principle of light.

And Milton's temper! His 'sedate and majestic patience'; his freedom from 'asperity'! If there is a defect which, above all others, is signal in Milton, which injures him even intellectually, which limits him as a poet, it is the defect common to him with the whole Puritan party to which he belonged,—the fatal defect of *temper*. He and they may have a thousand merits, but they are *unamiable*. Excuse them how one will, Milton's asperity and acerbity, his want of sweetness of temper, of the Shakspearian largeness and indulgence, are undeniable. Lord Macaulay in his Essay regrets that the prose writings of Milton should not be more read. 'They abound', he says in his rhetorical way, 'with passages, compared with which the finest declamations of Burke sink into insignificance.' At any rate, they enable us to judge of Milton's temper, of his freedom from asperity. Let us open the *Doctrine and Discipline of Divorce* and see how Milton treats an opponent.° 'How should he, a serving man both by nature and function, an idiot by breeding, and a solicitor by presumption, ever come to know or feel within himself what the meaning is of *gentle*?' What a gracious temper! 'At last, and in good hour, we come to his farewell, which is to be a concluding taste of his jabberment in law, the flashiest and the fustiest that ever corrupted in such an unswilled hogshead.' How 'sedate and majestic'!

Human progress consists in a continual increase in the number of those who, ceasing to live by the animal life alone and to feel the pleasures of sense only, come to participate in the intellectual life also, and to find enjoyment in the things of the mind. The enjoyment is not at first very discriminating. Rhetoric, brilliant writing, gives to such persons pleasure for its own sake; but it gives them pleasure, still more, when it is employed in commendation of a view of life which is on the whole theirs, and of men and causes with which they are naturally in sympathy. The immense popularity of Macaulay is due to his being pre-eminently fitted to give pleasure to all who are beginning to feel enjoyment in the things of the mind. It is said that the traveller in Australia,° visiting one settler's hut after another, finds again and again that the settler's third book, after the Bible and Shakspeare, is some work by Macaulay. Nothing can be more natural. The Bible and Shakspeare may be said to be imposed upon an Englishman as objects of his admiration; but as soon as the common Englishman, desiring culture, begins to choose for himself, he chooses Macaulay. Macaulay's view of things is, on the whole, the view of them which he feels to be his own also; the persons and causes praised are those which he himself is disposed to admire; the persons and causes blamed are those with which he himself is out of sympathy; and the rhetoric employed to praise or to blame them is animating and excellent. Macaulay is thus a great civiliser. In hundreds of men he hits their nascent taste for the things of the mind, possesses himself of it and stimulates it, draws it powerfully forth and confirms it.

But with the increasing number of those who awake to the intellectual life, the number of those also increases, who, having awoke to it, go on with it, follow where it leads them. And it leads them to see that it is their business to learn the real truth about the important men, and things, and books, which interest the human mind. For thus is gradually to be acquired a stock of sound ideas, in which the mind will habitually move, and which alone can give to our judgments security and solidity. To be satisfied with fine writing about the object of one's study, with having it praised or blamed in accordance with one's own likes or dislikes, with any conventional treatment of it whatever, is at this stage of growth seen to be futile. At this stage, rhetoric, even when it is so good as Macaulay's, dissatisfies. And the number of people who have reached this stage of mental growth is constantly, as things now are, increasing; increasing by the very same law of progress which plants the beginnings of mental life in more and more persons who, until now, have never known mental life at all. So that while the number of those who are delighted with rhetoric such as Macaulay's is always increasing, the number of those who are dissatisfied with it is always increasing too.

And not only rhetoric dissatisfies people at this stage, but convention-ality of any kind. This is the fault of Addison's Miltonic criticism,° once so celebrated; it rests almost entirely upon convention. Here is *Paradise Lost*, 'a work which does an honour to the English nation', a work claiming to be one of the great poems of the world, to be of the highest moment to us. 'The *Paradise Lost*', says Addison, 'is looked upon by the best judges as the greatest production, or at least the noblest work of genius, in our language, and therefore deserves to be set before an

English reader in its full beauty.' The right thing, surely, is for such a work to prove its own virtue by powerfully and delightfully affecting us as we read it, and by remaining a constant source of elevation and happiness to us for ever. But the *Paradise Lost* has not this effect certainly and universally; therefore Addison proposes to 'set before an English reader, in its full beauty', the great poem. To this end he has 'taken a general view of it under these four heads: the fable, the characters, the sentiments, and the language'. He has, moreover,

'endeavoured not only to prove that the poem is beautiful in general, but to point out its particular beauties and to determine wherein they consist. I have endeavoured to show how some passages are beautified by being sublime, others by being soft, others by being natural; which of them are recommended by the passion, which by the moral, which by the sentiment, and which by the expression. I have likewise endeavoured to show how the genius of the poet shines by a happy invention, or distant allusion, or a judicious imitation; how he has copied or improved Homer or Virgil, and raises his own imaginations by the use which he has made of several poetical passages in Scripture. I might have inserted also several passages in Tasso which our author has imitated; but as I do not look upon Tasso to be a sufficient voucher, I would not perplex my reader with such quotations as might do more honour to the Italian than the English poet.'

This is the sort of criticism which held our grandfathers and great-grandfathers spell-bound in solemn reverence. But it is all based upon convention, and on the positivism of the modern reader it is thrown away. Does the work which you praise, he asks, affect me with high pleasure and do me good, when I try it as fairly as I can? The critic who helps such a questioner is one who has sincerely asked himself, also, this same question; who has answered it in a way which agrees, in the main, with what the questioner finds to be his own honest experience in the matter, and who shows the reasons for this common experience. Where is the use of telling a man, who finds himself tired rather than delighted by *Paradise Lost*, that the incidents in that poem 'have in them all the beauties of novelty, at the same time that they have all the graces of nature'; that 'though they are natural, they are not obvious, which is the true character of all fine writing'? Where is the use of telling him that 'Adam and Eve are drawn with such sentiments as do not only interest the reader in their afflictions, but raise in him the most melting passions of humanity and commiseration'? His own experience, on the other hand, is that the incidents in *Paradise Lost* are such as awaken in him but the most languid interest; and that the afflictions and sentiments of Adam and Eve never melt or move him passionately at all. How is he advanced by hearing that 'it is not sufficient that the language of an epic poem be perspicuous, unless it be also sublime'; and that Milton's language is both? What avails it to assure him that 'the first thing to be considered in an epic poem is the fable, which is perfect or imperfect, according as the action which it relates is more or less so'; that 'this action should have three qualifications, should be but one action, an entire action, and a great action'; and that if we 'consider the action of the *Iliad*, *Æneid*, and *Paradise Lost*, in these three several lights', we shall find that Milton's poem does not 'fall short in the beauties which are essential to that kind of writing'? The patient whom Addison thus doctors will reply, that he does not care two straws whether the action of *Paradise Lost* satisfies the proposed test or no, if the poem does not give him pleasure. The truth is, Addison's criticism rests on certain conventions: namely, that incidents of a certain class *must* awaken keen interest; that sentiments of a certain kind *must* raise melting passions; that language of a certain strain, and an action with certain qualifications, *must* render a poem attractive and effective. Disregard the convention; ask solely whether the incidents *do* interest, whether the sentiments *do* move, whether the poem *is* attractive and effective, and Addison's criticism collapses.

Sometimes the convention is one which in theory ought, a man may perhaps admit, to be something more than a convention; but which yet practically is not. Milton's poem is of surpassing interest to us, says Addison, because in it, 'the principal actors are not only our progenitors but our representatives. We have an actual interest in everything they do, and no less than our utmost happiness is concerned, and lies at stake, in all their behaviour.' Of ten readers who may even admit that in theory this is so, barely one can be found whose practical experience tells him that Adam and Eve do really, as his representatives, excite his interest in this vivid manner. It is by a mere convention, then, that Addison supposes them to do so, and claims an advantage for Milton's poem from the supposition.

The theological speeches in the third book of *Paradise Lost* are not, in themselves, attractive poetry. But, says Addison:—

'The passions which they are designed to raise are a divine love and religious fear. The particular beauty of the speeches in the third book consists in that shortness and perspicuity of style in which the poet has couched the greatest mysteries of Christianity. . . . He has represented all the abstruse doctrines of predestination, free-will, and grace, as also the great points of incarnation and redemption (which naturally grow up in a poem that treats of the fall of man) with great energy of expression, and in a clearer and stronger light than I ever met with in any other writer.'

But nine readers out of ten feel that, as a matter of fact, their religious sentiments of 'divine love and religious fear' are wholly ineffectual even to reconcile them to the poetical tiresomeness of the speeches in question: far less can they render them interesting. It is by a mere convention, then, that Addison pretends that they do.

The great merit of Johnson's criticism on Milton is that from rhetoric and convention it is free. Mr Trevelyan says° that the enthusiasm of Macaulay's *Essay on Milton* is, at any rate, 'a relief from the perverted ability of that elaborate libel on our great epic poet, which goes by the name of Dr Johnson's *Life of Milton*'. This is too much in Lord Macaulay's own style. In Johnson's *Life of Milton* we have the straight-forward remarks, on Milton and his works, of a very acute and robust mind. Often they are thoroughly sound. 'What we know of Milton's character in domestic relations is that he was severe and arbitrary. His family consisted of women; and there appears in his books something like a Turkish contempt of females as subordinate and inferior beings.' Mr Trevelyan will forgive our saying that the truth is here much better hit than in Lord Macaulay's sentence telling us how Milton's 'conception of love unites all the voluptuousness of the Oriental haram, and all the gallantry of the chivalric tournament, with all the pure and quiet affection of an English fireside'. But Johnson's mind, acute and robust as it was, was at many points bounded, at many points warped. He was neither sufficiently disinterested, nor sufficiently flexible, nor sufficiently receptive, to be a satisfying critic of a poet like Milton. 'Surely no man could have fancied that he read Lycidas with pleasure had he not known the author!' Terrible sentence for revealing the deficiencies of the critic who utters it.

A completely disinterested judgment about a man like Milton is easier to a foreign critic than to an Englishman. From conventional obligation to admire 'our great epic poet' a foreigner is free. Nor has he any bias for or against Milton because he was a Puritan,—in his political and ecclesiastical doctrines to one of our great English parties a delight, to the other a bugbear. But a critic must have the requisite knowledge of the man and the works he is to judge; and from a foreigner—particularly perhaps from a Frenchman—one hardly expects such knowledge. M. Edmond Scherer,° however, whose essay on Milton lies before me, is an exceptional Frenchman. He is a senator of France and one of the directors of the *Temps* newspaper. But he was trained at Geneva, that home of large instruction and lucid intelligence. He was in youth the friend and hearer of Alexandre Vinet,°—one of the most salutary influences a man in our times can have experienced, whether he continue to think quite with Vinet or not. He knows thoroughly the language and literature of England, Italy, Germany, as well as of France. Well-informed, intelligent, disinterested, open-minded, sympathetic, M. Scherer has much in common with the admirable critic whom France has lost—Sainte-Beuve.° What he has not, as a critic, is Sainte-Beuve's elasticity and cheerfulness. He has not that gaiety, that radiancy, as of a man discharging with delight the very office for which he was born, which, in the *Causeries*, make Sainte-Beuve's touch so felicitous, his sentences so crisp, his effect so charming. But M. Scherer has the same open-mindedness as Sainte-Beuve, the same firmness and sureness of judgment; and having a much more solid acquaintance with foreign languages than Sainte-Beuve, he can much better appreciate a work like *Paradise Lost* in the only form in which it can be appreciated properly—in the original.

We will commence, however, by disagreeing with M. Scherer. He sees very clearly how vain is Lord Macaulay's sheer laudation of Milton, or Voltaire's sheer disparagement° of him. Such judgments, M. Scherer truly says, are not judgments at all. They merely express a personal sensation of like or dislike. And M. Scherer goes on to recommend, in the place of such 'personal sensations', the method of historical criticism—that great and famous power in the present day. He sings the praises of 'this method at once more conclusive and more equitable, which sets itself to understand things rather than to class them, to explain rather than to judge them; which seeks to account for a work from the genius of its author, and for the turn which this genius has taken from the circumstances amidst which it was developed';—the old story of 'the man and the *milieu*',° in short. 'For thus,' M. Scherer continues, 'out of these two things, the analysis of the writer's character and the study of his age, there spontaneously issues the right understanding of his work. In place of an appreciation thrown off by some chance comer, we have the work passing judgment, so to speak, upon itself, and assuming the rank which belongs to it among the productions of the human mind.'

The advice to study the character of an author and the circumstances in which he has lived, in order to account to oneself for his work, is excellent. But it is a perilous doctrine, that from such a study the right understanding of his work will 'spontaneously issue'. In a mind qualified in a certain manner it will—not in all minds. And it will be that mind's 'personal sensation'. It cannot be said that Macaulay had not studied the character of Milton, and the history of the times in which he lived. But a right understanding of Milton did not 'spontaneously issue' therefrom in the mind of Macaulay, because Macaulay's mind was that of a rhetorician, not of a disinterested critic. Let us not confound the method with the result intended by the method—right judgments. The critic who rightly appreciates a great man or a great work, and who can tell us faithfully—life being short, and art long,° and false information very plentiful—what we may expect from their study and what they can do for us; he is the critic we want, by whatever methods, intuitive or historical, he may have managed to get his knowledge.

M. Scherer begins with Milton's prose works, from which he translates many passages. Milton's sentences can hardly know themselves again in clear modern French, and with all their inversions and redundancies gone. M. Scherer does full justice to the glow and mighty eloquence with which Milton's prose, in its good moments, is instinct and alive; to the 'magnificences of his style', as he calls them:—

'The expression is not too strong. There are moments when, shaking from him the dust of his arguments, the poet bursts suddenly forth, and bears us away in a torrent of incomparable eloquence. We get, not the phrase of the orator, but the glow of the poet, a flood of images poured around his arid theme, a rushing flight carrying us above his paltry controversies. The polemical writings of Milton are filled with such beauties. The prayer which concludes the treatise on Reformation in England, the praise of zeal in the Apology for Smectymnuus, the portrait of Cromwell in the Second Defence of the English people, and, finally, the whole tract on the Liberty of Unlicensed Printing from beginning to end, are some of the most memorable pages in English literature, and some of the most characteristic products of the genius of Milton.'

Macaulay himself could hardly praise the eloquence of Milton's prose writings more warmly. But it is a very inadequate criticism which leaves the reader, as Macaulay's rhetoric would leave him, with the belief that the total impression to be got from Milton's prose writings is one of enjoyment and admiration. It is not; we are misled, and our time is wasted, if we are sent to Milton's prose works in the expectation of finding it so. Grand thoughts and beautiful language do not form the staple of Milton's controversial treatises, though they occur in them not unfrequently. But the total impression from those treatises is rightly given by M. Scherer:—

'In all of them the manner is the same. The author brings into play the treasures of his learning, heaping together testimonies from Scripture, passages from the Fathers, quotations from the poets; laying all antiquity, sacred and profane, under contribution; entering into subtle discussions on the sense of this or that Greek or Hebrew word. But not only by his undigested erudition and by his absorption in religious controversy does Milton belong to his age; he belongs to it, too, by the personal tone of his polemics. Morus and Salmasius had attacked his morals, laughed at his low stature, made unfeeling allusions to his loss of sight: Milton replies by reproaching them with the wages they have taken and with the servant-girls they have debauched. All this mixed with coarse witticisms, with terms of the lowest abuse. Luther and Calvin, those virtuosos of insult, had not gone further.'

No doubt there is, as M. Scherer says, 'something indescribably heroical and magnificent which overflows from Milton, even when he is engaged in the most miserable discussions', Still, for the mass of his prose treatises 'miserable discussions' is the final and right word. Nor, when Milton passed to his great epic, did he altogether leave the old man of these 'miserable discussions' behind him.

'In his soul he is a polemist and theologian—a Protestant Schoolman. He takes delight in the favourite dogmas of Puritanism: original sin, predestination, free-will. Not that even here he does not display somewhat of that independence which was in his nature. But his theology is, nevertheless, that of his epoch, tied and bound to the letter of Holy Writ, without grandeur, without horizons, without philosophy. He never frees himself from the bondage of the letter. He settles the most important questions by the authority of an obscure text, or a text isolated from its context. In a word, Milton is a great poet with a Salmasius or a Grotius bound up along with him; a genius nourished on the marrow of lions, of Homer, Isaiah, Virgil, Dante, but also, like the serpent of Eden, eating dust, the dust of dismal polemics. He is a doctor, a preacher, a man of didactics; and when the day shall arrive when he can at last realise the dreams of his youth and bestow on his country an epic poem, he will compose it of two elements, gold and clay, sublimity and scholasticism, and will bequeath to us a poem which is at once the most wonderful and the most insupportable poem in existence.'

From the first, two conflicting forces, two sources of inspiration, had contended with one another, says M. Scherer, for the possession of Milton,—the Renascence and Puritanism. Milton felt the power of both:—

'Elegant poet and passionate disputant, accomplished humanist and narrow sectary, admirer of Petrarch, of Shakspeare, and hair-splitting interpreter of Bible-texts, smitten with Pagan antiquity and smitten with the Hebrew genius; and all this at once, without effort, naturally;—an historical problem, a literary enigma!'

Milton's early poems, such as the *Allegro*, the *Penseroso*, are poems produced while a sort of equilibrium still prevailed in the poet's nature; hence their charm, and that of their youthful author:—

'Nothing morose or repellent, purity without excess of rigour, gravity without fanaticism. Something wholesome and virginal, gracious and yet strong. A son of the North who has passed the way of Italy; a last fruit of the Renascence, but a fruit filled with a savour new and strange!'

But Milton's days proceeded, and he arrived at the latter years of his life—a life which, in its outward fortunes, darkened more and more, *alla s'assombrissant de plus en plus*, towards its close. He arrived at the time when 'his friends had disappeared, his dreams had vanished, his eyesight was quenched, the hand of old age was upon him'. It was then that, 'isolated by the very force of his genius', but full of faith and fervour, he 'turned his eyes towards the celestial light' and produced *Paradise Lost*. In its form, M. Scherer observes, in its plan and distribution, the poem follows Greek and Roman models, particularly the *Æneid*. 'All in this respect is regular and classical; in this fidelity to the established models we recognise the literary superstitions of the Renascence.' So far as its form is concerned, *Paradise Lost* is, says M. Scherer, 'the copy of a copy, a tertiary formation. It is to the Latin epics what these are to Homer'.

The most important matter, however, is the contents of the poem, not the form. The contents are given by Puritanism. But let M. Scherer speak for himself:—

'*Paradise Lost* is an epic, but a theological epic; and the theology of the poem is made up of the favourite dogmas of the Puritans,—the Fall, justification, God's sovereign decrees. Milton, for that matter, avows openly that he has a thesis to maintain; his object is, he tells us at the outset, to "assert Eternal Providence and justify the ways of God to man". *Paradise Lost*, then, is two distinct things in one,—an epic and a theodicy. Unfortunately these two elements, which correspond to the two men of whom Milton was composed, and to the two tendencies which ruled his century, these two elements have not managed to get amalgamated. Far from doing so, they clash with one another, and from their juxtaposition there results a suppressed contradiction which extends to the whole work, impairs its solidity, and compromises its value.'

M. Scherer gives his reasons for thinking that the Christian theology is unmanageable in an epic poem, although the gods may come in very well in the *Iliad* and *Æneid*. Few will differ from him here, so we pass on. A theological poem is a mistake, says M. Scherer; but to call *Paradise Lost* a theological poem is to call it by too large a name. It is really a commentary on a biblical text,—the first two or three chapters of Genesis. Its subject, therefore, is a story, taken literally, which many of even the most religious people nowadays hesitate to take literally; while yet, upon our being able to take it literally, the whole real interest of the poem for us depends. Merely as matter of poetry, the story of the Fall has no special force or effectiveness; its effectiveness for us comes, and can only come, from our taking it all as the literal narrative of what positively happened.

Milton, M. Scherer thinks, was not strong in invention. The famous allegory of Sin and Death may be taken as a specimen of what he could do in this line, and the allegory of Sin and Death is uncouth and unpleasing. But invention is dangerous when one is dealing with a subject so grave, so strictly formulated by theology, as the subject of Milton's choice. Our poet felt this, and allowed little scope to free poetical invention. He adhered in general to data furnished by Scripture, and supplemented somewhat by Jewish legend. But this judicious self-limitation had, again, its drawbacks:—

'If Milton has avoided factitious inventions, he has done so at the price of another disadvantage; the bareness of his story, the epic poverty of his poem. It is not merely that the reader is carried up into the sphere of religious abstractions, where man loses power to see or breathe. Independently of this, everything is here too simple, both actors and actions. Strictly speaking, there is but one personage before us, God the Father; inasmuch as God cannot appear without effacing every one else, nor speak without the accomplishment of his will. The Son is but the Father's double. The angels and archangels are but his messengers, nay, they are less; they are but his decrees personified, the supernumeraries of a drama which would be transacted quite as well without them.

'Milton has struggled against these conditions of the subject which he had chosen. He has tried to escape from them, and has only made the drawback more visible. The long speeches with which he fills up the gaps of the action are sermons, and serve but to reveal the absence of action. Then as, after all, some action, some struggle, was necessary, the poet had recourse to the revolt of the angels. Unfortunately, such is the fundamental vice of the subject, that the poet's instrument has, one may say, turned against him. What his action has gained from it in movement it has lost in probability. We see a battle, indeed, but who can take either the combat or the combatants seriously? Belial shows his sense of this, when in the infernal council he rejects the idea of engaging in any conflict whatever, open or secret, with Him who is All-seeing and Almighty; and really one cannot comprehend how his mates should have failed to acquiesce in a consideration so evident. But, I repeat, the poem was not possible save at the price of this impossibility. Milton, therefore, has courageously made the best of it. He has gone with it all lengths, he has accepted in all its extreme consequences the most inadmissible of fictions. He has exhibited to us Jehovah apprehensive for his omnipotence, in fear of seeing his position turned, his residence surprised, his throne usurped. He has drawn the angels hurling mountains at one another's heads, and firing cannon at one another. He has shown us the victory doubtful until the Son appears armed with lightnings, and standing on a car horsed by four Cherubim.'

The fault of Milton's poem is not, says M. Scherer, that, with his Calvinism of the seventeenth century, Milton was a man holding other beliefs than ours. Homer, Dante, held other beliefs than ours:—

'But Milton's position is not the same as theirs. Milton has something he wants to prove, he supports a thesis. It was his intention, in his poem, to do duty as theologian as well as poet; at any rate, whether he meant it or not, *Paradise Lost* is a didactic work, and the form of it, therefore, cannot be separated from the substance. Now, it turns out that the idea of the poem will not bear examination; that its solution for the problem of evil is almost burlesque; that the character of its heroes, Jehovah and Satan, has no coherence; that what happens to Adam interests us but little; finally, that the action takes place in regions where the interests and passions of our common humanity can have no scope. I have already insisted on this contradiction in Milton's epic; the story on which it turns can have meaning and value only so long as it preserves its dogmatic weight, and, at the same time, it cannot preserve this without falling into theology,—that is to say, into a domain foreign to that of art. The subject of the poem is nothing if it is not real, and if it does not touch us as the turning-point of our destinies; and the more the poet seeks to grasp this reality, the more it escapes from him.'

In short, the whole poem of *Paradise Lost* is vitiated, says M. Scherer, 'by a kind of antinomy, by the conjoint necessity and impossibility of taking its contents literally'.

M. Scherer then proceeds to sum up. And in ending, after having once more marked his objections and accentuated them, he at last finds again that note of praise, which the reader will imagine him to have quite lost:—

'To sum up: *Paradise Lost* is a false poem, a grotesque poem, a tiresome poem; there is not one reader out of a hundred who can read the ninth and tenth books without smiling, or the eleventh and twelfth without yawning. The whole thing is without solidity; it is a pyramid resting on its apex, the most solemn of problems resolved by the most puerile of means. And, notwithstanding, *Paradise Lost* is immortal. It lives by a certain number of episodes which are for ever famous. Unlike Dante, who must be read as a whole if we want really to seize his beauties, Milton ought to be read only by passages. But these passages form part of the poetical patrimony of the human race.'

And not only in things like the address to light, or the speeches of Satan, is Milton admirable, but in single lines and images everywhere:—

'*Paradise Lost* is studded with incomparable lines. Milton's poetry is, as it were, the very essence of poetry. The author seems to think always in images, and these images are grand and proud like his soul, a wonderful mixture of the sublime and the picturesque. For rendering things he has the unique word, the word which is a discovery. Every one knows his *darkness visible*.'

M. Scherer cites other famous expressions and lines, so familiar that we need not quote them here. Expressions of the kind, he says, not only beautiful, but always, in addition to their beauty, striking one as the absolutely right thing (*toujours justes dans leur beauté*), are in *Paradise Lost* innumerable. And he concludes:—

'Moreover, we have not said all when we have cited particular lines of Milton. He has not only the image and the word, he has the period also, the large musical phrase, somewhat long, somewhat laden with ornaments and intricate with inversions, but bearing all along with it in its superb undulation. Lastly, and above all, he has a something indescribably serene and victorious, an unfailing level of style, power indomitable. He seems to wrap us in a fold of his robe, and to carry us away with him into the eternal regions where is his home.'

With this fine image M. Scherer takes leave of Milton. Yet the simple description of the man in Johnson's life of him touches us more than any image; the description of the old poet 'seen in a small house, neatly enough dressed in black clothes, sitting in a room hung with rusty green, pale but not cadaverous, with chalk stones in his hands. He said that, if it were not for the gout his blindness would be tolerable.'

But in his last sentences M. Scherer comes upon what is undoubtedly Milton's true distinction as a poet, his 'unfailing level of style'. Milton has always the sure, strong touch of the master. His power both of diction and of rhythm is unsurpassable, and it is characterised by being always present—not depending on an access of emotion, not intermittent, but, like the grace of Raphael, working in its possessor as a constant gift of nature. Milton's style, moreover, has the same propriety and soundness in presenting plain matters, as in the comparatively smooth task for a poet of presenting grand ones. His rhythm is as admirable where, as in the line—

'And Tiresias and Phineus, prophets old—'°

it is unusual, as in such lines as—

'With dreadful faces throng'd and fiery arms—'°

where it is simplest. And what high praise this is, we may best appreciate by considering the ever-recurring failure, both in rhythm and in diction, which we find in the so-called Miltonic blank verse of Thomson, Cowper, Wordsworth. What leagues of lumbering movement! what desperate endeavours, as in Wordsworth's

'And at the "Hoop" alighted, famous inn,'°

to render a platitude endurable by making it pompous! Shakspeare himself, divine as are his gifts, has not, of the marks of the master, this one: perfect sureness of hand in his style. Alone of English poets, alone in English art, Milton has it; he is our great artist in style, our one first-rate master in the grand style. He is as truly a master in this style as the great Greeks are, or Virgil, or Dante. The number of such masters is so limited that a man acquires a world-rank in poetry and art, instead of a mere local rank, by being counted among them. But Milton's importance to us Englishmen, by virtue of this distinction of his, is incalculable. The charm of a master's unfailing touch in diction and in rhythm, no one, after all, can feel so intimately, so profoundly, as his own countrymen. Invention, plan, wit, pathos, thought, all of them are in great measure capable of being detached from the original work itself, and of being exported for admiration abroad. Diction and rhythm are not. Even when a foreigner can read the work in its own language, they are not, perhaps, easily appreciable by him. It shows M. Scherer's thorough knowledge of English, and his critical sagacity also, that he has felt the force of them in Milton. We natives must naturally feel it yet more powerfully. Be it remembered, too, that English literature, full of vigour and genius as it is, is peculiarly impaired by gropings and inadequacies in form. And the same with English art. Therefore for the English artist in any line, if he is a true artist, the study of Milton may well have an indescribable attraction. It gives him lessons which nowhere else from an English-man's work can he obtain, and feeds a sense which English work, in general, seems bent on disappointing and baffling. And this sense is yet so deep-seated in human nature,—this sense of style,—that probably not for artists alone, but for all intelligent Englishmen who read him, its gratification by Milton's poetry is a large though often not fully recognised part of his charm, and a very wholesome and fruitful one.

As a man, too, not less than as a poet, Milton has a side of unsurpassable grandeur. A master's touch is the gift of nature. Moral qualities, it is commonly thought, are in our own power. Perhaps the germs of such qualities are in their greater or less strength as much a part of our natural constitution as the sense for style. The range open to our own will and power, however, in developing and establishing them, is evidently much larger. Certain high moral dispositions Milton had from nature, and he sedulously trained and developed them until they became habits of great power.

Some moral qualities seem to be connected in a man with his power of style. Milton's power of style, for instance, has for its great character *elevation*; and Milton's elevation clearly comes, in the main, from a moral quality in him,—his pureness. 'By pureness, by kindness!' says St Paul.° These two, pureness and kindness, are, in very truth, the two signal Christian virtues, the two mighty wings of Christianity, with which it winnowed and renewed, and still winnows and renews, the world. In kindness, and in all which that word conveys or suggests, Milton does not shine. He had the temper of his Puritan party. We often hear the boast, on behalf of the Puritans, that they produced 'our great epic poet'. Alas! one might not unjustly retort that they spoiled him. However, let Milton bear his own burden; in his temper he had natural affinities with the Puritans. He has paid for it by limitations as a poet. But, on the other hand, how high, clear, and splendid is his pureness; and how intimately does its might enter into the voice of his poetry! We have quoted some ill-conditioned passages from his prose, let us quote from it a passage of another stamp:—°

'And long it was not after, when I was confirmed in this opinion, that he, who would not be frustrate of his hope to write well hereafter in laudable things, ought himself to be a true poem; that is, a composition and pattern of the best and honourablest things; not presuming to sing high praises of heroic men, or famous cities, unless he have in himself the experience and the practice of all that which is praiseworthy. These reasonings, together with a certain niceness of nature, an honest haughtiness and self-esteem, either of what I was or what I might be (which let envy call pride), and lastly that modesty whereof here I may be excused to make some beseeming profession; all these uniting the supply of their natural aid together kept me still above low descents of mind. Next (for hear me out now, readers), that I may tell ye whither my younger feet wandered; I betook me among those lofty fables and romances which recount in solemn cantos the deeds of knighthood founded by our victorious kings, and from hence had in renown over all Christendom. There I read it in the oath of every knight, that he should defend to the expense of his best blood, or of his life if it so befell him, the honour and chastity of virgin or matron; from whence even then I learnt what a noble virtue chastity sure must be, to the defence of which so many worthies by such a dear adventure of themselves had sworn. Only this my mind gave me, that every free and gentle spirit, without that oath, ought to be born a knight, nor needed to expect the gilt spur, or the laying of a sword upon his shoulder, to stir him up both by his counsel and his arm to secure and protect the weakness of any attempted chastity.'

Mere fine professions are in this department of morals more common and more worthless than in any other. What gives to Milton's professions such a stamp of their own is their accent of absolute sincerity. In this elevated strain of moral pureness his life was really pitched; its strong, immortal beauty passed into the diction and rhythm of his poetry.

But I did not propose to write a criticism of my own upon Milton. I proposed to recite and compare the criticisms on him by others. Only one is tempted, after our many extracts from M. Scherer, in whose criticism of Milton the note of blame fills so much more place than the note of praise, to accentuate this note of praise, which M. Scherer touches indeed with justness, but hardly perhaps draws out fully enough or presses firmly enough. As a poet and as a man, Milton has a side of grandeur so high and rare, as to give him rank along with the half-dozen greatest poets who have ever lived, although to their masterpieces his *Paradise Lost* is, in the fulfilment of the complete range of conditions which a great poem ought to satisfy, indubitably inferior.

Nothing is gained by huddling on 'our great epic poet', in a promiscuous heap, every sort of praise. Sooner or later the question: How does Milton's masterpiece really stand to us moderns, what are we to think of it, what can we get from it? must inevitably be asked and answered. We have marked that side of the answer which is and will always remain favourable to Milton. The unfavourable side of the answer is supplied by M. Scherer. '*Paradise Lost* lives; but none the less is it true that its fundamental conceptions have become foreign to us, and that if the work subsists it is in spite of the subject treated by it.'

The verdict seems just, and it is supported by M. Scherer with considerations natural, lucid, and forcible. He, too, has his conventions when he comes to speak of Racine and Lamartine. But his judgments on foreign poets, on Shakspeare, Byron, Goethe, as well as on Milton, seem to me to be singularly uninfluenced by the conventional estimates of these poets, and singularly rational. Leaning to the side of severity, as is natural when one has been wearied by choruses of ecstatic and exaggerated praise, he yet well and fairly reports, I think, the real impression made by these great men and their works on a modern mind disinterested, intelligent, and sincere. The English reader, I hope, may have been interested in seeing how Milton and his *Paradise Lost* stand such a survey. And those who are dissatisfied with what has been thus given them may always revenge themselves by falling back upon their Addison, and by observing sarcastically that 'a few general rules extracted out of the French authors, with a certain cant of words, has sometimes set up an illiterate heavy writer for a most judicious and formidable critic'.°

A French Critic on Goethe

IT takes a long time to ascertain the true rank of a famous writer. A young friend of Joseph de Maistre, a M. de Syon, writing in praise of the literature of the nineteenth century as compared with that of the eighteenth, said of Chateaubriand, that 'the Eternal created Chateaubriand to be a guide to the universe'. Upon which judgment Joseph de Maistre comments thus:° 'Clear it is, my good young man, that you are only eighteen; let us hear what you have to say at forty.' '*On voit bien, excellent jeune homme, que vous avez dix-huit ans; je vous attends à quarante.*'

The same Joseph de Maistre has given° an amusing history of the rise of our own Milton's reputation:—

'No one had any suspicion of Milton's merits, when one day Addison took the speaking-trumpet of Great Britain (the instrument of loudest sound in the universe), and called from the top of the Tower of London: "Roman and Greek authors, give place!"

'He did well to take this tone. If he had spoken modestly, if he had simply said that there were great beauties in *Paradise Lost*, he would not have produced the slightest impression. But this trenchant sentence, dethroning Homer and Virgil, struck the English exceedingly. They said one to the other: "What, we possessed the finest epic poem in the world, and no one suspected it! What a thing is inattention! But now, at any rate, we have had our eyes opened." In fact, the reputation of Milton has become a national property, a portion of the Establishment, a Fortieth Article; and the English would as soon think of giving up Jamaica as of giving up the pre-eminence of their great poet.'

Joseph de Maistre goes on to quote a passage from a then recent English commentator on Milton,—Bishop Newton. Bishop Newton, it seems, declared that 'every man of taste and genius must admit *Paradise Lost* to be the most excellent of modern productions, as the Bible is the most perfect of the productions of antiquity'. In a note M. de Maistre adds: 'This judgment of the good bishop appears unspeakably ridiculous.'

Ridiculous, indeed! but a page or two later we shall find the clear-sighted critic himself almost as far astray as his 'good bishop' or as his 'good young man':—

'The strange thing is that the English, who are thorough Greek scholars, are willing enough to admit the superiority of the Greek tragedians over Shakspeare; but when they come to Racine, *who is in reality simply a Greek speaking French*, their standard of beauty all of a sudden changes, and Racine, who is at least the equal of the Greeks, has to take rank far below Shakspeare, who is inferior to them. This theorem in *trigonometry* presents no difficulties to the people of soundest understanding in Europe.'

So dense is the cloud of error here that the lover of truth and daylight will hardly even essay to dissipate it: he does not know where to begin. It is as when M. Victor Hugo gives his list of the sovereigns on the world's roll of creators and poets: 'Homer, Æschylus, Sophocles, Lucretius, Virgil, Horace, Dante, Shakspeare, *Rabelais, Molière, Corneille, Voltaire.*' His French audience rise and cry enthusiastically: '*And Victor Hugo!*' And really that is perhaps the best criticism on what he has been saying to them.

Goethe, the great poet of Germany, has been placed by his own countrymen now low, now high; and his right poetical rank they have certainly not yet succeeded in finding. Tieck, in his introduction° to the collected writings of Lenz, noticing Goethe's remark on Byron's *Manfred*,—that Byron had 'assimilated *Faust*, and sucked out of it the strangest nutriment to his hypochondria',—says tartly that Byron, when he himself talked about his obligations to Goethe, was merely using the language of compliment, and would have been highly offended if any one else had professed to discover them. And Tieck proceeds:—

'Everything which in the Englishman's poems might remind one of *Faust*, is in my opinion far above *Faust*; and the Englishman's feeling, and his incomparably more beautiful diction, are so entirely his own, that I cannot possibly believe him to have had *Faust* for his model.'

But then there comes a scion of the excellent stock of the Grimms, a Professor Herman Grimm, and lectures on Goethe° at Berlin, now that the Germans have conquered the French, and are the first military power in the world, and have become a great nation, and require a national poet to match; and Professor Grimm says of *Faust*, of which Tieck had spoken so coldly: 'The career of this, the greatest work of the greatest poet of all times and of all peoples, has but just begun, and we have been making only the first attempts at drawing forth its contents.'

If this is but the first letting out of the waters, the coming times may, indeed, expect a deluge.

Many and diverse must be the judgments passed upon every great poet, upon every considerable writer. There is the judgment of enthusiasm and admiration, which proceeds from ardent youth, easily fired, eager to find a hero and to worship him. There is the judgment of gratitude and sympathy, which proceeds from those who find in an author what helps them, what they want, and who rate him at a very high value accordingly. There is the judgment of ignorance, the judgment of incompatibility, the judgment of envy and jealousy. Finally, there is the systematic judgment, and this judgment is the most worthless of all. The sharp scrutiny of envy and jealousy may bring real faults to light. The judgments of incompatibility and ignorance are instructive, whether they reveal necessary clefts of separation between the experiences of different sorts of people, or reveal simply the narrowness and bounded view of those who judge. But the systematic judgment is altogether unprofitable. Its author has not really his eye upon the professed object of his criticism at all, but upon something else which he wants to prove by means of that object. He neither really tells us, therefore, anything about the object, nor anything about his own ignorance of the object. He never fairly looks at it, he is looking at something else. Perhaps if he looked at it straight and full, looked at it simply, he might be able to pass a good judgment on it. As it is, all he tells us is that he is no genuine critic, but a man with a system, an advocate.

Here is the fault of Professor Herman Grimm, and of his Berlin lectures on Goethe. The professor is a man with a system; the lectures are a piece of advocacy. Professor Grimm is not looking straight at 'the greatest poet of all times and of all peoples'; he is looking at the necessities, as to literary glory, of the new German empire.

But the definitive judgment on this great Goethe, the judgment of mature reason, the judgment which shall come 'at forty years of age', who may give it to us? Yet how desirable to have it! It is a mistake to think that the judgment of mature reason on our favourite author, even if it abates considerably our high-raised estimate of him, is not a gain to us. Admiration is positive, say some people, disparagement is negative; from what is negative we can get nothing. But is it no advantage, then, to the youthful enthusiast for Chateaubriand, to come to know that 'the Eternal did *not* create Chateaubriand to be a guide to the universe'? It is a very great advantage, because these over-charged admirations are always exclusive, and prevent us from giving heed to other things which deserve admiration. Admiration is salutary and formative, true; but things admirable are sown wide, and are to be gathered here and gathered there, not all in one place; and until we have gathered them wherever they are to be found, we have not known the true salutariness and formativeness of admiration. The quest is large; and occupation with the unsound or half sound, delight in the not good or less good, is a sore let and hindrance to us. Release from such occupation and delight sets us free for ranging farther, and for perfecting our sense of beauty. He is the happy man, who, encumbering himself with the love of nothing which is not beautiful, is able to embrace the greatest number of things beautiful in his love.

I have already spoken of the judgment of a French critic, M. Scherer, upon Milton. I propose now to draw attention to the judgment of the same critic upon Goethe. To set to work to discuss Goethe thoroughly, so as to arrive at the true definitive judgment respecting him, seems to me a most formidable enterprise. Certainly one should not think of attempting it within the limits of a single review-article. M. Scherer has devoted to Goethe not one article, but a series of articles.° I do not say that the adequate, definitive judgment on Goethe is to be found in these articles of M. Scherer. But I think they afford a valuable contribution towards it. M. Scherer is well-informed, clear-sighted, impartial. He is not warped by injustice and ill-will towards Germany, although the war has undoubtedly left him with a feeling of soreness. He is candid and cool, perhaps a little cold. Certainly he will not tell us that 'the Eternal created Goethe to be a guide to the universe'. He is free from all heat of youthful enthusiasm, from the absorption of a discoverer in his new discovery, from the subjugation of a disciple by the master who has helped and guided him. He is not a man with a system. And his point of view is in many respects that of an Englishman. We mean that he has the same instinctive sense rebelling against what is verbose, ponderous, roundabout, inane,—in one word, *niais* or silly,—in German literature,° just as a plain Englishman has.

This ground of sympathy between Englishmen and Frenchmen has not been enough remarked, but it is a very real one. They owe it to their having alike had a long-continued national life, a long-continued literary activity, such as no other modern nation has had. This course of practical experience does of itself beget a turn for directness and clearness of speech, a dislike for futility and fumbling, such as without it we shall rarely find general. Dr Wiese, in his recent useful work on English schools,° expresses surprise that the French language and literature should find more favour in Teutonic England than the German. But community of practice is more telling than community of origin. While English and French are printed alike, and while an English and a French sentence each of them says what it has to say in the same plain fashion, a German newspaper is still printed in black letter, and a German sentence is framed in the style of this which we quote from Dr Wiese himself:° 'Die Engländer einer grossen, in allen Erdtheilen eine Achtung gebietende Stellung einnehmenden Nation angehören!' The Italians are a Latin race, with a clear-cut language; but much of their modern prose has all the circuitousness and slowness of the German, and from the same cause: the want of the pressure of a great national life, with its practical discipline, its ever-active traditions; its literature, for centuries past, powerful and incessant. England has these in common with France.

M. Scherer's point of view, then, in judging the productions of German literature, will naturally, I repeat, coincide in several important respects with that of an Englishman. His mind will make many of the same instinctive demands as ours, will feel many of the same instinctive repugnances. We shall gladly follow him, therefore, through his criticism of Goethe's works. As far as possible he shall be allowed to speak for himself, as he was when we were dealing with his criticism on Milton. But as then, too, I shall occasionally compare M. Scherer's criticism on his author with the criticism of others. And I shall by no means attempt, on the present opportunity, a substantive criticism of my own, although I may from time to time allow myself to comment, in passing, upon the judgments of M. Scherer.

We need not follow M. Scherer in his sketch of Goethe's life. It is enough to remember that the main dates in Goethe's life are, his birth in 1749; his going to Weimar with the Grand Duke, Carl-August, in 1775; his stay in Italy from September 1786 to June 1788; his return in 1788 to Weimar; a severe and nearly fatal illness in 1801; the loss of Schiller in 1805, of Carl-August in 1828; his own death in 1832. With these dates fixed in our minds, we may come at once to the consideration of Goethe's works.

The long list begins, as we all know, with *Götz von Berlichingen* and *Werther*. We all remember how Mr Carlyle,° 'the old man eloquent', who in his younger days, fifty years ago, betook himself to Goethe for light and help, and found what he sought, and declared his gratitude so powerfully and well, and did so much to make Goethe's name a name of might for other Englishmen also, a strong tower into which the doubter and the despairer might run and be safe,—we all remember how Mr Carlyle has taught us to see in *Götz* and in *Werther* the double source from which have flowed those two mighty streams,—the literature of feudalism and romance, represented for us by Scott, and the literature of emotion and passion, represented for us by Byron.

M. Scherer's tone throughout is, we have said, not that of the ardent and grateful admirer, but of the cool, somewhat cold critic. He by no manner of means resembles Mr Carlyle. Already the cold tone appears in M. Scherer's way of dealing with Goethe's earliest productions.

M. Scherer seems to me to rate the force and interest of *Götz* too low. But his remarks on the derivedness of this supposed *source* are just. The Germans, he says, were bent, in their 'Sturm und Drang' period, on throwing off literary conventions, imitation of all sorts, and on being original. What they really did, was to fall from one sort of imitation, the imitation of the so-called classical French literature of the seventeenth century, into another.

'*Götz von Berlichingen* is a study composed after the dramatised chronicles of Shakspeare, and *Werther* is a product yet more direct of the sensibility and declamation brought into fashion by Jean Jacques Rousseau. All in these works is infantine, both the aim at being original, and the way of setting about it. It is exactly as it was with us about 1830. One imagines one is conducting an insurrection, making oneself independent; what one really does is to cook up out of season an old thing. Shakspeare had put the history of his nation upon the stage; Goethe goes for a subject to German history. Shakspeare, who was not fettered by the scenic conditions of the modern theatre, changed the place at every scene; *Götz* is cut up in the same fashion. I say nothing of the substance of the piece, of the absence of characters, of the nullity of the hero, of the commonplace of Weislingen "the inevitable traitor", of the melodramatic machinery of the secret tribunal. The style is no better. The astonishment is not that Goethe at twenty-five should have been equal to writing this piece; the astonishment is that after so poor a start he should have subsequently gone so far.'°

M. Scherer seems to me quite unjust, I repeat, to this first dramatic work of Goethe. Mr Hutton pronounces it 'far the most noble as well as the most powerful of Goethe's dramas'. And the merit which Mr Hutton finds in *Götz* is a real one; it is the work where Goethe, young and ardent, has most forgotten *himself* in his characters. 'There was something', says Mr Hutton° (and here he and M. Scherer are entirely in accord), 'which prevented Goethe, we think, from ever becoming a great dramatist. He could never lose himself sufficiently in his creations.' It is in *Götz* that he loses himself in them the most. *Götz* is full of faults, but there is a life and a power in it, and it is not dull. This is what distinguishes it from Schiller's *Robbers*. The *Robbers* is at once violent and tiresome. *Götz* is violent, but it is not tiresome.

Werther, which appeared a year later than *Götz*, finds more favour at M. Scherer's hands.° *Werther* is superior to *Götz*, he says, 'inasmuch as it is more modern, and is consequently alive, or, at any rate, has been alive lately. It has sincerity, passion, eloquence. One can still read it, and with emotion.' But then come the objections:—

'Nevertheless, and just by reason of its truth at one particular moment, *Werther* is gone by. It is with the book as with the blue coat and yellow breeches of the hero; the reader finds it hard to admit the pathetic in such accoutrement. There is too much enthusiasm for Ossian, too much absorption in nature, too many exclamations and apostrophes to beings animate and inanimate, too many torrents of tears. Who can forbear smiling as he reads the scene of the storm, where Charlotte first casts her eyes on the fields, then on the sky, and finally, laying her hand on her lover's, utters this one word: *Klopstock!* And then the cabbage-passage! . . . *Werther* is the poem of the German middle-class sentimentality of that day. It must be said that our sentimentality, even at the height of the *Héloïse* season, never reached the extravagance of that of our neighbours . . . Mdlle. Flachsland, who married Herder, writes to her betrothed that one night in the depth of the woods she fell on her knees as she looked at the moon, and that having found some glowworms she put them into her hair, being careful to arrange them in couples that she might not disturb their loves.'

One can imagine the pleasure of a victim of 'Kruppism and corporalism'° in relating that story of Mdlle. Flachsland. There is an even better story of the return of a Dr Zimmermann° to his home in Hanover, after being treated for hernia at Berlin; but for this story I must send the reader to M. Scherer's own pages.

After the publication of *Werther* began Goethe's life at Weimar. For ten years he brought out nothing except occasional pieces for the Court theatre, and occasional poems. True, he carried the project of his *Faust* in his mind, he planned *Wilhelm Meister*, he made the first draft of *Egmont*, he wrote *Iphigeneia* and *Tasso* in prose. But he could not make the progress he wished. He felt the need, for his work, of some influence which Weimar could not give. He became dissatisfied with the place, with himself, with the people about him. In the autumn of 1786 he disappeared from Weimar, almost by secret flight, and crossed the Alps into Italy. M. Scherer says truly° that this was the great event of his life.

Italy, Rome above all, satisfied Goethe, filled him with a sense of strength and joy. 'At Rome,' he writes from that city, 'he who has eyes to see, and who uses them seriously, becomes solid. The spirit receives a stamp of vigour; it attains to a gravity in which there is nothing dry or harsh,—to calm, to joy. For my own part, at any rate, I feel that I have never before had the power to judge things so justly, and I congratulate myself on the happy result for my whole future life.' So he wrote while he was in Rome.° And he told the Chancellor von Müller,° twenty-five years later, that from the hour when he crossed the Ponte Molle on his return to Germany, he had never known a day's happiness. 'While he spoke thus,' adds the Chancellor, 'his features betrayed his deep emotion.'

The Italy, from which Goethe thus drew satisfaction and strength, was

Græco-Roman Italy, pagan Italy. For mediæval and Christian Italy he had no heed, no sympathy. He would not even look at the famous church of St Francis at Assisi. 'I passed it by', he says, 'in disgust.' And he told a young Italian who asked him his opinion of Dante's great poem, that he thought the *Inferno* abominable, the *Purgatorio* dubious, and the *Paradiso* tiresome.°

I have not space to quote what M. Scherer says of the influence on Goethe's genius of his stay in Rome. We are more especially concerned with the judgments of M. Scherer on the principal works of Goethe as these works succeed one another. At Rome, or under the influence of Rome, *Iphigeneia* and *Tasso* were recast in verse, *Egmont* was resumed and finished, the chief portion of the first part of *Faust* was written. Of the larger works of Goethe in poetry, these are the chief. Let us see what M. Scherer has to say of them.

Tasso and *Iphigeneia*, says M. Scherer very truly,° mark a new phase in the literary career of Goethe:—

'They are works of finished style and profound composition. There is no need to inquire whether the *Iphigeneia* keeps to the traditional data of the subject; Goethe desired to make it Greek only by its sententious elevation and grave beauty. What he imitates are the conditions of art as the ancients understood them, but he does not scruple to introduce new thoughts into these mythological *motives*. He has given up the aim of rendering by poetry what is characteristic or individual; his concern is henceforth with the ideal, that is to say, with the transformation of things through beauty. If I were to employ the terms in use amongst ourselves, I should say that from romantic Goethe had changed to being classic; but, let me say again, he is classic only by the adoption of the elevated style, he imitates the ancients merely by borrowing their peculiar sentiment as to art, and within these bounds he moves with freedom and power. The two elements, that of immediate or passionate feeling, and that of well-considered combination of means, balance one another, and give birth to finished works. *Tasso* and *Iphigeneia* mark the apogee of Goethe's talent.'

It is curiously interesting to turn from this praise of *Tasso* and *Iphigeneia* to that by the late Mr Lewes,° whose *Life of Goethe*, a work in many respects of brilliant cleverness, will be in the memory of many of us. 'A marvellous dramatic poem!' Mr Lewes calls *Iphigeneia*. 'Beautiful as the separate passages are, admirers seldom think of passages, they think of the wondrous whole.' Of *Tasso*, Mr Lewes says: 'There is a calm, broad effulgence of light in it, very different from the concentrated lights of *effect* which we are accustomed to find in modern works. It has the clearness, unity, and matchless grace of a Raphael, not the lustrous warmth of a Titian, or the crowded gorgeousness of a Paul Veronese.'

Every one will remark the difference of tone between this criticism and M. Scherer's. Yet M. Scherer's criticism conveyed praise, and, for him, warm praise. *Tasso* and *Iphigeneia* mark, in his eyes, the period, the too short period, during which the forces of inspiration and of reflection, the poet in Goethe and the critic in him, the thinker and the artist, in whose conflict M. Scherer sees the history of our author's literary development, were in equilibrium.

Faust also, the first part of *Faust*, the only one which counts, belongs by its composition to this *Tasso* period. By common consent it is the best of Goethe's works. For while it had the benefit of his matured powers of thought, of his command over his materials, of his mastery in planning and expressing, it possesses by the nature of its subject an intrinsic richness, colour, and warmth. Moreover, from Goethe's long and early occupation with the subject, *Faust* has preserved many a stroke and flash out of the days of its author's fervid youth. To M. Scherer, therefore, as to the world in general, the first part of *Faust* seems Goethe's masterpiece. M. Scherer does not call *Faust* the greatest work of the greatest poet of all times and all peoples, but thus he speaks of it:—°

'Goethe had the good fortune early to come across a subject, which, while it did not lend itself to his faults, could not but call forth all the powers of his genius. I speak of *Faust*. Goethe had begun to occupy himself with it as early as 1774, the year in which *Werther* was published. Considerable portions of the First Part appeared in 1790; it was completed in 1808. We may congratulate ourselves that the work was already, at the time of his travels in Italy, so far advanced as it was; else there might have been danger of the author's turning away from it as from a Gothic, perhaps unhealthy, production. What is certain is, that he could not put into *Faust* his preoccupation with the antique, or, at any rate, he was obliged to keep this for the Second Part. The first *Faust* remained, whether Goethe would or no, an old story made young again, to serve as the poem of thought, the poem of modern life. This kind of adaptation had evidently great difficulties. It was impossible to give the story a satisfactory ending; the compact between the Doctor and the Devil could not be made good, consequently the original condition of the story was gone, and the drama was left without an issue. We must, therefore, take *Faust* as a work which is not finished, and which could not be finished. But, in compensation, the choice of this subject had all sorts of advantages for Goethe. In place of the somewhat cold symbolism for which his mind had a turn, the subject of *Faust* compelled him to deal with popular beliefs. Instead of obliging him to produce a drama with beginning, middle, and end, it allowed him to proceed by episodes and detached scenes. Finally, in a subject fantastic and diabolic there could hardly be found room for the imitation of models. Let me add, that in bringing face to face human aspiration represented by Faust and pitiless irony represented by Mephistopheles, Goethe found the natural scope for his keen observations on all things. It is unquestionable that

Faust stands as one of the great works of poetry; and, perhaps, the most wonderful work of poetry in our century. The story, the subject, do not exist as a whole, but each episode by itself is perfect, and the execution is nowhere defective. *Faust* is a treasure of poetry, of pathos, of the highest wisdom, of a spirit inexhaustible and keen as steel. There is not, from the first verse to the last, a false tone or a weak line.'

This praise is discriminating, and yet earnest, almost cordial. '*Faust* stands as one of the great works of poetry; and, perhaps, the most wonderful work of poetry in our century.' The *perhaps* might be away. But the praise is otherwise not coldly stinted, not limited ungraciously and unduly.

Goethe returned to 'the formless Germany',° to the Germanic north with its 'cold wet summers', of which he so mournfully complained. He returned to Weimar with its petty Court and petty town, its society which Carl-August himself, writing to Knebel, calls 'the most tiresome on the face of the earth', and of which the ennui drove Goethe sometimes to 'a sort of internal despair'. He had his animating friendship with Schiller. He had also his connection with Christiana Vulpius, whom he afterwards married. That connection both the moralist and the man of the world may unite in condemning. M. Scherer calls it 'a degrading connection with a girl of no education, whom Goethe established in his house to the great embarrassment of all his friends, whom he either could not or would not marry until eighteen years later, and who punished him as he deserved by taking a turn for drink,—a turn which their unfortunate son inherited'. In these circumstances was passed the second half of Goethe's life, after his return from Italy. The man of reflection, always present in him, but balanced for a while by the man of inspiration, became now, M. Scherer thinks, predominant. There was a *refroidisse-ment graduel*, a gradual cooling down, of the poet and artist.

The most famous works of Goethe which remain yet to be mentioned are *Egmont, Hermann and Dorothea, Wilhelm Meister*, the *Second Part of Faust*, and the *Gedichte*, or short poems. Of *Egmont* M. Scherer says:—°

'This piece also belongs, by the date of its publication, to the period which followed Goethe's stay in Rome. But in vain did Goethe try to transform it, he could not succeed. The subject stood in his way. We need not be surprised, therefore, if *Egmont* remains a mediocre performance, Goethe having always been deficient in dramatic faculty, and not in this case redeeming his defect by qualities of execution, as in *Iphigeneia*. He is too much of a generaliser to create a character, too meditative to create an action. *Egmont* must be ranked by the side of *Götz*; it is a product of the same order. The hero is not a living being; one does not know what he wants; the object of the conspiracy is not brought out. The unfortunate Count does certainly exclaim, as he goes to the scaffold, that he is dying for liberty, but nobody had suspected it until that moment. It is the same with the popular movement; it is insufficiently rendered, without breadth, without power. I say nothing of Machiavel, who preaches toleration to the Princess Regent and tries to make her understand the uselessness of persecution; nor of Claire, a girl sprung from the people, who talks like an epigram of the Anthology: "Neither soldiers nor lovers should have their arms tied." *Egmont* is one of the weakest among Goethe's weak pieces for the stage.'

But now, on the other hand, let us hear Mr Lewes:° 'When all is said, the reader thinks of Egmont and Clärchen, and flings criticism to the winds. These are the figures which remain in the memory; bright, genial, glorious creations, comparable to any to be found in the long galleries of art!' What a different tone!

Aristotle says, with admirable common-sense, that the determination of how a thing really is, is ὡς ἂν ὁ φρόνιμος ὁρίσειεν, 'as the judicious would determine'.° And would the judicious, after reading *Egmont*, determine with Mr Lewes, or determine with M. Scherer? Let us for the present leave the judicious to try, and let us pass to M. Scherer's criticism of *Hermann and Dorothea*. 'Goethe's epic poem', writes Schiller,° 'you have read; you will admit that it is the pinnacle of his and all our modern art.' In Professor Grimm's eyes, perhaps, this is but scant praise, but how much too strong is it for M. Scherer!°

'Criticism is considerably embarrassed in presence of a poem in many respects so highly finished as the antico-modern and heroico-middle-class idyll of Goethe. The ability which the author has spent upon it is beyond conception; and, the kind of poem being once allowed, the indispensable concessions having been once made, it is certain that the pleasure is doubled by seeing, at each step, difficulty so marvellously overcome. But all this cannot make the effort to be effort well spent, nor the kind of poem a true, sound and worthy kind. *Hermann and Dorothea* remains a piece of elegant cleverness, a wager laid and won, but for all that, a feat of ingenuity and nothing more. It is not quite certain that our modern society will continue to have a poetry at all; but most undoubtedly, if it does have one, it will be on condition that this poetry belongs to its time by its language, as well as by its subject. Has any critic remarked how Goethe's manner of proceeding is at bottom that of parody, and how the turn of a straw would set the reader laughing at these farm-horses transformed into coursers, these village innkeepers and apothecaries who speak with the magniloquence of a Ulysses or a Nestor? Criticism should have the courage to declare that all this is not sincere poetry at all, but solely the product of an exquisite dilettantism, and,—to speak the definitive judgment upon it,—a factitious work.'

Once again we will turn to Mr Lewes° for contrast:—

'Do not let us discuss whether *Hermann and Dorothea* is or is not an epic. It is a poem. Let us accept it for what it is,—a poem full of life, character, and beauty; of all idylls it is the most truly idyllic, of all poems describing country life and country people it is the most truthful. Shakspeare himself is not more dramatic in the presentation of character.'

It is an excellent and wholesome discipline for a student of Goethe to be brought face to face with such opposite judgments concerning his chief productions. It compels us to rouse ourselves out of the passiveness with which we in general read a celebrated work, to open our eyes wide, to ask ourselves frankly how, according to our genuine feeling, the truth stands. We all recollect Mr Carlyle on *Wilhelm Meister*,° 'the mature product of the first genius of our times':—

'Anarchy has now become peace; the once gloomy and perturbed spirit is now serene, cheerfully vigorous, and rich in good fruits. . . . The ideal has been built on the actual; no longer floats vaguely in darkness and regions of dreams, but rests in light, on the firm ground of human interest and business, as in its true scene, and on its true basis.'

Schiller, too, said° of *Wilhelm Meister*, that he 'accounted it the most fortunate incident in his existence to have lived to see the completion of this work'. And again: 'I cannot describe to you how deeply the truth, the beautiful vitality, the simple fulness of this work has affected me. The excitement into which it has thrown my mind will subside when I shall have thoroughly mastered it, and that will be an important crisis in my being.'

Now for the cold-water douche of our French critic:—°

'Goethe is extremely great, but he is extremely unequal. He is a genius of the first order, but with thicknesses, with spots, so to speak, which remain opaque and where the light does not pass. Goethe, to go farther, has not only genius, he has what we in France call *esprit*, he has it to any extent, and yet there are in him sides of commonplace and silliness. One cannot read his works without continually falling in with trivial admirations, solemn pieces of simplicity, reflections which bear upon nothing. There are moments when Goethe turns upon society and upon art a ken of astonishing penetration; and there are other moments when he gravely beats in an open door, or a door which leads nowhere. In addition, he has all manner of hidden intentions, he loves byways of effect, seeks to insinuate lessons, and so becomes heavy and fatiguing. There are works of his which one cannot read without effort. I shall never forget the repeated acts of self-sacrifice which it cost me to finish *Wilhelm Meister* and the *Elective Affinities*. As Paul de Saint-Victor has put it: "When Goethe goes in for being tiresome he succeeds with an astonishing perfection, he is the *Jupiter Pluvius* of ennui. The very height from which he pours it down, does but make its weight greater." What an insipid invention is the pedagogic city! What a trivial world is that in which the Wilhelms and the Philinas, the Eduards and the Ottilias, have their being! Mignon has been elevated into a poetic creation; but Mignon has neither charm, nor mystery, nor veritable existence; nor any other poetry belonging to her,—let us say it right out,—except the half-dozen immortal stanzas put into her mouth.'

And, as we brought Schiller to corroborate the praise of *Wilhelm Meister*, let us bring Niebuhr to corroborate the blame. Niebuhr calls *Wilhelm Meister* 'a menagerie of tame animals'.°

After this the reader can perhaps imagine, without any specimens of it, the sort of tone in which M. Scherer passes judgment° upon *Dichtung und Wahrheit*, and upon Goethe's prose in general. Even Mr Lewes declares° of Goethe's prose: 'He has written with a perfection no German ever achieved before, and he has also written with a feebleness which it would be gratifying to think no German would ever emulate again'.

Let us return, then, to Goethe's poetry. There is the continuation of *Faust* still to be mentioned. First we will hear Mr Carlyle. In *Helena* 'the design is', says Mr Carlyle,° 'that the story of *Faust* may fade away at its termination into a phantasmagoric region, where symbol and thing signified are no longer clearly distinguished', and that thus 'the final result may be curiously and significantly indicated rather than directly exhibited'. *Helena* is 'not a type of one thing, but a vague, fluctuating, fitful adumbration of many'. It is, properly speaking, 'what the Germans call a *Mährchen*, a species of fiction they have particularly excelled in'. As to its composition, 'we cannot but perceive it to be deeply studied, appropriate and successful'.

The 'adumbrative' style here praised, in which 'the final result is curiously and significantly indicated rather than directly exhibited', is what M. Scherer calls Goethe's 'last manner'.°

'It was to be feared that, as Goethe grew older and colder, the balance between those two elements of art, science and temperament, would not be preserved. This is just what happened, and hence arose Goethe's last manner. He had passed from representing characters to representing the ideal, he is now to pass from the ideal to the symbol. And this is quite intelligible; reflection, as it develops, leads to abstraction, and from the moment when the artist begins to prefer ideas to sensation he falls inevitably into allegory, since allegory is his only means for directly expressing ideas. Goethe's third epoch is characterised by three things: an ever-increasing devotion to the antique as to the supreme revelation of the beautiful, a disposition to take delight in aesthetic theories, and, finally, an irresistible desire for giving didactic intentions to art. This last tendency is evident in the continuation of *Wilhelm Meister*, and in the second *Faust*. We may say these two works are dead of a hypertrophy of reflection. They are a mere mass of symbols, hieroglyphics, sometimes even mystifications. There is something extraordinarily painful in seeing a genius so vigorous and a science so consummate thus mistaking the elementary conditions of poetry. The fault, we may add, is the fault of German art in general. The Germans have more ideas than their plasticity of temperament, evidently below par, knows how to deal with. They are wanting in the vigorous sensuousness, the concrete and immediate impression of things, which makes the artist, and which distinguishes him from the thinker.'

So much for Goethe's 'last manner' in general, and to serve as introduction to what M. Scherer has to say of the second *Faust*° more particularly:—

'The two parts of *Faust* are disparate. They do not proceed from one and the same conception. Goethe was like Defoe, like Milton, like so many others, who after producing a masterpiece have been bent on giving it a successor. Unhappily, while the first *Faust* is of Goethe's fairest time, of his most vigorous manhood, the second is the last fruit of his old age. Science, in the one, has not chilled poetic genius; in the other, reflection bears sway and produces all kind of symbols and abstractions. The beauty of the first comes in some sort from its very imperfection; I mean, from the incessant tendency of the sentiment of reality, the creative power, the poetry of passion and nature, to prevail over the philosophic intention and to make us forget it. Where is the student of poetry who, as he reads the monologues of Faust or the sarcasms of Mephistopheles, as he witnesses the fall and the remorse of Margaret, the most poignant history ever traced by pen, any longer thinks of the *Prologue in Heaven* or of the terms of the compact struck between Faust and the Tempter? In the second part it is just the contrary. The idea is everything. Allegory reigns there. The poetry is devoid of that simple and natural realism without which art cannot exist. One feels oneself in a sheer region of didactics. And this is true even of the finest parts,—of the third act, for example,—as well as of the weakest. What can be more burlesque than this Euphorion, son of Faust and Helen, who is found at the critical moment under a cabbage-leaf!—no, I am wrong, who descends from the sky "for all the world like a Phoebus", with a little cloak and a little harp, and ends by breaking his neck as he falls at the feet of his parents? And all this to represent Lord Byron, and, in his person, modern poetry, which is the offspring of romantic art! What decadence, good heavens! and what a melancholy thing is old age, since it can make the most plastic of modern poets sink down to these fantasticalities worthy of Alexandria!'

In spite of the high praise which he has accorded to *Tasso* and *Iphigeneia*, M. Scherer concludes,° then, his review of Goethe's productions thus:—

'Goethe is truly original and thoroughly superior only in his lyrical poems (the *Gedichte*), and in the first part of *Faust*. They are immortal works, and why? Because they issue from a personal feeling, and the spirit of system has not petrified them. And yet even his lyrical poems Goethe has tried to spoil. He went on correcting them incessantly; and, in bringing them to that degree of perfection in which we now find them, he has taken out of them their warmth.'

The worshipper of Goethe will ask with wrath and bitterness of soul whether M. Scherer has yet done. Not quite. We have still to hear some acute remarks on the pomposity of diction in our poet's stage pieces. The English reader will best understand, perhaps, the kind of fault meant, if we quote from the *Natural Daughter* a couple of lines not quoted, as it happens, by M. Scherer. The heroine has a fall from her horse, and the Court physician comes to attend her. The Court physician is addressed thus:—

> 'Erfahrner Mann, dem unseres König's Leben,
> Das unschätzbare Gut, vertraut ist. . .'°

'Experienced man, to whom the life of our sovereign, that inestimable treasure, is given in charge'. Shakspeare would have said *Doctor*. The German drama is full of this sort of roundabout, pompous language. 'Every one has laughed',° says M. Scherer, 'at the pomposity and periphrasis of French tragedy.' The heroic King of Pontus, in French tragedy, gives up the ghost with these words:—

> 'Dans cet embrassement dont la douceur me flatte,
> Venez, et recevez l'âme de Mithridate.'

'What has not been said,' continues M. Scherer, 'and justly said, against the artificial character of French tragedy?' Nevertheless, 'people do not enough remember that, convention being universally admitted in the seventeenth century, sincerity and even a relative simplicity remained possible' with an artificial diction; whereas Goethe did not find his artificial diction imposed upon him by conditions from without,—he made it himself, and of set purpose.

'It is a curious thing; this style of Goethe's has its cause just in that very same study which has been made such a matter of reproach against our tragedy-writers,—the study to maintain a pitch of general nobleness in all the language uttered. Everything with Goethe must be grave, solemn, sculptural. We see the influence of Winckelmann, and of his views on Greek art.'

Of Goethe's character, too, as well as of his talent, M. Scherer has something to say. English readers will be familiar enough with complaints of Goethe's 'artistic egotism',° of his tendency to set up his own intellectual culture as the rule of his life. The freshness of M. Scherer's repetition of these old complaints consists in his connecting them, as we have seen, with the criticism of Goethe's literary development. But M. Scherer has some direct blame° of defects in his author's character which is worth quoting:—

'It must fairly be confessed, the respect of Goethe for the mighty of this earth was carried to excesses which make one uncomfortable for him. One is confounded by these earnestnesses of servility. The King of Bavaria pays him a visit; the dear poet feels his head go round. The story should be read in the journal of the Chancellor von Müller:—Goethe after dinner became more and more animated and cordial. "It was no light matter", he said, "to work out the powerful impression produced by the King's presence, to assimilate it internally. It is difficult, in such circumstances, to keep one's balance and not to lose one's head. And yet the important matter is to extract from this apparition its real significance, to obtain a clear and distinct image of it."

'Another time he got a letter from the same sovereign; he talks of it to Eckermann with the same devout emotion—he "thanks Heaven for it as for a quite special favour". And when one thinks that the king in question was no other than that poor Louis of Bavaria, the ridiculous dilettante of whom Heine has made such fun! Evidently Goethe had a strong dose of what the English call "snobbishness". The blemish is the more startling in him, because Goethe is, in other respects, a simple and manly character. Neither in his person nor in his manner of writing was he at all affected; he has no self-conceit; he does not pose. There is in this particular all the difference in the world between him and the majority of our own French authors, who seem always busy arranging their draperies, and asking themselves how they appear to the world and what the gallery thinks of them.'

Goethe himself had in like manner called the French 'the women of Europe'.° But let us remark that it was not 'snobbishness' in Goethe, which made him take so seriously the potentate who loved Lola Montes;° it was simply his German 'corporalism'. A disciplinable and much-disciplined people, with little humour, and without the experience of a great national life, regards its official authorities in this devout and awe-struck way. To a German it seems profane and licentious to smile at his Dogberry.° He takes Dogberry seriously and solemnly, takes him at his own valuation.

We are all familiar with the general style of the critic who, as the phrase is, 'cuts up' his author. Such a critic finds very few merits and a great many faults, and he ends either with a phrase of condemnation, or with a phrase of compassion, or with a sneer. We saw, however, in the case of Milton, that one must not reckon on M. Scherer's ending in this fashion. After a course of severe criticism he wound up with earnest, almost reverential, praise. The same thing happens again in his treatment of Goethe. No admirer of Goethe will be satisfied with the treatment which hitherto we have seen Goethe receive at M. Scherer's hands. And the summing-up begins° in a strain which will not please the admirer much better:—

'To sum up, Goethe is a poet full of ideas and of observation, full of sense and taste, full even of feeling no less than of acumen, and all this united with an incomparable gift of versification. But Goethe has no artlessness, no fire, no invention; he is wanting in the dramatic fibre and cannot create; reflection, in Goethe, has been too much for emotion, the *savant* in him for poetry, the philosophy of art for the artist.'

And yet the final conclusion is this:—

'Nevertheless, Goethe remains one of the exceeding great among the sons of men. "After all," said he to one of his friends, "there are honest people up and down the world who have got light from my books; and whoever reads them, and gives himself the trouble to understand me, will acknowledge that he has acquired thence a certain inward freedom." I should like to inscribe these words upon the pedestal of Goethe's statue. No juster praise could be found for him, and in very truth there cannot possibly be for any man a praise higher or more enviable.'

And in an article on Shakspeare,° after a prophecy that the hour will come for Goethe, as in Germany it has of late come for Shakspeare, when criticism will take the place of adoration, M. Scherer, after insisting on those defects in Goethe of which we have been hearing so fully, protests that there are yet few writers for whom he feels a greater admiration than for Goethe, few to whom he is indebted for enjoyments more deep and more durable; and declares that Goethe, although he has not Shakspeare's power, is a genius more vast, more universal, than Shakspeare. He adds, to be sure, that Shakspeare had an advantage over Goethe in not outliving himself.

After all, then, M. Scherer is not far from being willing to allow, if any youthful devotee wishes to urge it, that 'the Eternal created Goethe to be a guide to the universe'. Yet he deals with the literary production of Goethe as we have seen. He is very far indeed from thinking it the performance 'of the greatest poet of all times and of all peoples'. And this is why I have thought M. Scherer's criticisms worthy of so much attention:—because a double judgment, somewhat of this kind, is the judgment about Goethe to which mature experience, the experience got 'by the time one is forty years old', does really, I think, bring us.

I do not agree with all M. Scherer's criticisms on Goethe's literary work. I do not myself feel, in reading the *Gedichte*, the truth of what M. Scherer says,—that Goethe has corrected and retouched them till he has taken all the warmth out of them. I do not myself feel the irritation in reading Goethe's Memoirs, and his prose generally, which they provoke in M. Scherer. True, the prose has none of those positive qualities of style which give pleasure, it is not the prose of Voltaire or Swift; it is loose, ill-knit, diffuse; it bears the marks of having been, as it mostly was, dictated,—and dictating is a detestable habit. But it is absolutely free from affectation; it lets the real Goethe reach us.

In other respects I agree in the main with the judgments passed by M. Scherer upon Goethe's works. Nay, some of them, such as *Tasso* and *Iphigeneia*, I should hesitate to extol so highly as he does. In that peculiar world of thought and feeling, wherein *Tasso* and *Iphigeneia* have their existence, and into which the reader too must enter in order to understand them, there is something factitious; something devised and determined by the thinker, not given by the necessity of Nature herself; something too artificial, therefore, too deliberately studied,—as the French say, *trop voulu*. They cannot have the power of works where we are in a world of thought and feeling not invented but natural,—of works like the *Agamemnon* or *Lear*. *Faust*, too, suffers by comparison with works like the *Agamemnon* or *Lear*. M. Scherer says, with perfect truth, that the first part of *Faust* has not a single false tone or weak line. But it is a work, as he himself observes, 'of episodes and detached scenes', not a work where the whole material together has been fused in the author's mind by strong and deep feeling, and then poured out in a single jet. It can never produce the single, powerful total-impression of works which have thus arisen.

The first part of *Faust* is, however, undoubtedly Goethe's best work. And it is so for the plain reason that, except his *Gedichte*, it is his most straightforward work in poetry. Mr Hayward's° is the best of the translations of *Faust* for the same reason,—because it is the most straightforward. To be simple and straightforward is, as Milton saw and said,° of the essence of first-rate poetry. All that M. Scherer says of the ruinousness, to a poet, of 'symbols, hieroglyphics, mystifications', is just. When Mr Carlyle praises the *Helena* for being 'not a type of one thing, but a vague, fluctuating, fitful adumbration of many', he praises it for what is in truth its fatal defect. The *Mährchen*, again, on which Mr Carlyle heaps such praise,° calling it 'one of the notablest performances produced for the last thousand years', a performance 'in such a style of grandeur and celestial brilliancy and life as the Western imagination has not elsewhere reached'; the *Mährchen*, woven throughout of 'symbol, hieroglyphic, mystification', is by that very reason a piece of solemn inanity, on which a man of Goethe's powers could never have wasted his time, but for his lot having been cast in a nation which has never lived.

Mr Carlyle has a sentence° on Goethe which we may turn to excellent account for the criticism of such works as the *Mährchen* and *Helena*:—

'We should ask', he says, 'what the poet's aim really and truly was, and how far this aim accorded, not with us and our individual crotchets and the crotchets of our little senate where we give or take the law, but with human nature and the nature of things at large; with the universal principles of poetic beauty, not as they stand written in our text-books, but in the hearts and imaginations of all men.'

To us it seems lost labour to inquire what a poet's *aim* may have been; but for aim let us read *work*, and we have here a sound and admirable rule of criticism. Let us ask how a poet's work accords, not with any one's fancies and crotchets, but 'with human nature and the nature of things at large, with the universal principles of poetic beauty as they stand written in the hearts and imaginations of all men', and we shall have the surest rejection of symbol, hieroglyphic, and mystification in poetry. We shall have the surest condemnation of works like the *Mährchen* and the second part of *Faust*.

It is by no means as the greatest of poets that Goethe deserves the pride and praise of his German countrymen. It is as the clearest, the largest, the most helpful thinker of modern times. It is not principally in his published works, it is in the immense Goethe-literature of letter, journal, and conversation, in the volumes of Riemer, Falk, Eckermann, the Chancellor von Müller, in the letters to Merck and Madame von Stein and many others, in the correspondence with Schiller, the correspondence with Zelter, that the elements for an impression of the truly great, the truly significant Goethe are to be found. Goethe is the greatest poet of modern times, not because he is one of the half-dozen human beings who in the history of our race have shown the most signal gift for poetry, but because, having a very considerable gift for poetry, he was at the same time, in the width, depth, and richness of his criticism of life, by far our greatest modern man. He may be precious and important to us on this account above men of other and more alien times, who as poets rank higher. Nay, his preciousness and importance as a clear and profound modern spirit, as a master-critic of modern life, must communicate a worth of their own to his poetry, and may well make it erroneously seem to have a positive value and perfectness as poetry, more than it has. It is most pardonable for a student of Goethe, and may even for a time be serviceable, to fall into this error. Nevertheless, poetical defects, where they are present, subsist, and are what they are.° And the same with defects of character. Time and attention bring them to light; and when they are brought to light, it is not good for us, it is obstructing and retarding, to refuse to see them. Goethe himself would have warned us against doing so. We can imagine, indeed, that great and supreme critic reading Professor Grimm's laudation of his poetical work with lifted eyebrows, and M. Scherer's criticisms with acquiescence.

Shall we say, however, that M. Scherer's tone in no way jars upon us, or that his presentation of Goethe, just and acute as is the view of faults both in Goethe's poetry and in Goethe's character, satisfies us entirely? By no means. One could not say so of M. Scherer's presentation of Milton; of the presentation of Goethe one can say so still less. Goethe's faults are shown by M. Scherer, and they exist. Praise is given, and the right praise. But there is yet some defect in the portraiture as a whole. Tone and perspective are somehow a little wrong; the distribution of colour, the proportions of light and shade, are not managed quite as they should be. One would like the picture to be painted over again by the same artist with the same talent, but a little differently. And meanwhile we instinctively, after M. Scherer's presentation, feel a desire for some last words of Goethe's own, something which may give a happier and more cordial turn to our thoughts, after they have been held so long to a frigid and censorious strain. And there rises to the mind this sentence: *'Die Gestalt dieser Welt vergeht*; und ich möchte mich nur mit dem beschäftigen, was bleibende Verhältnisse sind.'° *'The fashion of this world passeth away*; and I would fain occupy myself only with the abiding.' There is the true Goethe, and with that Goethe we would end!

But let us be thankful for what M. Scherer brings, and let us acknowledge with gratitude his presentation of Goethe to be, not indeed the definitive picture of Goethe, but a contribution, and a very able contribution, to that definitive picture. We are told that since the war of 1870 Frenchmen are abandoning literature for science. Why do they not rather learn of this accomplished senator of theirs, with his Geneva training, to extend their old narrow literary range a little, and to know foreign literatures as M. Scherer knows them?

Equality[1]

THERE is a maxim which we all know, which occurs in our copy-books, which occurs in that solemn and beautiful formulary against which the Nonconformist genius is just now so angrily chafing,—the Burial Service.° The maxim is this: 'Evil communications corrupt good manners.' It is taken from a chapter of the First Epistle to the Corinthians;° but originally it is a line of poetry, of Greek poetry. *Quid Athenis et Hierosolymis?* asks a Father;° what have Athens and Jerusalem to do with one another? Well, at any rate, the Jerusalemite Paul, exhorting his converts, enforces what he is saying by a verse of Athenian comedy,—a verse, probably, from the great master of that comedy, a man unsurpassed for fine and just observation of human life, Menander: Φθείρουσιν ἤθη χρήσθ᾽ ὁμιλίαι κακαί—'Evil communications corrupt good manners.'

In that collection of single, sententious lines, printed at the end of Menander's fragments, where we now find the maxim quoted by St Paul, there is another striking maxim, not alien certainly to the language of the Christian religion, but which has not passed into our copy-books: 'Choose equality and flee greed.'° The same profound observer, who laid down the maxim so universally accepted by us that it has become commonplace, the maxim that evil communications corrupt good manners, laid down also, as a no less sure result of the accurate study of human life, this other maxim as well: 'Choose equality and flee greed'— Ἰσότητα δ᾽ αἱροῦ καὶ πλεονεξίαν φύγε.

Pleonexia,° or greed, the wishing and trying for the bigger share, we know under the name of covetousness. We understand by covetousness something different from what *pleonexia* really means: we understand by it the longing for other people's goods: and covetousness, so understood, it is a commonplace of morals and of religion with us that we should shun. As to the duty of pursuing equality, there is no such consent amongst us. Indeed, the consent is the other way, the consent is against equality. Equality before the law we all take as a matter of course; that is not the equality which we mean when we talk of equality. When we talk of equality, we understand social equality; and for equality in this Frenchified sense of the term almost everybody in England has a hard word. About four years ago Lord Beaconsfield° held it up to reprobation in a

[1] Address delivered at the Royal Institution.

speech to the students at Glasgow;—a speech so interesting, that being asked soon afterwards to hold a discourse at Glasgow, I said that if one spoke there at all at that time it would be impossible to speak on any other subject but equality. However, it is a great way to Glasgow, and I never yet have been able to go and speak there.

But the testimonies against equality have been steadily accumulating from the date of Lord Beaconsfield's Glasgow speech down to the present hour. Sir Erskine May winds up his new and important *History of Democracy*° by saying: 'France has aimed at social equality. The fearful troubles through which she has passed have checked her prosperity, demoralized her society, and arrested the intellectual growth of her people.' Mr Froude,° again, who is more his own master than I am, has been able to go to Edinburgh and to speak there upon equality. Mr Froude told his hearers that equality splits a nation into a 'multitude of disconnected units', that 'the masses require leaders whom they can trust', and that 'the natural leaders in a healthy country are the gentry'. And only just before the *History of Democracy* came out, we had that exciting passage of arms between Mr Lowe and Mr Gladstone,° where equality, poor thing, received blows from them both. Mr Lowe declared° that 'no concession should be made to the cry for equality, unless it appears that the State is menaced with more danger by its refusal than by its admission. No such case exists now or ever has existed in this country.' And Mr Gladstone replied° that equality was so utterly unattractive to the people of this country, inequality was so dear to their hearts, that to talk of concessions being made to the cry for equality was absurd. 'There is no broad political idea', says Mr Gladstone quite truly, 'which has entered less into the formation of the political system of this country than the love of equality.' And he adds: 'It is not the love of equality which has carried into every corner of the country the distinct undeniable popular preference, wherever other things are equal, for a man who is a lord over a man who is not. The love of freedom itself is hardly stronger in England than the love of aristocracy.' Mr Gladstone goes on to quote a saying of Sir William Molesworth, that with our people the love of aristocracy 'is a religion'. And he concludes in his copious and eloquent way: 'Call this love of inequality by what name you please,—the complement of the love of freedom, or its negative pole, or the shadow which the love of freedom casts, or the reverberation of its voice in the halls of the constitution,—it is an active, living, and life-giving power, which forms an inseparable essential element in our political habits of mind, and asserts itself at every step in the processes of our system.'

And yet, on the other side, we have a consummate critic of life like Menander, delivering, as if there were no doubt at all about the matter, the maxim: 'Choose equality!' An Englishman with any curiosity must surely be inclined to ask himself how such a maxim can ever have got established, and taken rank along with 'Evil communications corrupt good manners.' Moreover, we see that among the French, who have suffered so grievously, as we hear, from choosing equality, the most gifted spirits continue to believe passionately in it nevertheless. 'The human ideal, as well as the social ideal, is', says George Sand,° 'to achieve equality.' She calls equality° 'the goal of man and the law of the future'. She asserts that France is the most civilised of nations, and that its pre-eminence in civilisation it owes to equality.

But Menander lived a long while ago, and George Sand was an enthusiast. Perhaps their differing from us about equality need not trouble us much. France, too, counts for but one nation, as England counts for one also. Equality may be a religion with the people of France, as inequality, we are told, is a religion with the people of England. But what do other nations seem to think about the matter?

Now, my discourse to-night is most certainly not meant to be a disquisition on law, and on the rules of bequest. But it is evident that in the societies of Europe, with a constitution of property such as that which the feudal Middle Age left them with,—a constitution of property full of inequality,—the state of the law of bequest shows us how far each society wishes the inequality to continue. The families in possession of great estates will not break them up if they can help it. Such owners will do all they can, by entail and settlement, to prevent their successors from breaking them up. They will preserve inequality. Freedom of bequest, then, the power of making entails and settlements, is sure, in an old European country like ours, to maintain inequality. And with us, who have the religion of inequality, the power of entailing and settling, and of willing property as one likes, exists, as is well known, in singular fulness,—greater fulness than in any country of the Continent. The proposal of a measure such as the Real Estates Intestacy Bill° is, in a country like ours, perfectly puerile. A European country like ours, wishing not to preserve inequality but to abate it, can only do so by interfering with the freedom of bequest. This is what Turgot,° the wisest of French statesmen, pronounced before the Revolution to be necessary, and what was done in France at the great Revolution. The *Code Napoléon*, the actual law of France, forbids entails altogether, and leaves a man free to dispose of but one-fourth of his property, of whatever kind, if he have three children or more, of one-third if he have two children, of one-half if he have but one child. Only in the rare case, therefore, of a man's having but one child, can that child take the whole of his father's property. If there are two children, two-thirds of the property must be equally divided between them; if there are more than two, three-fourths. In this way has France, desiring equality, sought to bring equality about.

Now the interesting point for us is, I say, to know how far other European communities, left in the same situation with us and with France, having immense inequalities of class and property created for them by the Middle Age, have dealt with these inequalities by means of the law of bequest. Do they leave bequest free, as we do? then, like us, they are for inequality. Do they interfere with the freedom of bequest, as France does? then, like France, they are for equality. And we shall be most interested, surely, by what the most civilised European communities do in this matter,—communities such as those of Germany, Italy, Belgium, Holland, Switzerland. And among those communities we are most concerned, I think, with such as, in the conditions of freedom and of self-government which they demand for their life, are most like ourselves. Germany, for instance, we shall less regard, because the conditions which the Germans seem to accept for their life are so unlike what we demand for ours; there is so much personal government there, so much *junkerism*, militarism, officialism; the community is so much more trained to submission than we could bear, so much more used to be, as the popular phrase is, sat upon. Countries where the community has more a will of its own, or can more show it, are the most important for our present purpose,—such countries as Belgium, Holland, Italy, Switzerland. Well, Belgium adopts purely and simply, as to bequest and inheritance, the provisions of the *Code Napoléon*. Holland adopts them purely and simply. Italy has adopted them substantially. Switzerland is a republic, where the general feeling against inequality is strong, and where it might seem less necessary, therefore, to guard against inequality by interfering with the power of bequest. Each Swiss canton has its own law of bequest. In Geneva, Vaud, and Zurich,—perhaps the three most distinguished cantons,—the law is identical with that of France. In Berne, one-third is the fixed proportion which a man is free to dispose of by will; the rest of his property must go among his children equally. In all the other cantons there are regulations of a like kind. Germany, I was saying, will interest us less than these freer countries. In Germany,— though there is not the English freedom of bequest, but the rule of the Roman law prevails, the rule obliging the parent to assign a certain portion to each child,—in Germany entails and settlements in favour of an eldest son are generally permitted. But there is a remarkable exception. The Rhine countries, which in the early part of this century were under French rule, and which then received the *Code Napoléon*, these countries refused to part with it when they were restored to Germany; and to this day Rhenish Prussia, Rhenish Hesse, and Baden, have the French law of bequest, forbidding entails, and dividing property in the way we have seen.

The United States of America° have the English liberty of bequest. But the United States are, like Switzerland, a republic, with the republican sentiment for equality. Theirs is, besides, a new society; it did not inherit the system of classes and of property which feudalism established in Europe. The class by which the United States were settled was not a class with feudal habits and ideas. It is notorious that to acquire great landed estates and to entail them upon an eldest son, is neither the practice nor the desire of any class in America. I remember hearing it said to an American in England: 'But, after all, you have the same freedom of bequest and inheritance as we have, and if a man tomorrow chose in your country to entail a great landed estate rigorously, what could you do?' The American answered: 'Set aside the will on the ground of insanity.'

You see we are in a manner taking the votes for and against equality. We ought not to leave out our own colonies. In general they are, of course, like the United States of America, new societies. They have the English liberty of bequest. But they have no feudal past, and were not settled by a class with feudal habits and ideas. Nevertheless it happens that there have arisen, in Australia, exceedingly large estates, and that the proprietors seek to keep them together. And what have we seen happen lately? An Act has been passed which in effect inflicts a fine upon every proprietor who holds a landed estate of more than a certain value. The measure has been severely blamed in England; to Mr Lowe° such a 'concession to the cry for equality' appears, as we might expect, pregnant with warnings. At present I neither praise it nor blame it; I simply count it as one of the votes for equality. And is it not a singular thing, I ask you, that while we have the religion of inequality, and can hardly bear to hear equality spoken of, there should be, among the nations of Europe which have politically most in common with us, and in the United States of America, and in our own colonies, this diseased appetite, as we must think it, for equality? Perhaps Lord Beaconsfield may not have turned your minds to this subject as he turned mine, and what Menander or George Sand happens to have said may not interest you much; yet surely, when you think of it, when you see what a practical revolt against inequality there is amongst so many people not so very unlike to ourselves, you must feel some curiosity to sift the matter a little further, and may be not ill-disposed to follow me while I try to do so.

I have received a letter from Clerkenwell,° in which the writer reproaches me for lecturing about equality at this which he calls 'the most aristocratic and exclusive place out'. I am here because your secretary invited me. But I am glad to treat the subject of equality before such an audience as this. Some of you may remember that I have roughly divided° our English society into Barbarians, Philistines, Populace, each of them with their prepossessions, and loving to hear what gratifies them. But I remarked at the same time, that scattered throughout all these classes were a certain number of generous and humane souls, lovers of man's perfection, detached from the prepossessions of the class to which they might naturally belong, and desirous that he who speaks to them should, as Plato says,° not try to please his fellow-servants, but his true and legitimate masters—the heavenly Gods. I feel sure that among the members and frequenters of an institution like this, such humane souls are apt to congregate in numbers. Even from the reproach which my Clerkenwell friend brings against you of being too aristocratic, I derive some comfort. Only I give to the term *aristocratic* a rather wide extension. An accomplished American, much known and much esteemed in this country, the late Mr Charles Summer, says° that what particularly struck him in England was the large class of gentlemen as distinct from the nobility, and the abundance amongst them of serious knowledge, high accomplishment, and refined taste,—taste fastidious perhaps, says Mr Sumner, to excess, but erring on virtue's side. And he goes on: 'I do not know that there is much difference between the manners and social observances of the highest classes of England and those of the corresponding classes of France and Germany; but in the rank immediately below the highest,—as among the professions, or military men, or literary men,—there you will find that the Englishmen have the advantage. They are better educated and better bred, more careful in their personal habits and in social conventions, more refined.' Mr Sumner's remark is just and important; this large class of gentlemen in the professions, the services, literature, politics,—and a good contingent is now added from business also,—this large class, not of the nobility, but with the accomplishments and taste of an upper class, is something peculiar to England. Of this class I may probably assume that my present audience is in large measure composed. It is aristocratic in this sense, that it has the tastes of a cultivated class, a certain high standard of civilisation. Well, it is in its effects upon *civilisation* that equality interests me. And I speak to an audience with a high standard of civilisation. If I say that certain things in certain classes do not come up to a high standard of civilisation, I need not prove how and why they do not; you will feel instinctively whether they do or no. If they do not, I need not prove that this is a bad thing, that a high standard of civilisation is desirable; you will instinctively feel that it is. Instead of calling this 'the most aristocratic and exclusive place out', I conceive of it as a *civilised* place; and in speaking about civilisation half one's labour is saved when one speaks about it among those who are civilised.

Politics are forbidden here; but equality is not a question of English politics. The abstract right to equality may, indeed, be a question of speculative politics. French equality appeals to this abstract natural right as its support. It goes back to a state of nature where all were equal, and supposes that 'the poor consented', as Rousseau says,° 'to the existence of rich people', reserving always a natural right to return to the state of nature. It supposes that a child has a natural right to his equal share in his father's goods. The principle of abstract right, says Mr Lowe,° has never been admitted in England, and is false. I so entirely agree with him, that I run no risk of offending by discussing equality upon the basis of this principle. So far as I can sound human consciousness, I cannot, as I have often said,° perceive that man is really conscious of any abstract natural rights at all. The natural right to have work found for one to do, the natural right to have food found for one to eat—rights sometimes so confidently and so indignantly asserted—seem to me quite baseless. It cannot be too often repeated: peasants and workmen have no natural rights, not one. Only we ought instantly to add, that kings and nobles have none either. If it is the sound English doctrine that all rights are created by law and are based on expediency, and are alterable as the public advantage may require, certainly that orthodox doctrine is mine. Property is created and maintained by law. It would disappear in that state of private war and scramble which legal society supersedes. Legal society creates, for the common good, the right of property; and for the common good that right is by legal society limitable. That property should exist, and that it should be held with a sense of security and with a power of disposal, may be taken, by us here at any rate, as a settled matter of expediency. With these conditions a good deal of inequality is inevitable. But that the power of disposal should be practically *unlimited*, that the inequality should be *enormous*, or that the degree of inequality admitted at one time should be admitted *always*,—this is by no means so certain. The right of bequest was in early times, as Sir Henry Maine and Mr Mill° have pointed out, seldom recognised. In later times it has been limited in many countries in the way that we have seen; even in England itself it is not formally quite unlimited. The question is one of expediency. It is assumed, I grant, with great unanimity amongst us, that our signal inequality of classes and property is expedient for our civilisation and welfare. But this assumption, of which the distinguished personages who adopt it seem so sure that they think it needless to produce grounds for it, is just what we have to examine.

Now, there is a sentence of Sir Erskine May, whom I have already quoted, which will bring us straight to the very point that I wish to raise. Sir Erskine May, after saying, as you have heard, that France has pursued social equality, and has come to fearful troubles, demoralisation, and intellectual stoppage by doing so, continues thus:° 'Yet is she high, if not the first, in the scale of civilised nations.' Why, here is a curious thing, surely! A nation pursues social equality, supposed to be an utterly false and baneful ideal; it arrives, as might have been expected, at fearful misery and deterioration by doing so; and yet, at the same time, it is high, if not the first, in the scale of civilised nations. What do we mean by *civilised*? Sir Erskine May does not seem to have asked himself the question, so we will try to answer it for ourselves. Civilisation is the humanisation of man in society. To be humanised is to comply with the true law of our human nature: *servare modum, finemque tenere, Naturamque sequi*, says Lucan;° 'to keep our measure, and to hold fast our end, and to follow Nature'. To be humanised is to make progress towards this, our true and full humanity. And to be civilised is to make progress towards this in civil society; in that civil society 'without which', says Burke,° 'man could not by any possibility arrive at the perfection of which his nature is capable, nor even make a remote and faint approach to it'. To be the most civilised of nations, therefore, is to be the nation which comes nearest to human perfection, in the state which that perfection essentially demands. And a nation which has been brought by the pursuit of social equality to moral deterioration, intellectual stoppage, and fearful troubles, is perhaps the nation which has come nearest to human perfection in that state which such perfection essentially demands! Michelet himself, who would deny the demoralisation and the stoppage, and call the fearful troubles a sublime expiation for the sins of the whole world, could hardly say more for France than this. Certainly Sir Erskine May never intended to say so much. But into what a difficuly has he somehow run himself, and what a good action would it be to extricate him from it!° Let us see whether the performance of that good action may not also be a way of clearing our minds as to the uses of equality.

When we talk of man's advance towards his full humanity, we think of an advance, not along one line only, but several. Certain races and nations, as we know, are on certain lines pre-eminent and representative. The Hebrew nation was pre-eminent on one great line. 'What nation', it was justly asked by their lawgiver,° 'hath statutes and judgments so righteous as the law which I set before you this day? Keep therefore and do them; for this is your wisdom and your understanding in the sight of the nations which shall hear all these statutes and say: Surely this great nation is a wise and understanding people!' The Hellenic race was pre-eminent on other lines. Isocrates could say° of Athens: 'Our city has left the rest of the world so far behind in philosophy and eloquence, that those educated by Athens have become the teachers of the rest of mankind; and so well has she done her part, that the name of Greeks seems no longer to stand for a race but to stand for intelligence itself, and they who share in our culture are called Greeks even before those who are merely of our own blood.' The power of intellect and science, the power of beauty, the power of social life and manners,°—these are what Greece so felt, and fixed, and may stand for. They are great elements in our humanisation. The power of conduct is another great element; and this was so felt and fixed by Israel that we can never with justice refuse to permit Israel, in spite of all his shortcomings, to stand for it.

So you see that in being humanised we have to move along several lines, and that on certain lines certain nations find their strength and take a lead. We may elucidate the thing yet further. Nations now existing may be said to feel or to have felt the power of this or that element in our humanisation so signally that they are characterised by it. No one who knows this country would deny that it is characterised, in a remarkable degree, by a sense of the power of conduct. Our feeling for religion is one part of this; our industry is another. What foreigners so much remark in us,—our public spirit, our love, amidst all our liberty, for public order and for stability,—are parts of it too. Then the power of beauty was so felt by the Italians that their art revived, as we know, the almost lost idea of beauty, and the serious and successful pursuit of it. Cardinal Antonelli,° speaking to me about the education of the common people in Rome, said that they were illiterate indeed, but whoever mingled with them at any public show, and heard them pass judgment on the beauty or ugliness of what came before them,—'è brutto', 'è bello',—would find that their judgment agreed admirably, in general, with just what the most cultivated people would say. Even at the present time, then, the Italians are pre-eminent in feeling the power of beauty. The power of knowledge, in the same way, is eminently an influence with the Germans. This by no means implies, as is sometimes supposed, a high and fine general culture. What it implies is a strong sense of the necessity of knowing *scientifically*,° as the expression is, the things which have to be known by us; of knowing them systematically, by the regular and right process, and in the only real way. And this sense the Germans especially have. Finally, there is the power of social life and manners. And even the Athenians themselves, perhaps, have hardly felt this power so much as the French.

Voltaire, in a famous passage° where he extols the age of Louis the Fourteenth and ranks it with the chief epochs in the civilisation of our race, has to specify the gift bestowed on us by the age of Louis the Fourteenth, as the age of Pericles, for instance, bestowed on us its art and literature, and the Italian Renascence its revival of art and literature. And Voltaire shows all his acuteness in fixing on the gift to name. It is not the sort of gift which we expect to see named. The great gift of the age of Louis the Fourteenth to the world, says Voltaire, was this: *l'esprit de société*, the spirit of society, the social spirit. And another French writer, looking for the good points in the old French nobility, remarks that this at any rate is to be said in their favour: they established a high and charming ideal of social intercourse and manners, for a nation formed to profit by such an ideal, and which has profited by it ever since. And in America, perhaps, we see the disadvantages of having social equality before there has been any such high standard of social life and manners formed.

We are not disposed in England, most of us, to attach all this importance to social intercourse and manners. Yet Burke says:° 'There ought to be a system of manners in every nation which a well-formed mind would be disposed to relish.' And the power of social life and manners is truly, as we have seen, one of the great elements in our humanisation. Unless we have cultivated it, we are incomplete. The impulse for cultivating it is not, indeed, a moral impulse. It is by no means identical with the moral impulse to help our neighbour and to do him good. Yet in many ways it works to a like end. It brings men together, makes them feel the need of one another, be considerate of one another, understand one another. But, above all things, it is a promoter of equality. It is by the humanity of their manners that men are made equal. 'A man thinks to show himself my equal', says Goethe, 'by being *grob*,— that is to say, coarse and rude; he does not show himself my equal, he shows himself *grob*.' But a community having humane manners is a community of equals, and in such a community great social inequalities have really no meaning, while they are at the same time a menace and an embarrassment to perfect ease of social intercourse. A community with the spirit of society is eminently, therefore, a community with the spirit of equality. A nation with a genius for society, like the French or the

Athenians, is irresistibly drawn towards equality. From the first moment
when the French people, with its congenital sense for the power of social
intercourse and manners, came into existence, it was on the road to
equality. When it had once got a high standard of social manners
abundantly established, and at the same time the natural, material
necessity for the feudal inequality of classes and property pressed upon it
no longer, the French people introduced equality and made the French
Revolution. It was not the spirit of philanthropy which mainly impelled
the French to that Revolution, neither was it the spirit of envy, neither
was it the love of abstract ideas, though all these did something towards
it; but what did most was the spirit of society.

The well-being of the many comes out more and more distinctly, in
proportion as time goes on, as the object we must pursue. An individual
or a class, concentrating their efforts upon their own well-being
exclusively, do but beget troubles both for others and for themselves also.
No individual life can be truly prosperous, passed, as Obermann says,° in
the midst of men who suffer; *passée au milieu des générations qui souffrent*.
To the noble soul, it cannot be happy; to the ignoble, it cannot be secure.
Socialistic and communistic schemes have generally, however, a fatal
defect; they are content with too low and material a standard of well-
being. That instinct of perfection, which is the master-power in humanity, always rebels at this, and frustrates the work. Many are to be made
partakers of well-being, true; but the ideal of well-being is not to be, on
that account, lowered and coarsened. M. de Laveleye,° the political
economist, who is a Belgian and a Protestant, and whose testimony
therefore we may the more readily take about France, says that France,
being the country of Europe where the soil is more divided than
anywhere except in Switzerland and Norway, is at the same time the
country where material well-being is most widely spread, where wealth
has of late years increased most, and where population is least outrun-
ning the limits which, for the comfort and progress of the working classes
themselves, seem necessary. This may go for a good deal. It supplies an
answer to what Sir Erskine May says about the bad effects of equality
upon French prosperity. But I will quote to you from Mr Hamerton°
what goes, I think, for yet more. Mr Hamerton is an excellent observer
and reporter, and has lived for many years in France. He says of the
French peasantry that they are exceedingly ignorant. So they are. But he
adds: 'They are at the same time full of intelligence; their manners are
excellent, they have delicate perceptions, they have tact, they have a
certain refinement which a brutalised peasantry could not possibly have.
If you talk to one of them at his own home, or in his field, he will enter

into conversation with you quite easily, and sustain his part in a perfectly becoming way, with a pleasant combination of dignity and quiet humour. The interval between him and a Kentish labourer is enormous.'

This is indeed worth your attention. Of course all mankind are, as Mr Gladstone says,° of our own flesh and blood. But you know how often it happens in England that a cultivated person, a person of the sort that Mr Charles Sumner describes, talking to one of the lower class, or even of the middle class, feels, and cannot but feel, that there is somehow a wall of partition between himself and the other, that they seem to belong to two different worlds. Thoughts, feelings, perceptions, susceptibilities, language, manners,—everything is different. Whereas, with a French peasant, the most cultivated man may find himself in sympathy, may feel that he is talking to an equal. This is an experience which has been made a thousand times, and which may be made again any day. And it may be carried beyond the range of mere conversation, it may be extended to things like pleasures, recreations, eating and drinking, and so on. In general the pleasures, recreations, eating and drinking of English people, when once you get below that class which Mr Charles Sumner calls the class of gentlemen, are to one of that class unpalatable and impossible. In France there is not this incompatibility. Whether he mix with high or low, the gentleman feels himself in a world not alien or repulsive, but a world where people make the same sort of demands upon life, in things of this sort, which he himself does. In all these respects France is the country where the people, as distinguished from a wealthy refined class, most lives what we call a humane life, the life of a civilised man.

Of course, fastidious persons can and do pick holes in it. There is just now, in France,° a *noblesse* newly revived, full of pretension, full of airs and graces and disdains; but its sphere is narrow, and out of its own sphere no one cares very much for it. There is a general equality in a humane kind of life. This is the secret of the passionate attachment with which France inspires all Frenchmen, in spite of her fearful troubles, her checked prosperity, her disconnected units, and the rest of it. There is so much of the goodness and agreeableness of life there, and for so many. It is the secret of her having been able to attach so ardently to her the German and Protestant people of Alsace,° while we have been so little able to attach the Celtic and Catholic people of Ireland. France brings the Alsatians into a social system so full of the goodness and agreeableness of life; we offer to the Irish no such attraction. It is the secret, finally, of the prevalence which we have remarked in other continental countries of a legislation tending, like that of France, to social equality. The social system which equality creates in France is, in the eyes of others, such a giver of the goodness and agreeableness of life, that they seek to get the goodness by getting the equality.

Yet France has had her fearful troubles, as Sir Erskine May justly says. She suffers too, he adds, from demoralisation and intellectual stoppage. Let us admit, if he likes, this to be true also. His error is that he attributes all this to equality. Equality, as we have seen, has brought France to a really admirable and enviable pitch of humanisation in one important line. And this, the work of equality, is so much a good in Sir Erskine May's eyes, that he has mistaken it for the whole of which it is a part, frankly identifies it with civilisation, and is inclined to pronounce France the most civilised of nations.

But we have seen how much goes to full humanisation, to true civilisation, besides the power of social life and manners. There is the power of conduct, the power of intellect and knowledge, the power of beauty. The power of conduct is the greatest of all. And without in the least wishing to preach, I must observe, as a mere matter of natural fact and experience, that for the power of conduct France has never had anything like the same sense which she has had for the power of social life and manners. Michelet, himself a Frenchman, gives us° the reason why the Reformation did not succeed in France. It did not succeed, he says, because *la France ne voulait pas de réforme morale*—moral reform France would not have; and the Reformation was above all a moral movement. The sense in France for the power of conduct has not greatly deepened, I think, since. The sense for the power of intellect and knowledge has not been adequate either. The sense for beauty has not been adequate. Intelligence and beauty have been, in general, but so far reached as they can be and are reached by men who, of the elements of perfect humanisation, lay thorough hold upon one only,—the power of social intercourse and manners. I speak of France in general; she has had, and she has, individuals who stand out and who form exceptions. Well then, if a nation laying no sufficient hold upon the powers of beauty and knowledge, and a most failing and feeble hold upon the power of conduct, comes to demoralisation and intellectual stoppage and fearful troubles, we need not be inordinately surprised. What we should rather marvel at is the healing and bountiful operation of Nature, whereby the laying firm hold on one real element in our humanisation has had for France results so beneficent.

And thus, when Sir Erskine May gets bewildered between France's equality and fearful troubles on the one hand, and the civilisation of France on the other, let us suggest to him that perhaps he is bewildered by his data because he combines them ill. France has not exemplary disaster and ruin as the fruits of equality, and at the same time, and independently of this, an exemplary civilisation. She has a large measure of happiness and success as the fruits of equality, and she has a very large measure of dangers and troubles as the fruits of something else.

We have more to do, however, than to help Sir Erskine May out of his scrape about France. We have to see whether the considerations which we have been employing may not be of use to us about England.

We shall not have much difficulty in admitting whatever good is to be said of ourselves, and we will try not to be unfair by excluding all that is not so favourable. Indeed, our less favourable side is the one which we should be the most anxious to note, in order that we may mend it. But we will begin with the good. Our people has energy and honesty as its good characteristics. We have a strong sense for the chief power in the life and progress of man,—the power of conduct. So far we speak of the English people as a whole. Then we have a rich, refined, and splendid aristocracy. And we have, according to Mr Charles Sumner's acute and true remark, a class of gentlemen, not of the nobility, but well-bred, cultivated, and refined, larger than is to be found in any other country. For these last we have Mr Sumner's testimony. As to the splendour of our aristocracy, all the world is agreed. Then we have a middle class and a lower class; and they, after all, are the immense bulk of the nation.

Let us see how the civilisation of these classes appears to a Frenchman, who has witnessed, in his own country, the considerable humanisation of these classes by equality. To such an observer our middle class divides itself into a serious portion and a gay or rowdy portion; both are a marvel to him. With the gay or rowdy portion we need not much concern ourselves; we shall figure it to our minds sufficiently if we conceive it as the source of that war-song° produced in these recent days of excitement:

'We don't want to fight, but by jingo, if we do,
We've got the ships, we've got the men, and we've got the money too.'

We may also partly judge its standard of life, and the needs of its nature, by the modern English theatre, perhaps the most contemptible in Europe. But the real strength of the English middle class is in its serious portion. And of this a Frenchman, who was here some little time ago as the correspondent, I think, of the *Siècle* newspaper, and whose letters were afterwards published in a volume, writes as follows. He had been attending some of the Moody and Sankey meetings,° and he says: 'To

understand the success of Messrs Moody and Sankey, one must be
familiar with English manners, one must know the mind-deadening
influence of a narrow Biblism, one must have experienced the sense of
acute ennui, which the aspect and the frequentation of this great division
of English society produce in others, the want of elasticity and the
chronic ennui which characterise this class itself, petrified in a narrow
Protestantism and in a perpetual reading of the Bible.'

You know the French;—a little more Biblism, one may take leave to
say, would do them no harm. But an audience like this,—and here, as I
said, is the advantage of an audience like this,—will have no difficulty in
admitting the amount of truth which there is in the Frenchman's picture.
It is the picture of a class which, driven by its sense for the power of
conduct, in the beginning of the seventeenth century entered,—as I have
more than once said,° and as I may more than once have occasion in
future to say,—*entered the prison of Puritanism, and had the key turned upon
its spirit there for two hundred years*. They did not know, good and earnest
people as they were, that to the building up of human life there belong all
those other powers also,—the power of intellect and knowledge, the
power of beauty, the power of social life and manners. And something, by
what they became, they gained, and the whole nation with them; they
deepened and fixed for this nation the sense of conduct. But they created
a type of life and manners, of which they themselves indeed are slow to
recognise the faults, but which is fatally condemned by its hideousness,
its immense ennui, and against which the instinct of self-preservation in
humanity rebels.

Partisans fight against facts in vain. Mr Goldwin Smith,° a writer of
eloquence and power, although too prone to acerbity, is a partisan of the
Puritans, and of the Nonconformists who are the special inheritors of the
Puritan tradition. He angrily resents the imputation upon that Puritan
type of life, by which the life of our serious middle class has been formed,
that it was doomed to hideousness, to immense ennui. He protests that it
had beauty, amenity, accomplishment. Let us go to facts. Charles the
First, who, with all his faults, had the just idea that art and letters are
great civilisers, made, as you know, a famous collection of pictures,—our
first National Gallery. It was, I suppose, the best collection at that time
north of the Alps. It contained nine Raphaels, eleven Correggios,
twenty-eight Titians. What became of that collection? The journals of
the House of Commons° will tell you. There you may see the Puritan
Parliament disposing of this Whitehall or York House collection as
follows: 'Ordered, that all such pictures and statues there as are without
any superstition, shall be forthwith sold. . . . Ordered, that all such

pictures there as have the representation of the Second Person in Trinity upon them, shall be forthwith burnt. Ordered, that all such pictures there as have the representation of the Virgin Mary upon them, shall be forthwith burnt.' There we have the weak side of our parliamentary government and our serious middle class. We are incapable of sending Mr Gladstone to be tried at the Old Bailey because he proclaims his antipathy to Lord Beaconsfield. A majority in our House of Commons is incapable of hailing, with frantic laughter and applause, a string of indecent jests against Christianity and its Founder.° But we are not, or were not, incapable of producing a Parliament which burns or sells the masterpieces of Italian art. And one may surely say of such a Puritan Parliament, and of those who determine its line for it, that they had not the spirit of beauty.

What shall we say of amenity? Milton was born a humanist, but the Puritan temper, as we know, mastered him. There is nothing more unlovely and unamiable than Milton the Puritan disputant. Some one answers his *Doctrine and Discipline of Divorce*. 'I mean not', rejoins Milton,° 'to dispute philosophy with this pork, who never read any.' However, he does reply to him, and throughout the reply Milton's great joke is, that his adversary, who was anonymous, is a serving-man. 'Finally, he winds up his text with much doubt and trepidation; for it may be his trenchers were not scraped, and that which never yet afforded corn of savour to his noddle,—the salt-cellar,—was not rubbed; and therefore, in this haste, easily granting that his answers fall foul upon each other, and praying you would not think he writes as a prophet, but as a man, he runs to the black jack, fills his flagon, spreads the table, and serves up dinner.' There you have the same spirit of urbanity and amenity, as much of it, and as little, as generally informs the religious controversies of our Puritan middle class to this day.

But Mr Goldwin Smith insists, and picks out his own exemplar of the Puritan type of life and manners; and even here let us follow him. He picks out the most favourable specimen he can find,—Colonel Hutchinson, whose well-known memoirs, written by his widow, we have all read with interest. 'Lucy Hutchinson', says Mr Goldwin Smith,° 'is painting what she thought a perfect Puritan would be; and her picture presents to us not a coarse, crop-eared, and snuffling fanatic, but a highly accomplished, refined, gallant, and most amiable, though religious and seriously minded, gentleman.' Let us, I say, in this example of Mr Goldwin Smith's own choosing, lay our finger upon the points where this type deflects from the truly humane ideal.

Mrs Hutchinson relates° a story which gives us a good notion of what the amiable and accomplished social intercourse, even of a picked Puritan family, was. Her husband was governor of Nottingham. He had occasion, she says, 'to go and break up a private meeting in the cannoneer's chamber'; and in the cannoneer's chamber 'were found some notes concerning pædobaptism, which, being brought into the governor's lodgings, his wife having perused them and compared them with the Scriptures, found not what to say against the truths they asserted concerning the misapplication of that ordinance to infants'. Soon afterwards she expects her confinement, and communicates the cannoneer's doubts about pædobaptism to her husband. The fatal cannoneer makes a breach in him too. 'Then he bought and read all the eminent treatises on both sides, which at that time came thick from the presses, and still was cleared in the error of the pædobaptists.' Finally, Mrs Hutchinson is confined. Then the governor 'invited all the ministers to dinner, and propounded his doubt and the ground thereof to them. None of them could defend their practice with any satisfactory reason, but the tradition of the Church from the primitive times, and their main buckler of federal holiness, which Tombs and Denne had excellently overthrown. He and his wife then, professing themselves unsatisfied, desired their opinions.' With the opinions I will not trouble you, but hasten to the result: 'Whereupon that infant was not baptised.'

No doubt to a large division of English society at this very day, that sort of dinner and discussion, and, indeed, the whole manner of life and conversation here suggested by Mrs Hutchinson's narrative, will seem both natural and amiable, and such as to meet the needs of man as a religious and social creature. You know the conversation which reigns in thousands of middle-class families at this hour, about nunneries, teetotalism, the confessional, eternal punishment, ritualism, disestablishment. It goes wherever the class goes which is moulded on the Puritan type of life. In the long winter evenings of Toronto° Mr Goldwin Smith has had, probably, abundant experience of it. What is its enemy? The instinct of self-preservation in humanity. Men make crude types and try to impose them, but to no purpose. '*L'homme s'agite, Dieu le mène*', says Bossuet.° 'There are many devices in a man's heart; nevertheless the counsel of the Eternal, that shall stand.'° Those who offer us the Puritan type of life offer us a religion not true, the claims of intellect and knowledge not satisfied, the claim of beauty not satisfied, the claim of manners not satisfied. In its strong sense for conduct that life touches truth; but its other imperfections hinder it from employing even this sense aright. The type mastered our nation for a time. Then came the reaction. The nation said: 'This type, at any rate, is amiss; we are not going to be all like *that*!' The type retired into our middle class, and fortified itself there. It seeks to endure, to emerge, to deny its own imperfections, to impose itself again;—impossible! If we continue to live, we must outgrow it. The very class in which it is rooted, our middle class, will have to acknowledge the type's inadequacy, will have to acknowledge the hideousness, the immense ennui of the life which this type has created, will have to transform itself thoroughly. It will have to admit the large part of truth which there is in the criticisms of our Frenchman, whom we have too long forgotten.

After our middle class he turns his attention to our lower class. And of the lower and larger portion of this, the portion not bordering on the middle class and sharing its faults, he says: 'I consider this multitude to be absolutely devoid, not only of political principles, but even of the most simple notions of good and evil. Certainly it does not appeal, this mob, to the principles of '89, which you English make game of; it does not insist on the rights of man; what it wants is beer, gin, and *fun*.'[1]

That is a description of what Mr Bright would call° the residuum, only our author seems to think the residuum a very large body. And its condition strikes him with amazement and horror. And surely well it may. Let us recall Mr Hamerton's account of the most illiterate class in France; what an amount of civilisation they have notwithstanding! And this is always to be understood, in hearing or reading a Frenchman's praise of England. He envies our liberty, our public spirit, our trade, our stability. But there is always a reserve in his mind. He never means for a moment that he would like to change with us. Life seems to him so much better a thing in France for so many more people, that, in spite of the fearful troubles of France, it is best to be a Frenchman. A Frenchman might agree with Mr Cobden,° that life is good in England for those people who have at least £5000 a year. But the civilisation of that immense majority who have not £5000 a year, or £500, or even £100,— of our middle and lower class,—seems to him too deplorable.

And now what has this condition of our middle and lower classes to tell us about equality? How is it, must we not ask, how is it that, being without fearful troubles, having so many achievements to show and so much success, having as a nation a deep sense for conduct, having signal energy and honesty, having a splendid aristocracy, having an exceptionally large class of gentlemen, we are yet so little civilised? How is it that our middle and lower classes, in spite of the individuals among them who are raised by happy gifts of nature to a more humane life, in spite of the

[1] So in the original.

seriousness of the middle class, in spite of the honesty and power of true work, the *virtus verusque labor*,° which are to be found in abundance throughout the lower, do yet present, as a whole, the characters which we have seen?

And really it seems as if the current of our discourse carried us of itself to but one conclusion. It seems as if we could not avoid concluding, that just as France owes her fearful troubles to other things and her civilisedness to equality, so we owe our immunity from fearful troubles to other things, and our uncivilisedness to inequality. 'Knowledge is easy', says the wise man,° 'to him that understandeth'; easy, he means, to him who will use his mind simply and rationally, and not to make him think he can know what he cannot, or to maintain, *per fas et nefas*,° a false thesis with which he fancies his interests to be bound up. And to him who will use his mind as the wise man recommends, surely it is easy to see that our shortcomings in civilisation are due to our inequality; or in other words, that the great inequality of classes and property, which came to us from the Middle Age and which we maintain because we have the religion of inequality, that this constitution of things, I say, has the natural and necessary effect, under present circumstances, of materialising our upper class, vulgarising our middle class, and brutalising our lower class. And this is to fail in civilisation.

For only just look how the facts combine themselves. I have said little as yet about our aristocratic class, except that it is splendid. Yet these, 'our often very unhappy brethren', as Burke calls them,° are by no means matter for nothing but ecstasy. Our charity ought certainly, Burke says, to 'extend a due and anxious sensation of pity to the distresses of the miserable great'. Burke's extremely strong language about their miseries and defects I will not quote. For my part, I am always disposed to marvel that human beings, in a position so false, should be so good as these are. Their reason for existing was to serve as a number of centres in a world disintegrated after the ruin of the Roman Empire, and slowly reconstituting itself. Numerous centres of material force were needed, and these a feudal aristocracy supplied. Their large and hereditary estates served this public end. The owners had a positive function, for which their estates were essential. In our modern world the function is gone; and the great estates, with an infinitely multiplied power of ministering to mere pleasure and indulgence, remain. The energy and honesty of our race does not leave itself without witness in this class, and nowhere are there more conspicuous examples of individuals raised by happy gifts of nature far above their fellows and their circumstances. For distinction of all kinds this class has an esteem. Everything which succeeds they tend to welcome, to win over, to put on their side; genius may generally make, if it will, not bad terms for itself with them. But the total result of the class, its effect on society at large and on national progress, are what we must regard. And on the whole, with no necessary function to fulfil, never conversant with life as it really is, tempted, flattered, and spoiled from childhood to old age, our aristocratic class is inevitably materialised, and the more so the more the development of industry and ingenuity augments the means of luxury. Every one can see how bad is the action of such an aristocracy upon the class of newly enriched people, whose great danger is a materialistic ideal, just because it is the ideal they can easiest comprehend. Nor is the mischief of this action now compensated by signal services of a public kind. Turn even to that sphere which aristocracies think specially their own, and where they have under other circumstances been really effective,—the sphere of politics. When there is need, as now, for any large forecast of the course of human affairs, for an acquaintance with the ideas which in the end sway mankind, and for an estimate of their power, aristocracies are out of their element, and materialised aristocracies most of all. In the immense spiritual movement of our day, the English aristocracy, as I have elsewhere said,° always reminds me of Pilate confronting the phenomenon of Christianity. Nor can a materialised class have any serious and fruitful sense for the power of beauty. They may imagine themselves to be in pursuit of beauty; but how often, alas, does the pursuit come to little more than dabbling a little in what they are pleased to call art, and making a great deal of what they are pleased to call love!

Let us return to their merits. For the power of manners an aristocratic class, whether materialised or not, will always, from its circumstances, have a strong sense. And although for this power of social life and manners, so important to civilisation, our English race has no special natural turn, in our aristocracy this power emerges and marks them. When the day of general humanisation comes, they will have fixed the standard of manners. The English simplicity, too, makes the best of the English aristocracy more frank and natural than the best of the like class anywhere else, and even the worst of them it makes free from the incredible fatuities and absurdities of the worst. Then the sense of conduct they share with their countrymen at large. In no class has it such trials to undergo; in none is it more often and more grievously overborne. But really the right comment on this is the comment of Pepys° upon the evil courses of Charles the Second and the Duke of York and the court of that day: 'At all which I am sorry; but it is the effect of idleness, and having nothing else to employ their great spirits upon.'

Heaven forbid that I should speak in dispraise of that unique and most English class which Mr Charles Sumner extols—the large class of gentlemen, not of the landed class or of the nobility, but cultivated and refined. They are a seemly product of the energy and of the power to rise in our race. Without, in general, rank and splendour and wealth and luxury to polish them, they have made their own the high standard of life and manners of an aristocratic and refined class. Not having all the dissipations and distractions of this class, they are much more seriously alive to the power of intellect and knowledge, to the power of beauty. The sense of conduct, too, meets with fewer trials in this class. To some extent, however, their contiguousness to the aristocratic class has now the effect of materialising them, as it does the class of newly enriched people. The most palpable action is on the young amongst them, and on their standard of life and enjoyment. But in general, for this whole class, established facts, the materialism which they see regnant, too much block their mental horizon, and limit the possibilities of things to them. They are deficient in openness and flexibility of mind, in free play of ideas, in faith and ardour. Civilised they are, but they are not much of a civilising force; they are somehow bounded and ineffective.

So on the middle class they produce singularly little effect. What the middle class sees is that splendid piece of materialism, the aristocratic class, with a wealth and luxury utterly out of their reach, with a standard of social life and manners, the offspring of that wealth and luxury, seeming utterly out of their reach also. And thus they are thrown back upon themselves—upon a defective type of religion, a narrow range of intellect and knowledge, a stunted sense of beauty, a low standard of manners. And the lower class see before them the aristocratic class, and its civilisation, such as it is, even infinitely more out of *their* reach than out of that of the middle class; while the life of the middle class, with its unlovely types of religion, thought, beauty, and manners, has naturally, in general, no great attractions for them either. And so they too are thrown back upon themselves; upon their beer, their gin, and their *fun*. Now, then, you will understand what I meant by saying that our inequality materialises our upper class, vulgarises our middle class, brutalises our lower.

And the greater the inequality the more marked is its bad action upon the middle and lower classes. In Scotland the landed aristocracy fills the scene, as is well known, still more than in England; the other classes are more squeezed back and effaced. And the social civilisation of the lower middle class and of the poorest class, in Scotland, is an example of the consequences. Compared with the same class even in England, the

Scottish lower middle class is most visibly, to vary Mr Charles Sumner's phrase, *less* well-bred, *less* careful in personal habits and in social conventions, *less* refined. Let any one who doubts it go, after issuing from the aristocratic solitudes which possess Loch Lomond, let him go and observe the shopkeepers and the middle class in Dumbarton, and Greenock, and Gourock, and the places along the mouth of the Clyde. And for the poorest class, who that has seen it can ever forget the hardly human horror, the abjection and uncivilisedness of Glasgow?

What a strange religion, then, is our religion of inequality! Romance often helps a religion to hold its ground, and romance is good in its way; but ours is not even a romantic religion. No doubt our aristocracy is an object of very strong public interest. The *Times* itself° bestows a leading article by way of epithalamium on the Duke of Norfolk's marriage. And those journals of a new type,° full of talent, and which interest me particularly because they seem as if they were written by the young lion° of our youth,—the young lion grown mellow and, as the French say, *viveur*, arrived at his full and ripe knowledge of the world, and minded to enjoy the smooth evening of his days,—those journals, in the main a sort of social gazette of the aristocracy, are apparently not read by that class only which they most concern, but are read with great avidity by other classes also. And the common people too have undoubtedly, as Mr Gladstone says, a wonderful preference for a lord. Yet our aristocracy, from the action upon it of the Wars of the Roses, the Tudors, and the political necessities of George the Third, is for the imagination a singularly modern and uninteresting one. Its splendour of station, its wealth, show, and luxury, is then what the other classes really admire in it; and this is not an elevating admiration. Such an admiration will never lift us out of our vulgarity and brutality, if we chance to be vulgar and brutal to start with; it will rather feed them and be fed by them. So that when Mr Gladstone invites us to call our love of inequality 'the complement of the love of freedom or its negative pole, or the shadow which the love of freedom casts, or the reverberation of its voice in the halls of the constitution', we must surely answer that all this mystical eloquence is not in the least necessary to explain so simple a matter; that our love of inequality is really the vulgarity in us, and the brutality, admiring and worshipping the splendid materiality.

Our present social organisation, however, will and must endure until our middle class is provided with some better ideal of life than it has now. Our present organisation has been an appointed stage in our growth; it has been of good use, and has enabled us to do great things. But the use is at an end, and the stage is over. Ask yourselves if you do not sometimes feel in yourselves a sense, that in spite of the strenuous efforts for good of so many excellent persons amongst us, we begin somehow to flounder and to beat the air; that we seem to be finding ourselves stopped on this line of advance and on that, and to be threatened with a sort of standstill. It is that we are trying to live on with a social organisation of which the day is over. Certainly equality will never of itself alone give us a perfect civilisation. But, with such inequality as ours, a perfect civilisation is impossible.

To that conclusion, facts, and the stream itself of this discourse, do seem, I think, to carry us irresistibly. We arrive at it because they so choose, not because we so choose. Our tendencies are all the other way. We are all of us politicians, and in one of two camps, the Liberal or the Conservative. Liberals tend to accept the middle class as it is, and to praise the nonconformists; while Conservatives tend to accept the upper class as it is, and to praise the aristocracy. And yet here we are at the conclusion, that whereas one of the great obstacles to our civilisation is, as I have often said, British nonconformity, another main obstacle to our civilisation is British aristocracy! And this while we are yet forced to recognise excellent special qualities as well as the general English energy and honesty, and a number of emergent humane individuals, in both nonconformists and aristocracy. Clearly such a conclusion can be none of our own seeking.

Then again, to remedy our inequality, there must be a change in the law of bequest, as there has been in France; and the faults and inconveniences of the present French law of bequest are obvious.It tends to over-divide property; it is unequal in operation, and can be eluded by people limiting their families; it makes the children, however ill they may behave, independent of the parent. To be sure, Mr Mill and others have shown° that a law of bequest fixing the maximum, whether of land or money, which any one individual may take by bequest or inheritance, but in other respects leaving the testator quite free, has none of the inconveniences of the French law, and is in every way preferable. But evidently these are not questions of practical politics. Just imagine Lord Hartington° going down to Glasgow, and meeting his Scotch Liberals there, and saying to them: 'You are ill at ease, and you are calling for change, and very justly. But the cause of your being ill at ease is not what you suppose. The cause of your being ill at ease is the profound imperfectness of your social civilisation. Your social civilisation is indeed such as I forbear to characterise. But the remedy is not disestablishment. The remedy is social equality. Let me direct your attention to a reform in the law of bequest and entail.' One can hardly speak of such a thing without laughing. No, the matter is at present one for the thoughts of those who think. It is a thing to be turned over in the minds of those who, on the one hand, have the spirit of scientific inquirers, bent on seeing things as they really are; and, on the other hand, the spirit of friends of the humane life, lovers of perfection. To your thoughts I commit it. And perhaps, the more you think of it, the more you will be persuaded that Menander showed his wisdom quite as much when he said *Choose equality*, as when he assured us that *Evil communications corrupt good manners*.

PRACTICAL people talk with a smile of Plato and of his absolute ideas; and it is impossible to deny that Plato's ideas do often seem unpractical and impracticable, and especially when one views them in connexion with the life of a great work-a-day world like the United States. The necessary staple of the life of such a world Plato regards with disdain; handicraft and trade and the working professions he regards with disdain; but what becomes of the life of an industrial modern community if you take handicraft and trade and the working professions out of it? The base mechanic arts and handicrafts, says Plato,° bring about a natural weakness in the principle of excellence in a man, so that he cannot govern the ignoble growths in him, but nurses them, and cannot understand fostering any other. Those who exercise such arts and trades, as they have their bodies, he says, marred by their vulgar businesses, so they have their souls, too, bowed and broken by them. And if one of these uncomely people has a mind to seek self-culture and philosophy, Plato compares him to a bald little tinker, who has scraped together money, and has got his release from service, and has had a bath, and bought a new coat, and is rigged out like a bridegroom about to marry the daughter of his master who has fallen into poor and helpless estate.

Nor do the working professions fare any better than trade at the hands of Plato. He draws for us° an inimitable picture of the working lawyer, and of his life of bondage; he shows how this bondage from his youth up has stunted and warped him, and made him small and crooked of soul, encompassing him with difficulties which he is not man enough to rely on justice and truth as means to encounter, but has recourse, for help out of them, to falsehood and wrong. And so, says Plato, this poor creature is bent and broken, and grows up from boy to man without a particle of soundness in him, although exceedingly smart and clever in his own esteem.

One cannot refuse to admire the artist who draws these pictures. But we say to ourselves that his ideas show the influence of a primitive and obsolete order of things, when the warrior caste and the priestly caste were alone in honour, and the humble work of the world was done by slaves. We have now changed all that; the modern majesty consists in work, as Emerson declares;° and in work, we may add, principally of such plain and dusty kind as the work of cultivators of the ground, handicraftsmen, men of trade and business, men of the working professions. Above all is this true in a great industrious community such as that of the United States.

Now education, many people go on to say, is still mainly governed by the ideas of men like Plato, who lived when the warrior caste and the priestly or philosophical class were alone in honour, and the really useful part of the community were slaves. It is an education fitted for persons of leisure in such a community. This eduction passed from Greece and Rome to the feudal communities of Europe, where also the warrior caste and the priestly caste were alone held in honour, and where the really useful and working part of the community, though not nominally slaves as in the pagan world, were practically not much better off than slaves, and not more seriously regarded. And how absurd it is, people end by saying, to inflict this education upon an industrious modern community, where very few indeed are persons of leisure, and the mass to be considered has not leisure, but is bound, for its own great good, and for the great good of the world at large, to plain labour and to industrial pursuits, and the education in question tends necessarily to make men dissatisfied with these pursuits and unfitted for them!

That is what is said. So far I must defend Plato, as to plead that his view of education and studies is in the general, as it seems to me, sound enough, and fitted for all sorts and conditions of men, whatever their pursuits may be. 'An intelligent man', says Plato,° 'will prize those studies which result in his soul getting soberness, righteousness, and wisdom, and will less value the others.' I cannot consider *that* a bad description of the aim of education, and of the motives which should govern us in the choice of studies, whether we are preparing ourselves for a hereditary seat in the English House of Lords or for the pork trade in Chicago.

Still I admit that Plato's world was not ours, that his scorn of trade and handicraft is fantastic, that he had no conception of a great industrial community such as that of the United States, and that such a community must and will shape its education to suit its own needs. If the usual education handed down to it from the past does not suit it, it will certainly before long drop this and try another. The usual education in the past has been mainly literary. The question is whether the studies which were long supposed to be the best for all of us are practically the best now; whether others are not better. The tyranny of the past, many think, weighs on us injuriously in the predominance given to letters in education. The question is raised whether, to meet the needs of our modern life, the predominance ought not now to pass from letters to science; and naturally the question is nowhere raised with more energy than here in the United States. The design of abasing what is called 'mere literary instruction and education', and of exalting what is called 'sound, extensive, and practical scientific knowledge',° is, in this intensely modern world of the United States, even more perhaps than in Europe, a very popular design, and makes great and rapid progress.

I am going to ask whether the present movement for ousting letters from their old predominance in education, and for transferring the predominance in education to the natural sciences, whether this brisk and flourishing movement ought to prevail, and whether it is likely that in the end it really will prevail. An objection may be raised which I will anticipate. My own studies have been almost wholly in letters, and my visits to the field of the natural sciences have been very slight and inadequate, although those sciences have always strongly moved my curiosity. A man of letters, it will perhaps be said, is not competent to discuss the comparative merits of letters and natural science as means of education. To this objection I reply, first of all, that his incompetence, if he attempts the discussion but is really incompetent for it, will be abundantly visible; nobody will be taken in; he will have plenty of sharp observers and critics to save mankind from that danger. But the line I am going to follow is, as you will soon discover, so extremely simple, that perhaps it may be followed without failure even by one who for a more ambitious line of discussion would be quite incompetent.

Some of you may possibly remember a phrase of mine° which has been the object of a good deal of comment; an observation to the effect that in our culture, the aim being *to know ourselves and the world*, we have, as the means to this end, *to know the best which has been thought and said in the world*. A man of science, who is also an excellent writer and the very prince of debaters, Professor Huxley, in a discourse at the opening of Sir Josiah Mason's college at Birmingham, laying hold of this phrase, expanded it by quoting some more words of mine, which are these: 'The civilised world is to be regarded as now being, for intellectual and spiritual purposes, one great confederation, bound to a joint action and working to a common result; and whose members have for their proper outfit a knowledge of Greek, Roman, and Eastern antiquity, and of one another. Special local and temporary advantages being put out of account, that modern nation will in the intellectual and spiritual sphere make most progress, which most thoroughly carries out this programme.'

Now on my phrase, thus enlarged, Professor Huxley remarks° that when I speak of the above-mentioned knowledge as enabling us to know ourselves and the world, I assert *literature* to contain the materials which suffice for thus making us know ourselves and the world. But it is not by any means clear, says he, that after having learnt all which ancient and modern literatures have to tell us, we have laid a sufficiently broad and deep foundation for that criticism of life, that knowledge of ourselves and the world, which constitutes culture. On the contrary, Professor Huxley declares that he finds himself 'wholly unable to admit that either nations or individuals will really advance, if their outfit draws nothing from the stores of physical science. An army without weapons of precision, and with no particular base of operations, might more hopefully enter upon a campaign on the Rhine, than a man, devoid of a knowledge of what physical science has done in the last century, upon a criticism of life.'

This shows how needful it is for those who are to discuss any matter together, to have a common understanding as to the sense of the terms they employ,—how needful, and how difficult. What Professor Huxley says, implies just the reproach which is so often brought against the study of *belles lettres*, as they are called: that the study is an elegant one, but slight and ineffectual; a smattering of Greek and Latin and other ornamental things, of little use for any one whose object is to get at truth, and to be a practical man. So, too, M. Renan talks of° the 'superficial humanism' of a school-course which treats us as if we were all going to be poets, writers, preachers, orators, and he opposes this humanism to positive science, or the critical search after truth. And there is always a tendency in those who are remonstrating against the predominance of letters in education, to understand by letters *belles lettres*, and by *belles lettres* a superficial humanism, the opposite of science or true knowledge.

But when we talk of knowing Greek and Roman antiquity, for instance, which is the knowledge people have called the humanities, I for my part mean a knowledge which is something more than a superficial humanism, mainly decorative. 'I call all teaching *scientific*', says Wolf, the critic of Homer,° 'which is systematically laid out and followed up to its original sources. For example: a knowledge of classical antiquity is scientific when the remains of classical antiquity are connectedly studied in the original languages.' There can be no doubt that Wolf is perfectly right; that all learning is scientific which is systematically laid out and followed up to its original sources, and that a genuine humanism is scientific.

When I speak of knowing Greek and Roman antiquity, therefore, as a help to knowing ourselves and the world, I mean more than a knowledge of so much vocabulary, so much grammar, so many portions of authors in the Greek and Latin languages. I mean knowing the Greeks and Romans, and their life and genius, and what they were and did in the world; what we get from them, and what is its value. That, at least, is the ideal; and when we talk of endeavouring to know Greek and Roman antiquity, as a help to knowing ourselves and the world, we mean endeavouring so to know them as to satisfy this ideal, however much we may still fall short of it.

The same also as to knowing our own and other modern nations, with the like aim of getting to understand ourselves and the world. To know the best that has been thought and said by the modern nations, is to know, says Professor Huxley,° 'only what modern *literatures* have to tell us; it is the criticism of life contained in modern literature'. And yet 'the distinctive character of our times', he urges, 'lies in the vast and constantly increasing part which is played by natural knowledge'. And how, therefore, can a man, devoid of knowledge of what physical science has done in the last century, enter hopefully upon a criticism of modern life?

Let us, I say, be agreed about the meaning of the terms we are using. I talk of knowing the best which has been thought and uttered in the world; Professor Huxley says this means knowing *literature*. Literature is a large word; it may mean everything written with letters or printed in a book. Euclid's *Elements* and Newton's *Principia* are thus literature. All knowledge that reaches us through books is literature. But by literature Professor Huxley means *belles lettres*. He means to make me say, that knowing the best which has been thought and said by the modern nations is knowing their *belles lettres* and no more. And this is no sufficient equipment, he argues, for a criticism of modern life. But as I do not mean, by knowing ancient Rome, knowing merely more or less of Latin *belles lettres*, and taking no account of Rome's military, and political, and legal, and administrative work in the world; and as, by knowing ancient Greece, I understand knowing her as the giver of Greek art, and the guide to a free and right use of reason and to scientific method, and the founder of our mathematics and physics and astronomy and biology,—I understand knowing her as all this, and not merely knowing certain Greek poems, and histories, and treatises, and speeches,—so as to the knowledge of modern nations also. By knowing modern nations, I mean not merely knowing their *belles lettres*, but knowing also what has been done by such men as Copernicus, Galileo, Newton, Darwin. 'Our ancestors learned',° says Professor Huxley, 'that the earth is the centre of the visible universe, and that man is the cynosure of things terrestrial; and more especially was it inculcated that the course of nature had no fixed order, but that it could be, and constantly was, altered.' But for us now, continues Professor Huxley, 'the notions of the beginning and the end of the world entertained by our forefathers are no longer credible. It is very certain that the earth is not the chief body in the material universe, and that the world is not subordinated to man's use. It is even more certain that nature is the expression of a definite order, with which nothing interferes.' 'And yet,' he cries, 'the purely classical education advocated by the representatives of the humanists in our day gives no inkling of all this!'

In due place and time I will just touch upon that vexed question of classical education; but at present the question is as to what is meant by knowing the best which modern nations have thought and said. It is not knowing their *belles lettres* merely which is meant. To know Italian *belles lettres* is not to know Italy, and to know English *belles lettres* is not to know England. Into knowing Italy and England there comes a great deal more, Galileo and Newton amongst it. The reproach of being a superficial humanism, a tincture of *belles lettres*, may attach rightly enough to some other disciplines; but to the particular discipline recommended when I proposed knowing the best that has been thought and said in the world, it does not apply. In that best I certainly include what in modern times has been thought and said by the great observers and knowers of nature.

There is, therefore, really no question between Professor Huxley and me as to whether knowing the great results of the modern scientific study of nature is not required as a part of our culture, as well as knowing the products of literature and art. But to follow the processes by which those results are reached, ought, say the friends of physical science, to be made the staple of education for the bulk of mankind. And here there does arise a question between those whom Professor Huxley calls with playful sarcasm 'the Levites of culture',° and those whom the poor humanist is sometimes apt to regard as its Nebuchadnezzars.°

The great results of the scientific investigation of nature we are agreed upon knowing, but how much of our study are we bound to give to the processes by which those results are reached? The results have their visible bearing on human life. But all the processes, too, all the items of fact, by which those results are reached and established, are interesting. All knowledge is interesting to a wise man, and the knowledge of nature is interesting to all men. It is very interesting to know, that, from the albuminous white of the egg, the chick in the egg gets the materials for its flesh, bones, blood, and feathers; while, from the fatty yolk of the egg, it gets the heat and energy which enable it at length to break its shell and begin the world. It is less interesting, perhaps, but still it is interesting, to know that when a taper burns, the wax is converted into carbonic acid and water. Moreover, it is quite true that the habit of dealing with facts, which is given by the study of nature, is, as the friends of physical science praise it for being, an excellent discipline. The appeal, in the study of nature, is constantly to observation and experiment; not only is it said that the thing is so, but we can be made to see that it is so. Not only does a man tell us that when a taper burns the wax is converted into carbonic acid and water, as a man may tell us, if he likes,° that Charon is punting his ferry-boat on the river Styx, or that Victor Hugo is a sublime poet, or Mr Gladstone the most admirable of statesmen; but we are made to see that the conversion into carbonic acid and water does actually happen. This reality of natural knowledge it is, which makes the friends of physical science contrast it, as a knowledge of things, with the humanist's knowledge, which is, say they, a knowledge of words. And hence Professor Huxley is moved to lay it down that,° 'for the purpose of attaining real culture, an exclusively scientific education is at least as effectual as an exclusively literary education'. And a certain President° of the Section for Mechanical Science in the British Association is, in Scripture phrase, 'very bold', and declares that if a man, in his mental training, 'has substituted literature and history for natural science, he has chosen the less useful alternative'. But whether we go these lengths or not, we must all admit that in natural science the habit gained of dealing with facts is a most valuable discipline, and that every one should have some experience of it.

More than this, however, is demanded by the reformers. It is proposed to make the training in natural science the main part of education, for the great majority of mankind at any rate. And here, I confess, I part company with the friends of physical science, with whom up to this point I have been agreeing. In differing from them, however, I wish to proceed with the utmost caution and diffidence. The smallness of my own acquaintance with the disciplines of natural science is ever before my mind, and I am fearful of doing these disciplines an injustice. The ability and pugnacity of the partisans of natural science make them formidable persons to contradict. The tone of tentative inquiry, which befits a being of dim faculties and bounded knowledge, is the tone I would wish to take and not to depart from. At present it seems to me, that those who are for giving to natural knowledge, as they call it, the chief place in the education of the majority of mankind, leave one important thing out of their account: the constitution of human nature. But I put this forward on the strength of some facts not at all recondite, very far from it; facts capable of being stated in the simplest possible fashion, and to which, if I so state them, the man of science will, I am sure, be willing to allow their due weight.

Deny the facts altogether, I think, he hardly can. He can hardly deny, that when we set ourselves to enumerate the powers° which go to the building up of human life, and say that they are the power of conduct, the power of intellect and knowledge, the power of beauty, and the power of social life and manners,—he can hardly deny that this scheme, though drawn in rough and plain lines enough, and not pretending to scientific exactness, does yet give a fairly true representation of the matter. Human nature is built up by these powers; we have the need for them all. When we have rightly met and adjusted the claims of them all, we shall then be in a fair way for getting soberness and righteousness, with wisdom. This is evident enough, and the friends of physical science would admit it.

But perhaps they may not have sufficiently observed another thing: namely, that the several powers just mentioned are not isolated, but there is, in the generality of mankind, a perpetual tendency to relate them one to another in divers ways. With one such way of relating them I am particularly concerned now. Following our instinct for intellect and knowledge, we acquire pieces of knowledge; and presently, in the generality of men, there arises the desire to relate these pieces of knowledge to our sense for conduct, to our sense for beauty,—and there is weariness and dissatisfaction if the desire is baulked. Now in this desire lies, I think, the strength of that hold which letters have upon us.

All knowledge is, as I said just now, interesting; and even items of knowledge which from the nature of the case cannot well be related, but must stand isolated in our thoughts, have their interest. Even lists of exceptions have their interest. If we are studying Greek accents, it is interesting to know that *pais* and *pas*, and some other monosyllables of the same form of declension, do not take the circumflex upon the last syllable of the genitive plural, but vary, in this respect, from the common rule. If we are studying physiology, it is interesting to know that the pulmonary artery carries dark blood and the pulmonary vein carries bright blood, departing in this respect from the common rule for the division of labour between the veins and the arteries. But every one knows how we seek naturally to combine the pieces of our knowledge together, to bring them under general rules, to relate them to principles; and how unsatisfactory and tiresome it would be to go on for ever learning lists of exceptions, or accumulating items of fact which must stand isolated.

Well, that same need of relating our knowledge, which operates here within the sphere of our knowledge itself, we shall find operating, also, outside that sphere. We experience, as we go on learning and know-ing,—the vast majority of us experience,—the need of relating what we have learnt and known to the sense which we have in us for conduct, to the sense which we have in us for beauty.

A certain Greek prophetess of Mantineia in Arcadia, Diotima by name, once explained to the philosopher Socrates° that love, and impulse, and bent of all kinds, is, in fact, nothing else but the desire in men that good should for ever be present to them. This desire for good, Diotima assured Socrates, is our fundamental desire, of which fundamental desire every impulse in us is only some one particular form. And therefore this fundamental desire it is, I suppose,—this desire in men that good should be for ever present to them,—which acts in us when we feel the impulse for relating our knowledge to our sense for conduct and to our sense for beauty. At any rate, with men in general the instinct exists. Such is human nature. And the instinct, it will be admitted, is innocent, and human nature is preserved by our following the lead of its innocent instincts. Therefore, in seeking to gratify this instinct in question, we are following the instinct of self-preservation° in humanity.

But, no doubt, some kinds of knowledge cannot be made to directly serve the instinct in question, cannot be directly related to the sense for beauty, to the sense for conduct. These are instrument-knowledges; they lead on to other knowledges, which can. A man who passes his life in instrument-knowledges is a specialist. They may be invaluable as instruments to something beyond, for those who have the gift thus to employ them; and they may be disciplines in themselves wherein it is useful for every one to have some schooling. But it is inconceivable that the generality of men should pass all their mental life with Greek accents or with formal logic. My friend Professor Sylvester,° who is one of the first mathematicians in the world, holds transcendental doctrines as to the virtue of mathematics, but those doctrines are not for common men. In the very Senate House and heart of our English Cambridge I once ventured,° though not without an apology for my profaneness, to hazard the opinion that for the majority of mankind a little of mathematics, even, goes a long way. Of course this is quite consistent with their being of immense importance as an instrument to something else; but it is the few who have the aptitude for thus using them, not the bulk of mankind.

The natural sciences do not, however, stand on the same footing with these instrument-knowledges. Experience shows us that the generality of men will find more interest in learning that, when a taper burns, the wax is converted into carbonic acid and water, or in learning the

explanation of the phenomenon of dew, or in learning how the circulation of the blood is carried on, than they find in learning that the genitive
plural of *pais* and *pas* does not take the circumflex on the termination.
And one piece of natural knowledge is added to another, and others are
added to that, and at last we come to propositions so interesting as Mr
Darwin's famous proposition° that 'our ancestor was a hairy quadruped
furnished with a tail and pointed ears, probably arboreal in his habits'.
Or we come to propositions of such reach and magnitude as those which
Professor Huxley delivers,° when he says that the notions of our forefathers about the beginning and the end of the world were all wrong,
and that nature is the expression of a definite order with which nothing
interferes.

Interesting, indeed, these results of science are, important they are,
and we should all of us be acquainted with them. But what I now wish you
to mark is, that we are still, when they are propounded to us and we
receive them, we are still in the sphere of intellect and knowledge. And
for the generality of men there will be found, I say, to arise, when they
have duly taken in the proposition that their ancestor was 'a hairy
quadruped furnished with a tail and pointed ears, probably arboreal in
his habits', there will be found to arise an invincible desire to relate this
proposition to the sense in us for conduct, and to the sense in us for
beauty. But this the men of science will not do for us, and will hardly even
profess to do. They will give us other pieces of knowledge, other facts,
about other animals and their ancestors, or about plants, or about stones,
or about stars; and they may finally bring us to those great 'general
conceptions of the universe, which are forced upon us all', says Professor
Huxley, 'by the progress of physical science'.° But still it will be *knowledge*
only which they give us; knowledge not put for us into relation with our
sense for conduct, our sense for beauty, and touched with emotion by
being so put; not thus put for us, and therefore, to the majority of
mankind, after a certain while, unsatisfying, wearying.

Not to the born naturalist, I admit. But what do we mean by a born
naturalist? We mean a man in whom the zeal for observing nature is so
uncommonly strong and eminent, that it marks him off from the bulk of
mankind. Such a man will pass his life happily in collecting natural
knowledge and reasoning upon it, and will ask for nothing, or hardly
anything, more. I have heard it said that the sagacious and admirable
naturalist whom we lost not very long ago,° Mr Darwin, once owned to a
friend that for his part he did not experience the necessity for two things
which most men find so necessary to them,—religion and poetry; science
and the domestic affections, he thought, were enough. To a born

naturalist, I can well understand that this should seem so. So absorbing is his occupation with nature, so strong his love for his occupation, that he goes on acquiring natural knowledge and reasoning upon it, and has little time or inclination for thinking about getting it related to the desire in man for conduct, the desire in man for beauty. He relates it to them for himself as he goes along, so far as he feels the need; and he draws from the domestic affections all the additional solace necessary. But then Darwins are extremely rare. Another great and admirable master of natural knowledge, Faraday,° was a Sandemanian. That is to say, he related his knowledge to his instinct for conduct and to his instinct for beauty, by the aid of that respectable Scottish sectary, Robert Sandeman. And so strong, in general, is the demand of religion and poetry to have their share in a man, to associate themselves with his knowing, and to relieve and rejoice it, that, probably, for one man amongst us with the disposition to do as Darwin did in this respect, there are at least fifty with the disposition to do as Faraday.

Education lays hold upon us, in fact, by satisfying this demand. Professor Huxley holds up to scorn° mediaeval education, with its neglect of the knowledge of nature, its poverty even of literary studies, its formal logic devoted to 'showing how and why that which the Church said was true must be true'. But the great mediaeval Universities were not brought into being, we may be sure, by the zeal for giving a jejune and contemptible education. Kings have been their nursing fathers, and queens have been their nursing mothers, but not for this. The mediaeval Universities came into being, because the supposed knowledge, delivered by Scripture and the Church, so deeply engaged men's hearts, by so simply, easily, and powerfully relating itself to their desire for conduct, their desire for beauty. All other knowledge was dominated by this supposed knowledge and was subordinated to it, because of the surpassing strength of the hold which it gained upon the affections of men, by allying itself profoundly with their sense for conduct, their sense for beauty.

But now, says Professor Huxley, conceptions of the universe fatal to the notions held by our forefathers have been forced upon us by physical science. Grant to him that they are thus fatal, that the new conceptions must and will soon become current everywhere, and that every one will finally perceive them to be fatal to the beliefs of our forefathers. The need of humane letters, as they are truly called, because they serve the paramount desire in men that good should be for ever present to them,— the need of humane letters, to establish a relation between the new conceptions, and our instinct for beauty, our instinct for conduct, is only the more visible. The Middle Age could do without humane letters, as it could do without the study of nature, because its supposed knowledge was made to engage its emotions so powerfully. Grant that the supposed knowledge disappears, its power of being made to engage the emotions will of course disappear along with it,—but the emotions themselves, and their claim to be engaged and satisfied, will remain. Now if we find by experience that humane letters have an undeniable power of engaging the emotions, the importance of humane letters in a man's training becomes not less, but greater, in proportion to the success of modern science in extirpating what it calls 'mediaeval thinking'.°

Have humane letters, then, have poetry and eloquence, the power here attributed to them of engaging the emotions, and do they exercise it? And if they have it and exercise it, *how* do they exercise it, so as to exert an influence upon man's sense for conduct, his sense for beauty? Finally, even if they both can and do exert an influence upon the senses in question, how are they to relate to them the results,—the modern results,—of natural science? All these questions may be asked. First, have poetry and eloquence the power of calling out the emotions? The appeal is to experience. Experience shows that for the vast majority of men, for mankind in general, they have the power. Next, do they exercise it? They do. But then, *how* do they exercise it so as to affect man's sense for conduct, his sense for beauty? And this is perhaps a case for applying the Preacher's words: 'Though a man labour to seek it out, yet he shall not find it; yea, farther, though a wise man think to know it, yet shall he not be able to find it.'[1] Why should it be one thing, in its effect upon the emotions, to say, 'Patience is a virtue', and quite another thing, in its effect upon the emotions, to say with Homer,

τλητὸν γὰρ Μοῖραι θυμὸν θέσαν ἀνθρώποισιν—[2]

'for an enduring heart have the destinies appointed to the children of men'? Why should it be one thing, in its effect upon the emotions, to say with the philosopher Spinoza,° *Felicitas in eo consistit quod homo suum esse conservare potest*—'Man's happiness consists in his being able to preserve his own essence', and quite another thing, in its effect upon the emotions, to say with the Gospel,° 'What is a man advantaged, if he gain the whole world, and lose himself, forfeit himself?' How does this difference of effect arise? I cannot tell, and I am not much concerned to know; the important thing is that it does arise, and that we can profit by it. But how, finally, are poetry and eloquence to exercise the power of relating the modern results of natural science to man's instinct for

[1] *Ecclesiastes*, viii. 17. [2] *Iliad*, xxiv. 49.

conduct, his instinct for beauty? And here again I answer that I do not know *how* they will exercise it, but that they can and will exercise it I am sure. I do not mean that modern philosophical poets and modern philosophical moralists are to come and relate for us, in express terms, the results of modern scientific research to our instinct for conduct, our instinct for beauty. But I mean that we shall find, as a matter of experience, if we know the best that has been thought and uttered in the world, we shall find that the art and poetry and eloquence of men who lived, perhaps, long ago, who had the most limited natural knowledge, who had the most erroneous conceptions about many important matters, we shall find that this art, and poetry, and eloquence, have in fact not only the power of refreshing and delighting us, they have also the power,—such is the strength and worth, in essentials, of their authors' criticism of life,—they have a fortifying, and elevating, and quickening, and suggestive power, capable of wonderfully helping us to relate the results of modern science to our need for conduct, our need for beauty. Homer's conceptions of the physical universe were, I imagine, grotesque; but really, under the shock of hearing from modern science that 'the world is not subordinated to man's use, and that man is not the cynosure of things terrestrial', I could, for my own part, desire no better comfort than Homer's line which I quoted just now:

τλητὸν γὰρ Μοῖραι θυμὸν θέσαν ἀνθρώποισιν—

'for an enduring heart have the destinies appointed to the children of men'!

And the more that men's minds are cleared, the more that the results of science are frankly accepted, the more that poetry and eloquence come to be received and studied as what in truth they really are,—the criticism of life by gifted men, alive and active with extraordinary power at an unusual number of points;—so much the more will the value of humane letters, and of art also, which is an utterance having a like kind of power with theirs, be felt and acknowledged, and their place in education be secured.

Let us therefore, all of us, avoid indeed as much as possible any invidious comparison between the merits of humane letters, as means of education, and the merits of the natural sciences. But when some President of a Section for Mechanical Science insists on making the comparison, and tells us that 'he who in his training has substituted literature and history for natural science has chosen the less useful alternative', let us make answer to him that the student of humane letters only, will, at least, know also the great general conceptions brought in by

modern physical science; for science, as Professor Huxley says, forces
them upon us all. But the student of the natural sciences only, will, by our
very hypothesis, know nothing of humane letters; not to mention that in
setting himself to be perpetually accumulating natural knowledge, he
sets himself to do what only specialists have in general the gift for doing
genially. And so he will probably be unsatisfied, or at any rate incomplete, and even more incomplete than the student of humane letters only.

I once mentioned in a school-report, how a young man in one of our
English training colleges having to paraphrase the passage in *Macbeth*°
beginning,

> 'Can'st thou not minister to a mind diseased?'

turned this line into, 'Can you not wait upon the lunatic?' And I remarked
what a curious state of things it would be, if every pupil of our national
schools knew, let us say, that the moon is two thousand one hundred and
sixty miles in diameter, and thought at the same time that a good
paraphrase for

> 'Can'st thou not minister to a mind diseased?'

was, 'Can you not wait upon the lunatic?' If one is driven to choose, I
think I would rather have a young person ignorant about the moon's
diameter, but aware that 'Can you not wait upon the lunatic?' is bad, than
a young person whose education had been such as to manage things the
other way.

Or to go higher than the pupils of our national schools. I have in my
mind's eye a member of our British Parliament° who comes to travel here
in America, who afterwards relates his travels, and who shows a really
masterly knowledge of the geology of this great country and of its mining
capabilities, but who ends by gravely suggesting that the United States
should borrow a prince from our Royal Family, and should make him
their king, and should create a House of Lords of great landed proprietors after the pattern of ours; and then America, he thinks, would
have her future happily and perfectly secured. Surely, in this case, the
President of the Section for Mechanical Science would himself hardly
say that our member of Parliament, by concentrating himself upon
geology and mineralogy, and so on, and not attending to literature and
history, had 'chosen the more useful alternative'.

If then there is to be separation and option between humane letters on
the one hand, and the natural sciences on the other, the great majority of
mankind, all who have not exceptional and overpowering aptitudes for
the study of nature, would do well, I cannot but think, to choose to be

educated in humane letters rather than in the natural sciences. Letters will call out their being at more points, will make them live more.

I said that before I ended I would just touch on the question of classical education, and I will keep my word. Even if literature is to retain a large place in our education, yet Latin and Greek, say the friends of progress, will certainly have to go. Greek is the grand offender in the eyes of these gentlemen. The attackers of the established course of study think that against Greek, at any rate, they have irresistible arguments. Literature may perhaps be needed in education, they say; but why on earth should it be Greek literature? Why not French or German? Nay, 'has not an Englishman models in his own literature of every kind of excellence?'° As before, it is not on any weak pleadings of my own that I rely for convincing the gainsayers; it is on the constitution of human nature itself, and on the instinct of self-preservation in humanity. The instinct for beauty is set in human nature, as surely as the instinct for knowledge is set there, or the instinct for conduct. If the instinct for beauty is served by Greek literature and art as it is served by no other literature and art, we may trust to the instinct of self-preservation in humanity for keeping Greek as part of our culture. We may trust to it for even making the study of Greek more prevalent than it is now. Greek will come, I hope, some day to be studied more rationally than at present; but it will be increasingly studied as men increasingly feel the need in them for beauty, and how powerfully Greek art and Greek literature can serve this need. Women will again study Greek, as Lady Jane Grey° did; I believe that in that chain of forts, with which the fair host of the Amazons are now engirdling our English universities,° I find that here in America,° in colleges like Smith College in Massachusetts, and Vassar College in the State of New York, and in the happy families of the mixed universities out West, they are studying it already.

Defuit una mihi symmetria prisca,—'The antique symmetry was the one thing wanting to me,' said Leonardo da Vinci;° and he was an Italian. I will not presume to speak for the Americans, but I am sure that, in the Englishman, the want of this admirable symmetry of the Greeks is a thousand times more great and crying than in any Italian. The results of the want show themselves most glaringly, perhaps, in our architecture, but they show themselves, also, in all our art. *Fit details strictly combined, in view of a large general result nobly conceived;* that is just the beautiful *symmetria prisca* of the Greeks, and it is just where we English fail, where all our art fails. Striking ideas we have, and well-executed details we have; but that high symmetry which, with satisfying and delightful effect, combines them, we seldom or never have. The glorious beauty of the

Acropolis at Athens did not come from single fine things stuck about on that hill, a statue here, a gateway there;—no, it arose from all things being perfectly combined for a supreme total effect. What must not an Englishman feel about our deficiencies in this respect, as the sense for beauty, whereof this symmetry is an essential element, awakens and strengthens within him! what will not one day be his respect and desire for Greece and its *symmetria prisca*, when the scales drop from his eyes as he walks the London streets, and he sees such a lesson in meanness as the Strand, for instance, in its true deformity! But here we are coming to our friend Mr Ruskin's province,° and I will not intrude upon it, for he is its very sufficient guardian.

And so we at last find, it seems, we find flowing in favour of the humanities the natural and necessary stream of things, which seemed against them when we started. The 'hairy quadruped furnished with a tail and pointed ears, probably arboreal in his habits', this good fellow carried hidden in his nature, apparently, something destined to develop into a necessity for humane letters. Nay, more; we seem finally to be even led to the further conclusion that our hairy ancestor carried in his nature, also, a necessity for Greek.

And therefore, to say the truth, I cannot really think that humane letters are in much actual danger of being thrust out from their leading place in education, in spite of the array of authorities against them at this moment. So long as human nature is what it is, their attractions will remain irresistible. As with Greek, so with letters generally: they will some day come, we may hope, to be studied more rationally, but they will not lose their place. What will happen will rather be that there will be crowded into education other matters besides, far too many; there will be, perhaps, a period of unsettlement and confusion and false tendency; but letters will not in the end lose their leading place. If they lose it for a time, they will get it back again. We shall be brought back to them by our wants and aspirations. And a poor humanist may possess his soul in patience, neither strive nor cry,° admit the energy and brilliancy of the partisans of physical science, and their present favour with the public, to be far greater than his own, and still have a happy faith that the nature of things works silently on behalf of the studies which he loves, and that, while we shall all have to acquaint ourselves with the great results reached by modern science, and to give ourselves as much training in its disciplines as we can conveniently carry, yet the majority of men will always require humane letters; and so much the more, as they have the more and the greater results of science to relate to the need in man for conduct, and to the need in him for beauty.

Emerson

FORTY years ago, when I was an undergraduate at Oxford,° voices were in the air there which haunt my memory still. Happy the man who in that susceptible season of youth hears such voices! they are a possession to him for ever.° No such voices as those which we heard in our youth at Oxford are sounding there now. Oxford has more criticism now, more knowledge, more light; but such voices as those of our youth it has no longer. The name of Cardinal Newman is a great name to the imagination still; his genius and his style are still things of power. But he is over eighty years old; he is in the Oratory at Birmingham; he has adopted, for the doubts and difficulties which beset men's minds to-day, a solution which, to speak frankly, is impossible. Forty years ago he was in the very prime of life; he was close at hand to us at Oxford; he was preaching in St Mary's pulpit every Sunday; he seemed about to transform and to renew what was for us the most national and natural institution in the world, the Church of England. Who could resist the charm of that spiritual apparition, gliding in the dim afternoon light through the aisles of St Mary's, rising into the pulpit, and then, in the most entrancing of voices, breaking the silence with words and thoughts which were a religious music,—subtle, sweet, mournful? I seem to hear him still, saying:° 'After the fever of life, after wearinesses and sicknesses, fightings and despondings, languor and fretfulness, struggling and failing, struggling and succeeding; after all the changes and chances of this troubled, unhealthy state,—at length comes death, at length the white throne of God, at length the beatific vision.' Or, if we followed him back to his seclusion at Littlemore,° that dreary village by the London road, and to the house of retreat and the church which he built there,—a mean house such as Paul might have lived in when he was tent-making at Ephesus,° a church plain and thinly sown with worshippers,—who could resist him there either, welcoming back to the severe joys of church-fellowship, and of daily worship and prayer, the firstlings of a generation which had well-nigh forgotten them? Again I seem to hear him:° 'The season is chill and dark, and the breath of the morning is damp, and worshippers are few; but all this befits those who are by profession penitents and mourners, watchers and pilgrims. More dear to them that loneliness, more cheerful that severity, and more bright that gloom, than all those aids and appliances of luxury by which men nowadays attempt to make prayer less disagreeable to them. True faith does not covet comforts; they who realize that awful day when they shall see Him face to face, whose eyes are as a flame of fire, will as little bargain to pray pleasantly now, as they will think of doing so then.'

Somewhere or other I have spoken° of those 'last enchantments of the Middle Age' which Oxford sheds around us, and here they were! But there were other voices sounding in our ear besides Newman's. There was the puissant voice of Carlyle, so sorely strained, over-used and misused since, but then fresh, comparatively sound, and reaching our hearts with true, pathetic eloquence. Who can forget the emotion of receiving in its first freshness such a sentence as that sentence of Carlyle upon Edward Irving,° then just dead: 'Scotland sent him forth a herculean man; our mad Babylon wore and wasted him with all her engines,—and it took her twelve years!' A greater voice still,—the greatest voice of the century,—came to us in those youthful years through Carlyle: the voice of Goethe. To this day,—such is the force of youthful associations,—I read the *Wilhelm Meister* with more pleasure in Carlyle's translation than in the original. The large, liberal view of human life in *Wilhelm Meister*, how novel it was to the Englishman in those days! and it was salutary, too, and educative for him, doubtless, as well as novel. But what moved us most in *Wilhelm Meister* was that which, after all, will always move the young most,—the poetry, the eloquence. Never surely was Carlyle's prose so beautiful and pure as in his rendering of the Youths' dirge over Mignon!°—'Well is our treasure now laid up, the fair image of the past. Here sleeps it in the marble, undecaying; in your hearts, also, it lives, it works. Travel, travel, back into life! Take along with you this holy earnestness, for earnestness alone makes life eternity.' Here we had the voice of the great Goethe;—not of the stiff, and hindered, and frigid, and factitious Goethe who speaks to us too often from those sixty volumes of his, but of the great Goethe, and the true one.

And besides those voices, there came to us in that old Oxford time a voice, also, from this side of the Atlantic,—a clear and pure voice, which for my ear, at any rate, brought a strain as new, and moving, and unforgettable, as the strain of Newman, or Carlyle, or Goethe. Mr Lowell has well described° the apparition of Emerson to your young generation here, in that distant time of which I am speaking, and his workings upon them. He was your Newman, your man of soul and genius visible to you in the flesh, speaking to your bodily ears, a present object for your heart and imagination. That is surely the most potent of all influences! nothing can come up to it. To us at Oxford Emerson was but a voice speaking from three thousand miles away. But so well he spoke, that from that time forth Boston Bay and Concord were names invested, to my ear, with a sentiment akin to that which invests for me the names of Oxford and of Weimar; and snatches of Emerson's strain fixed themselves in my mind as imperishably as any of the eloquent words which I have been just now quoting. 'Then dies the man in you; then once more perish the buds of art, poetry, and science, as they have died already in a thousand thousand men.'° 'What Plato has thought, he may think; what a saint has felt, he may feel; what at any time has befallen any man, he can understand.'° 'Trust thyself! every heart vibrates to that iron string. Accept the place the divine providence has found for you, the society of your contemporaries, the connexion of events. Great men have always done so, and confided themselves childlike to the genius of their age; betraying their perception that the Eternal was stirring at their heart, working through their hands, predominating in all their being. And we are now men, and must accept in the highest spirit the same transcendent destiny; and not pinched in a corner, not cowards fleeing before a revolution, but redeemers and benefactors, pious aspirants to be noble clay plastic under the Almighty effort, let us advance and advance on Chaos and the Dark!'° These lofty sentences of Emerson, and a hundred others of like strain, I never have lost out of my memory; I never *can* lose them.

At last I find myself in Emerson's own country, and looking upon Boston Bay. Naturally I revert to the friend of my youth. It is not always pleasant to ask oneself questions about the friends of one's youth; they cannot always well support it. Carlyle, for instance, in my judgment, cannot well support such a return upon him. Yet we should make the return; we should part with our illusions, we should know the truth. When I come to this country, where Emerson now counts for so much and where such high claims are made for him, I pull myself together, and ask myself what the truth about this object of my youthful admiration really is. Improper elements often come into our estimate of men. We have lately seen a German critic° make Goethe the greatest of all poets, because Germany is now the greatest of military powers, and wants a poet to match. Then, too, America is a young country; and young countries, like young persons, are apt sometimes to evince in their literary judgments a want of scale and measure. I set myself, therefore, resolutely to come at a real estimate of Emerson, and with a leaning even to strictness rather than to indulgence. That is the safer course. Time has no indulgence; any veils of illusion which we may have left around an object because we loved it, Time is sure to strip away.

I was reading the other day a notice of Emerson° by a serious and interesting American critic. Fifty or sixty passages in Emerson's poems, says this critic,—who had doubtless himself been nourished on Emerson's writings, and held them justly dear,—fifty or sixty passages from Emerson's poems have already entered into English speech as matter of familiar and universally current quotation. Here is a specimen of that personal sort of estimate° which, for my part, even in speaking of authors dear to me, I would try to avoid. What is the kind of phrase of which we may fairly say that it has entered into English speech as matter of familiar quotation? Such a phrase, surely, as the 'Patience on a monument'° of Shakspeare; as the 'Darkness visible'° of Milton; as the 'Where ignorance is bliss'° of Gray. Of not one single passage in Emerson's poetry can it be truly said that it has become a familiar quotation like phrases of this kind. It is not enough that it should be familiar to his admirers, familiar in New England, familiar even throughout the United States; it must be familiar to all readers and lovers of English poetry. Of not more than one or two passages in Emerson's poetry can it, I think, be truly said, that they stand ever-present in the memory of even many lovers of English poetry. A great number of passages from his poetry are no doubt perfectly familiar to the mind and lips of the critic whom I have mentioned, and perhaps of a wide circle of American readers. But this is a very different thing from being matter of universal quotation, like the phrases of the legitimate poets.

And, in truth, one of the legitimate poets, Emerson, in my opinion, is not. His poetry is interesting, it makes one think; but it is not the poetry of one of the born poets. I say it of him with reluctance, although I am sure that he would have said it of himself; but I say it with reluctance, because I dislike giving pain to his admirers, and because all my own wish, too, is to say of him what is favourable. But I regard myself, not as speaking to please Emerson's admirers, not as speaking to please myself; but rather, I repeat, as communing with Time and Nature concerning the productions of this beautiful and rare spirit, and as resigning what of him is by their unalterable decree touched with caducity, in order the better to mark and secure that in him which is immortal.

Milton says that poetry° ought to be simple, sensuous, impassioned. Well, Emerson's poetry is seldom either simple, or sensuous, or impassioned. In general it lacks directness; it lacks concreteness; it lacks energy. His grammar is often embarrassed; in particular, the want of clearly-marked distinction between the subject and the object of his sentence is a frequent cause of obscurity in him. A poem which shall be a plain, forcible, inevitable whole he hardly ever produces. Such good work as the noble lines graven on the Concord Monument° is the exception with him; such ineffective work as the 'Fourth of July Ode' or the 'Boston Hymn' is the rule. Even passages and single lines of thorough plainness and commanding force are rare in his poetry. They exist, of course; but when we meet with them they give us a slight shock of surprise, so little has Emerson accustomed us to them. Let me have the pleasure of quoting one or two of these exceptional passages:—

> So nigh is grandeur to our dust,
> So near is God to man,
> When Duty whispers low, *Thou must*,
> The youth replies, *I can*.°

Or again this:—

> Though love repine and reason chafe,
> There came a voice without reply:
> ''Tis man's perdition to be safe,
> When for the truth he ought to die.'°

Excellent! but how seldom do we get from him a strain blown so clearly and firmly! Take another passage where his strain has not only clearness, it has also grace and beauty:—

> And ever, when the happy child
> In May beholds the blooming wild,
> And hears in heaven the bluebird sing,
> 'Onward', he cries, 'your baskets bring!
> In the next field is air more mild,
> And o'er yon hazy crest is Eden's balmier spring.'°

In the style and cadence here there is a reminiscence, I think, of Gray; at any rate, the pureness, grace and beauty of these lines are worthy even of Gray. But Gray holds his high rank as a poet, not merely by the beauty and grace of passages in his poems; not merely by a diction generally pure in an age of impure diction: he holds it, above all, by the power and skill with which the evolution of his poems is conducted. Here is his grand superiority to Collins, whose diction in his best poem, the 'Ode to Evening', is purer than Gray's; but then the 'Ode to Evening' is like a river which loses itself in the sand, whereas Gray's best poems have an evolution sure and satisfying. Emerson's 'Mayday', from which I just now quoted, has no real evolution at all; it is a series of observations. And, in general, his poems have no evolution. Take for example his 'Titmouse'. Here he has an excellent subject; and his observation of Nature, moreover, is always marvellously close and fine. But compare what he makes of his meeting with his titmouse with what Cowper or Burns makes of the like kind of incident! One never quite arrives at learning what the titmouse actually did for him at all, though one feels a strong interest and desire to learn it; but one is reduced to guessing, and cannot be quite sure that after all one has guessed right. He is not plain and concrete enough,—in other words, not poet enough,—to be able to tell us. And a failure of this kind goes through almost all his verse, keeps him amid symbolism, and allusion, and the fringes of things, and in spite of his spiritual power deeply impairs his poetic value. Through the inestimable virtue of concreteness, a simple poem like 'The Bridge' of Longfellow or the 'School Days' of Mr Whittier is of more poetic worth, perhaps, than all the verse of Emerson.

I do not, then, place Emerson among the great poets. But I go further, and say that I do not place him among the great writers, the great men of letters. Who are the great men of letters? They are men like Cicero, Plato, Bacon, Pascal, Swift, Voltaire,—writers with, in the first place, a genius and instinct for style; writers whose prose is by a kind of native necessity true and sound. Now the style of Emerson, like the style of his transcendentalist friends° and of the 'Dial' so continually,—the style of Emerson is capable of falling into a strain like this, which I take from the beginning of his 'Essay on Love': 'Every soul is a celestial Venus to every other soul. The heart has its sabbaths and jubilees, in which the world appears as a hymeneal feast, and all natural sounds and the circle of the seasons are erotic odes and dances.' Emerson altered this sentence in the later editions. Like Wordsworth, he was in later life fond of altering; and in general his later alterations, like those of Wordsworth, are not improvements. He softened the passage in question, however, though without really mending it. I quote it in its original and strongly-marked form. Arthur Stanley used to relate° that, about the year 1840, being in conversation with some Americans in quarantine at Malta and thinking to please them he declared his warm admiration for Emerson's 'Essays', then recently published. However, the Americans shook their heads, and told him that for home taste Emerson was decidedly too *greeny*. We will hope, for their sakes, that the sort of thing they had in their heads was such writing as I have just quoted. Unsound it is indeed, and in a style almost impossible to a born man of letters.

It is a curious thing, that quality of style which marks the great writer, the born man of letters. It resides in the whole tissue of his work, and of his work regarded as a composition for literary purposes. Brilliant and powerful passages in a man's writings do not prove his possession of it; it lies in their whole tissue. Emerson has passages of noble and pathetic eloquence, such as those which I quoted at the beginning; he has

passages of shrewd and felicitous wit; he has crisp epigrams; he has
passages of exquisitely touched observation of nature. Yet he is not a
great writer; his style has not the requisite wholeness of good tissue.
Even Carlyle is not, in my judgment, a great writer. He has surpassingly
powerful qualities of expression, far more powerful than Emerson's, and
reminding one of the gifts of expression of the great poets,—of even
Shakspeare himself. What Emerson so admirably says° of Carlyle's
'devouring eyes and portraying hand', 'those thirsty eyes, those portrait-
eating, portrait-painting eyes of thine, those fatal perceptions', is
thoroughly true. What a description is Carlyle's° of the first publisher of
Sartor Resartus, 'to whom the idea of a new edition of *Sartor* is frightful, or
rather ludicrous, unimaginable'; of this poor Fraser, in whose 'wonderful world of Tory pamphleteers, Conservative Younger-brothers,
Regent Street Loungers, Crockford Gamblers, Irish Jesuits, drunken
reporters, and miscellaneous unclean persons (whom nitre and much
soap will not wash clean) not a soul has expressed the smallest wish that
way!' What a portrait, again, of the well-beloved John Sterling!° 'One,
and the best, of a small class extant here, who, nigh drowning in a black
wreck of Infidelity (lighted up by some glare of Radicalism only, now
growing *dim* too), and about to perish, saved themselves into a Col-
eridgian Shovel-Hattedness.' What touches in the invitation of Emerson
to London!° 'You shall see blockheads by the million; Pickwick himself
shall be visible,—innocent young Dickens, reserved for a questionable
fate. The great Wordsworth shall talk till you yourself pronounce him to
be a bore. Southey's complexion is still healthy mahogany brown, with a
fleece of white hair, and eyes that seem running at full gallop. Leigh
Hunt, man of genius in the shape of a cockney, is my near neighbour,
with good humour and no common-sense; old Rogers with his pale head,
white, bare, and cold as snow, with those large blue eyes, cruel,
sorrowful, and that sardonic shelf chin.' How inimitable it all is! And
finally, for one must not go on for ever, this version of a London Sunday,°
with the public-houses closed during the hours of divine service! 'It is
silent Sunday; the populace not yet admitted to their beer-shops, till the
respectabilities conclude their rubric mummeries,—a much more auda-
cious feat than beer.' Yet even Carlyle is not, in my judgment, to be called
a great writer; one cannot think of ranking him with men like Cicero and
Plato and Swift and Voltaire. Emerson freely promises to Carlyle
immortality for his histories.° They will not have it. Why? Because the
materials furnished to him by that devouring eye of his and that
portraying hand were not wrought in and subdued by him to what his
work, regarded as a composition for literary purposes, required. Occur-

ring in conversation, breaking out in familiar correspondence, they are magnificent, inimitable; nothing more is required of them; thus thrown out anyhow, they serve their turn and fulfil their function. And therefore I should not wonder if really Carlyle lived, in the long run, by such an invaluable record as that correspondence between him and Emerson, of which we owe the publication to Mr Charles Norton,°—by this and not by his works, as Johnson lives in Boswell, not by his works.° For Carlyle's sallies, as the staple of a literary work, become wearisome; and as time more and more applies to Carlyle's works its stringent test, this will be felt more and more. Shakspeare, Molière, Swift,—they too had, like Carlyle, the devouring eye and the portraying hand. But they are great literary masters, they are supreme writers, because they knew how to work into a literary composition their materials, and to subdue them to the purposes of literary effect. Carlyle is too wilful for this, too turbid, too vehement.

You will think I deal in nothing but negatives. I have been saying that Emerson is not one of the great poets, the great writers. He has not their quality of style. He is, however, the propounder of a philosophy. The Platonic dialogues afford us the example of exquisite literary form and treatment given to philosophical ideas. Plato is at once a great literary man and a great philosopher. If we speak carefully, we cannot call Aristotle or Spinoza or Kant great literary men, or their productions great literary works. But their work is arranged with such constructive power that they build a philosophy and are justly called great philosophical writers. Emerson cannot, I think, be called with justice a great philosophical writer. He cannot build; his arrangement of philosophical ideas has no progress in it, no evolution; he does not construct a philosophy. Emerson himself knew the defects of his method, or rather want of method, very well; indeed, he and Carlyle criticise themselves and one another in a way which leaves little for any one else to do in the way of formulating their defects. Carlyle formulates perfectly the defects of his friend's poetic and literary production when he says of the 'Dial':° 'For me it is too ethereal, speculative, theoretic; I will have all things condense themselves, take shape and body, if they are to have my sympathy.' And, speaking of Emerson's orations he says:° 'I long to see some concrete Thing, some Event, Man's Life, American Forest, or piece of Creation, which this Emerson loves and wonders at, well *Emersonised*,—depictured by Emerson, filled with the life of Emerson, and cast forth from him, then to live by itself. If these orations balk me of this, how profitable soever they may be for others, I will not love them.' Emerson himself formulates° perfectly the defect of his own philosophical productions when he speaks of his 'formidable tendency to the lapidary style. I build my house of boulders.' 'Here I sit and read and write,' he says again, 'with very little system, and as far as regards composition, with the most fragmentary result; paragraphs incompressible, each sentence an infinitely repellent particle.' Nothing can be truer; and the work of a Spinoza or Kant, of the men who stand as great philosophical writers, does not proceed in this wise.

Some people will tell you that Emerson's poetry indeed is too abstract, and his philosophy too vague, but that his best work is his *English Traits*.° The *English Traits* are, beyond question, very pleasant reading. It is easy to praise them, easy to commend the author of them. But I insist on always trying Emerson's work by the highest standards. I esteem him too much to try his work by any other. Tried by the highest standards, and compared with the work of the excellent markers and recorders of the traits of human life,—of writers like Montaigne, La Bruyère, Addison,— the *English Traits* will not stand the comparison. Emerson's observation has not the disinterested quality of the observation of these masters. It is the observation of a man systematically benevolent, as Hawthorne's observation in *Our Old Home*° is the work of a man chagrined. Hawthorne's literary talent is of the first order. His subjects are generally not to me subjects of the highest interest; but his literary talent is of the first order, the finest, I think, which America has yet produced,—finer, by much, than Emerson's. Yet *Our Old Home* is not a masterpiece any more than *English Traits*. In neither of them is the observer disinterested enough. The author's attitude in each of these cases can easily be understood and defended. Hawthorne was a sensitive man, so situated in England that he was perpetually in contact with the British Philistine; and the British Philistine is a trying personage. Emerson's systematic benevolence comes from what he himself calls somewhere his 'persistent optimism'; and his persistent optimism is the root of his greatness and the source of his charm. But still let us keep our literary conscience true, and judge every kind of literary work by the laws really proper to it. The kind of work attempted in the *English Traits* and in *Our Old Home* is work which cannot be done perfectly with a bias such as that given by Emerson's optimism or by Hawthorne's chagrin. Consequently, neither *English Traits* nor *Our Old Home* is a work of perfection in its kind.

Not with the Miltons and Grays, not with the Platos and Spinozas, not with the Swifts and Voltaires, not with the Montaignes and Addisons, can we rank Emerson. His work of various kinds, when one compares it with the work done in a corresponding kind by these masters, fails to stand the comparison. No man could see this clearer than Emerson himself. It is hard not to feel despondency when we contemplate our failures and shortcomings: and Emerson, the least self-flattering and the most modest of men, saw so plainly what was lacking to him that he had his moments of despondency. 'Alas, my friend,' he writes in reply to Carlyle,° who had exhorted him to creative work,—'Alas, my friend, I can do no such gay thing as you say. I do not belong to the poets, but only to a low department of literature,—the reporters; suburban men.' He deprecates° his friend's praise; praise 'generous to a fault', he calls it; praise 'generous to the shaming of me,—cold, fastidious, ebbing person that I am. Already in a former letter you had said too much good of my poor little arid book, which is as sand to my eyes. I can only say that I heartily wish the book were better; and I must try and deserve so much favour from the kind gods by a bolder and truer living in the months to come,—such as may perchance one day relax and invigorate this cramp hand of mine. When I see how much work is to be done; what room for a poet, for any spiritualist, in this great, intelligent, sensual and avaricious America,—I lament my fumbling fingers and stammering tongue.' Again, as late as 1870, he writes to Carlyle:° 'There is no example of constancy like yours, and it always stings my stupor into temporary recovery and wonderful resolution to accept the noble challenge. But "the strong hours conquer us"; and I am the victim of miscellany,— miscellany of designs, vast debility, and procrastination.' The forlorn note belonging to the phrase 'vast debility', recalls that saddest and most discouraged of writers, the author of *Obermann*, Senancour, with whom Emerson has in truth a certain kinship. He has in common with Senancour his pureness, his passion for nature, his single eye; and here we find him confessing, like Senancour, a sense in himself of sterility and impotence.

And now I think I have cleared the ground. I have given up to envious Time° as much of Emerson as Time can fairly expect ever to obtain. We have not in Emerson a great poet, a great writer, a great philosophy-maker. His relation to us is not that of one of those personages; yet it is a relation of, I think, even superior importance. His relation to us is more like that of the Roman Emperor Marcus Aurelius. Marcus Aurelius is not a great writer, a great philosophy-maker; he is the friend and aider of those who would live in the spirit.° Emerson is the same. He is the friend and aider of those who would live in the spirit. All the points in thinking which are necessary for this purpose he takes; but he does not combine them into a system, or present them as a regular philosophy. Combined in a system by a man with the requisite talent for this kind of thing, they would be less useful than as Emerson gives them to us; and the man with the talent so to systematise them would be less impressive than Emerson. They do very well as they now stand;—like 'boulders', as he says;—in 'paragraphs incompressible, each sentence an infinitely repellent particle'. In such sentences his main points recur again and again, and become fixed in the memory.

We all know them. First and foremost, character. Character is everything.° 'That which all things tend to educe,—which freedom, cultivation, intercourse, revolutions, go to form and deliver,—is character.' Character and self-reliance. 'Trust thyself! every heart vibrates to that iron string.'° And yet we have our being in a *not ourselves*.° 'There is a power above and behind us, and we are the channels of its communications.'° But our lives must be pitched higher. 'Life must be lived on a higher plane; we must go up to a higher platform, to which we are always invited to ascend; there the whole scene changes.'° The good we need is for ever close to us, though we attain it not. 'On the brink of the waters of life and truth, we are miserably dying.'° This good is close to us, moreover, in our daily life, and in the familiar, homely places. 'The unremitting retention of simple and high sentiments in obscure duties',°—that is the maxim for us. 'Let us be poised and wise, and our own to-day. Let us treat the men and women well,—treat them as if they were real; perhaps they are. Men live in their fancy, like drunkards whose hands are too soft and tremulous for successful labour. I settle myself ever firmer in the creed, that we should not postpone and refer and wish, but do broad justice where we are, by whomsoever we deal with; accepting our actual companions and circumstances, however humble or odious, as the mystic officials to whom the universe has delegated its whole pleasure for us.'° 'Massachusetts, Connecticut River, and Boston Bay, you think paltry places, and the ear loves names of foreign and classic topography. But here we are; and if we will tarry a little we may come to learn that here is best. See to it only that thyself is here.'° Furthermore, the good is close to us *all*. 'I resist the scepticism of our education and of our educated men. I do not believe that the differences of opinion and character in men are organic. I do not recognise, beside the class of the good and the wise, a permanent class of sceptics, or a class of conservatives, or of malignants, or of materialists. I do not believe in two classes.'° 'Every man has a call of the power to do something unique.'° Exclusiveness is deadly. 'The exclusive in social life does not see that he excludes himself from enjoyment in the attempt to appropriate it. The exclusionist in religion does not see that he shuts the door of heaven on himself in striving to shut out others. Treat men as pawns and ninepins, and you shall suffer as well as they. If you leave out their heart you shall lose your own.'° 'The selfish man suffers more from his selfishness than he from whom that selfishness withholds some important benefit.'° A sound nature will be inclined to refuse ease and self-indulgence. 'To live with some rigour of temperance, or some extremes of generosity, seems to be an asceticism which common good-nature would appoint to those who are at ease and in plenty, in sign that they feel a brotherhood with the great multitude of suffering men.'° Compensation, finally, is the great law of life; it is everywhere, it is sure, and there is no escape from it. This is that 'law alive and beautiful, which works over our heads and under our feet. Pitiless, it avails itself of our success when we obey it, and of our ruin when we contravene it. Men are all secret believers in it. It rewards actions after their nature. The reward of a thing well done is to have done it.'° 'The thief steals from himself, the swindler swindles himself. You must pay at last your own debt.'°

This is tonic indeed! And let no one object that it is too general; that more practical, positive direction is what we want; that Emerson's optimism, self-reliance, and indifference to favourable conditions for our life and growth, have in them something of danger. 'Trust thyself'; 'what attracts my attention shall have it'°; 'though thou shouldst walk the world over, thou shalt not be able to find a condition inopportune or ignoble';° 'what we call vulgar society is that society whose poetry is not yet written, but which you shall presently make as enviable and renowned as any'.° With maxims like these, we surely, it may be said, run some risk of being made too well satisfied with our own actual self and state, however crude and imperfect they may be. 'Trust thyself'? It may be said that the common American or Englishman is more than enough disposed already to trust himself. I often reply, when our sectarians are praised for following conscience: Our people are very good in following their conscience; where they are not so good is in ascertaining whether their conscience tells them right. 'What attracts my attention shall have it'? Well, that is our people's plea when they run after the Salvation Army, and desire Messrs Moody and Sankey.° 'Thou shalt not be able to find a condition inopportune or ignoble'? But think of the turn of the good people of our race for producing a life of hideousness and immense ennui; think of that specimen of your own New England life which Mr Howells° gives us in one of his charming stories which I was reading lately; think of the life of that ragged New England farm in the *Lady of the Aroostook*; think of Deacon Blood, and Aunt Maria, and the straight-backed chairs with black horse-hair seats, and Ezra Perkins with perfect self-reliance depositing his travellers in the snow! I can truly say that in

the little which I have seen of the life of New England, I am more struck
with what has been achieved than with the crudeness and failure. But no
doubt there is still a great deal of crudeness also. Your own novelists say
there is, and I suppose they say true. In the New England, as in the Old,
our people have to learn, I suppose, not that their modes of life are
beautiful and excellent already; they have rather to learn that they must
transform them.

To adopt this line of objection to Emerson's deliverances would,
however, be unjust. In the first place, Emerson's points are in themselves
true, if understood in a certain high sense; they are true and fruitful. And
the right work to be done, at the hour when he appeared, was to affirm
them generally and absolutely. Only thus could he break through the
hard and fast barrier of narrow, fixed ideas which he found confronting
him, and win an entrance for new ideas. Had he attempted developments
which may now strike us as expedient, he would have excited fierce
antagonism, and probably effected little or nothing. The time might
come for doing other work later, but the work which Emerson did was
the right work to be done then.

In the second place, strong as was Emerson's optimism, and
unconquerable as was his belief in a good result to emerge from all which
he saw going on around him, no misanthropical satirist ever saw
shortcomings and absurdities more clearly than he did, or exposed them
more courageously. When he sees 'the meanness', as he calls it, 'of
American politics', he congratulates Washington on being 'long already
happily dead', on being 'wrapt in his shroud and for ever safe'.° With how
firm a touch he delineates the faults of your two great political parties of
forty years ago! The Democrats,° he says, 'have not at heart the ends
which give to the name of democracy what hope and virtue are in it. The
spirit of our American radicalism is destructive and aimless; it is not
loving; it has no ulterior and divine ends, but is destructive only out of
hatred and selfishness. On the other side, the conservative party,
composed of the most moderate, able, and cultivated part of the
population, is timid, and merely defensive of property. It vindicates no
right, it aspires to no real good, it brands no crime, it proposes no
generous policy. From neither party, when in power, has the world any
benefit to expect in science, art, or humanity, at all commensurate with
the resources of the nation.' Then with what subtle though kindly irony
he follows the gradual withdrawal in New England,° in the last half
century, of tender consciences from the social organisations,—the bent
for experiments such as that of Brook Farm and the like,—follows it in all
its 'dissidence of dissent and Protestantism of the Protestant religion'!

He even loves to rally the New Englander on his philanthropical activity, and to find his beneficence and its institutions a bore! 'Your miscellaneous popular charities, the education at college of fools, the building of meeting-houses to the vain end to which many of these now stand, alms to sots, and the thousandfold relief societies,—though I confess with shame that I sometimes succumb and give the dollar, yet it is a wicked dollar which by and by I shall have the manhood to withhold.'° 'Our Sunday schools and churches and pauper societies are yokes to the neck. We pain ourselves to please nobody. There are natural ways of arriving at the same ends at which these aim, but do not arrive.'° 'Nature does not like our benevolence or our learning much better than she likes our frauds and wars. When we come out of the caucus, or the bank, or the Abolition Convention, or the Temperance meeting, or the Transcendental Club, into the fields and woods, she says to us: "So hot, my little Sir?" '°

Yes, truly, his insight is admirable; his truth is precious. Yet the secret of his effect is not even in these; it is in his temper. It is in the hopeful, serene, beautiful temper wherewith these, in Emerson, are indissolubly joined; in which they work, and have their being.° He says himself: 'We judge of a man's wisdom by his hope, knowing that the perception of the inexhaustibleness of nature is an immortal youth.'° If this be so, how wise is Emerson! for never had man such a sense of the inexhaustibleness of nature, and such hope. It was the ground of his being; it never failed him. Even when he is sadly avowing the imperfection of his literary power and resources, lamenting his fumbling fingers and stammering tongue, he adds: 'Yet, as I tell you, I am very easy in my mind and never dream of suicide. My whole philosophy, which is very real, teaches acquiescence and optimism. Sure I am that the right word will be spoken though I cut out my tongue.'° In his old age, with friends dying and life failing, his tone of cheerful, forward-looking hope is still the same: 'A multitude of young men are growing up here of high promise, and I compare gladly the social poverty of my youth with the power on which these draw.'° His abiding word for us, the word by which being dead he yet speaks to us, is this: 'That which befits us, embosomed in beauty and wonder as we are, is cheerfulness and courage, and the endeavour to realize our aspirations. Shall not the heart, which has received so much, trust the Power by which it lives?'°

One can scarcely overrate the importance of thus holding fast to happiness and hope. It gives to Emerson's work an invaluable virtue. As Wordsworth's poetry is, in my judgment, the most important work done in verse, in our language, during the present century, so Emerson's

Essays are, I think, the most important work done in prose. His work is more important than Carlyle's. Let us be just to Carlyle, provoking though he often is. Not only has he that genius of his which makes Emerson say truly of his letters, that 'they savour always of eternity'.° More than this may be said of him. The scope and upshot of his teaching are true; 'his guiding genius', to quote Emerson again, is really 'his moral sense, his perception of the sole importance of truth and justice'.° But consider Carlyle's temper, as we have been considering Emerson's; take his own account of it:° 'Perhaps London is the proper place for me after all, seeing all places are *im*proper: who knows? Meanwhile, I lead a most dyspeptic, solitary, self-shrouded life; consuming, if possible in silence, my considerable daily allotment of pain; glad when any strength is left in me for working, which is the only use I can see in myself,—too rare a case of late. The ground of my existence is black as death; too black, when all *void* too; but at times there paint themselves on it pictures of gold, and rainbow, and lightning; all the brighter for the black ground, I suppose. Withal, I am very much of a fool.'—No, not a fool, but turbid and morbid, wilful and perverse. 'We judge of a man's wisdom by his hope.'

Carlyle's perverse attitude towards happiness cuts him off from hope. He fiercely attacks the desire for happiness; his grand point in *Sartor*,° his secret in which the soul may find rest, is that one shall cease to desire happiness, that one should learn to say to oneself: 'What if thou wert born and predestined not to be happy, but to be unhappy!' He is wrong; Saint Augustine is the better philosopher, who says: 'Act we *must* in pursuance of what gives us most delight.' Epictetus° and Augustine can be severe moralists enough; but both of them know and frankly say that the desire for happiness is the root and ground of man's being. Tell him and show him that he places his happiness wrong, that he seeks for delight where delight will never be really found; then you illumine and further him. But you only confuse him by telling him to cease to desire happiness, and you will not tell him this unless you are already confused yourself.

Carlyle preached the dignity of labour, the necessity of righteousness, the love of veracity, the hatred of shams. He is said by many people to be a great teacher, a great helper for us, because he does so. But what is the due and eternal result of labour, righteousness, veracity?—Happiness. And how are we drawn to them by one who, instead of making us feel that with them is happiness, tells us that perhaps we were predestined not to be happy but to be unhappy?

You will find, in especial, many earnest preachers of our popular religion to be fervent in their praise and admiration of Carlyle. His insistence on labour, righteousness, and veracity pleases them; his contempt for happiness pleases them too. I read the other day a tract against smoking, although I do not happen to be a smoker myself. 'Smoking', said the tract, 'is liked because it gives agreeable sensations. Now it is a positive objection to a thing that it gives agreeable sensations. An earnest man will expressly avoid what gives agreeable sensations.' Shortly afterwards I was inspecting a school, and I found the children reading a piece of poetry° on the common theme that we are here to-day and gone to-morrow. I shall soon be gone, the speaker in this poem was made to say,—

> And I shall be glad to go,
> For the world at best is a weary place,
> And my pulse is getting low.

How usual a language of popular religion that is, on our side of the Atlantic at any rate! But then our popular religion, in disparaging happiness here below, knows very well what it is after. It has its eye on a happiness in a future life above the clouds, in the New Jerusalem, to be won by disliking and rejecting happiness here on earth. And so long as this ideal stands fast, it is very well. But for very many it now stands fast no longer; for Carlyle, at any rate, it had failed and vanished. Happiness in labour, righteousness and veracity,—in the life of the spirit,—here was a gospel still for Carlyle to preach, and to help others by preaching. But he baffled them and himself by preferring the paradox that we are not born for happiness at all.

Happiness in labour, righteousness and veracity; in all the life of the spirit; happiness and eternal hope;—that was Emerson's gospel. I hear it said that Emerson was too sanguine; that the actual generation in America is not turning out so well as he expected. Very likely he was too sanguine as to the near future; in this country it is difficult not to be too sanguine. Very possibly the present generation may prove unworthy of his high hopes; even several generations succeeding this may prove unworthy of them. But by his conviction that in the life of the spirit is happiness, and by his hope that this life of the spirit will come more and more to be sanely understood, and to prevail, and to work for happiness,—by this conviction and hope Emerson was great, and he will surely prove in the end to have been right in them. In this country it is difficult, as I said, not to be sanguine. Very many of your writers are over-sanguine, and on the wrong grounds. But you have two men who in what they have written show their sanguineness in a line where courage and hope are just, where they are also infinitely important, but where they are not easy. The two men are Franklin and Emerson.[1] These two are, I think, the most distinctively and honourably American of your writers; they are the most original and the most valuable. Wise men everywhere know that we must keep up our courage and hope; they know that hope is, as Wordsworth well says,—°

> The paramount *duty* which Heaven lays,
> For its own honour, on man's suffering heart.

But the very word *duty* points to an effort and a struggle to maintain our hope unbroken. Franklin and Emerson maintained theirs with a convincing ease, an inspiring joy. Franklin's confidence in the happiness with which industry, honesty, and economy will crown the life of this work-day world, is such that he runs over with felicity. With a like felicity does Emerson run over, when he contemplates the happiness eternally attached to the true life in the spirit. You cannot prize him too much, nor heed him too diligently. He has lessons for both the branches of our race. I figure him to my mind as visible upon earth still, as still standing here by Boston Bay, or at his own Concord, in his habit as he lived,° but of heightened stature and shining feature, with one hand stretched out towards the East, to our laden and labouring England; the other towards the ever-growing West, to his own dearly-loved America,—'great, intelligent, sensual, avaricious America'.° To us he shows for guidance his lucid freedom, his cheerfulness and hope; to you his dignity, delicacy, serenity, elevation.

---

[1] I found with pleasure that this conjunction of Emerson's name with Franklin's had already occurred to an accomplished writer and delightful man, a friend of Emerson, left almost the sole survivor, alas! of the famous literary generation of Boston,—Dr Oliver Wendell Holmes. Dr Holmes has kindly allowed me to print here the ingenious and interesting lines,° hitherto unpublished, in which he speaks of Emerson thus:—

> Where in the realm of thought, whose air is song,
> Does he, the Buddha of the West, belong?
> He seems a wingéd Franklin, sweetly wise,
> Born to unlock the secrets of the skies;
> And which the nobler calling—if 'tis fair
> Terrestrial with celestial to compare—
> To guide the storm-cloud's elemental flame,
> Or walk the chambers whence the lightning came
> Amidst the sources of its subtile fire,
> And steal their effluence for his lips and lyre?

Civilisation in the United States

Two or three years ago I spoke in this Review° on the subject of America; and after considering the institutions and the social condition of the people of the United States, I said that what, in the jargon of the present day, is called 'the political and social problem', does seem to be solved there with remarkable success. I pointed out the contrast which in this respect the United States offer to our own country, a contrast, in several ways, much to their advantage. But I added that the solution of the political and social problem, as it is called, ought not so to absorb us as to make us forget the human problem; and that it remained to ask how the human problem is solved in the United States. It happened that Sir Lepel Griffin, a very acute and distinguished Indian official, had just then been travelling in the United States, and had published his opinion, from what he saw of the life there, that there is no country calling itself civilised where one would not rather live than in America, except Russia. Certainly then, I said, one cannot rest satisfied, when one finds such a judgment passed on the United States as this, with admiring their institutions and their solid social condition, their freedom and equality, their power, energy, and wealth. One must, further, go on to examine what is done there towards solving the human problem, and must see what Sir Lepel Griffin's objection comes to.

And this examination I promised that I would one day make. However, it is so delicate a matter to discuss how a sensitive nation solves the human problem, that I found myself inclined to follow the example of the Greek moralist Theophrastus,° who waited, before composing his famous *Characters*, until he was ninety-nine years old. I thought I had perhaps better wait until I was about that age, before I discussed the success of the Americans in solving the human problem. But ninety-nine is a great age; it is probable that I may never reach it,° or even come near it. So I have determined, finally, to face the question without any such long delay, and thus I come to offer to the readers of this Review the remarks following. With the same frankness with which I discussed here the solution of the political and social problem by the people of the United States, I shall discuss their success in solving the human problem.

Perhaps it is not likely that any one will now remember what I said three years ago here about the success of the Americans in solving the political and social problem. I will sum it up in the briefest possible manner. I said that the United States had constituted themselves in a modern age; that their institutions complied well with the form and pressure of those circumstances and conditions which a modern age presents. Quite apart from all question how much of the merit for this may be due to the wisdom and virtue of the American people, and how much to their good fortune, it is undeniable that their institutions do work well and happily. The play of their institutions suggests, I said, the image of a man in a suit of clothes which fits him to perfection, leaving all his movements unimpeded and easy; a suit of clothes loose where it ought to be loose, and sitting close where its sitting close is an advantage; a suit of clothes able, moreover, to adapt itself naturally to the wearer's growth, and to admit of all enlargements as they successively arise.

So much as to the solution, by the United States, of the political problem. As to the social problem, I observed that the people of the United States were a community singularly free from the distinction of classes, singularly homogeneous; that the division between rich and poor was consequently less profound there than in countries where the distinction of classes accentuates that division. I added that I believed there was exaggeration in the reports of their administrative and judicial corruption; and altogether, I concluded, the United States, politically and socially, are a country living prosperously in a natural modern condition, and conscious of living prosperously in such a condition. And being in this healthy case, and having this healthy consciousness, the community there uses its understanding with the soundness of health; it in general, as to its own political and social concerns, sees clear and thinks straight. Comparing the United States with ourselves, I said that while they are in this natural and healthy condition, we on the contrary are so little homogeneous, we are living with a system of classes so intense, with institutions and a society so little modern, so unnaturally complicated, that the whole action of our minds is hampered and falsened by it; we are in consequence wanting in lucidity, we do not see clear or think straight, and the Americans have here much the advantage of us.

Yet we find an acute and experienced Englishman saying that there is no country, calling itself civilised, where one would not rather live than in the United States, except Russia! The civilisation of the United States must somehow, if an able man can think thus, have shortcomings, in spite of the country's success and prosperity. What is civilisation? It is the humanisation of man in society,° the satisfaction for him, in society, of the true law of human nature. Man's study, says Plato,° is to discover the right answer to the question *how to live?* our aim, he says, is very and true life. We are more or less civilised as we come more or less near to this aim, in that social state which the pursuit of our aim essentially demands. But several elements or powers, as I have often insisted,° go to build up a complete human life. There is the power of conduct, the power of intellect and knowledge, the power of beauty, the power of social life and manners; we have instincts responding to them all, requiring them all. And we are perfectly civilised only when all these instincts in our nature, all these elements in our civilisation, have been adequately recognised and satisfied. But of course this adequate recognition and satisfaction of all the elements in question is impossible; some of them are recognised more than others, some of them more in one community, some in another; and the satisfactions found are more or less worthy.

And meanwhile, people use the term *civilisation* in the loosest possible way, for the most part attaching to it, however, in their own mind some meaning connected with their own preferences and experiences. The most common meaning thus attached to it is perhaps that of a satisfaction, not of all the main demands of human nature, but of the demand for the comforts and conveniences of life, and of this demand as made by the sort of person who uses the term.

Now we should always attend to the common and prevalent use of an important term. Probably Sir Lepel Griffin had this notion of the comforts and conveniences of life much in his thoughts when he reproached American civilisation with its shortcomings. For men of his kind, and for all that large number of men, so prominent in this country and who make their voice so much heard, men who have been at the public schools and universities, men of the professional and official class, men who do the most part of our literature and our journalism, America is not a comfortable place of abode. A man of this sort has in England everything in his favour; society appears organised expressly for his advantage. A Rothschild or a Vanderbilt can buy his way anywhere, and can have what comforts and luxuries he likes whether in America or in England. But it is in England that an income of from three or four to fourteen or fifteen hundred° a year does so much for its possessor, enables him to live with so many of the conveniences of far richer people. For his benefit, his benefit above all, clubs are organised and hansom cabs ply; service is abundant, porters stand waiting at the railway stations. In America all luxuries are dear except oysters and ice; service is in general scarce and bad; a club is a most expensive luxury; the cab-rates are prohibitive—more than half of the people who in England would use cabs must in America use the horse-cars,° the tram. The charges of tailors and mercers are about a third higher than they are with us. I mention only a few striking points as to which there can be no dispute, and in which a man of Sir Lepel Griffin's class would feel the great difference between America and England in the conveniences at his command. There are a hundred other points one might mention, where he would feel the same thing. When a man is passing judgment on a country's civilisation, points of this kind crowd to his memory, and determine his sentence.

On the other hand, for that immense class of people, the great bulk of the community, the class of people whose income is less than three or four hundred a year, things in America are favourable. It is easier for them there than in the Old World to rise and to make their fortune; but I am not now speaking of that. Even without making their fortune, even with their income below three or four hundred a year, things are favourable to them in America, society seems organised there for their benefit. To begin with, the humbler kind of work is better paid in America than with us, the higher kind worse. The official, for instance, gets less, his office-keeper gets more. The public ways are abominably cut up by rails and blocked with horse-cars; but the inconvenience is for those who use private carriages and cabs, the convenience is for the bulk of the community who but for the horse-cars would have to walk. The ordinary railway cars are not delightful, but they are cheap, and they are better furnished and in winter are warmer than third-class carriages in England. Luxuries are, as I have said, very dear—above all, European luxuries; but a working man's clothing is nearly as cheap as in England, and plain food is on the whole cheaper. Even luxuries of a certain kind are within a labouring man's easy reach. I have mentioned ice, I will mention fruit also. The abundance and cheapness of fruit is a great boon to people of small incomes in America. Do not believe the Americans when they extol their peaches as equal to any in the world, or better than any in the world; they are not to be compared to peaches grown under glass. Do not believe that the American Newtown pippins appear in the New York and Boston fruit-shops as they appear in those of London and Liverpool; or that the Americans have any pear to give you like the Marie Louise. But what labourer, or artisan, or small clerk, ever gets hot-house peaches, or Newtown pippins, or Marie Louise pears? Not such good pears, apples, and peaches as those, but pears, apples, and peaches by no means to be despised, such people and their families do in America get in plenty.

Well, now, what would a philosopher or a philanthropist say in this case? which would he say was the more civilised condition—that of the country where the balance of advantage, as to the comforts and conveniences of life, is greatly in favour of the people with incomes below three hundred a year, or that of the country where it is greatly in favour of those with incomes above that sum?

Many people will be ready to give an answer to that question without the smallest hesitation. They will say that they are, and that all of us ought to be, for the greatest happiness of the greatest number.° However, the question is not one which I feel bound now to discuss and answer. Of course, if happiness and civilisation consist in being plentifully supplied with the comforts and conveniences of life, the question presents little difficulty. But I believe neither that happiness consists, merely or mainly, in being plentifully supplied with the comforts and conveniences of life, nor that civilisation consists in being so supplied; therefore I leave the question unanswered.

I prefer to seek for some other and better tests by which to try the civilisation of the United States. I have often insisted on the need of more equality in our own country, and on the mischiefs caused by inequality over here. In the United States there is not our intense division of classes, our inequality; there is great equality. Let me mention two points in the system of social life and manners over there in which this equality seems to me to have done good. The first is a mere point of form, but it has its significance. Every one knows it is the established habit with us in England, if we write to people supposed to belong to the class of gentlemen, of addressing them by the title of *Esquire*, while we keep *Mr* for people not supposed to belong to that class. If we think of it, could one easily find a habit more ridiculous, more offensive? The title of *Esquire*, like most of our titles, comes out of the great frippery shop of the Middle Age; it is alien to the sound taste and manner of antiquity, when men said *Pericles* and *Camillus*.° But unlike other titles, it is applied or withheld quite arbitrarily. Surely, where a man has no specific title proper to him, the one plain title of *Master* or *Mr* is enough, and we need not be encumbered with a second title of *Esquire*, now quite unmeaning, to draw an invidious and impossible line of distinction between those who are gentlemen and those who are not; as if we actually wished to provide a source of embarrassment for the sender of a letter, and of mortification for the receiver of it.

The French, those great authorities in social life and manners, find *Mr* enough, and the Americans are more and more, I am glad to say, following the French example. I only hope they will persevere, and not be seduced by *Esquire* being 'so English, you know'. And I do hope, moreover, that we shall one day take the same course and drop our absurd *Esquire*.°

The other point goes deeper. Much may be said against the voices and intonation of American women. But almost every one acknowledges that there is a charm in American women—a charm which you find in almost all of them, wherever you go. It is the charm of a natural manner, a manner not self-conscious, artificial, and constrained. It may not be a beautiful manner always, but it is almost always a natural manner, a free and happy manner; and this gives pleasure. Here we have, undoubtedly, a note of civilisation, and an evidence, at the same time, of the good effect of equality upon social life and manners. I have often heard it observed that a perfectly natural manner is as rare among Englishwomen of the middle classes as it is general among American women of like condition with them. And so far as the observation is true, the reason of its truth no doubt is, that the Englishwoman is living in presence of an upper class, as it is called—in presence, that is, of a class of women recognised as being the right thing in style and manner, and whom she imagines criticising *her* style and manner, finding this or that to be amiss with it, this or that to be vulgar. Hence self-consciousness and constraint in her. The American woman lives in presence of no such class; there may be circles trying to pass themselves off as such a class, giving themselves airs as such, but they command no recognition, no authority. The American woman in general is perfectly unconcerned about their opinion, is herself, enjoys her existence, and has consequently a manner happy and natural. It is her great charm; and it is moreover, as I have said, a real note of civilisation, and one which has to be reckoned to the credit of American life, and of its equality.

But we must get nearer still to the heart of the question raised as to the character and worth of American civilisation. I have said how much the word civilisation really means—the humanisation of man in society; his making progress there towards his true and full humanity. Partial and material achievement is always being put forward as civilisation. We hear a nation called highly civilised by reason of its industry, commerce, and wealth, or by reason of its liberty or equality, or by reason of its numerous churches, schools, libraries, and newspapers. But there is something in human nature, some instinct of growth, some law of perfection, which rebels against this narrow account of the matter. And perhaps what human nature demands in civilisation, over and above all those obvious things which first occur to our thoughts—what human nature, I say, demands in civilisation, if it is to stand as a high and satisfying civilisation, is best described by the word *interesting*. Here is the extraordinary charm of the old Greek civilisation—that it is so *interesting*. Do not tell me only, says human nature, of the magnitude of your industry and commerce; of the beneficence of your institutions, your freedom, your equality; of the great and growing number of your churches and schools, libraries and newspapers; tell me also if your civilisation—which is the grand name you give to all this development—tell me if your civilisation is *interesting*.

An American friend of mine, Professor Norton,° has lately published the early letters of Carlyle. If any one wants a good antidote to the unpleasant effect left by Mr Froude's *Life of Carlyle*, let him read those letters. Not only of Carlyle will those letters make him think kindly, but they will also fill him with admiring esteem for the qualities, character, and family life, as there delineated, of the Scottish peasant. Well, the Carlyle family were numerous, poor, and struggling. Thomas Carlyle, the eldest son, a young man in wretched health and worse spirits, was fighting his way in Edinburgh. One of his younger brothers talked of emigrating. 'The very best thing he could do!' we should all say. Carlyle dissuades him.° 'You shall never,' he writes, 'you shall never seriously meditate crossing the great Salt Pool to plant yourself in the Yankee-land. That is a miserable fate for any one, at best; never dream of it. Could you banish yourself from all that is interesting to your mind, forget the history, the glorious institutions, the noble principles of old Scotland—that you might eat a better dinner, perhaps?'

There is our word launched—the word *interesting*. I am not saying that Carlyle's advice was good, or that young men should not emigrate. I do but take note, in the word *interesting*, of a requirement, a cry of aspiration, a cry not sounding in the imaginative Carlyle's own breast only, but sure of a response in his brother's breast also, and in human nature.

Amiel,° that contemplative Swiss whose journals the world has been reading lately, tells us that 'the human heart is, as it were, haunted by confused reminiscences of an age of gold; or rather, by aspirations towards a harmony of things which every-day reality denies to us'. He says that the splendour and refinement of high life is an attempt by the rich and cultivated classes to realise this ideal, and is 'a form of poetry'. And the interest which this attempt awakens in the classes which are not rich or cultivated, their indestructible interest in the pageant and fairy tale, as to them it appears, of the life in castles and palaces, the life of the great, bears witness to a like imaginative strain in them also, a strain tending after the elevated and the beautiful. In short, what Goethe describes as 'was uns alle bändigt, *das Gemeine*°—that which holds us all in bondage, the common and ignoble', is, notwithstanding its admitted prevalence, contrary to a deep-seated instinct of human nature and repelled by it. Of civilisation, which is to humanise us in society, we demand, before we will consent to be satisfied with it—we demand, however much else it may give us, that it shall give us, too, the *interesting.*

Now, the great sources of the *interesting* are distinction and beauty: that which is elevated, and that which is beautiful. Let us take the beautiful first, and consider how far it is present in American civilisation. Evidently this is that civilisation's weak side. There is little to nourish and delight the sense of beauty there. In the long-settled States east of the Alleghanies the landscape in general is not interesting, the climate harsh and in extremes. The Americans are restless, eager to better themselves and to make fortunes; the inhabitant does not strike his roots lovingly down into the soil, as in rural England. In the valley of the Connecticut you will find farm after farm which the Yankee settler has abandoned in order to go West, leaving the farm to some new Irish immigrant. The charm of beauty which comes from ancientness and permanence of rural life the country could not yet have in a high degree, but it has it in an even less degree than might be expected. Then the Americans come originally, for the most part, from that great class in English society° amongst whom the sense for conduct and business is much more strongly developed than the sense for beauty. If we in England were without the cathedrals, parish churches, and castles of the catholic and feudal age, and without the houses of the Elizabethan age, but had only the towns and buildings which the rise of our middle class has created in the modern age, we should be in much the same case as the Americans. We should be living with much the same absence of training for the sense of beauty through the eye, from the aspect of outward things. The American cities have hardly anything to please a trained or a natural sense for beauty. They have buildings which cost a great deal of money and produce a certain effect—buildings, shall I say, such as our Midland Station at St Pancras; but nothing such as Somerset House° or Whitehall. One architect of genius they had—Richardson.° I had the pleasure to know him; he is dead, alas! Much of his work was injured by the conditions under which he was obliged to execute it; I can recall but one building, and that of no great importance, where he seems to have had his own way, to be fully himself; but that is indeed excellent. In general, where the Americans succeed best in their architecture—in that art so indicative and educative of a people's sense for beauty—is in the fashion of their villa-cottages in wood. These are often original and at the same time very pleasing, but they are pretty and coquettish, not beautiful. Of the really beautiful in the other arts, and in literature, very little has been produced there as yet. I asked a German portrait-painter, whom I found painting and prospering in America, how he liked the country? 'How *can* an artist like it?' was his answer. The American artists live chiefly in Europe; all Americans of cultivation and wealth visit Europe more and more constantly. The mere nomenclature of the country acts upon a cultivated person like the incessant pricking of pins. What people in whom the sense for beauty and fitness was quick could have invented, or could tolerate, the hideous names ending in *ville*, the Briggsvilles, Higginsvilles, Jacksonvilles,° rife from Maine to Florida; the jumble of unnatural and inappropriate names everywhere? On the line from Albany to Buffalo you have, in one part, half the names in the classical dictionary° to designate the stations; it is said that the folly is due to a surveyor who, when the country was laid out, happened to possess a classical dictionary; but a people with any artist-sense would have put down that surveyor. The Americans meekly retain his names; and indeed his strange Marcellus or Syracuse is perhaps not much worse than their congenital Briggsville.

So much as to beauty, and as to the provision, in the United States, for the sense of beauty. As to distinction, and the interest which human nature seeks from enjoying the effect made upon it by what is elevated, the case is much the same. There is very little to create such an effect, very much to thwart it. Goethe says somewhere that 'the thrill of awe is the best thing humanity has':—

Das Schaudern ist der Menschheit bestes Theil.°

But, if there be a discipline in which the Americans are wanting, it is the discipline of awe and respect. An austere and intense religion imposed on their Puritan founders the discipline of respect, and so provided for them the thrill of awe; but this religion is dying out. The Americans have produced plenty of men strong, shrewd, upright, able, effective; very few who are highly distinguished. Alexander Hamilton is indeed a man of rare distinction; Washington, though he has not the high mental distinction of Pericles or Caesar, has true distinction of style and character. But these men belong to the pre-American age. Lincoln's recent American biographers declare that Washington is but an Englishman, an English officer; the typical American, they say, is Abraham Lincoln. Now Lincoln is shrewd, sagacious, humorous, honest, courageous, firm; he is a man with qualities deserving the most sincere esteem and praise, but he has not distinction.

In truth everything is against distinction in America, and against the sense of elevation to be gained through admiring and respecting it. The glorification of 'the average man', who is quite a religion with statesmen and publicists there, is against it. The addiction to 'the funny man', who is a national misfortune there, is against it. Above all, the newspapers are against it.

It is often said that every nation has the government it deserves. What is much more certain is that every nation has the newspapers it deserves. The newspaper is the direct product of the want felt; the supply answers closely and inevitably to the demand. I suppose no one knows what the American newspapers are, who has not been obliged, for some length of time, to read either those newspapers or none at all. Powerful and valuable contributions occur scattered about in them. But on the whole, and taking the total impression and effect made by them, I should say that if one were searching for the best means to efface and kill in a whole nation the discipline of respect, the feeling for what is elevated, one could not do better than take the American newspapers. The absence of truth and soberness in them, the poverty in serious interest, the personality and sensation-mongering, are beyond belief. There are a few newspapers which are in whole, or in part, exceptions. The *New York Nation*, a weekly paper, may be paralleled with the *Saturday Review* as it was in its old and good days; but the *New York Nation* is conducted by a foreigner,° and has an extremely small sale. In general, the daily papers are such that when one returns home one is moved to admiration and thankfulness not only at the great London papers, like the *Times* or the *Standard*, but quite as much at the great provincial newspapers too— papers like the *Leeds Mercury* and the *Yorkshire Post* in the north of England, like the *Scotsman* and the *Glasgow Herald* in Scotland.

The Americans used to say to me that what they valued was news, and that this their newspapers gave them. I at last made the reply: 'Yes, news for the servants' hall!' I remember that a New York newspaper,° one of the first I saw after landing in the country, had a long account, with the prominence we should give to the illness of the German Emperor or the arrest of the Lord Mayor of Dublin, of a young woman who had married a man who was a bag of bones, as we say, and who used to exhibit himself as a skeleton; of her growing horror in living with this man, and finally of her death. All this in the most minute detail, and described with all the writer's powers of rhetoric. This has always remained by me as a specimen of what the Americans call news.

You must have lived amongst their newspapers to know what they are. If I relate some of my own experiences, it is because these will give a clear enough notion of what the newspapers over there are, and one remembers more definitely what has happened to oneself. Soon after arriving in Boston, I opened a Boston newspaper and came upon a column headed: 'Tickings'. By *tickings* we are to understand news conveyed through the tickings of the telegraph. The first 'ticking' was: 'Matthew Arnold is sixty-two years old'—an age, I must just say in passing, which I had not then reached. The second 'ticking' was: 'Wales says, Mary is a darling'; the meaning being, that the Prince of Wales expressed great admiration for Miss Mary Anderson. This was at Boston, the American Athens. I proceeded to Chicago. An evening paper was given me soon after I arrived; I opened it, and found under a large-type heading, '*We have seen him arrive*', the following picture of myself: 'He has harsh features, supercilious manners, parts his hair down the middle, wears a single eyeglass and ill-fitting clothes.' Notwithstanding this rather unfavourable introduction I was most kindly and hospitably received at Chicago. It happened that I had a letter for Mr Medill, an elderly gentleman° of Scotch descent, the editor of the chief newspaper in those parts, the *Chicago Tribune*. I called on him, and we conversed amicably together. Some time afterwards, when I had gone back to England, a New York paper° published a criticism of Chicago and its people, purporting to have been contributed by me to the *Pall Mall Gazette* over here. It was a poor hoax, but many people were taken in and were excusably angry, Mr Medill of the *Chicago Tribune* amongst the number. A friend telegraphed to me to know if I had written the criticism. I, of course, instantly telegraphed back that I had not written a syllable of it. Then a Chicago paper is sent to me; and what I have the pleasure of reading, as the result of my contradiction, is this: 'Arnold denies; Mr Medill [my old friend] refuses to accept Arnold's disclaimer; says Arnold is a cur.'

I once declared that in England the born lover of ideas and of light could not but feel that the sky over his head is of brass and iron.° And so I say that, in America, he who craves for the *interesting* in civilisation, he who requires from what surrounds him satisfaction for his sense of beauty, his sense for elevation, will feel the sky over his head to be of brass and iron. The human problem, then, is as yet solved in the United States most imperfectly; a great void exists in the civilisation over there: a want of what is elevated and beautiful, of what is interesting.

The want is grave; it was probably, though he does not exactly bring it out, influencing Sir Lepel Griffin's feelings when he said that America is one of the last countries in which one would like to live. The want is such as to make any educated man feel that many countries, much less free and prosperous than the United States, are yet more truly civilised; have more which is interesting, have more to say to the soul; are countries, therefore, in which one would rather live.

The want is graver because it is so little recognised by the mass of Americans; nay, so loudly denied by them. If the community over there perceived the want and regretted it, sought for the right ways of remedying it, and resolved that remedied it should be; if they said, or even if a number of leading spirits amongst them said: 'Yes, we see what is wanting to our civilisation, we see that the average man is a danger, we see that our newspapers are a scandal, that bondage to the common and ignoble is our snare; but under the circumstances our civilisation could not well have been expected to begin differently. What you see are *beginnings*, they are crude, they are too predominantly material, they omit much, leave much to be desired—but they could not have been otherwise, they have been inevitable, and we will rise above them'; if the Americans frankly said this, one would have not a word to bring against it. One would *then* insist on no shortcoming, one would accept their admission that the human problem is at present quite insufficiently solved by them, and would press the matter no further. One would congratulate them on having solved the political problem and the social problem so successfully, and only remark, as I have said already, that in seeing clear and thinking straight on *our* political and social questions, we have great need to follow the example they set us on theirs.

But now the Americans seem, in certain matters, to have agreed, as a people, to deceive themselves, to persuade themselves that they have what they have not, to cover the defects in their civilisation by boasting, to fancy that they well and truly solve, not only the political and social problem, but the human problem too. One would say that they do really hope to find in tall talk and inflated sentiment a substitute for that real sense of elevation which human nature, as I have said, instinctively craves—and a substitute which may do as well as the genuine article. The thrill of awe, which Goethe pronounces to be the best thing humanity has, they would fain create by proclaiming themselves at the top of their voices to be 'the greatest nation upon earth', by assuring one another, in the language of their national historian,° that American democracy proceeds in its ascent 'as uniformly and majestically as the laws of being, and is as certain as the decrees of eternity'.

Or, again, far from admitting that their newspapers are a scandal, they assure one another that their newspaper press is one of their most signal distinctions. Far from admitting that in literature they have as yet produced little that is important, they play at treating American literature as if it were a great independent power; they reform the spelling of the English language by the insight of their average man. For every English writer they have an American writer to match. And him good Americans read; the Western States are at this moment being nourished and formed, we hear, on the novels of a native author called Roe,° instead of those of Scott and Dickens. Far from admitting that their average man is a danger, and that his predominance has brought about a plentiful lack of refinement, distinction and beauty, they declare in the words of my friend Colonel Higginson, a prominent critic at Boston, that 'Nature said, some years since: "Thus far the English is my best race, but we have had Englishmen enough; put in one drop more of nervous fluid and make the American." And with that drop a new range of promise opened on the human race, and a lighter, finer, more highly organised type of mankind was born.'° Far from admitting that the American accent, as the pressure of their climate and of their average man has made it, is a thing to be striven against, they assure one another that it is the right accent, the standard English speech of the future. It reminds me of a thing in Smollet's dinner-party of authors.° Seated by 'the philosopher who is writing a most orthodox refutation of Bolingbroke, but in the meantime has just been presented to the Grand Jury as a public nuisance for having blasphemed in an alehouse on the Lord's day'—seated by this philosopher is 'the Scotchman who is giving lectures on the pronunciation of the English language'.

The worst of it is, that all this tall talk and self-glorification meets with hardly any rebuke from sane criticism over there. I will mention, in regard to this, a thing which struck me a good deal. A Scotchman who has made a great fortune at Pittsburgh, a kind friend of mine, one of the most hospitable and generous of men, Mr Andrew Carnegie, published a year or two ago a book called *Triumphant Democracy*, a most splendid picture of American progress. The book is full of valuable information, but religious people thought that it insisted too much on mere material progress, and did not enough set forth America's deficiencies and dangers. And a friendly clergyman in Massachusetts, telling me how he regretted this, and how apt the Americans are to shut their eyes to their own dangers, put into my hands a volume written by a leading minister among the Congregationalists, a very prominent man, which he said supplied a good antidote to my friend Mr Carnegie's book. The volume is entitled *Our Country*.° I read it through. The author finds in evangelical Protestantism, as the orthodox Protestant sects present it, the grand remedy for the deficiencies and dangers of America. On this I offer no criticism; what struck me, and that on which I wish to lay stress, is, the writer's entire failure to perceive that such self-glorification and self-deception as I have been mentioning is one of America's dangers, or even that it *is* self-deception at all. He himself shares in all the self-deception of the average man among his countrymen, he flatters it. In the very points where a serious critic would find the Americans most wanting he finds them superior; only they require to have a good dose of evangelical Protestantism still added. 'Ours is the elect nation', preaches this reformer of American faults°—'ours is the elect nation for the age to come. We are the chosen people.' Already, says he, we are taller and heavier than other men, longer lived than other men, richer and more energetic than other men, above all, 'of finer nervous organisation' than other men. Yes, this people, who endure to have the American newspaper for their daily reading, and to have their habitation in Briggsville, Jacksonville, and Marcellus—this people is of finer, more delicate nervous organisation than other nations! It is Colonel Higginson's 'drop more of nervous fluid', over again. This 'drop' plays a stupendous part in the American rhapsody of self-praise. Undoubtedly the Americans are highly nervous, both the men and the women. A great Paris physician says that he notes a distinct new form of nervous disease, produced in American women by worry about servants. But this nervousness, developed in the race out there by worry, overwork, want of exercise, injudicious diet, and a most trying climate—this morbid nervousness our friends ticket as the fine susceptibility of genius, and cite it as a proof of their distinction, of their superior capacity for civilisation! 'The roots of civilisation are the nerves,' says our Congregationalist instructor° again; 'and, other things being equal, the finest nervous organisation will produce the highest civilisation. Now, the finest nervous organisation is ours.'

The new West promises to beat in the game of brag even the stout champions I have been quoting. Those belong to the old Eastern States; and the other day there was sent to me a Californian newspaper which calls all the Easterners 'the unhappy denizens of a forbidding clime', and adds: 'The time will surely come when all roads will lead to California. Here will be the home of art, science, literature, and profound knowledge.'

Common-sense criticism, I repeat, of all this hollow stuff there is in America next to none. There are plenty of cultivated, judicious, delightful individuals there. They are our hope and America's hope; it is through their means that improvement must come. They know perfectly well how false and hollow the boastful stuff talked is; but they let the storm of self-laudation rage, and say nothing. For political opponents and their doings there are in America hard words to be heard in abundance; for the real faults in American civilisation, and for the foolish boasting which prolongs them, there is hardly a word of regret or blame, at least in public. Even in private, many of the most cultivated Americans shrink from the subject, are irritable and thin-skinned when it is canvassed. Public treatment of it, in a cool and sane spirit of criticism, there is none. In vain I might plead that I had set a good example of frankness, in confessing over here, that, so far from solving our problems successfully, we in England find ourselves with an upper class materialised, a middle class vulgarised, and a lower class brutalised.° But it seems that nothing will embolden an American critic to say firmly and aloud to his countrymen and to his newspapers, that in America they do not solve the human problem successfully, and that with their present methods they never can. Consequently the masses of the American people do really come to believe all they hear about their finer nervous organisation, and the rightness of the American accent, and the importance of American literature; that is to say, they see things not as they are,° but as they would like them to be; they deceive themselves totally. And by such self-deception they shut against themselves the door to improvement, and do their best to make the reign of *das Gemeine* eternal. In what concerns the solving of the political and social problem they see clear and think straight; in what concerns the higher civilisation they live in a fool's paradise. This it is which makes a famous French critic speak of 'the hard unintelligence of the people of the United States'—*la dure inintelligence des Américains du Nord*°—of the very people who in general pass for being specially intelligent—and so, within certain limits, they are. But they have been so plied with nonsense and boasting that outside those limits, and where it is a question of things in which their civilisation is weak, they seem, very many of them, as if in such things they had no power of perception whatever, no idea of a proper scale, no sense of the difference between good and bad. And at this rate they can never, after solving the political and social problem with success, go on to solve happily the human problem too, and thus at last to make their civilisation full and interesting.

To sum up, then. What really dissatisfies in American civilisation is the want of the *interesting*, a want due chiefly to the want of those two great elements of the interesting, which are elevation and beauty. And the want of these elements is increased and prolonged by the Americans being assured that they have them when they have them not. And it seems to me that what the Americans now most urgently require, is not so much a vast additional development of orthodox Protestantism, but rather a steady exhibition of cool and sane criticism by their men of light and leading° over there. And perhaps the very first step of such men should be to insist on having for America, and to create if need be, better newspapers.

To us, too, the future of the United States is of incalculable importance. Already we feel their influence much, and we shall feel it more. We have a good deal to learn from them; we shall find in them, also, many things to beware of, many points in which it is to be hoped our democracy may not be like theirs. As our country becomes more democratic, the malady here may no longer be that we have an upper class materialised, a middle class vulgarised, and a lower class brutalised. But the predominance of the common and ignoble, born of the predominance of the average man, is a malady too. That the common and ignoble is human nature's enemy, that, of true human nature, distinction and beauty are needs, that a civilisation is insufficient where these needs are not satisfied, faulty where they are thwarted, is an instruction of which we, as well as the Americans, may greatly require to take fast hold, and not to let go. We may greatly require to keep, as if it were our life, the doctrine that we are failures after all, if we cannot eschew vain boasting and vain imaginations, eschew what flatters in us the common and ignoble, and approve things that are truly excellent.°

I have mentioned evangelical Protestantism. There is a text which evangelical Protestantism—and for that matter Catholicism too—translates wrong and takes in a sense too narrow. The text is that well-known one;° 'Except a man be born again he cannot see the kingdom of God.' Instead of *again*, we ought to translate *from above*; and instead of taking the kingdom of God in the sense of a life in Heaven above, we ought to take it, as its speaker meant it, in the sense of the reign of saints, a renovated and perfected human society on earth, the ideal society of the future. In the life of such a society, in the life *from above*, the life born of inspiration or *the spirit*—in that life elevation and beauty are not everything; but they are much, and they are indispensable. Humanity cannot reach its ideal while it lacks them: 'Except a man be born *from above*, he cannot have part in the society of the future.'

NOTES

A list of Abbreviations will be found at the beginning of the book.

1 *Mycerinus*. Arnold noted his source as Herodotus ii. 133 in *1849* but in *1853* substituted a conflation of passages from Herodotus ii. 129 and 133: 'After Cephren, Mycerinus, son of Cheops, reigned over Egypt. He abhorred his father's courses, and judged his subjects more justly than any of their kings had done.—To him there came an oracle from the city of Buto, to the effect that he was to live but six years longer, and to die in the seventh year from that time.' The rest of the story records Mycerinus' sense of injustice and decision to defy the gods by revelling night and day 'that he might make his six years into twelve and so prove the oracle false'. Dr Arnold, who set 'Mycerinus' as the subject for a prize poem at Rugby in 1831, died from a heart attack in June 1842 at the age of forty-seven. This event and Arnold's learning that he had inherited his father's weakness of heart may account for the bitterness in the stanzaic part of the poem.

l. 10 *self-governed* (and l. 28 'self-mastered'). Obedience to a divinely sanctioned law now called in question by seemingly arbitrary 'tyrannous necessity' (l. 42).

3 ll. 103–6. Herodotus ii. 78 (of Egyptian customs): 'At rich men's banquets, after dinner a man carries round an image of a corpse in a coffin . . . this he shows to each of the company, saying "Drink and make merry, but look on this; for such shalt thou be when thou art dead."'

3 ll. 107–11: perhaps reflecting Arnold's recent serious reading of Stoicism (see 'To a Friend' and 'Resignation', headnn., pp. 513, 511).

4 *Stagirius*. Entitled 'Stagyrus' in *1849*, 'Desire' in *1855*, and given the present title in *1877* with the note; 'Stagirius was a young monk to whom St Chrysostom addresses three books, and of whom these books give an account. They will be found in the first volume of the Benedictine edition of St Chrysostom's works.' Arnold came on 'Stagyre' when reading in 1848 St Marc Girardin's 1843 analysis of the parallel between modern romantic malaise and the writings of the Church Fathers (see *CL*, p. 68). The title 'Desire' links the poem with a favourite passage in George Sand's *Lélia* (1833), ch. lxvii, on 'le sanglot désespéré du désir impuissant' which accompanies man's perennial search for truth (see *CPW* viii. 221 and for his dissatisfaction with the poem, headn. to 'The Voice' below).

6 *The Voice*. The 'voice' may be John Henry Newman's, who resigned his living at St Mary's, Oxford on 18 September 1843, and whose sermons Arnold 'for a long time attended' being 'powerfully attracted' (Tom Arnold, *Passages in a Wandering Life* (London, 1900, p. 57). In his essay on

'Emerson', written over forty years after the poem's publication in *1849*, Arnold still recalled 'that . . . most entrancing of voices, breaking the silence with words and thoughts which were a religious music,—subtle, sweet, mournful' (*CPW* x. 165). He told Palgrave in 1869 (Russell, p. 43): 'In the "Voice" the falsetto rages too furiously; I can do nothing with it; ditto in "Stagirius".'

7 *A Question. To Fausta*: first published in *1849* as 'To Fausta' and given its present title in *1877*. 'Fausta' is Arnold's eldest sister Jane, known in the family as 'K'; see 'Resignation. To Fausta' (p. 39). Swinburne noticed the 'echo of Shelley's voice in its fainter but not least exquisite modulation' (*Miscellanies* (London, 1886), p. 112).

8 *Shakespeare*. Arnold's sonnet, dated 1 August 1844 in his fair copy for his sister Jane and first published in *1849*, reflects the sense of Shakespeare's inscrutability which he shared with others after the appearance of various attempts at a biography in the 1830s and 1840s. He told Clough in early December 1847: 'I keep saying Shakespeare, Shakespeare, you are as obscure as life is' (*CL*, p. 63). His use of the mountain as a symbol (ll. 3–6) parallels Goethe's Shakespeare as a Mont Blanc among poets (J. P. Eckermann, *Gespräche mit Goethe* (Berlin, 1916) under 2 January 1824) and Emerson's 'The Poet' (1884), who stands 'out of our limitations, like a Chimborazo' (Emerson's *Works*, Riverside edn. (London, 1883), iii. 14–15). Arnold's irregular sonnet consists of an Italian octet—with a third rhyme in pp. 6–7—and a sestet comprising a Shakespearian quatrain and a couplet.

l. 5: echoing Cowper's 'Light Shining out of Darkness' 1–3: 'God moves in a mysterious way / His wonders to perform; / He plants his footsteps in the sea'.

Written in Emerson's 'Essays'. Arnold probably wrote this sonnet, published in *1849*, shortly after the publication of Emerson's *Essays* (2nd series, 1844); he recalls his early feeling for Emerson in *Discourses in America* (1885), 'Forty years ago, when I was an undergraduate at Oxford, voices were in the air . . . a voice, also, from this side of the Atlantic,—a clear and pure voice . . . as new, and moving, and unforgettable, as the strain of Newman, or Carlyle, or Goethe' (*CPW* x. 165, 167). With ll. 11–13 compare Emerson's 'History' (1841): 'What Plato has thought, he [man] may think; what a saint has felt, he may feel; what at any time has befallen any man, he can understand' (*Works*, Riverside edn., ii. 9).

l. 1. *Hamlet* I. ii. 133–4: 'How weary, stale, flat and unprofitable / Seem to me all the uses of this world.'

l. 14. Arnold's first attempt, 'O barren boast and joyless Mockery' (Yale MS), seems to justify the scorn felt for the enthusiast; the revision 'leaves it open whether the world's pessimism may not be due to inertia' (*Poems*, p. 43).

9 *In Harmony with Nature*. Written probably in 1844–7 and first published in *1849*—with the title 'To an Independent Preacher, who preached that we

should be "In Harmony with Nature" '—as one of a group of sonnets printed together (including the two preceding poems). Arnold seems to contradict 'Quiet Work' ('One lesson, Nature, let me learn of thee', p. 53), but note that (*a*) l. 3 'When true' indicates that we can be 'in harmony with Nature' by learning a moral lesson from it, and (*b*) Arnold is concerned here with the distinction between the 'law for the man' and the 'law for the thing', as in *Literature and Dogma* (1873: *CPW* vi. 389, 391–2) on the 'pitfalls in that word *Nature* . . . do you mean that we are to give full swing to our inclinations . . . the constitution of things turns out to be somehow or other against it . . . the free development of our *apparent* self, has to undergo a profound modification from the law of our higher *real* self, the law of righteousness'. Arnold first reprinted the poem, with its present title, in *1877*.

l. 13. 'It is shameful for man to begin and to end where irrational animals do, but rather he ought to begin where they begin, and to end where nature ends in us, and nature ends in contemplation and understanding' (Epictetus, *Discourses* i, ch. 6: transl. G. Long (London, 1877), p. 21).

In Utrumque Paratus. The alternatives of the title are the world as emanation (man has descended) or as eternal matter achieving consciousness in man (man has ascended) and derive from Arnold's reading in 1846 of Plotinus, whose idea in the *Enneads* of the One all-pure, the world as emanation, and the reascent of the soul by purification he reproduces in ll. 1–5. After the poem's publication in *1849* he first reprinted it in *1869* with an altered version of ll. 36–42 (see *Poems*, p. 46), thereafter retaining it in its original form in all subsequent collections.

10 l. 31 (and l. 37 below). Confusing, since the 'brother world' dreams because it does not possess man's consciousness, whereas the 'chief dreamer' is man, who deludes himself that he is 'monarch' of all and the world exists for him: cf. 'Empedocles on Etna' I. ii. 177–81 (p. 86).

10 *The New Sirens*. The history of this poem and its revisions spans some thirty years from its publication in 1849 as an expression of Arnold's youthful enthusiasm for George Sand in the mid 1840s to its first reprinting as a tribute to her in *Macmillan's Magazine* in December 1876 (she died on 9 June 1876). He told C. E. Swinerton in 1885 (unpublished letter) that he composed it 'when I was only twenty-four', i.e. 1847 (perhaps not long after visiting George Sand at Nohant in July 1846). In March 1849 he agreed with Clough's criticism that it was 'a mumble—and I have doctored it and looked at it so long that I am now powerless respecting it' (*CL*, p. 107); on 1 September 1875 he told G. Smith: 'I don't thoroughly understand it myself, but I believe it is very fine and Rossetti and his school say it is the best thing I ever wrote' (Buckler, p. 97); in 1876, having revised the poem yet again before republication, he added the note: 'To a work of his youth, a work produced in long-past days of ardour and emotion, an author can never be very hard-hearted.' He explains in the 1885 letter: 'the mythological Sirens were . . . beautiful and charming creatures, whose charm and song allured men to their ruin. The idea of the poem is to consider as their successors Beauty and Pleasure as they meet in actual life, and distract us

from spiritual Beauty, though in themselves transitory and unsatisfying.' His distinction between Old and New Sirens was probably first inspired by various debates in George Sand's *Lélia* (1833) between Lélia, Sténio, and Pulchérie, which turn on the relative claims of sensuality, romantic passion, stoicism, and renunciation. He recalls his early feelings in his 1877 commemorative essay for her: 'agitations more or less stormy . . . of youth . . . days of Valentine, days of Lélia, days never to return' (*CPW* vii. 220). He set out his 'argument' for the poem at length for Clough, in March 1849 (*CL*, pp. 105–7; repr. *Poems*, pp. 48–50).

11 ll. 31–2. The laurel belongs to poetry and knowledge, the myrtle to love.

l. 40. *ceiled*: ceilinged.

12 l. 44. *Odyssey* xi. 39 ff., 184 ff.

18 *Horatian Echo. (To an Ambitious Friend)*. Written in 1847 (autograph signed and dated in Yale Papers) but not published till July 1887 (*Century Guild Hobby Horse*, repr. *1890* etc.) as a poem 'which has lain by me for years, discarded because of an unsatisfactory stanza' (see ll. 7–12) and 'a relic of youth . . . quite artificial in sentiment but . . . some tolerable lines, perhaps' (Arnold's letters of April and June 1887, published in *Century Guild Hobby Horse*, 18 (1890), 47–55, 49, 51). Horace replaced Béranger in Arnold's esteem during 1847–8: 'Horace whom he [Béranger] resembles had to write only for a circle of highly cultivated désillusionés roués in a sceptical age: we have the sceptical age but a far different and wider audience: voilà pourquoi, with all his genius, there is something "fade" about Béranger's Epicureanism' (Letter to Clough 29 September 1848, *CL*, pp. 92–3: part quoted in headn. to 'Switzerland', p. 514). Stanzas 2 and 3 glance at the Chartist agitations of 1847. The 'ambitious friend' could perhaps be John Blackett (1821–56), writer for the Whig *Globe* and invited to contest in 1846 a Newcastle upon Tyne constituency (*Poems*, p. 58). The opening stanza has a light *vers-de-société* flavour which Arnold rarely permits himself in his poetry.

ll. 1–6. See Milton, Sonnet xviii. 7–8: 'Let *Euclid* rest and *Archimedes* pause, / And what the *Swede* intend and what the *French*', and Milton's source, Horace, *Odes* II. xi. 1–4: 'What the warlike Cantabrian is plotting, Quinctus Hirpinus, and the Scythian, divided from us by the intervening Adriatic, cease to inquire.'

19 ll. 7–12. Aimed at both populace and aristocracy but in the first draft (see headn.) aimed at the crowd alone: 'Him not the noisy swarming race / Mounting in power long denied— / Who will not mount in peace—but love / At such despotic length to prove / That right is on their side'.

l. 24. *Eugenia*: used again in 'Philomela', l. 28 (p. 216).

20 *Fragment of an 'Antigone'*. Published in *1849* and reflecting Arnold's current feeling for Sophocles (see also 'To a Friend', p. 53), 'What is valuable is not so much his contributions to psychology and the anatomy of sentiment, as the grand moral effects produced by *style* . . .' But my Antigone supports me and in some degree subjugates destiny' (*CL*, p. 101); *Antigone* was performed in translation with music by Mendelssohn in London in 1844–5 and

1847. This and the following piece hint that, possibly with this stimulus, Arnold was interested in writing a play in the manner of Greek tragedy long before *Merope* (1858: see further headn. to 'Empedocles on Etna', p.520). The 'Chorus' here does not imitate any chorus in Sophocles' *Antigone* but follows his version of the legend.

21 ll. 59–64. Euripus is the strait separating Boeotia from Euboea; Parnes is a mountain forming part of the boundary with Attica; Tanagra is a Boeotian town on one bank of the Asopus.

l. 64. 'Orion, the Wild Huntsman of Greek legend, and in this capacity appearing in both earth and sky' (footnote in *1869*). Orion, loved by the Dawn Goddess (Eos, Aurora) was, according to one tradition, slain accidentally by an arrow from Artemis' bow while swimming off Delos.

22 l. 86. The river Ismenus flows through Thebes to Lake Hylica.

l. 90. *tired son*: Hercules, son of Zeus and Alcmene, two of whose twelve labours for Eurystheus were to fetch the golden apples of the Hesperides and to bring Cerberus from Hades (ll. 91–3).

l. 92. 'Erythia, the legendary region round the Pillars of Hercules, probably took its name from the redness of the West under which the Greeks saw it' (footnote in *1869*).

ll. 98–103. Zeus snatched Hercules to heaven from his funeral pyre on Mt. Oeta, which he mounted when agonized by Nessus' shirt (innocently sent by his wife Deianira); Oeta is in southern Thessaly where the river Spercheios flows into the Malic Gulf; Trachis, Hercules' home for a time, was situated there.

23 *Fragment of a Chorus of a 'Dejaneira'*. First published in *1867* but probably written in 1847–8 with the preceding poem which it follows in *1869* etc. Swinburne hoped 'to have more of the tragedy in time, that must be a noble statue which could match this massive fragment' (review of *1867* in *Essays and Studies* (London, 1875), p. 160). Dejaneira, innocently responsible for the death of her husband Hercules (see preceding poem, ll. 98–103 n.), killed herself in despair. She appears in Sophocles' *Trachiniae*.

l. 16. Those approaching the oracle first bathed and donned fresh linen.

24 *The Strayed Reveller*. Arnold's title poem for *1849* was probably written 1847–8 when, besides exploring classical themes and styles as in the two preceding poems, he was haunted by the 'natural magic' of Maurice de Guérin's 'Le Centaure': 'extraordinary delicacy of organisation and suceptibility to impressions; in exercising it the poet . . . aspires to be a sort of Aeolian harp . . . To assist at the evolution of the whole world is his craving' ('Maurice de Guérin', *E in C* I (1865): *CPW* iii. 30). The qualities are dramatized in the 'reveller', whom Arnold places in a setting from the *Odyssey* which he uses for his characteristic balancing of different visions (other instances are in 'Resignation', 'The Scholar-Gipsy', and 'Bacchanalia' pp. 39, 208, 264). The irregular chant is an individual attempt to suggest the effect of Greek lyric measures, though L. Binyon noted an

influence from Goethe's and Heine's free-verse poems, which were 'modelled on a misunderstanding of Greek lyrics' (*Tradition and Reaction in Modern Poetry* (London, 1923), p. 12). The unusual wealth of vivid pictorial detail is mainly drawn from Arnold's extensive recent reading of Eastern travel books, including A. Burnes's *Travels into Bokhara* (1834), also used for 'The Sick King in Bokhara' and 'Sohrab and Rustum' (pp. 32, 186).

25 l. 38. *Iacchus*: minor deity often identified with Dionysus. The vine crown and thyrsis (l. 33) are Bacchic insignia.

26 ll. 78–9. 'son of a nymph and satyr . . . loved by Bacchus . . . Upon him the god bestowed a vine that . . . still . . . takes from the boy its name' (Ovid, *Fasti* iii. 409–12).

28 ll. 135–42. *Tiresias*: the blind soothsayer who foretold the fate of Oedipus and the defeat of Thebes by the Epigoni.

29 l. 183. The lower Oxus. The Chorasmii inhabited its banks and islands.

    l. 206. *Happy Islands*: according to Pindar located in the 'stream of Ocean reached by the Argonauts' (*Pythian Odes* iv. 251).

30 ll. 220–2. Hera blinded Tiresias in a rage; Zeus in mitigation extended his life to seven generations.

    ll. 223–32. The Centaurs and Lapiths fought at Pirithous' wedding-feast; Theseus was present. Hercules fought the Centaurs when in pursuit of the Erymanthian boar.

31 ll. 258–60. The Argonauts sailed 'over an unknown sea in that first ship to seek the bright gleaming fleece of gold' (Ovid, *Metamorphoses* vi. 720–1).

    l. 261. The satyr who was companion and instructor to the young Dionysus.

32 l. 287. *much enduring*: recurrent epithet for Ulysses in the *Odyssey*.

    The Sick King in Bokhara. Arnold found the story for this poem, along with its topographical and pictorial details, in A. Burnes's *Travels into Bokhara* (see headn. to 'The Strayed Reveller, above); according to his 1869 letter to F. T. Palgrave the poem, published in *1849*, was 'the first thing of mine dear old Clough thoroughly liked' (Russell, p. 42). The subject interested Arnold because of its suggestion that the moral law may transcend expediency and yet be sanctioned by the individual conscience. The mullah, who breaks the law but 'in the presence of the king, stated his crime and demanded justice according to the Koran' (see Burnes, i. 307–9), is seen as a Raskolnikov figure who insists on retribution to purge his sense of guilt in Trilling (pp. 104–6).

    l. 12. *Ferdousi*: a Persian epic poet whose Shāhnāma includes the story of Sohrab and Rustum.

33 l. 31. A mullah is a Muhammadan theologian.

    l. 89. Tennyson's 'The Lady of Shalott' (1832, rev. 1842), l. 168: 'But Lancelot mused a little space'.

37 l. 167. The Shiahs are an Islamic sect (the vizier is a Sunni).

    l. 173. *kaffirs*: infidels.

38 ll. 186–8. Echoing Horace, *Odes* III. i. 5–6: 'The rule of dreaded kings is over their own peoples; but over the kings themselves is the rule of Jove.'

39 *Resignation. To Fausta.* Arnold must have begun the poem shortly after July 1843, the date of the second walk celebrated in the poem, and then revised and added to it until its publication in *1849*: it shows dependence on Goethe's *Wilhelm Meister* which he read in Carlyle's translation in 1845, while at Oxford, together with influences from his more general reading in the later 1840s, including the Stoics, Spinoza, *Obermann* (see p. 66). Like Wordsworth's 'Tintern Abbey', to which it reads in some respects as a reply, it is a poem addressed to a favourite sister (see 'A Question. To Fausta', headn., p. 506) and is about revisiting a place and the reflections aroused by the two occasions. Arnold's view that nature seems 'to bear rather than rejoice' (l. 270) contradicts Wordsworth's claim that nature's privilege is 'Through all the years of this our life, to lead / From joy to joy' (ll. 124–5). In 1842, the year before the second walk, Jane's broken engagement to George Cotton in May, and her subsequent depression, was followed in the June by Dr Arnold's sudden death (see headn. to 'Mycerinus', p. 505). The name 'Fausta' perhaps suggest 'a female Faust' who 'desires poignant experience to relieve the dulness of her life' (Trilling, p. 99), but see *Poems*, p. 88: 'with the Latin meaning of the word in mind, it may hint a gentle rebuke—in spite of her sense of frustration Jane is "fortunate" '. Jane is silent in the poem (l. 30: 'your thoughts I scan'), which is a characteristically Arnoldian 'dialogue of the mind with itself' and his most substantial achievement to date. The handling of the octosyllabics and some of the pictorial details argue the influence of Scott's introductions to the various Cantos of *Marmion* (1808), especially the tone of reflective reminiscence in the introduction to Canto 4.

l. 1. 'I knew but to obtain or die' (Byron's 'The Giaour' (1813), 111).

40 l. 21. *Past straits*: past difficulties, and also narrow gorges already passed through.

l. 23. *unblamed serenity*: detachment as in Stoic or Hindu philosophy.

ll. 40–1. 'Those who have been long familiar with the English Lake-Country will find no difficulty in recalling, from the description in the text, the roadside inn of Wythburn on the descent from Dunmail Rise towards Keswick; its sedentary landlord of twenty years ago, and the passage over the Wythburn Fells to Watendlath' (footnote in *1869*, with 'thirty' substituted for 'twenty' in *1877*; for a reconstruction of the whole journey see *Poems*, p. 90). A memorial stone at Wythburn commemorates 'the two walks from hence over the Armboth Fells July 1833–43'.

41 ll. 86–107. Cf. Clough's 'Blank Misgivings' (dated February 1841) ix. 1–21, beginning 'Once more the wonted road I tread / Once more dark heavens above me spread'.

42 ll. 116–19. Stressing their 'torpid life' as in Wordsworth's 'Gipsies', ll. 1–8.

43 ll. 144–69. The poet's 'high station' (l. 64) parallels Goethe's description of the poet as 'exalted above all this [continual agitation]', *Wilhelm Meister*, ii, ch. 2, transl. Carlyle (1824: Traill, iii. 112).

46 l. 253. The image is of the Tree of Life; cf. Carlyle's 'Tree Igdrasil', *On Heroes and Hero-Worship* (1841: Traill, v. 20–1).

l. 265. *turf we tread.* Echoes Wordsworth's 'The Brothers' (1800), l. 14.

l. 268. *strange-scrawled rocks*: striated rocks.

l. 278. *infects.* Suggests some sympathy with the 'intemperate' Romantic expectations frustrated by Necessity.

The Forsaken Merman. This familiar and much anthologized poem was a favourite with reviewers of *1849* and many of Arnold's fellow writers including Clough, Froude, Palgrave, and Tennyson (with whose youthful poem 'The Mermaid' ll. 2–8, 23–34 cf. ll. 41, 51). Arnold may first have come on the story in Mary Howitt's 1847 translation of Hans Andersen's *The True Story of my Life* (Andersen visited England in June 1847) which refers to the Danish ballad of Agnes and the Merman, but he uses many details from the fuller version in George Borrow's review of J. M. Thiele's *Danske Folkesagn* (*Universal Review*, 2 (1825), 563–4 (see *Poems*, p. 101); there is also a debt in ll. 112–19 to Byron's *Manfred* I. i. 76–87). The themes of loss and separation because of rival claims and incompatible longings (the colourful Romantic undersea world set against the serene but austere white-walled town on the shore) suggest an emotional association with the 'Marguerite' poems written 1847–8 (see headn. to 'Switzerland', p. 514). The irregular chant, consisting of verse paragraphs in which the lines vary in number, length, stress, and rhyme, is individual to Arnold (see headnn. to 'The Strayed Reveller' and 'The Youth of Nature', pp. 509, 532).

49 l. 96. *spindle drops*: shuttle falls *1849–1877*. Clough pointed out the error in March 1849, when Arnold replied 'I will look in the technological dict:' (*CL*, p. 107).

50 *The World and the Quietist.* This poem and the four following sonnets, published together in *1849*, are addressed to Clough and constitute in effect a running debate with him *c.* March–*c.* August 1848 when Arnold found himself at odds with Clough's enthusiasm for the revolutionary movements of that year, over-fastidious conscience, and, as here, want of sympathy for his own attraction to Indian 'quietism' inspired by the *Bhagavadgita*, which he urged on Clough in 1848 and was 'disappointed the Oriental wisdom pleased you not' (*CL*, p. 90). Clough's review of *1853* censured Arnold's 'dismal cycle of rehabilitated Hindoo-Greek theosophy' since 'for the present age, the lessons of reflectiveness and caution do not appear to be more needful than . . . calls to action' (*Correspondence of AHC* (1869) i. 377–8). Arnold refers to 'the Indian virtue of detachment' in 'The Function of Criticism at the Present Time' (*E in C* I) when speaking of 'the dizzying and attractive effect' of the 'rush and roar of practical life . . . only by remaining collected, and refusing to tend himself to the view of the practical man . . . [can] the critic . . . do the practical man any service' (*CPW* iii. 274–5). 'Critias', prominent in the revolutionary 'oligarchy' following Athens' collapse after the Peloponnesian War (404–403 BC), figures in Plato's dialogues; especially relevant here are the discussions in the *Charmides*

concerning the relative wisdom of 'pure' versus 'practical' and 'applied' knowledge.

51 ll. 25–7. Herodotus v. 105; Darius made his servant repeat thrice at dinner-time 'Master, remember the Athenians', Arnold's point being that he felt the full extent of his power only when reminded of the single check to it.

To a Republican Friend, 1848. See preceding headn. Clough claimed in March 1848: 'if it were not for all these blessed revolutions I should sink into hopeless lethargy' but believed 'the millenium as Matt, says, won't come this bout'; Arnold teasingly addressed a letter to 'Citizen Clough, Oriel Lyceum, Oxford' (*Correspondence of AHC* i. 119, 216, 243). The 'great ones' of l. 10 are probably seminal thinkers familiar to Arnold, e.g. Carlyle, Emerson, George Sand.

52 *To a Republican Friend, Continued.* See preceding headn.

l. 4. Quoted by Arnold twenty years later with the comment: 'lovers of France as we are . . . this still seems to be the true criticism on her' (*CPW* vii. 48).

Religious Isolation. To the Same Friend. See headn. to 'The World and the Quietist' (p. 512). The sonnet is addressed to Clough *c.* July 1848 on his decision to resign his Oriel fellowship on conscientious grounds. Arnold thought him 'the most conscientious man I know; but sometimes morbidly so' (*CPW*, p. 130).

53 l. 9. The need to act according to the 'inner light' of conscience.

To a Friend. See headn. to 'The World and the Quietist' (p. 512). Probably written August 1848 when Arnold was reading Epictetus ('that halting slave' l. 6).

l. 1. *these bad days.* 1848: 'French Revolutions, Chartisms . . . that make the heart sick in these bad days' (Carlyle, *Past and Present* (1848): Traill, x. 36). A notably awkward line, rivalled only by 'Austerity of Poetry', l. 8 (p. 239): 'A prop gave way! crash fell a platform! lo . . .'.

l. 2. *the old man*: Homer.

l. 3 'The name Europe (Εὐρώπη, *the wide prospect*) probably describes the appearance of the European coast to the Greeks on the coast of Asia Minor opposite. The name Asia, again, comes, it has been thought, from the muddy fens of the rivers of Asia Minor, much as the Caster or Maeander, which struck the imaginations of the Greeks near them' (footnote in *1869*; before this simply 'Εὐρώπη').

l. 4 *Smyrna*: one of seven cities claiming to be Homer's birthplace.

ll. 6–8. The Stoic philosopher Epictetus taught at Nicopolis after his expulsion by Domitian in AD 89. He was born a slave; his lameness was said to result from injury received before he was freed.

ll. 9–14. Sophocles was born at Colonus in 496 BC and died when nearly ninety. *Oedipus Coloneus* was his last play.

Quiet Work. Probably written in 1848, the year of the 'discords' (ll. 9-10) also referred to in the two preceding sonnets. Published as the first poem in *1849* with the title 'Sonnet' and given its present title in *1869*. With the attitude to Nature cf. 'Self-Dependence' (p. 73).

54 l. 11. *sleepless ministers*: the stars; cf. Keats's 'nature's patient sleepless eremite' in his 'Bright Star!' sonnet, published in 1848.

55 *Switzerland*. Arnold first used this title in his third collection—*1853*—for a series of poems, some already published separately in *1852* and *1849*, addressed to 'Marguerite', a girl whose identity is still a matter of speculation. The nine poems in the series, all referring directly to her and all except one composed during the period from late 1847 to early 1850, were never all published together because Arnold frequently altered the selection and ordering of the poems before finally settling on the arrangement in *1877* (see below). They appear together here in chronological order of composition (except 'The Terrace at Berne' (p. 237), written some ten years later). They unfold a narrative of meetings and partings which run *pari passu* with his visits to Switzerland in September 1848 and September 1849, recorded in two letters to Clough which contain the only external evidence available of 'Marguerite's' existence and her connection with Switzerland: 29 September 1848, 'Tomorrow I . . . get to Thun, linger one day at the Hotel Bellevue for the sake of the blue eyes of one of its inmates . . . then proceed by slow steps down the Rhine to Cologne, thence to . . . England'; 23 September 1849, from Thun: 'I am here in a curious and not altogether comfortable state: however tomorrow I carry my aching head to the mountains' (*CL*, pp. 91, 110; p. 283 above); Arnold then wrote out an early version of 'Parting' ll. 23-4 (see below). 'Marguerite's' appearance, her setting at Thun, and the emotions associated with her are described with a degree of circumstantial detail making it difficult to regard the poems as a fictional working-up of events, a view necessary for the recent argument that 'Marguerite' was Mary Claude, a family friend living in England and not, as the poems suggest, a French girl living in Switzerland and separated from the poet by distance, 'estranging sea' ('To Marguerite—Continued', l. 24, 'Parting', l. 65) and year-long intervals between his visits abroad to meet her. (For debate about this see Miriam Allott, 'Arnold and Marguerite— Continued', and Park Honan, 'The Character of Marguerite in Arnold's *Switzerland*', in *Victorian Poetry*, vol. 23, no. 2, summer 1985, pp. 125-43, 145-59.) The themes of loss and separation, conflict of feeling caused by incompatible longings and the presence of Switzerland itelf, to which Arnold was powerfully drawn by Senancour's *Obermann*, link the series with other poems of the late 1840s and early 1850s, notably 'Empedocles on Etna', 'Tristram and Iseult', and shorter pieces including 'The Forsaken Merman', 'Youth's Agitations', and 'Self-Dependence' (pp. 76, 109, 46, 74, 73). The following is the order of the poems in *1877*, together with their earlier titles and dates of first publication: 1. 'Meeting' ('The Lake' *1852*); 2. 'Parting' (*1852*); 3. 'A Farewell' (*1852*); 4. 'Isolation. To Marguerite' ('To Marguerite' *1857*); 5. 'To Marguerite—Continued' ('To Marguerite, in

Returning a Volume of the Letters of Ortis' *1852*, 'To Marguerite' *1853*, 'Isolation' *1857*); 6. 'Absence' (*1852*); 7. 'The Terrace at Berne' (*1867*).

A Memory-Picture. Arnold's list of poems to be composed in 1849 (Yale MS) indicates this poem with 'Thun and vividness of sight and memory compared: sight would be less precious if memory could equally realise for us.' This, the earliest of the 'Marguerite' poems, is the only one published in *1849*, where it was entitled 'To my Friends, who ridiculed a tender Leave-taking.' It was 'Switzerland' 1 in *1853–69* but excluded from the series in *1877* when it became an 'Early Poem'. 'Marguerite's' pallor, ash-blonde hair, blue eyes, and mingled gaiety and gravity recur in 'Meeting', 'Parting', and 'A Farewell'. Her reference to 'last year . . . the coming year . . . next year' (ll. 17–20) hints at a meeting the previous year (see 'Rachel', headn., p. 554).

56 *Meeting*. Seemingly the meeting referred to in Arnold's September 1848 letter to Clough; see headnn. to 'Switzerland' and 'A Memory-Picture' above. The poem, first published as 'The Lake' in *1852* and given its present title in *1869*, was 'Switzerland' 2 in *1853–9*, 1 in *1877*.

l. 2. Thun, a fashionable resort at the time, is situated on the Aar where the river emerges from Lake Thun to flow down to Berne. The setting is recalled in greater detail in 'A Farewell' and 'The Terrace at Berne'.

57 l. 11. See 'To Marguerite—Continued', l. 22: 'A God, a God their severance ruled'.

Parting. Associated with the distress hinted at in Arnold's September 1849 letter to Clough on his 'uncomfortable state' and intention to 'carry my aching head to the mountains': see headn. to 'Switzerland' (p. 514). The metrical changes seem to represent wavering between desire to give free play to emotion ('storm-winds of Autumn') and the desire for peace and detachment (the stillness of 'the white peaks in air'). There are several verbal links with 'Obermann' (p. 66), e.g. with l. 3 at l. 1; ll. 33–4 at ll. 29–30 and 56; and l. 9 at l. 7.

59 *Isolation. To Marguerite*. Not published until *1857* (as 'To Marguerite') but Arnold must have written this and the following poem, 'To Marguerite—Continued' (*1852*), at about the same time, using the same stanza in both to continue the argument about human isolation in the emotional aftermath of the September 1849 'parting' lamented in the preceding poem; 'We were apart' (l. 1) refers to the interval between September 1848 and September 1849. It has been suggested (*Poems*, p. 127) that Arnold delayed publication because stanzas 3–5 might read oddly so soon after his marriage in 1852 (see headn. to 'Faded Leaves', p. 528). 'Switzerland' 6 in *1857–69*, 4 in *1877*.

60 l. 11. Our feelings are seen as ebbing and swelling under the influence of the heart. The moon image prompts the reference to the story of Luna and Endymion in ll. 19–30.

61 *To Marguerite—Continued*. 'Yes' at the opening of this poem—one of

Arnold's best-known lyrics—links it with the concluding lines of its companion, 'Isolation. To Marguerite' (see above), and with the theme of the book referred to in its original title in *1852*, 'To Marguerite, in Returning a Volume of the Letters of Ortis'. The hero of Foscolo's *Ultime Letteri di Jacopo Ortis* (1802) is an unhappy lover and Wertherian misfit who finds the universe incomprehensible and commits suicide in despair. It was perhaps a copy of the recent 1847 reprint of the 1839 French translation by A. Dumas that Arnold read and returned to 'Marguerite' (note 'Daughter of France . . . France thy home' in 'The Terrace at Berne, l. 18, p. 237). The 'sea of life' (l. 1) and the idea of voyaging across it is recurrent in Arnold (e.g. 'Human Life', l. 27 (p. 72), 'A Summer Night', ll. 52–3 (p. 151), the central image in 'The Future', pp. 148–50), one of its most arresting expressions occurring in this poem's final line. 'Switzerland' 5 in *1853*, 6 in *1854*, 7 in *1857–69*, 5 in *1877*.

l. 2. *echoing*: making communications confused and uncertain.

l. 6. *endless*: unending; cf. 'Isolation. To Marguerite', ll. 40–1: 'isolation without end / Prolonged'.

ll. 7–12. The islands, subject to the same laws as human beings, are enchanted by moonlight and nightingales just as humans fall under the enchantment of love.

ll. 22–4. Conflates Horace's *Epodes* xiv. 6: ''tis the god, yea 'tis the god, that forbids me' and, for 'estranging', *Odes* I. iii. 21–3, which has 'Oceano dissociabili', known at Rugby through A. C. Tait's translation 'the estranging main'.

A Farewell. Occasioned by the final parting from 'Marguerite' at Thun, seemingly after returning from the mountain expedition mentioned in the letter to Clough of 23 September 1849 (see headn. to 'Parting', p. 515). The poem was first published in *1852* and became 'Switzerland' 5 in *1854–7*, 4 in *1869*, 3 in *1877*.

ll. 5–6. The roofed bridge and poplars along the Frutigen road were features of Thun in the 1840s. See 'Meeting', l. 2 n. (p. 515).

63 l. 56. *affinities*: A spiritual attraction believed to exist between persons; the subjects of the affinity (*OED* cites from 1868).

64 ll. 73–6. Cf. Clough's astronomical image in his letter to Arnold of 23 June 1849: 'Our orbits . . . early in August might . . . cross, and we two . . . undeviating stars salute each other once again for a moment across the infinite spaces' (*CL*, p. 108).

l. 81. *boon*: gracious, benign, as in 'Thyrsis', l. 175 (p. 245): 'To a boon southern country he is fled'.

Absence. Expressing the conflict of feeling when passion for a lost love survives alongside the stirring of a new attraction, here probably inspired by Frances Lucy Wightman with whom Arnold became emotionally involved in the summer of 1850 and later married; 'this fair stranger's eyes of grey' (l. 1) is echoed in 'Separation', ll. 3–16, a poem in the 'Faded Leaves' sequence associated with her (headn., pp. 528–9). The 'light' of ll. 13–17 is knowledge of the self and of the world; the image, which has a Goethean flavour, is of a man struggling through a storm to a light which will mean safety. The confused feelings are indicated by 'bear' (l. 16), which suggests that the poet is drawn to the storm, and by his plea to 'Marguerite', the cause of the storm, to stay with him till the light is reached. The poem, published in *1852*, became 'Switzerland' 6 in *1853*, 7 in *1854*, 8 in *1857*, 5 in *1869*, and 6 again in *1877*.

65 *A Dream*. The unusually vivid pictorial detail suggests that Arnold is describing an actual dream. The images of Swiss mountain freshness contrasted with the heat and noise of the 'cities of the plain' reflect how yearningly he looked back to Thun and its associations after the separation. 'Martin' and 'Olivia' have not been identified but 'Martin' may be Wyndham Slade, Arnold's travelling companion in Switzerland in September 1849. The scarlet berries of the ash, harvested gourds, and Indian corn (ll. 9, 17–19) point to late summer or autumn. The 'river of Life' (l. 31) is a favourite metaphor with Arnold; see 'The Buried Life', ll. 3–63 n. (p. 534). The poem, first published in *1853*, became 'Switzerland' 3 in *1853–7*, was omitted altogether from *1869* and *1877*, but was restored as an 'Early Poem' in *1881*.

ll. 26–7. A neatly mannered 'epic' touch: cf. Virgil's Sibyl, *Aeneid* vi. 48–50: 'her bosom heaves, her heart swells with wild frenzy, and she is taller to behold, nor has her voice a mortal ring' and Milton's *Comus* (1637), l. 297: 'Their port was more than human as they stood'.

66 ll. 36–7. Cf. 'The Future', ll. 50–65 (p. 149) and n. (p. 533).

Stanzas in Memory of the Author of 'Obermann'. November 1849. This key poem in Arnold's early poetic career was first published in *1852* but without the lengthy note on Senancour attached to it in *1869* (first used for 'Obermann Once More', *1868*) which opens with a reference to Senancour's lack of recognition apart from 'some of the most remarkable spirits of this century, such as George Sand and Sainte-Beuve', followed by a biographical sketch. Etienne Pivert de Senancour, born in France in 1770, was educated for the priesthood, broke away to live in Switzerland, married there, and later returned to live in France as a 'man of letters, but with hardly any fame or success'; he died 1846, at Sèvres near Paris, having requested as his only epitaph 'Éternité, deviens mon asile'. The rest of the note, celebrating Senancour's qualities, includes the following reflections:

> ... though *Obermann*, a collection of letters from Switzerland treating almost entirely of nature and the human soul may be called a work of sentiment, Senancour has a gravity and severity which distinguish him from all other writers of the sentimental school ... of all writers he is the most perfectly isolated and the least attitudinising ... The stir of all the main forces, by which modern life is, and has been, impelled, lives in the letters of Obermann; the dissolving agencies of the eighteenth century, the fiery storm of the French Revolution, the first faint dawn and promise of that new world which our own time is but now faintly bringing to light—all these are to be felt, almost to be touched, there. To me, indeed,

it will always seem that the impressiveness of this production can hardly
be rated too high . . .

The poem's connection with the 'Switzerland' series is suggested by
Arnold's *1869* note to l. 5, which links its composition with the journey
anticipated in 'Parting' and his September 1849 letter to Clough quoting
ll. 143–7 (see headnn. to 'Switzerland' and 'Parting', pp. 514 and 515).
'Once more' (l. 12) alludes to his first journey in September 1848. Sainte-
Beuve included his translation of the poem in his *Chateaubriand et son groupe
littéraire* . . . (1860) as 'une immortelle couronne funèbre'. For their
correspondence see Bonnerot, pp. 520–80 and for Arnold's 1869 article on
Senancour *CPW* v. 295–303.

l. 3. *rack*. 'un mot de la vieille langue anglaise pour signifier les *bords*, la *frange*
d'un nuage qui passe. Shakespeare s'en sert dans la "Tempête" ' (Arnold, 8
March 1855: Bonnerot, p. 525).

l. 5. 'The Baths of Leuk. This poem was conceived, and partly composed, in
the valley going down from the foot of the Gemmi Pass towards the Rhone'
(footnote in *1869*).

l. 18. cast: turning the ship's head away from the wind to ease its passage.

67 l. 27. 'les glaciers *prêtent, donnent quelquechose* de l'âme de leurs neiges au
livre d'Obermann' (Arnold to Sainte-Beuve: see Bonnerot, p. 526).

l. 44. Luke 23:34.

ll. 49–50. 'Written in November, 1849' (footnote in *1852–69*). Wordsworth
had been dead more than two years when 'Obermann' was first published.

ll. 51–4. Goethe died in 1832; in 'Memorial Verses' (pp. 137–9) he is again
linked with Wordsworth. With the latter's averting his eyes 'from half of
human fate' cf. 'Heinrich Heine', *E in C* I (1865): 'Wordsworth . . . plunged
himself into the inward life, he voluntarily cut himself off from the modern
spirit' (*CPW* iii. 121).

68 ll. 65–6. Goethe was forty in 1789; see 'the fiery storm of the French
Revolution' in Arnold's note on Senancour (headn., p. 517).

69 ll. 91–2. Achilles in *Iliad* xxi. 166–7: 'Why lamentest thou thus? Patroclus
also died, who was better than thou.'

l. 114. Jaman overlooks Vevey on the Lake of Geneva (also known as Lake
Leman, l. 121).

70 l. 132. The poetic half of the divided self: also refers to the emotions
surrounding 'Marguerite'. With the 'unknown Power' (ll. 133–4) cf. 'Meeting', l. 13 and 'Human Life', l. 26 (pp. 57, 72).

ll. 143–4. Quoted in Arnold's letter of 23 September 1849 to Clough, whom
he describes as 'the only living one almost that I know of "The children of
the second birth" ' (*CL*, pp. 109–10; p. 283 above). See John 3: 3, 7.

71 l. 156. *Unspotted*. James 1:27.

l. 180. For Senancour's burial place see headn. (p. 517).

ll. 183–4. See Arnold to Clough, 23 September 1849: 'I have never yet succeeded . . . in consciously mastering myself . . . at the critical point I am too apt to hoist up the mainsail to the wind and let her drive' (*CL*, p. 110; p. 283 above); 'The lines are Arnold's farewell to youth, insouciance and Marguerite—and also, in the long run, to the writing of poetry' (*Poems*, p. 144).

72 *Human Life*. First published in *1852* but an 'Early Poem' in *1877*; perhaps written when the separation from 'Marguerite' was still recent—see l. 16; with the central image of the 'sea of life' cf. 'To Marguerite—Continued' and headn. above. Stanzas 2 and 3 seem contradictory, but the implication is that our freedom of action is circumscribed, not that it is illusory. The unusual stanza ($a^3\ b^4$ c a c b^5) was used by Arnold only here.

l. 5. *inly written chart*: conscience.

l. 23. *prore*: prow.

l. 27. *stem*: hold to a fixed course.

73 *Self-Dependence*. Published in *1852* but perhaps written when memories of 'Marguerite' were still fresh; as in 'Human Life' above, Arnold seems to be recalling his return by sea to England after the parting at Thun. Basil Willey found the poem 'as concise a summary of Stoic teaching as one could hope to find' (*The English Moralists* (London, 1965), pp. 66–7).

l. 17. Pascal, *Pensées* (Paris, 1882), p. 91: 'Le silence éternel de ces espaces infinis m'effraie'; the poem's closing lines on the need for self-knowledge parallel *Pensées*, p. 81 (Pascal is referred to in 'Eugénie de Guérin', *E in C* I (1865), *CPW* iii. 89).

74 *Destiny*. Not reprinted by Arnold after *1852*: 'Probably a "Marguerite" poem cancelled by Arnold as duplicating the fuller self-analysis of "A Farewell" 16–17 [p. 62]' (*Poems*, p. 150).

Youth's Agitations. Arnold's list of poems to be composed in 1849 (Yale MS) points to this and the sonnet 'The World's Triumphs' as the second and third of a projected group of five sonnets (the others, if written, have not survived). Brought together in *1869* and classified as 'Early Poems' in *1877*, these were Arnold's only Shakespearian sonnets. Each was first published in *1852* as 'Sonnet' and acquired its present title in *1867*. Arnold's mingled relief and regret as youth is felt slipping away is again expressed in his letter to 'K', 25 January 1851: 'The aimless and settled, but also open and liberal state of our youth we *must* perhaps all leave . . . but with most of us it is a melancholy passage from which we emerge shorn of so many beams that we are almost tempted to quarrel with the law of nature which imposes it on us' (*L* i. 14).

l. 3. *tedious vain expense*: Shakespeare, Sonnet 131: 'The expense of spirit in a waste of shame'.

75 *The World's Triumphs*. See 'Youth's Agitations' above. The title refers to the world that triumphs over idealism and youthful enthusiasm.

76 *Empedocles on Etna. A Dramatic Poem.* Arnold's—and arguably his age's—
major long poem took over from his long-projected tragedy on the subject of
Lucretius, whom he was reading closely from 1845. Some of the stanzas in
this uncompleted work became 'Empedocles on Etna', which was included
in a numbered list of suggestions for other poems (possibly to compose in
1849), the first being 'Empedocles—refusal of limitation by the religious
sentiment' (Yale MS); the description fits the philosophical chant in I. ii. 77–
426 but not the completed 'dramatic poem', suggesting that the original
poem outgrew Arnold's first conception (drawing on material intended for
'Lucretius') so that he decided to give it dramatic form. J. C. Shairp's letter
to Clough of 30 June 1849 refers to Arnold's 'working at an "Empedocles".
I wish Matt would give up that old greek form' (*Correspondence of AHC* i.
270). Also at this time Arnold made notes on the life of Empedocles from the
introduction to S. Karsten's *Philosophorum Graecorum Veterum . . . Operum
Reliquiae* 11 (Amsterdam, 1838), pp. 3–78 (reproduced in *Commentary*,
pp. 289–90—Karsten's material is extensively drawn on in the poem), and
wrote out an analysis of his protagonist's character and motives (Yale MS):

He is a philosopher.

He has not the religious consolation of other men, facile because
adapted to their weakness, or because shared by all around and charging
the atmosphere they breathe.

He sees things as they are—the world as it is—God as he is: in their
stern simplicity.

The sight is a severe and mind-tasking one: to know the mysteries
which are communicated to others by fragments, in parables.

But he started towards it in hope: his first glimpses of it filled him with
joy; he had friends who shared his hope and joy and communicated to him
theirs: even now he does not deny that the sight is capable of affording
rapture and the purest peace.

But his friends are dead: the world is all against him, and incredulous of
the truth: his mind is overtasked by the effort to hold fast so great and
severe a truth in solitude; the atmosphere he breathes not being modified
by the presence of human life, is too rare for him. He perceives still the
truth of the truth, but cannot be transported and rapturously agitated by
its grandeur: his spring and elasticity of mind are gone: he is clouded,
oppressed, dispirited, without hope and energy.

Before he becomes the victim of depression and overtension of mind, to
the utter deadness to joy, grandeur, spirit, and animated life, he desires to
die; to be reunited with the universe, before by exaggerating his human
side he has become utterly estranged from it.

This resembles Arnold's description in 'On the Modern Element in
Literature' (1857) of the 'modern feeling' in Lucretius (*CPW* i. 32–4) and
his essay on Senancour ('Obermann', headn., p. 517; on Arnold's distinction between I. i as 'modern feeling' and I. ii as 'modern thought' see W. E.
Houghton's interpretation of the poem in *Victorian Studies*, 1 (1958), 311–
36). Arnold's ideas in the poem are drawn mainly from Empedocles,
Lucretius, Senancour's *Obermann*, and Spinoza; the songs of Callicles (who

recalls the Youth in 'The Strayed Reveller' (p. 24) draw on Pindar, Hesiod, and Ovid; the dramatic form and ordering of the action owe something to Byron's *Manfred*, whose heading 'A Dramatic Poem' Arnold borrowed for his own subtitle. Arnold published the poem in *1852* but suppressed it in *1853*, to which he attached the famous preface in explanation (p. 172). He reproduced the poem entire in *1867*—having retained Callicles' songs in other collections—'at the request of a man of genius . . . Mr Robert Browning' (footnote in *1867*).

Persons and Scene. Callicles is Arnold's own conception but the name appears in Pindar and means 'fair fame'. The three scenes of the poem are set in the cool forest region of Etna in the early morning (I. i), on the boundary of this region and the unsheltered upper slopes at noon (I. ii), and the barren summit at evening (II. [i]). The journey to the summit is the journey from youth to middle age, elasticity of spirit to world-weariness, a balance of mental powers to 'enslavement' by thought. Arnold found his topographical details in A. de Quatrefages, 'Souvenirs d'un naturaliste: les Côtes de Sicile. V. Etna', *Revue des deux mondes*, NS 19 (1847), 5–36. The details of Empedocles' life and character are freely drawn from Karsten (see headn. above and *Poems*, pp. 157–63).

I. i. 20. *sick or sorry.* 'the old, ideal, limited, pagan world never . . . *was* sick or sorry, never at least shows itself to us as sick or sorry' (Arnold, 'Pagan and Medieval Religious Sentiment', *E in C* I (1865): *CPW* iii. 227–8).

77   I. i. 31. *Peisianax.* A nobleman who once entertained Empedocles (Karsten, ii. 34–5).

78   I. i. 85–98. Callicles shadows Empedocles as Byron's Abbot proposes to follow Manfred at a distance and try to save him (*Manfred* III. i. 160–7; see headn. above).

80   I. i. 151–3. Cf. Arnold in 1849 on the 'root of failure, powerlessness, and ennui . . . in the constitution of Senancour's . . . nature . . . so . . . we should err in attributing to any outward circumstances the whole of the discouragement by which he is pervaded' (*CPW* iii. 372–3; and see headn., p. 520).

81   I. ii. 4–8. Cf. the Alpine setting in 'Obermann Once More', ll. 21–2 (p. 253).

I. ii. 8. The stage direction recalls Byron's 'The Shepherd's pipe in the distance is heard' (*Manfred* I. ii. 47).

82   I. ii. 36–76. Reproduced in *1855* as 'The Harp-player on Etna I. 'The Last Glen'.

83   I. ii. 51. *muffle.* Cf. 'The Scholar-Gipsy', l. 69 (p. 209) and Arnold's July 1865 letter about the vegetation of Lombardy 'not muffling and cooling the ground itself in the way I love' (*L* i. 28).

I. ii. 57–76. Chiron like Empedocles was famed as a teacher and healer; his traditional lore for Achilles contrasts here with the philosophical instruction about to be given to Pausanias. He gave a 'spear of ash' (*Iliad* xvi. 140–1) to Achilles' father, hence 'ashes for spears' (l. 65).

I. ii. 77–426. Empedocles' attempt hitherto to free Pausanias from superstitious fears has been Lucretian, but his long philosophical chant is primarily Stoical in its insistence on recognizing and accepting the limits of human freedom. The stanza consists of two hexameters broken to form a quatrain rhyming a b a b and an unbroken hexameter which rhymes with the corresponding unbroken hexameter in the following stanza.

84 I. ii. 101. *who tries to see least ill*: who, following the less 'degrading' alternative, assumes that man has a limited freedom of action. The subsequent comment (l. 102) is Epicurean: the evils of life are due to fear, belief in the gods is due to fear, fear leads to superstition.

I. ii. 111. Glossed by l. 216 below: 'How man may here best live no care too great to explore'. Empedocles is 'son of wise Anchitus' (Karsten (see headn., p. 520), ii. 92).

85 I. ii. 122–6. A mocking allusion to the miracles attributed to himself (cf. I. i. 108–11, 117–18).

I. ii. 128–30. On the 'burden of ourselves' see Arnold's *1853* Preface, 'the calm, the cheerfulness, the disinterested objectivity have disappeared; the dialogue of the mind with itself has commenced' (p. 172) and on the hidden self Lucretius, *De Rerum Natura* iii. 273–5: 'this nature lies . . . hidden in the most secret recess and there is nothing in our body more deeply seated than this' (cf. 'The Buried Life', ll. 55–6, p. 154).

86 I. ii. 152–6. 'Epicureanism is Stoical, and there is no theory of life but is' (Yale MS). 'Thou hast no *right* to bliss' (l. 160) is Carlylean: 'What Act of Legislature was there that *thou* shouldst be Happy?' (*Sartor Resartus*: Traill, i. 153).

87 I. ii. 197–201. 'The Spirit of the world enounces the problems which this or that generation of Men is to work: men do not fix for themselves' (Yale MS).

I. ii. 217–20. See cancelled motto to 'The Future' (headn., p. 533).

88 I. ii. 237–8. General Confession. *The Book of Common Prayer*: 'We have left undone those things which we ought to have done; And we have done those things which we ought not to have done.'

I. ii. 241. 'We learn too late the truth of *Cause and Effect*: that this or that operation will bring forth this or that result, and not another . . . that there is no pity—no compromise' (Yale MS).

89 I. ii. 266. 'The confusion and sinfulness of men which *we* are avoiding will continue to throw obstacles in our way even when we are cured' (Yale MS).

I. ii. 287–301. 'We learn not to *abuse* or *storm at* the Gods or Fate: knowing this mere madness: as there is nothing wilfully operating against us . . . the power we would curse is the same with ourselves' (Yale MS).

91 I. ii. 332–6. Cf. 'The Second Best', ll. 5–8 (*Poems*, p. 296).

92 I. ii. 362–6. See Arnold's letter to 'K' quoted in headn. to 'Youth's Agitations' (p. 519).

I. ii. 386. Carlyle's famous recipe for content in *Sartor Resartus* (1838): 'the fraction of Life can be increased in value not so much by increasing your Numerator as by lessening your Denominator' (Traill, i. 152–3).

93 I. ii. 422–3. See 'A Summer Night', ll. 91–2 (p. 152).

I. ii. 427–60. Printed separately as 'Cadmus and Harmonia' in *1853*, *1854*, *1857*. Lyrical relief after the philosopher's instruction of Pausanias: Callicles' songs aim at the effects of the chorus in Greek tragedy as Arnold describes them in the preface to *Merope* (see *Poems*, p. 698, ll. 739–42), 'to combine, to harmonise, to deepen the feelings excited' by the action and provide 'relief and relaxation from their intensity'. Cadmus was the founder of Thebes and Zeus gave him Harmonia for his wife; Arnold conflates details of their story from Ovid's *Metamorphoses* iv. 563–605 and Pindar's *Pythian Odes* iii. 86–96. His Empedocles is fleeing from men as Cadmus fled from Thebes.

94 I. ii. 440. The Theban Sphinx.

I. ii. 445–7. Sophocles, *Trachiniae* 112ff.

I. ii. 446. The children were Autonoe, Semele, Agave, and Polydorus; Apollodorus records their misfortunes in *Bibliotheca* iii. 4.

I. ii. 457–60. Arnold echoes *Metamorphoses* iv. 600–3, but alters Ovid's account of the snakes being content 'rembering what they were'; he sees them, significantly, as unhappy if they did not 'wholly forget' (l. 458).

96 II. [i.] 11–19. *Obermann* has 'Je suis las de mener une vie si vaine' and, as often, 'Je suis seul . . . me voilà dans le monde, errant, solitaire' (Lettres XLI, XXII); see also *Manfred* I. ii. 25–7.

II. [i.] 21–2. The 'something' is identified in ll. 235–76 with the hostility of the age, loss of youth, and tyranny of the 'imperious lonely thinking-power'.

II. [i.] 25–6. The elements are 'friends' in Epictetus' *Discourses* iii, ch. 13 and the roots of all things in Empedocles (Karsten (see headn., p. 520) ii. 96); see also Lucretius, *De Rerum Natura* ii. 1112–15 (copied out by Arnold, Yale MS).

II. [i.] 29–32. 'I am sure that in the air of the present times il nous manquent d'aliment . . . we deteriorate in spite of our struggles' (Arnold to Clough, June 1852, *CL*, i. 125); see also his letter of 23 September 1849 (pp. 283–4 above).

II. [i.] 37–88. Printed separately in 1855 as 'The Harp-player on Etna. II Typho'. Empedocles interprets the myth—freely adapted from Pindar's *Pythian Odes* i. 1–28—at ll. 89–107 below.

97 II. [i.] 54. 'Mount Haemus, so called, said the legend, from Typho's blood spilt on it in his last battle with Zeus, when the giant's strength failed, owing to the Destinies having a short time before given treacherously to him, for his refreshment, perishable fruits. See Apollodorus, *Bibliotheca*, book i, chap. vi' (footnote in *1885*).

II. [i.] 67–83. See Pindar, *Pythian Odes* i. 5–12.

98 II. [i.] 90–1. Cf. the 'heads o'ertaxed' and 'palsied hearts' of modern life in 'The Scholar-Gipsy', l. 205 (p. 213).

II. [i.] 92–4. Reflecting current fears about the individual lost in the mass, e.g. Carlyle's 'Signs of the Times' (1829), Mill's 'Civilisation' (1836).

II. [i.] 95–8. The eruptions of Mt. Etna were supposedly caused by the rage and sufferings of Enceladus, the Titan imprisoned under the mountain by the victorious Olympians.

II. [i.] 109–20 (and 191–3 below). Inspired by Prospero's farewell to his powers in *The Tempest*.

99 II. [i.] 121–90. Published separately in *1855* as 'The Harp-player on Etna. III Marsyas'. Pan's jealousy and the presence of the Muses are Arnold's additions to the myth of Apollo and Marsyas; the song is about the price of being a poet.

II. [i.] 128–48. Pan, inventor of the syrinx or shepherd's flute, was jealous of the music of Apollo's lyre (l. 28); the fir was sacred to him (l. 48). Marsyas was associated with Phrygia, where flutes were used in the cult of Cybele (l. 131); the Meander is a Phrygian river flowing on the south side of Mt. Messogis (ll. 136–8).

100 II. [i.] 191–3. see II. [i.] 109–20 n. above. The laurel was sacred to Apollo.

101 II. [i.] 220–9. Cf. 'Obermann', ll. 93–6 (p. 69).

104 II. [i.] 313. Stromboli is one of the volcanic Lipari islands north-east of Sicily.

105 II. [i.] 353–4. The All is the noumen behind phenomenal appearance.

II. [i.] 368. In Karsten (see headn., p. 520) ii. 508–9 Empedocles holds that human beings 'for thirty thousand years . . . are forced to migrate through various shapes of the body . . . when . . . they have absolved their . . . punishment . . . at last they emerge to their heavenly home'.·

II. [i.] 371–2. See I. ii. 128–30 n.

106 II. [i.] 404–16. Empedocles leaps to his death half-exultantly; see Arnold's analysis of his character (headn., p. 520).

107 II. [i.] 417–68. Reprinted in *1855* as 'The Harp-player on Etna. IV Apollo'; in *Selected Poems* (1878) as 'Apollo Musagetes'. Mostly adapted from Hesiod's *Theogony*, with an echo at ll. 436–56 of Aeschylus, *Prometheus Unbound* 114–15 and at ll. 417–20 of the Fourth Spirit's Song (*Manfred* I. i. 88–95).

II. [i.] 427. *Thisbe*. A Boeotian town between Mt. Helicon and the Gulf of Corinth, famed for its wild pigeons (ll. 431–2).

108 II. [i.] 457–68. Echoing Hesiod's invocation of the Muses who 'uttering their immortal voice' celebrate Zeus and the other immortals, the sun, the moon, 'Earth, too, and great Oceanus, and dark Night, and the holy race of all the other deathless ones' (*Theogony* 36–9, 11–21, 43–56).

109 *Tristram and Iseult*. The first modern treatment of the Tristram legend in English, published in *1852*. Arnold told Herbert Hill on 5 November 1852 (*TLS* 19 May 1932), 'I read the story . . . some years ago at Thun in an article

in a French Review on the romance literature: I had never met with it before, and it fastened on me: when I got back to England I looked at the Morte d'Arthur and took what I could, but the poem was in the main formed, and I could not well disturb it.' This probably refers to the second rather than the first of his visits to Thun in September *1848* and September *1849*, when a tale of doomed love might have 'fastened' on him most strongly (see headnn. to 'Switzerland' and 'Parting', pp. 514 and 515; Introduction, p. xvii). The poem was not wholly 'formed' in Thun; there were additions to parts I and III for inclusion in *1853* and revisions in *1853* and *1854*. The 'article' was Théodore de la Villemarqué's 'Les Poèmes gallois et les romans de la Table-Ronde', of which the first instalment—alone used by Arnold—appeared in *Revue de Paris*, 3rd series, 24 (*1841*), 266–82. The poem draws on (i) la Villemarqué's summary of the legend (pp. 274–5); (ii) for the Merlin episode in part III, the Vulgate Merlin quoted in notes to the Preface of Southey's *1817* edition of Malory's *The Birth, Lyf, and Actes of Kyng Arthur*; (iii) Malory's *Morte d'Arthur* itself for background material and incidental phrases. The following note was first attached to the poem in *1853* to meet complaints about the poem's obscurity made by Clough and others after *1852*; see Arnold to Clough, 25 August 1853: 'Froude recommends prefacing Tristram and Iseult with an extract from Dunlop's Hist. of fiction . . . in preference to telling it in my own words' (*CL*, p. 140). The note is a patchwork of sentences from Dunlop's *History of Fiction* (3rd edn., 1845, pp. 85–7), not read by Arnold before 1853; he regretted not having known his story of 'the *sails* . . . a beautiful ending, which I should perhaps have used, had I known of it, but I did not' (letter to Swinburne, 26 July 1882, *TLS* 12 August 1920). The note runs:

'In the court of his uncle King Marc, the king of Cornwall, who at this time resided at the castle of Tyntagel, Tristram became expert in all knightly exercises.—The king of Ireland, at Tristram's solicitations, promised to bestow his daughter Iseult in marriage on King Marc. The mother of Iseult gave to her daughter's confidante a philtre, or love-potion, to be administered on the night of her nuptials. Of this beverage Tristram and Iseult, on their voyage to Cornwall, unfortunately partook. Its influence, during the remainder of their lives, regulated the affections and destiny of the lovers.—

After the arrival of Tristram and Iseult in Cornwall, and the nuptials of the latter with King Marc, a great part of the romance is occupied with their contrivances to procure secret interviews.—Tristram, being forced to leave Cornwall, on account of the displeasure of his uncle, repaired to Brittany, where lived Iseult with the White Hands.—He married her—more out of gratitude than love.—Afterwards he proceeded to the dominions of Arthur, which became the theatre of unnumbered exploits.

Tristram, subsequent to these events, returned to Brittany, and to his long-neglected wife. There, being wounded and sick, he was soon reduced to the lowest ebb. In this situation, he despatched a confidant to the queen of Cornwall, to try if he could induce her to follow him to Brittany, etc.'—Dunlop's *History of Fiction*.

I. *Tristram*. For Tristram's life story Arnold uses his own invention filled out by la Villemarqué's account in *Revue de Paris*, pp. 274–5, in which Tristram dies of chagrin after his wife tells him that the Queen of Cornwall has refused to come. Arnold rejects this ending, depicting the second Iseult as the mother of two children by Tristram and a pattern of wifely patience and sympathy.

I. 4. Keats, 'Eve of St Agnes', ll. 322–4: 'the frost-wind blows / Like Love's alarum pattering the sharp sleet / Against the window-panes'. Cf. I. 83, 'sharp patters the rain'.

I. 9–93. This and the octosyllabic link passages between Tristram's dreams are recited by a medieval Breton bard.

I. 9. Keats's 'La Belle Dame Sans Merci', ll. 5–6: 'O what can ail thee, knight-at-arms! / So haggard and so woe-begone'.

111   I. 74. Added in *1853*, probably after encountering Dunlop's description 'Yseult with the White Hands' (see headn., p. 525).

114   I. 203. 'sire Launcelot broughte sire Tristram and la beale Iseud unto Joyous gard that was his own Castel' (Southey's Malory ii. 86).

115   I. 232–3. 'Youth and Calm', l. 23 (p. 131): '*Calm's not life's crown, though calm is well.*'

I. 234–42. Malory's Tristram plays no part in these campaigns; Arnold expands la Villemarqué by using Malory on King Arthur's struggle against the Emperor of Rome.

119   II. *Iseult of Ireland*. The meeting of the lovers at Tristram's death-bed is Arnold's invention (in la Villemarqué Iseult dies after arriving too late). He was 'by no means satisfied with . . . the second part' (letter to Herbert Hill, quoted in headn., pp. 524–5).

122   II. 84. *deep draughts of death*. Borrowed directly from Southey's Malory i. 247 (death of Elizabeth, wife of Melodias, in childbirth).

II. 101–11. 'A subject for a Pre-Raphaelite painting' (*Poems*, p. 224).

123   II. 131–46. Added in *1869*; in *1852* constituting the first paragraph of 'Lines Written by a Deathbed' (see 'Youth and Calm', headn., p. 528).

II. 147–51. Suggested by Byron's *The Siege of Corinth* (1816), ll. 620–7, included in Arnold's selection *The Poetry of Byron* (1881).

124   II. 152–4. Arnold's arras owes something to both Keats's 'The Eve of St Agnes', ll. 358–9 and Tennyson's 'The Palace of Art' (1832), ll. 61–4.

II. 163–4. Tennyson's 'The Lady of Shalott', l. 163, 'Who is this? and what is here?' (see 'The Sick King in Bokhara', l. 89 n., p. 510).

II. 164–7. Resembling 'The Church of Brou', part III (pp. 170–1).

II. 174. A favourite rhythm, as in 'To Marguerite—Continued', l. 24: 'The unplumb'd, salt, estranging sea' (p. 61).

125   II. 192–3. Spoken by the Breton bard. The change of perspective—the

lovers 'lived and loved / A thousand years ago'—recalls the close of 'The Eve of St Agnes'.

III. *Iseult of Brittany*. The second Iseult is Arnold's own invention but he took her story about Merlin and Vivian from the Vulgate Merlin quoted by Southey (see headn., p. 525). He told Herbert Hill: 'The story of Merlin, of which I am particularly fond, was brought in on purpose to relieve the poem, which would else I thought have ended so sadly; but perhaps the new element introduced is too much' (letter quoted in headn.)—it was perhaps 'brought in' after he had consulted Malory on his return to England from Thun. The story tells of Merlin's fatal enchantment by Vivian, whose spell he already finds so powerful that he succumbs to magic 'such as he himself had taught her . . . and when he awoke and looked about him, it seemed that he was enclosed in the strongest tower in the world' (Southey's Malory i. xiv–xlvi). The second Iseult's tale of doomed passion is in keeping with the *Liebestod* of her husband and the first Iseult, but Arnold's 'particular fondness' for it seems to indicate its special significance for himself (see headn.).

III. 7. *a green circular hollow*. In la Villemarqué's 'Visite au tombe de Merlin' the Val-des-Fées is 'un immense amphithéâtre' (*Revue de Paris*, 2nd series, 41 (1837), 45–62).

III. 22. *fell-fare*: more commonly fieldfare; a species of thrush.

128 III. 112–50. Omitted in *1853*, *1854*, perhaps because the digression seemed too long. Supposedly the reflections of the Breton narrator (see the 'medievalizing' in ll. 133–50), it is clearly autobiographical: 'The misery of the present age is not in the intensity of men's suffering—but in their incapacity to suffer, enjoy, feel at all, wholly and profoundly . . . the eternal tumult of the world mingling, breaking in upon, hurrying away all. Deep suffering is the consciousness of oneself no less than deep enjoyment' (Yale MS). Cf. 'Empedocles on Etna' II. [i.] 21–2; 'The Scholar-Gipsy', ll. 142–5 (pp. 96, 212).

III. 122. *bloom*. See 'Sohrab and Rustum', l. 856 (p. 206) and 'The Youth of Nature', ll. 53–5 n. (p. 532).

II. 143–5. Suetonius, *De Vita Caesarum* i. 7 (of Caesar when quaestor in Spain): 'noticing a statue of Alexander the Great [in Gades] . . . he heaved a sigh, and as if out of patience with his own incapacity in having as yet done nothing at a time of life when Alexander had already brought the world to his feet, he straightway asked for his discharge, to grasp the first opportunity for greater enterprises at Rome.'

III. 149. *Soudan's realm*: the Persian empire of Darius.

129 III. 156. *Broce-liande*: legendary region adjoining Brittany, where in the Arthurian cycle Merlin was enchanted by Vivian; his tomb is supposedly in one of its forests.

III. 179. Cf. the description of Marguerite, 'Parting', ll. 17–18; 'what voice is this I hear, / Buoyant as morning, and as morning clear?' (p. 57).

130 III. 224. 'And always Merlyn lay aboute the lady to have her maydenhode, and she was ever passynge weary of his love' (Southey's Malory i. 91).

131 *Youth and Calm*. Probably originally conceived as part of 'Tristram and Iseult': on its first publication in *1852* the poem formed part of 'Lines written by a Death-bed', of which ll. 1–16 were in *1869* inserted in 'Tristram and Iseult' II to describe 'Iseult of Ireland' (see II. 131–46 and n., p. 526). The poem, linked in theme with 'Youth's Agitations' and 'Growing Old' (pp. 74, 267), was printed with them in *1867* and became an 'Early Poem' in *1877*.

l. 21. Cf. 'Tristram and Iseult' II. 165–7 and 'The Church of Brou' III. 16 (pp. 124, 170).

ll. 23–5. Sainte-Beuve on Senancour and the 'crise antérieure à toute maturité' (1832): 'ils devient plus calme, plus capable de cette regulière stabilité qui n'est pas le bonheur au fond, mais qui le simule à la longue même à nos propres yeux' (*Portraits contemporains* i (Paris, rev. edn. 1868), 168).

Faded Leaves. Arnold first gave this title in *1855* to a series of five poems— four of them published separately in *1852*—inspired by difficulties in his courtship of Frances Lucy Wightman whom he met in 1849 or early 1850 and married 10 June 1851 (see 'Absence', headn., p. 516). Tom's obituary notice for Arnold (*Manchester Guardian*, 18 May 1888) refers to its not having been 'all prosperous sailing in his love . . . of one such counterblast which drove him out of England and towards the Alps the lovely stanzas "Vain is the effort to forget" are the record' (see (iv) 'On the Rhine' below). According to family tradition, Justice Wightman, unsure of Arnold's financial prospects, forbade the two to meet; 'Calais Sands' (p. 134), dated August 1850 but not included in the series, recalls Arnold's delay at Calais to catch sight of Frances Lucy before travelling on to the Rhine. Her father agreed to the engagement after Lord Landsdowne had appointed Arnold an Inspector of Schools, which became his lifelong career. The order of the poems in the series, which seems to be chronological, is followed here.

(i) *The River*. Published in a shorter version in *1852*. Justice Wightman (see n. above) had a house at Hampton-on-Thames by the river (l. 1) which in l. 25 acquires properties from Wordsworth's 'The River Duddon' (1820) xxxiv. 5: 'Still glides the Stream, and shall for ever glide'. Miss Wightman shares her mixture of gaiety and gravity with 'Marguerite' (with ll. 5–6 cf. 'A Memory-Picture', ll. 27–8 and 'Parting', l. 39, pp. 55, 58).

132 (ii) *Too Late*. Written in August 1850; see headn. above. In l. 4: 'the twin soul which halves their own' ('halves' meaning 'be the other half of'), Arnold may be running together memories of the origin of love in Plato's *Symposium* and Horace's 'half of my own soul' in *Odes* I. iii. 8 (also recalled in 'To Marguerite—Continued', ll. 22–4; see p. 61 and n., p. 516).

(iii) *Separation*. Written in August 1850 but not published till *1855*; see headn. (p. 528). With l. 6 cf. Arnold's 1849 project for 'A Memory-Picture' (headn., p. 515).

133 ll. 15–16. 'Marguerite' had blue eyes and ash-blonde hair; see 'Absence' and 'Calais Sands' (pp. 64, 134).

(iv) *On the Rhine*. Written August or September 1850; the 'iron knot' (ll. 7–8) probably means the poor prospects which kept the lovers apart (see headn., p. 528).

134 (v) *Longing*. Probably written August or September 1850, or perhaps a little later; Arnold and Frances Lucy resumed their correspondence by the end of the year.

Calais Sands. Arnold's manuscript is dated 'By the seashore near Calais. August 1850', but he did not publish the piece (much revised) until *1867*, 'perhaps because it was too intimate a self-revelation' (*Poems*, p. 243). See headn. to 'Faded Leaves' (p. 528).

l. 5. Ardres, a small town ten miles south-east of Calais, is near the 'Field of the Cloth of Gold', the meeting-place in 1520 of Henry VIII and Francis I of France.

135 *Dover Beach*. Arnold's most famous lyric was not published until *1867* but is agreed to have been written in the summer of 1851 at the time of his union with Frances Lucy Wightman. A pencilled autograph of ll. 1–28 ending 'And naked shingles etc.' (implying that ll. 29–37 were written earlier) appears on the back of notes about Empedocles from Karsten (see 'Empedocles on Etna' headn., p. 520) which can be plausibly dated 1849–50. Existing records of Arnold's visits to Dover with his wife after their marriage on 10 June 1851 suggest the actual occasion of the poem (his accounts for 19–30 June include 'Journey to Dover and back—£9–9–0'), the implication here being that he wrote ll. 29–37 at Dover on his first visit there after his marriage, adding ll. 1–28 shortly afterwards in London. His concern with the decay of orthodox religious belief links 'Dover Beach' with 'Stanzas from the Grande Chartreuse' (pp. 159–65), which is associated with his honeymoon visit to the monastery on 7 September 1851. The lyric consists of four unequal irregularly rhymed verse paragraphs; the lines vary between two and five stresses, but more than half are five-stressed.

ll. 1–14. Characteristically Arnoldian setting of sea and moonlight with some overtones from the passage in Senancour's *Obermann* singled out in Arnold's 1869 essay on Senancour ('Obermann' headn., p. 517) for its 'poetic emotion and deep feeling for nature', and beginning: 'je me suis plaçai sur le sable où venait expirer les vagues. L'air était calme . . . La lune . . . parut' (Lettre IV. i. 22). In ll. 1–6 the effect owes much to the high proportion of monosyllables and simplicity of the key epithets 'calm', 'fair', 'tranquil'. In l. 6 the sweetness of the air is felt before the melancholy sound of the sea is is first heard.

ll. 3–4. *the light / Gleams and is gone*. The movement is in keeping with the observation.

136 l. 13. *tremulous cadence slow*: Miltonic placing of epithets, as in 'vast edges drear' (l. 27).

ll. 14–18. The waves inevitably remind mankind of time and change as, Arnold believes, they reminded Sophocles 'long ago': there is no direct reference in the plays but the feeling is similar in *Trachiniae* 112ff.

l. 20. See 'Stanzas from the Grande Chartreuse', ll. 80–2 (p. 161) where the Greek's musings over two dead faiths resemble the elegiac tone of Arnold's reflections at Dover.

ll. 21–5. 'Obermann Once More', ll. 189–90: 'But slow that tide of common thought, / Which bathed our life, retired' (p. 258).

l. 23. 'This difficult line means, in general, that at high tide the sea envelops the land closely. Its forces are "gathered" up (to use Wordsworth's term for it) like the "folds" of bright clothing ("girdle") which have been compressed ("furled"). At ebb tide, as the sea retreats, it is unfurled and spread out' (G. H. Ford, *The Norton Anthology of English Literature* (rev. edn. (1968), ii. 1039).

ll. 24–8. 'Probably the most musically expressive passage in all Arnold's poetry and a valid poetic equivalent for his feelings of loss, exposure, dismay' (*Poems*, p. 256).

ll. 29–34. 'The beauty of the moonlit scene, lingeringly dwelt on in l. 32, is a deceptive enchantment if it leads us to suppose that the universe is anything but indifferent to man' (*Poems*, p. 256). Cf. 'The Youth of Man', ll. 28–31 (p. 144).

ll. 35–7. The source is Thucydides on the Battle of Epipolae (413 BC). His *History of the Peloponnesian War* (vii, ch. 44) describes the confusion of the Athenians 'in a battle by night' where friend could not be distinguished from foe and 'they not only became panic-stricken but came to blows with one another'. The image of the night battle, probably familiar to everyone who read 'Greats' at Oxford, was used by Clough in 'The Bothie of Toper-na-Fuosich' (1848) ix. 51–4 and by Newman in his January 1839 sermon (*University Sermons* (London, 1843), p. 193; perhaps affecting Arnold's lines): 'Controversy, at least in this age, does not lie between the hosts of heaven, Michael and his angels on the one side, and the powers of evil on the other; but it is a sort of night battle, where each fights for himself, and friend and foe stand together.'

137 *Stanzas in Memory of Edward Quillinan*. Edward Quillinan (1791–1851), minor poet and translator of Camoens, married Wordsworth's daughter Dora in 1841; she died in 1847, leaving their two daughters Rotha and Jemima. Arnold wrote out the poem in Rotha Quillinan's album, dating it 27 December 1851, and included it in *1853*; he also selected ll. 14–20 ('Good' for 'Sweet' in l. 16) as an inscription, dated 14 August 1869, for the second volume of *1869*, Jemima's gift to her sister Rotha.

l. 2. Dora's death in 1847 profoundly affected her husband and her father.
Memorial Verses. April 1850. This and the following poem, 'The Youth of

Nature', Arnold's tributes to Wordsworth who died 23 April 1850, combine elegy and literary criticism as in 'Obermann', 'Haworth Churchyard', and 'Heine's Grave' (pp. 66, 216, 230). The present poem, published in *Fraser's Magazine* in June 1850 and reprinted in *1852*, was requested by Wordsworth's son-in-law Edward Quillinan (see preceding headn.) who described it in 1851 as '*very* classical . . . a triple Epicede on . . . Wordsworth and Goethe, and on Byron, who, I think . . . is not tall enough for the other two' (*Correspondence of H. C. Robinson with the Wordsworth Circle*, ed. E. J. Morley (Oxford, 1927) ii. 769). Byron, also mentioned in 'Stanzas from the Grande Chartreuse' and 'Haworth Churchyard', remains a pervasive influence in Arnold's work and forms the subject of a late essay (*E in C* II (1888): *CPW* ix. 217). Arnold also returns in the 1880s to the celebration of Wordsworth's healing and 'rejuvenating power' (ll. 49–57), emphasizing the 'extraordinary power with which he feels the joy offered us in nature . . . Wordsworth's poetry, when he is at his best, is . . . as inevitable as Nature herself' ('Wordsworth', *CPW* ix. 51–2; cf. 'The Youth of Nature', ll. 53–5, p. 141). On Wordsworth's limitations, unsuitable material for a panegyric, see the passage linking him with Goethe, 'Obermann', ll. 53–4 (p. 67).

ll. 1–2. Goethe died in 1832, Byron at Missalonghi in Greece in 1824.

138 l. 8. 'Goethe lays his finger on . . . [Byron's] real source of weakness . . . "The moment he reflects he is a child." ' ('Byron', *E in C* I (1888): *CPW* i. 227).

ll. 12–14. Byron's 'On this day I complete my thirty-sixth year', ll. 9–10: 'The fire which on my bosom preys / Is lone as some Volcanic isle. . .' (quoted on the Byronic 'Titanism of the Celt', *On the Study of Celtic Literature* (1867): *CPW* iii. 372).

ll. 19–22. 'this strange disease of modern life' ('The Scholar-Gipsy', l. 203, p. 213).

ll. 29–33. Cf. Virgil, *Georgics* ii. 490–2.

ll. 43–4. Carlyle's 'Characteristics' (1831): 'How changed in these new days . . . Not Godhead, but an iron, ignoble circle of Necessity embraces all things . . . Doubt storms in . . . through every avenue, inquiries of the deepest, painfulest sort . . . sceptical, suicidal cavillings' (Traill, xxviii. 30).

139 l. 56. Cf. 'Empedocles on Etna' II. [i.] 221–2 (p. 101).

l. 72. The Rotha flows by Grasmere Churchyard.

The Youth of Nature. Probably begun in the same period as the preceding poem (see headn.); the 'pure June-night' (l. 6) must belong to 1850 when Arnold made a summer visit to the Lake District (not to 1851 when he married Miss Wightman—see 'Faded Leaves' headn., p. 528): the reference in ll. 2–9 suggests that Wordsworth's death was still recent. Wordsworth is buried in the churchyard at Grasmere and Arnold's boat is on the lake; 'if he looks in the direction of Ambleside haze hides Rydal Water, but Fairfield rises on the left. Arnold was used to looking across at Fairfield from his home at Fox How' (*Poems*, p. 260). He finished the poem in January 1852:

'windy bright day—walked alone along Rydal and Grasmere ... finished Wordsworth; pindaric' (unpublished diary). 'Pindaric' is used here to mean an unrhymed lyric consisting of a varying number of verse-paragraphs each with an unfixed number of lines as in 'The Youth of Man' and 'The Future' (headn. below), the only other unrhymed lyrics of this kind in *1852*. The line is normally three-stressed but has between six and nine syllables; some lines with fewer or more than three stresses occur at irregular intervals. Nature as a 'Power' in this poem and 'The Youth of Man' (headn. below) suggest Arnold's recent reading of *The Prelude* (1850).

140   ll. 15–17. In Wordsworth's 'The Brothers' (1800): Egremont is a small town through which the river from Ennerdale Lake flows to the sea; for the Pillar see ll. 366–8; 'one particular rock / That rises like a column from the vale, / Where by our shepherds it is called THE PILLAR'.

l. 18. The name of the shepherd's cottage in Wordsworth's 'Michael', ll. 136–9.

l. 24. She lived on 'the Banks of Tone' (Wordsworth's 'Ruth' (1800), l. 214).

ll. 28–35. Wordsworth outlived all the other Romantic poets and became a high Tory in his later years.

l. 36 Tilphusa was near Thebes.

ll. 41–5. Tiresias fled with the Thebans after their defeat by the Epigoni; according to one tradition he died in this manner.

141   ll. 53–5. Cf. 'Memorial Verses', ll. 45–50 (pp. 138–9) and Carlyle's 'Man of Letters' as 'the light of the world; the world's Priest' (*On Heroes and Hero-Worship* (1841): Traill, v. 157); 'bloom' (l. 54) is a favourite word in Arnold suggesting ephemeral delicacy and freshness.

ll. 59–74. Arnold's question whether he would experience the beauty of nature had not a great poet first revealed it to him is answered by Nature at ll. 79–86 (beauty is experienced by all but in different degrees, even by poets) and ll. 103–6 (our knowledge and appreciation depend on how far we really know ourselves; cf. 'The Buried Life', pp. 153–5).

ll. 76–7. Ida is both a mountain range in Mysia associated with the worship of Cybele and a mountain in Crete where Rhea hid the infant Zeus to save him from Cronos. Rhea was identified with the Asiatic 'Magna Mater' by the Greeks of Asia Minor.

142   ll. 110–14. Recalling 'A Memory-Picture' (headn., p. 515).

ll. 115–16. 'No' is the expected answer, but see 'Epilogue to Lessing's "Laocoön"' where the poet—unlike the painter or musician—'must life's *movement* tell' (l. 140, p. 250).

143   ll. 133–4. See 'the mateless, the one' (l. 118).

The Youth of Man. 'Pindaric "sink o youth in thy soul"' (l. 116)—Arnold's note in his 1852 diary on this sequel to 'The Youth of Nature', completed on 4 January 1852 and published in *1852*. With l. 27 cf. Empedocles' warning ('Empedocles on Etna' I. ii. 177–81) ending: 'No, we are strangers here; the world is from of old'; and with ll. 29–31 the deceptive enchantment of the beauty of nature in 'Dover Beach', ll. 30–4 (pp. 86, 136).

144 ll. 51–60. Cf. Goethe's invocation, 'Eins und Alles', l. 7: 'Come, soul of the world, and flood us through'; and with l. 58 'The Youth of Nature', ll. 51–2: 'The complaining millions of men / Darken in labour and pain' (p. 141).

146 ll. 112–18. Reprinted separately as 'Power of Youth' in *1853*; ll. 116–18 recall the 'unlit gulph' in 'The Youth of Nature', l. 102 (p. 142) and the Stoic notion of the 'buried self' ('Empedocles on Etna' I. ii. 129–30, p. 85).

Lines Written in Kensington Gardens. The invocation of Nature linked with fear of having missed 'life' (ll. 37–40) is close to the preceding poem 'The Youth of Man', also published in *1852*. Arnold's '*1852* temper' desires feeling without agitation, calm without indifference, but as in 'Youth's Agitations' (p. 74) fears discontent from this as much as from 'youth's hurrying fever'. With the Spinozist prayer to Nature at ll. 37–44 and l. 37, 'Calm soul of all things' cf. 'The Youth of Man', ll. 51–60 (pp. 144–5).

147 ll. 13–16. Echoes Werther's description of 'the swarming little world' among the grass stalks (Goethe, *Werther* (1774): *Werke* xvi (Stuttgart, 1828) 8).

l. 36. 'every day, all these appear, live, go thro: their stages whether I see them or no. I, in an unnatural state of effort and personal wrapped-upness do not see them' (prose argument for this line, Yale MS).

ll. 43–4. The literal meaning coexists with the longing still to experience and feel intensely.

148 *The Future*. Published in *1852*; for the style see headnn. to 'The Youth of Nature' and 'The Youth of Man' (pp. 531–2 and 532), and on Arnold's images of river and sea for human life headnn. to 'To Marguerite—Continued', 'Human Life', and 'The Buried Life' (pp. 515–16, 519, and 534). In *1853* and *1854* the title is followed by a motto of Arnold's own composition:

For Nature hath long kept this inn, The Earth
And many a guest hath she therein received.

149 ll. 36–7. Genesis 26: 1–67.

ll. 41–9. Exodus 3: 1–6. Carlyle refers to the wonder of the 'first Pagan Thinker among rude men' (*On Heroes and Hero-Worship* (1841): Traill, v. 7, 9).

ll. 50–65. See 'A Dream', ll. 36–7. Probably a recollection of part of a poem by Dr Arnold, including the lines, 'A straight embankéd Line / Confines thee, wont to trace at Will erewhile / Thy own free margin; and the Haunts of Men / Thy Spotless Waves defile . . .' See further *Poems*, 280 and cf. 'The Buried Life', ll. 97–8 (p. 155).

l. 57 *shot*. The image is of textiles woven to give a changing colour effect.

150 l. 82. *the man*. Mankind, but the thought is also turning to the end of an individual life.

A Summer Night. The two moonlit nights (ll. 1–10, 13–21) occurred before *1852* (date of the poem's publication): the first belongs to May 1851 when, shortly after their engagement, Arnold went late at night to gaze up at Frances Lucy's window (unpublished 1851 diary and see headn. to 'Faded Leaves', p. 528); the second perhaps refers to unhappy earlier stages of their courtship when he tried to catch sight of her at Calais (see headn. to 'Calais Sands', p. 529). The 'moon-blanched' scene is a favourite setting—cf. headn. to 'Dover Beach' above; its calm is contrasted with the conflict of the 'divided self' (ll. 27–33; cf. 'Empedocles on Etna' II. [i.] 220–34 and 'Obermann', ll. 95–6, pp. 101–2, 69). The alternative ways of life are illustrated in the following two verse paragraphs (ll. 37–50, 51–73).

151 ll. 37–8. Similar images of imprisonment in 'Memorial Verses', ll. 45–6 and 'Growing Old', ll. 23–4 (pp. 138–9, 268).

153 *The Buried Life*. Published in *1852* but not reprinted until *1885* (possibly because of the references to early personal experience in ll. 1–25); but the poem is important as expressing certain of Arnold's key ideas about the nature of the real or true self. It links two favourite ideas, the river of life and the Stoic notion of an inner or buried self (see n. to ll. 30–63 below), together with the distinction which haunted him between our apparent freedom of behaviour and its determination by forces beyond our control (ll. 43–4). Other characteristic ideas include the modern world as a sort of Waste Land, noisy, hot, full of 'sick hurry' under a glaring noonday sun (ll. 79–80, 82, 91). *Commentary* sees an 'obvious relation, in its opening motive, to the Marguerite series and, in the second half, to *Dover Beach*' (p. 195). Arnold's pessimism about the chance of understanding between lovers (ll. 12–15: cf. 'Isolation. To Marguerite', p. 59) shifts to the possibility of self-knowledge at rare moments of happiness in love (ll. 77–87).

ll. 30–63. Risks confusion: in ll. 38–40 the river is subterranean, agreeing with ll. 55–6 where shafts cannot reach the buried self or stream; but ll. 43–4 suggest the everyday self is the surface of a stream broken by eddies, while the buried self is the deeper level of the same stream the direction of whose flow is unmistakable.

154 l. 55. 'Empedocles on Etna' I. ii. 129–30; 'the wiser wight / In his own bosom delves' (p. 85).

ll. 57–60. 'We have been on a thousand lines and on each have shown spirit talent even geniality but hardly for an hour between birth and death have we been on our own one natural line, have we been ourselves, have we breathed freely' (prose argument of these lines, Yale MS).

l. 67. i.e., we wish no more to be racked.

155 ll. 88–90. An ordinary river when self-knowledge is momentarily achieved.

Self-Deception. Another of Arnold's *1852* poems on his 'master-feeling' (l. 23), a 'natural craving . . . for . . . a distinct seeing of my own way as far as my own nature is concerned' (*CL*, p. 110); but 'not a piece [that] at all satisfies me' (Russell, p. 43).

156 *Despondency*. Published in *1852* and probably written 1849–1852. With l. 2 cf. 'To Marguerite—Continued', l. 1: 'Yes! in the sea of life enisled' and with ll. 5–8 cf. 'Self-Deception', ll. 21–2: 'And on earth we wander, groping, reeling; / Powers stir in us, stir and disappear' (pp. 61, 156).

The Neckan. Arnold printed this after 'The Forsaken Merman' (1852) as a companion piece from the time of its first publication in *1853*. It is 'a song about a song' in the manner of Heine, whose 'sweetest note, his plaintive note' and the 'rapidity and grace' of his use of the ballad form Arnold warmly praised in his 'Heinrich Heine' (*E in C* I (1865): *CPW* iii. 124, 131). The story of a non-human partner who longs to possess a soul, common in Scandinavian and German folklore, was widely drawn on for tales and ballads (for example by Goethe and La Motte Fouqué). Arnold's source is the tale of the Neckan and the priest in Benjamin Thorpe's *Northern Mythology* (1851) ii. 80, which he varied by (*a*) making the Neckan not happy first and last, as in the Swedish tale, but melancholy throughout, thus sustaining the 'plaintive' note, and (*b*) omitting in *1853* the miracle of the flowering staff; he first added this (ll. 53–6) in *1869*, when he also added ll. 61–4, thus making his Neckan less rejoiced at his salvation than grieved by the thought of human unkindness.

159 *A Caution to Poets*. Used as the prefatory poem (untitled) for vol. i of *1869* but first published with this title in *1867* and reprinted in *1868* and *1877* also with the present title. It was written as early as 14 December 1852, when Arnold quoted it in a letter to Clough (p. 287): 'There is an oracular quatrain for you, terribly true' (*CL*, p. 126). His prose argument in the Yale MS, 'What it gives us no pleasure to conceive or make, it will give the world no pleasure to contemplate', echoes Horace's celebrated statement in *Ars Poetica* 101–5 (including 'If you wish me to weep, you must first weep yourself') and Goethe's expansion of the theme in 'Antik und Modern' (1818: *Werke* xxxix (Stuttgart, 1830), 76–7; see *Poems*, p. 301).

Stanzas from the Grande Chartreuse. Arnold wrote his 'Stanzas' in the period between 7 September 1851, the date of his honeymoon visit to the Grande Chartreuse, and April 1855, the date of their publication in *Fraser's Magazine*. They were first reprinted in *1867*, as with 'Dover Beach' which was written in the same period and was similarly concerned with the decay of orthodox religious belief. Other links with poems written in 1849–52 include the occasional echoes of 'Tristram and Iseult' (with ll. 181–6 cf. II. 153–6, 186–9 and also 'The Church of Brou' I. 9–12, 15–18; pp. 124, 165); the stanza form, used elsewhere by Arnold only in three poems of these years ('Isolation. To Marguerite', 'To Marguerite—Continued', and 'Morality', pp. 59, 61, *Poems*, 273); and the sympathy with Romantic melancholy expressed in the poem.

l. 2. *the crocus*: the autumn crocus.

l. 10. The Guiers Mort, arising near the Grande Chartreuse, joins the Guiers Vif in the valley.

160 l. 27. *this outbuilding*: the infirmary outside the main gate.

l. 30. The first settlement was by St Bruno, the founder of the Carthusians, in 1084.

ll. 40–2. Perhaps referring to the tablet of the Pax, which is kissed by the priest and then by each member of the congregation; the Host would not be passed hand to hand nor Holy Communion received standing.

ll. 47–8. Carthusians are in fact buried in their habits on a bare plank of wood.

ll. 56–60. Discreet allusion to Chartreuse, the famous liqueur manufactured by the monks.

161 l. 67. Including Carlyle, Goethe, Senancour, Spinoza, some of the Stoic philosophers, and Lucretius.

ll. 79–88. The exiled Greek, emancipated from Greek paganism and gazing at the 'Runic stone', laments two dead faiths as Arnold regrets that the Protestant and Catholic forms of Christian orthodoxy are alike incredible (ll. 79–84). Arnold's celebrated lines on the age as a spiritual No Man's Land (ll. 85–8: 'Wandering between two worlds, one dead, / The other powerless to be born') express a commonplace among reflecting minds in his circle: see Tom Arnold to Clough on the 'age of transition' being for the thinking few 'nothing but sadness and isolation' (*Correspondence of AHC* i. 180) and cf. J. S. Mill, 'The Spirit of the Age' (1831): 'Mankind have outgrown old institutions and doctrines and have not yet acquired new ones' (*Essays on Literature and Society*, ed. J. B. Schneewind (Toronto, 1965), p. 30); Carlyle's 'Characteristics' has the same theme, 'the Old has passed away: but, alas, the New appears not in its stead' (Traill, xxviii. 29–30).

l. 87. Matthew 8:20.

162 l. 99. A sciolist is a pretender to knowledge.

l. 109. Presumably the 'Sons of the world' (l. 161)

163 ll. 133–42. Echoing Carlyle's 'Characteristics' (1831: Traill, xxviii 31): 'behold a Byron in melodious tones "cursing his day" . . . Hear a Shelley filling the earth with inarticulate wail; like the infinite, inarticulate grief and weeping of forsaken infants.'

ll. 145–50. See headn. to 'Obermann' (p. 517).

ll. 163–8. Grudging but only partly ironic.

164 l. 169. The shadow of Romantic melancholy.

165 l. 210. The region surrounding the monastery was known as the *désert*; 'a general term, probably from the Celtic *dysart*, a holy, retired place . . . so used by Senancour for any Alpine solitude' (Sells, pp. 295–6).

The Church of Brou. Arnold was planning this poem as early as 1851 ('La châtelaine architecte' appears in his tentative list of poems for *1852*) but postponed publication till *1853* because of the closeness of part III to 'Tristram and Iseult'. Part I suggests influences from Heine's metre and manner in his 'romances' and perhaps also from Tennyson, since the stanza

is that of 'The Lord of Burleigh' (1842). Part III has a strong Keatsian
flavour (ll. 16–17n). Arnold took the story from Edgar Quinet's 'Des Arts de
la Renaissance et de l'Eglise de Brou', *Mélanges* (1839): *Œuvres complètes*
(Paris, 1857) vi. 351–63). Quinet included few historical details (hence the
poem's factual inaccuracies), but Arnold was attracted by his reflections: 'ce
n'est pas seulement la duchesse et le duc de Savoie qui dorment là dans ce
cercueil, c'est l'ancienne foi, c'est l'ancien amour; c'est la poussière de
toutes les croyances tombés'; 'l'idée que l'homme pût être separé par la mort
de ce qu'il avait aimé n'avait pas encore approché de l'âme humaine' (op. cit.
vi. 355). After publishing the whole poem in *1853* and subsequent collections, Arnold omitted parts I and II in *1877* because of their inaccuracies
(Buckler, pp. 41–2; E. Cook, *Literary Recreations* (London, 1919), p. 295 n.),
printing part III on its own as 'A Tomb among the Mountains'; but in *1881*
he restored the poem to its original shape, classifying it as an 'Early Poem'.
See also R. H. Super, 'The Church of Brou and its Occupants', *Arnoldian*,
12 (1985), 1–2.

I. 6. *Savoy's Duke*. Philibert II of Savoy (unnamed by Quinet) died 1504 of a
disease contracted while hunting.

I. 12. *Duchess Marguerite*. Marguerite d'Autriche (1480–1530), daughter of
the Emperor Maximilian, married Philibert II in 1501.

I. 15–18. Close to 'Stanzas from the Grande Chartreuse', ll. 181–6 and
'Tristram and Iseult' II. 153–6, 186–9 (pp. 164, 124), all three poems
probably written within a short interval.

I. 17. *prickers*: mounted attendants at a hunt.

166 I. 24. Almost as clumsy as 'Austerity of Poetry', l. 8: 'A prop gave way! crash
fell a platform! lo. . .'.

167 I. 61–4. The Dukes of Savoy were also Counts of Bresse; the church is in
flat country close to the town of Bourg-en-Bresse. 'Maud' is invented (the
duchess was Marguerite de Bourbon).

168 I. 95–6, 98–100. Details from Quinet, *Œuvres* (see headn.) vi. 357, 360.

I. 104–8. The duchess's tomb is on the right of the choir; the effigies of the
duke and duchess are on separate tombs.

169 II. 1–8. In Arnold's stanza the style of Tennyson's 'The Lady of Shalott'
(1832, revised 1842) 'is suggested but not actually reproduced' (G. C.
Macaulay, *Poems by Matthew Arnold* (London, 1896), p. 112).

170 II. 37. Quinet has the phrase 'leurs rocs d'albâtre ciselés et brodés (*Œuvres*
(see headn.) vi. 357).

III. 1–15. Inspired by Quinet's reflections, *Œuvres* (see headn.) vi. 358,
360–1, beginning: 'Ah! que la vieille société se couche ici sans regret dans
son tombeau!' and including (of the duchess): 'c'est de cette heure seulement que commence pour elle le vrai mariage dans son duché éternel . . .
son époux . . . ne poursuit plus le sanglier dans le forêt . . . elle ne l'attendra
plus vainement jusqu'à la nuit, en sanglotant à la fenêtre de son tour'.

III. 16–17. Quinet's moving description of the sleeping figures on their marble tomb, lit by 'la lumière transfigurée des vitraux' (*Œuvres* (see headn., p. 537) vi. 361) unites with Arnold's recollections of Keats's 'The Eve of St Agnes', ll. 217–21: 'Full on this casement shone the wintry moon / And threw warm gules on Madeline's fair breast / . . . Rose bloom fell on her hands, together prest, / And on her silver cross soft amethyst'.

171 III. 30–1. *Paradise Lost* iii. 515 (of Jacob's dream): 'And waking cri'd, This is the gate of Heav'n'.

III. 46. 'Entendez-vous aussi sur votre dais la pluie de l'éternel Amour' (Quinet (see headn., p. 537) vi. 361).

172 *Preface to Poems (1853, 1854).* The Preface to the *Poems* of 1853—Arnold's third volume of poems, but the first to bear his name on the title-page—had a twofold purpose: to confute some of the reviewers of his earlier volumes and to make a case for 'Sohrab and Rustum', the principal poem of the new volume (see headnote, pp. 541–3). His second book had appeared almost coincidentally with a pretentious work by a Glasgow factory man, Alexander Smith (1830–67), who was hailed as a new Keats and whose verse was several times reviewed along with Arnold's—even by Arnold's good friend Clough—as more inspiriting for the present age. (It should be remembered that when Arnold published his first book of poems he had already lived longer than Keats lived.) The rhetoric of the reviews was often silly and not hard for Arnold to demolish; meanwhile, though 'Sohrab and Rustum' is nowhere mentioned in the Preface, the discussion of the eternal timelessness of an ancient subject, of epic plot and style, was clearly meant to prepare the case for the new poem.

With respect to 'Empedocles on Etna' (1852), however, the Preface is less than candid. That poem had not been designed on the lines of an Aristotelian tragedy; it had been in fact an attempt to deal with the same religious problems that were treated in its two contemporaries, Tennyson's *In Memoriam* and Browning's *Christmas Eve and Easter Day* (both 1850). Though it ends on a note of exaltation, its readers were impressed with what they took to be its pessimism and gloom; not one of its reviewers grasped its point. And it is undoubtedly because of this misunderstanding on the part of his readers (though Arnold never said this) that he withdrew his most important poem from the public eye for fourteen years.

The Preface (with its brief appendix of 1854) is essentially Aristotelian in doctrine, with a strong admixture of Wordsworth, of Goethe, of other nineteenth-century continental critics. It can easily be misread to suggest that a poet should not draw his subjects from contemporary life, but that is not what Arnold said: he said that poets' 'business is not to praise their age', not to 'inflate themselves with a belief in the pre-eminent importance and greatness of their own times'. To this position Arnold would remain constant, as well as to the position that poetry should be conceptual, not merely ornamental; objective rather than self-expressive. Arnold's entire career, however, was an address to the spiritual problems of his age; his best-known critical essay was on 'The Function of Criticism *at the Present Time*', for example. And as he looked over his poems in their first collected edition of 1869 (see Introduction, p. xiii), he commented that they 'represent, on the whole, the main movement of mind of the last quarter of a century'. (R. H. S.)

says Aristotle: *Poetics* iv. 1–5.

as Hesiod says: *Theogony* 55; and see also 102–3.

173 *is not enough*: an echo of Horace, *Ars Poetica* 333–4.

says Schiller: Preface to *Die Braut von Messina*, para. 4.

derive enjoyment. See Aristotle, *Poetics* iv. 3 and xi. 6: 'No less universal is the pleasure felt in things imitated. . . . Objects which in themselves we view with pain, we delight to contemplate when reproduced with minute fidelity' (tr. S. H. Butcher).

by an intelligent critic: anonymous review of Edwin Arnold's *Poems Narrative and Lyrical*. Matthew Arnold attributed the review to R. S. Rintoul, editor of the *Spectator*.

174 *They are actions*. 'Tragedy is an imitation . . . of an action and of life, and life consists in action. . . . Character comes in as subsidiary to the actions. Hence the incidents and the plot are the end of a tragedy' (Aristotle, *Poetics* vi. 9–10, tr. S. H. Butcher).

will the latter imagine: an echo of Goethe's *Dichtung und Wahrheit*, part II, ch. vii, nearly halfway through.

select an excellent action: See Aristotle, *Poetics* xiii. 1–2, 5; ix. 2–4, 9.

primary human affections: an echo of Wordsworth's language in the Preface to the Second Edition of *Lyrical Ballads* (1800), paras. 6, 15: the poet deals with 'the primary laws of our nature', 'the great and universal passions of men'.

Achilles, Prometheus, Clytemnestra, Dido: the heroes of Homer's epic *Iliad* and Aeschylus' tragedy *Prometheus Bound*, the heroines of Aeschylus' tragedy *Agamemnon* (first part of the Oresteia trilogy) and Virgil's epic *Aeneid*.

175 *Hermann . . . The Excursion*: a domestic epic by Goethe (1797), Byron's 'pageant of his bleeding heart' (1812–18), a narrative poem by Alphonse de Lamartine (1836), and Wordsworth's long philosophical poem (1814).

the grand style. Eight years later, in *On Translating Homer: Last Words*, Arnold defined this term: 'The grand style arises in poetry, *when a noble nature, poetically gifted, treats with simplicity or with severity a serious subject*.' He deprecated definition, however: 'One must feel it in order to know what it is' (*CPW* i. 188).

176 *Orestes, or Merope, or Alcmaeon*. All three are mentioned by Aristotle as subjects of Greek tragedy, but Orestes is the only one who is the subject of ancient tragedies now extant. In 1858 Arnold published his own tragedy of *Merope*.

stood in his memory. Aristotle also speaks of the advantage of the dramatist's using a story already known to the audience, though not in Arnold's terms (*Poetics* ix. 6–8).

the Persae: a tragedy based on the Athenians' defeat of the Persian invasion under Xerxes, an event that occurred only eight years before the play was produced.

expression of Polybius. Polybius, the Greek historian of the second century BC, commonly uses the term 'pragmatic' to define the writing of history. Goethe, in 'Über den sogenannten Dilettantismus', speaks of 'pragmatische Poesie' (*Werke* (Stuttgart, 1833) xliv. 282).

treatise of Aristotle: *Poetics* xvii. 1–2. In one of his commonplace books Arnold cites a passage from Voltaire that makes the same point (*Note-Books*, ed. Lowry, p. 533). And so does Goethe in *Dichtung und Wahrheit*, part II, ch. vii, nearly halfway through.

177 *what Menander meant*. Plutarch tells this story about the fourth-century (BC) Greek comic poet (*Moralia* 347e–f).

in a representative history: in 'Theories of Poetry and a New Poet', *North British Review*, 19 (1853), 338, an anonymous review (by David Masson) of Alexander Smith's 'A Life Drama'.

178 *something incommensurable*: reported in J. P. Eckermann, *Gespräche mit Goethe*, 3 January 1830.

says Goethe: in 'Über den sogenannten Dilettantismus' (1799): *Werke* (Stuttgart, 1833) xliv. 262–3.

179 *rien à dire*: 'He says all he wishes to say, but unfortunately he has nothing to say.' This is a fair summary of Alfred Crampon's 'Les Fantaisistes', a review of Gautier's *Émaux et Camées* and other books (*Revue des deux mondes*, NS 16 (1852), 582–97) but the precise words are not there.

formed in the school of Shakspeare. For a preliminary statement of these ideas, see Arnold's letter to Clough of 28 October 1852 (pp. 285–6). Clough used the ideas in his review of Arnold's 1852 volume and Alexander Smith's 'A Life Drama' (*North American Review*, 77 (1853), 1–30). 'Isabella' is less highly regarded today than when Arnold wrote, and when it was chosen, for example, as the subject of one of John Everett Millais' Pre-Raphaelite paintings (1849). David Masson, in his anonymous review of Smith, makes the comparison of *Endymion* to *The Fairy Queen*.

eye of the mind: an echo of Wordsworth's 'They flash upon that inward eye', from 'I wandered lonely as a cloud'.

in the Decameron: Boccaccio, *Decameron* iv. 5.

180 *Mr Hallam*: Henry Hallam, *Introduction to the Literature of Europe in the Fifteenth, Sixteenth, and Seventeenth Centuries* (2nd edn., London, 1843) iii. 91–2.

M. Guizot meant: François Guizot, *Shakspeare et son temps* (Paris, 1852), p. 114 (with special reference, however, to Shakespeare's youthful *Sonnets*).

181 *laws of her country*. The modern reader is more likely to see the *Antigone* as a study of the corrupting effect of dictatorial power upon the dictator, and therefore a subject of deep modern significance.

as Pittacus said: 'it is hard to be good'; recorded by Diogenes Laertius (i. 76) and by Plato (*Protagoras* 343c). Pittacus, a seventh-century (BC) statesman of Lesbos, was traditionally one of the Seven Sages of ancient Greece.

182 *interpreting their age*. Alexander Smith looked forward to 'a mighty Poet, whom this age shall choose / To be its spokesman to all coming times' ('A Life Drama', sc. ii).

by Goethe and by Niebuhr: Goethe, for example, in a conversation with Eckermann on 29 January 1826; the historian Niebuhr in a letter to F. K. von Savigny, 19 February 1830: 'How barren and dumb is our literature now! How apathetic are all hearts!' (*Life and Letters of Barthold George Niebuhr* (New York, 1852), p. 519).

183 *Jupiter hostis*. 'Your fiery words don't frighten me; the gods frighten me, and Jupiter if he is my enemy' (Virgil, *Aeneid* xii. 894–5).

says Goethe: 'Über den sogenannten Dilettantismus' (1799: *Werke* (Stuttgart, 1833) xliv. 281). The term 'regulative laws' is from the same work (p. 278).

184 *An objection*: in the review of Arnold's 1853 *Poems* (*Spectator*, 3 December 1853, Supplement, p. 5).

Prometheus or Joan of Arc. When he republished this Preface in 1857 Arnold made his pairings symmetrical by writing 'Alcestis or Joan of Arc'—Alcestis, the heroine of a tragedy by Euripides.

It has been said: in reviews of Arnold's poems by his friends J. D. Coleridge (*Christian Remembrancer*, 27 (1854), 318–20, and J. A. Froude, *Westminster Review*, 61 (1854), 158–9.

185 *imitate them*: G. H. Lewes, 'Schools of Poetry', *The Leader*, 4 (1853), 1147.

186 *Sohrab and Rustum. An Episode.* This is the long narrative poem with which Arnold replaced 'Empedocles on Etna', the 'dramatic poem' dropped from the 1853 edition of his 1852 collection. It is composed in accordance with the 'classical' principles of his 1853 Preface, especially its emphasis on the all-importance of the choice of a subject, the necessity of accurate construction, and the subordinated character of expression (see pp. 174, 177, 179, 181). He was generally satisfied with it, telling Clough in May 1853: 'I have just got through a thing which pleases me better than anything I have yet done', and again in November 1853: 'Homer *animates*—Shakspeare *animates*—in its poor way I think Sohrab *animates*—the Gipsy Scholar at best awakens a pleasing melancholy' (*CL*, pp. 136, 146; see also pp. 292–3 above). Reviews of *1853* were divided on the poem's merits, often objecting to the frequency of the epic similes. Broadly, modern judgment (i) agrees with Coventry Patmore's in 1854 that though the tone is Virgilian, 'the poem fixes our attention rather as a vivid reproduction of Homer's manner and spirit, than as a new and independent creation'; (ii) finds, against the 1853 principles, that it is the poem's subjectivity that guarantees a limited success (see Trilling, pp. 134–5: 'The strong son is slain by the mightier father . . . Sohrab draws his father's spear from his own side to let out his life . . . we must wonder if Arnold . . . is not, in a psychical sense, doing the same

thing'); and (iii) sees the Oxus coda (ll. 875–92) as one of his finest poetic passages (Introduction, p. xviii). Arnold's main source was Sainte-Beuve's notice (*Constitutionnel*, 11 February 1850) of J. Mohl's edition of 'Le Livre des Rois, par Firdousi', containing long extracts from Mohl's French translation; it was reprinted in *Causeries du Lundi*, 1 (Paris, December 1850), and acknowledged enthusiastically in Arnold's letter to Sainte-Beuve, 6 January 1854 (Bonnerot, p. 518). For further names and details he drew extensively on Sir John Malcolm's *History of Persia* (1815); Alexander Burnes's *Travels into Bokhara* (1834), already used for 'The Sick King in Bokhara' and 'The Strayed Reveller' (pp. 32–9, 24–32); and James Atkinson's *The Sháh Námeh of the Persian Poet Firdousi, translated and abridged in Prose and Verse with Notes and Illustrations* (1832), which also reprints Atkinson's full-length *Soohrab, A Poem from the Original Persian of Firdousi* (second edition, revised and enlarged, Calcutta, 1828), a work clearly affecting incidental details in Arnold's handling of the story. Atkinson's annotations contain numerous Homeric analogues and parallels to illuminate his prefatory statement, 'The author of the Sháh Námeh has usually been·called the Homer of the East', which parallels Sainte-Beuve's on Firdousi as 'l'Homère de son Pays'. The *Iliad* appears regularly in Arnold's monthly reading lists from December 1852 until June 1853. The poem's wide range of Homeric parallels, reworkings of its source materials, and conflation of Homeric and Miltonic influences can only be glanced at in the space available for these notes (a detailed commentary appears in *Poems*, pp. 319–55). Arnold first attached his note to the poem in 1854, reprinting it in the following shorter form in *1869* (there was no note in *1853* or *1857*; the lengthy extract from Sainte-Beuve following in *1854* the quotation from Malcolm is reprinted in *Poems*, Appendix C, pp. 676–97):

The story of Sohrab and Rustum is told in Sir John Malcolm's *History of Persia*, as follows:

'The young Sohrab was the fruit of one of Roostum's early amours. He had left his mother, and sought fame under the banners of Afrasiab, whose armies he commanded, and soon obtained a renown beyond that of all contemporary heroes but his father. He had carried death and dismay into the ranks of the Persians, and had terrified the boldest warriors of that country, before Rustum encountered him, which at last that hero resolved to do, under a feigned name. They met three times. The first time they parted by mutual consent, though Sohrab had the advantage. The second time the youth obtained a victory, but granted life to his unknown father. The third was fatal to Sohrab, who, when writhing in the pangs of death, warned his conqueror to shun the vengeance that is inspired by parental woes, and bade him dread the rage of the mighty Roostum, who must soon learn that he had slain his son Sohrab. These words, we are told, were as death to the aged hero; and when he recovered from a trance, he called in despair for proofs of what Sohrab had said. The afflicted and dying youth tore open his mail, and showed his father a seal which his mother had placed on his arm when she discovered to him the secret of his birth, and bade him seek his father. The sight of his own signet rendered Roostum quite frantic: he cursed himself, attempting to put an end to his existence, and was only prevented by the efforts of his expiring son. After Sohrab's death, he burnt his tents and all his goods, and carried the corpse to Seistan, where it was interred. The army of Turan was, agreeably to the last request of Sohrab, permitted to cross the Oxus unmolested. . . . To reconcile us to the improbability of this tale, we are informed that Roostum could have no idea his son was in existence. The mother of Sohrab had written to him her child was a daughter, fearing to lose her darling infant if she revealed the truth; and Roostum, as before stated, fought under a feigned name, an usage not uncommon in the chivalrous combats of those days.'

l. 3. Sohrab is champion of the Tartars in Malcolm and Atkinson, of the Turks in Sainte-Beuve.

11. 14–15. Burnes records the periodical flooding of the Oxus from 'the melting of the snows' on the 'high plain of Pamere' (*Travels* (see headn., p. 542) ii. 192, 207–8).

187 ll. 38–40. Peran-Wisa is 'the Nestor of the Tartars' in Malcolm (op. cit. i. 40 n.). See ll. 242–62 n. below.

l. 74. 'I think "if this one desire indeed rules all" *is* rather Tennysonian—at any rate it is not good' (Arnold to Clough, 30 November 1853 (*CL*, p. 145; on Tennyson see further ll. 860–3 n., p. 544).

188 ll. 110–16. Combining influences from *Iliad* ii. 459–66 (simile of cranes followed by roll-call of Greek chieftains) and *Paradise Lost* vii. 425–31: 'Milton is . . . sufficiently great . . . to imitate. The cranes are not taken direct from him . . . but the passage is, no doubt, an imitation of his manner. So with many others' (Arnold to J. D. Coleridge, 22 November 1853, *Life of JDC* i. 210–11).

190 ll. 178–9. Sainte-Beuve's comparison of Rustum with Achilles, 'il entre dans un colère d'Achille' (*Causeries* (see headn., p. 542), p. 347), is a hint followed in ll. 202–6 (recalling Achilles' welcome to the envoys, *Iliad* ix. 193–204) and again at ll. 233–5, 242–62, 730–6 (see notes below).

191 l. 230. See Malcolm (quoted in headn. above). The ruse is absent from Sainte-Beuve, prominent in Atkinson.

ll. 233–5. Priam to Achilles (*Iliad* xxiv. 486–9): 'Remember thy father, O Achilles . . . whose years are even as mine'; l. 241 echoes the Homeric epithet 'man-slaying' for Achilles' hands (e.g. *Iliad* xxiv. 474 (see ll. 178–9 n. above).

ll. 242–62. Atkinson sees Gudurz as the Nestor of the Persians (*Sháh-Námeh* (see headn., p. 542), p. 572 n.; see ll. 38–40 n. above).

193 ll. 284–8. 'The Gulf of Persia has several pearl fisheries, particularly near the island of Bahrein' (Malcolm, op. cit. ii. 513).

ll. 302–8. A vivid mid-Victorian image which Arnold failed to 'orientalise': 'I took a great deal of trouble to orientalise them [the poem's similes]' (Letter of 26 November 1853: *L* i. 32).

195 ll. 370–2. An instance of the poem's frequent use of dramatic irony; see l. 381 (Sohrab's 'I am no girl'—Rustum thinks he has a daughter) and ll. 679–81 n. below.

ll. 390–7. 'We lie outstretched on a vast wave of this starlit sea of life, balancing backwards and forwards with it: we desire the shore, but we shall reach it only when our wave reaches it' (Yale MS; cf. *Iliad* xiv. 16–19).

196 ll. 409–15. Burnes describes the vast cedar tree 'floated down with the inundations of the river from the Hemilaya' (*Travels* i. 50; cf. Satan's spear, *Paradise Lost* i. 292–4).

ll. 439–47. The combat takes place during two days in Firdousi, Sohrab's speech occurring before its renewal on the second day.

197 ll. 448–526. Drawing freely on scenes of verbal and physical combat in the *Iliad*, e.g. between Aeneas and Meriones, Achilles and Hector, Patroclus and Sarpedon (xvi. 617–18, 633–7; xxii. 261; xvi. 426–30).

198 ll. 527–39. Homeric vaunt over the defeated, e.g. Achilles on Hector (*Iliad* xxii. 331–3, 335–6).

199 ll. 541–53. Imitating Patroclus' dying speech (*Iliad* xv. 844–54).

ll. 556–75. The poem's most ambitious attempt at the long Miltonic simile; the image of the afflicted eagles crying for their young is from Aeschylus, *Agamemnon* 49 ff.

201 ll. 619–23. Wordsworthian.

ll. 634–8. Virgil, *Aeneid* ix. 435–8 (the dying Euryalus): 'as when a purple flower, severed by the plough, droops in death'.

202 ll. 658–60. Follows Malcolm; in Sainte-Beuve Rustum gives Tahmineh an onyx, in Atkinson an amulet in the form of a gold bracelet.

ll. 679–81. Conflates for dramatic irony the story of the 'seal' and the legend of Rustum's father Zal. He was exposed to die at birth, his father being persuaded that the infant was not his own, but survived to father 'the champion of the world'.

203 ll. 730–6. Achilles' horses similarly mourn for Patroclus (*Iliad* xvii. 426–40; see ll. 178–9 n., p. 543).

205 ll. 783–9. Sohrab's burial in Seistan is from Malcolm (quoted in headn., p. 543); mound and pillar from *Iliad* xvi. 674–5.

ll. 815–19. 2 Samuel 18:33 (David's lament for Absalom).

206 ll. 830–4. Arnold's invention. Rustum outlives Kai-Kosru in Malcolm and Atkinson.

ll. 853–6. Echoing the death of Hector (*Iliad* xxii. 361–3).

l. 856. *bloom*. See 'The Youth of Nature', ll. 53–5 n. (p. 532).

207 ll. 860–3. Tennyson's 'Morte d'Arthur' (1842), l. 221: 'So like a shatter'd column lay the king'. Cf. ll. 891–2 n. (p.545).

ll. 869–74. Cf. the fires of the Trojan encampment at *Iliad* vii. 542–65, partly translated by Arnold (*On Translating Homer* (1861): *CPW* i. 120).

ll. 875–92. Maud Bodkin on Arnold's famous coda: 'the poet or his reader, dreaming on the river that breaks at last into the free ocean, sees in this image his own life and death . . . in accordance with a deep organic need for release from conflict and tension' (*Archetypal Patterns in Poetry* (1934), p. 66. See also Introduction, p. xviii). Burnes traces the course of the Oxus in *Travels* ii. 186–7.

l. 878. Shelley's 'Alastor', ll. 272–4: 'At length upon the lone Chorasmian shore / He paused, a wide and melancholy waste / Of putrid marshes' (cf. 'The Strayed Reveller', l. 183 n., p. 510).

ll. 890–1. Cf. 'Empedocles on Etna' II. [i.] 315 (p. 104).

ll. 891–2. Tennyson's 'Morte d'Arthur', ll. 242–3: 'And on a sudden, lo! the level lake, / And the long glories of the winter moon'; Arnold told Clough on 25 November 1853: 'I think the likeness [between 'Sohrab and Rustum' and 'Morte d'Arthur'] where there is likeness (except in the last lines which I own are a regular slip), proceeds from our both having imitated Homer' (*CL*, p. 145, p. 293 above; cf. ll. 860–3 n., p. 544).

208 *The Scholar-Gipsy*. Arnold took the story for his famous poem from Joseph Glanvill's *The Vanity of Dogmatising* (1661), of which he bought a copy in 1844; his description 'Our friend, the Gipsy-Scholar' in 'Thyrsis' (ll. 28–30, p. 241), the companion poem and elegy for Clough published thirteen years later, is an indication of the figure's special meaning for them both in their Oxford days. According to his diary entries, Arnold was contemplating the poem as early as 1848 (for evidence about the period of composition see *Poems*, p. 356) but he published it in *1853*, with a note pieced together from passages in Glanvill:

> 'There was very lately a lad in the University at Oxford, who was by his poverty forced to leave his studies there; and at last to join himself to a company of vagabond gipsies. Among these extravagant people, by the insinuating subtilty of his carriage, he quickly got so much of their love and their esteem as that they discovered to him their mystery. After he had been a pretty while well exercised in the trade, there chanced to ride by a couple of scholars, who had formerly been of his acquaintance. They quickly spied out their old friend among the gipsies; and he gave them an account of the necessity which drove him to that kind of life, and told them that the people he went with were not imposters as they were taken for, but that they had a traditional kind of learning among them, and could do wonders by the power of imagination, their fancy binding that of others: that himself had learned much of their art, and when he had compassed the whole secret, he intended, he said, to leave their company, and give the world an account of what he had learned'—Glanvill's *Vanity of Dogmatising*, 1661.

The poem, which Arnold classified as an elegy, is a lament for youth's wholeheartedness and energy 'which are sapped by life in the world, i.e. the scholar-gipsy is a Callicles miraculously preserved from turning into an Empedocles . . . to oppose the ideal and the actual is the method adopted here for a criticism of Victorian civilization (which is also a Romantic criticism of life for its failure to match expectation)' (*Poems*, p. 356). Arnold rated 'The Scholar-Gipsy' below 'Sohrab and Rustum' for its failure to 'animate' and capacity at best to arouse 'a pleasing melancholy' (*CL*, p. 146; see 'Sohrab and Rustum' headn., p. 541) but writes of it warmly to his brother Tom, 15 May 1857: 'that life at Oxford, the *freest* and most delightful part, perhaps, of my life, when with you and Clough and Walrond I shook off all the bonds and formalities of the place, and enjoyed the spring of life and unforgotten Oxfordshire and Berkshire country. Do you remember a poem of mine called 'The Scholar-Gipsy'? It was meant to fix the remembrance of those delightful wanderings of ours in the Cumnor Hills.' (Mrs H. Ward, *A Writer's Recollections* (London, 1918), p. 54.) The stanza is original but influenced by that of the 'Ode to a Nightingale', which evolved from Keats's experiments with the sonnet; it consists of the sestet followed by the first half of the octet of a Petrarchan sonnet, the sixth line being shortened to three stresses. Keats's influence diffusely pervades the poem— see for instance ll. 11–20—with specific echoes of 'Ode to Autumn' at ll. 21–3, 101–4 and of 'Ode to a Nightingale' at ll. 88, 131, 141.

l. 1. *shepherd*: Clough; see 'Thyrsis', l. 35 (p. 241).

l. 2. *wattled cotes*. Milton's *Comus*, l. 344: 'The folded flocks pen'd in their watled cotes'.

l. 10. i.e. the quest of the scholar-gipsy and also for the simple integrity of his quest for the truth.

ll. 34–5. Glanvill (see headn., p. 545), p. 96, omitted from Arnold's note above.

209 l. 50. Moments of inspiration.

l. 54. He is glimpsed in lonely places and by the young and happy, the simple and innocent, attributes also belonging to himself; the poet is thus seen in Gray's *Elegy*, whose contrast between two ways of life helps to shape Arnold's poem.

l. 57. *the Hurst*: Cumnor Hurst; mentioned 'Thyrsis', ll. 216–17 (p. 246).

210 l. 95. *lasher*: a weir or, as here, the pool below it.

211 l. 120. 'The lightning-spark of Thought . . . heaven-kindled' (Carlyle, 'Characteristics' (1839): Traill, xxviii, 11); Acts 2: 3.

212 l. 147. *teen*: grief, woe.

l. 149. *Genius*: Tutelary spirit guiding the individual through life.

l. 164. See ll. 201–3 n. (p. 547).

l. 167. *fluctuate*. See 'A Summer Night', ll. 27, 31–3 (p. 151) and Arnold to Clough, 30 November 1853: 'You are too content to *fluctuate* . . . That is why, with you, I feel it necessary to stiffen myself—and hold fast to my rudder' (*CL*, p. 146). Cf. in ll. 232 ff. the 'grave Tyrian trader' who 'snatched his rudder' and fled the threat to his peace of mind.

213 ll. 182–9. The 'one' may be either Goethe, whose *Dichtung und Wahrheit* Arnold had been reading recently, or Tennyson, created Poet Laureate in November 1850, for whose *In Memoriam* (1850) ll. 185–90 are apt and whose phrase 'intellectual throne' from 'The Palace of Art' (l. 216) is borrowed in l. 184. 'This for our wisest!' (l. 191), suggesting admiration mingled with rueful irony, is in keeping with Arnold's general response to Tennyson rather than to Goethe, for whom he felt unqualified respect (cf. 'Memorial Verses', ll. 15–23, p. 138).

ll. 201–3. See headn. (p. 545) and the image of sickness in 'Memorial Verses', ll. 19–22 (p. 138).

l. 205. 'Empedocles on Etna' II. [i.] 90–1: 'The brave impetuous heart yields everywhere / To the subtle, contriving head' (p. 98).

214 l. 217. *pales*: stakes of a fence.

l. 220. *dingles*: wooded dells.

ll. 232–50. Derives from Herodotus iv. 196 on the trading transactions between the voyaging Carthaginians and the 'shy' inhabitants dwelling 'beyond the Pillars of Hercules'. E. K. Brown notes: 'The Tyrian trader's flight before the clamorous spirited Greeks is exactly analogous to the scholar-gipsy's flight before the drink and clatter of the smock-frock'd boors or before the bathers in the abandoned lasher or before the Oxford riders blithe. Both flights express a desire for calm, a desire for aloofness' (*Revue Anglo-Américaine*, 12 (1934–5), 224–5).

215 *Requiescat.* This widely anthologized poem was published in *1853* (reprinted *1854*, *1857*, *1869*, etc.) but the circumstances of its composition and the identity of the lady remain unknown. 'The poem may be an exercise in the manner of Wordsworth's "A slumber did my spirit seal" and its dismissal in *1877* to "Early Poems" perhaps indicates that A. thought less of it than do many of its critics' (*Poems*, p. 371).

ll. 5–7. 'the girl forced to play her part in a busy worldly life she detests is from Victorian stock' (*Poems*, p. 371, which cites in illustration Mary Claude's 'An Angel's Tears', *Twilight Thoughts* (London, 1848), p. 75).

l. 13. *cabin'd. Macbeth* III. iv. 24: 'cabin'd, cribb'd, confin'd, bound in'.

Philomela. The name is associated in classical legend with the nightingale; but there are two versions of which Arnold follows the less familiar. In the Greek story Tereus, King of Thrace, married Prokne but fell in love with her sister Philomela and after seducing, or raping, her, cut out her tongue and hid her away; she embroidered her tale in a piece of needlework and sent it to her sister, who revenged herself on her husband by serving up their son Itys to him at a banquet; before Tereus could catch up with the sisters to kill them Prokne was changed into a nightingale and Philomela into a swallow. In Latin versions Prokne is the swallow and Philomela the nightingale. See H. J. Rose, *A Handbook of Greek Mythology* (London, 1928), pp. 262–3: 'the Greek account is better for it explains why the nightingale always sings mournfully (she is lamenting her child), and why the swallow chatters (she has no tongue, and keeps trying to tell her story)'. Arnold followed the Greek version in his manuscript—which includes after l. 21 'Dost thou still reach / Thy husband, weak avenger, through thyself?'—and hence should have called his poem 'Prokne'. He dropped the lines before publishing the poem in *1853*, realizing the firmness of the popular identification of Philomela with the nightingale, but the rest of his poem still implies that it was Prokne who lost her tongue (see l. 21). The reference to the 'sweet, tranquil Thames' (l. 12) may link the poem with Arnold's frequent visits to his father-in-law's house at Hampton-on-Thames (see 'The River', l. 1 (p. 131) and n., p. 528). The poem carries overtones from Arnold's earlier songs about the triumph and pain of poetic creativity: with l. 4 cf. 'The Strayed Reveller', ll. 210-11 (p. 29): 'But oh, what labour! / O prince, what pain', and note the autobiographical flavour in ll. 10-15.

216 ll. 27-8. Daulis is in Phocis, through which the River Cephisus runs to Lake Copais.

Haworth Churchyard. April 1855. Written shortly after Charlotte Brontë's death and published in *Fraser's Magazine* for May 1855 as a commemorative tribute also saluting Emily, Anne, and Branwell Brontë and Harriet Martineau, who at the time seemed mortally ill (see ll. 30-3). Her recovery and survival until 1876 accounts for the long-delayed first reprinting of the poem, with many revisions, in *1877*: the passage following l. 73, removed from the 1855 version for inclusion in *1867* as 'Early Death and Fame' (*Poems*, 430), was not restored; forty-six further lines were dropped (reprinted in *Poems*, pp. 425-6); and an epilogue was added (ll. 125-38). From the error about the Brontës' graves (ll. 88-9) Arnold seemingly did not visit the church or churchyard when staying at Haworth to inspect the Wesleyan school there on 6 May 1852: his reply on 1 June 1855 to Mrs Gaskell's congratulatory letter runs: 'I am almost sorry you told me about the place of their burial. It really seems to put the finishing touch to the strange cross-grained character of the fortunes of that ill-fated family that they should even be placed after death in the wrong, uncongenial spot' (*Bulletin of John Rylands Library*, 19 (1935), 135-6): Charlotte, Emily, and Branwell are buried in a vault inside the church; Anne is buried at Scarborough. The metre, which he says in the letter 'must have interfered with many people's enjoyment . . . but I could not manage to say what I wished *as* I wished in any other', is that of his earlier elegiac 'pindarics' celebrating literary figures (see headnn. to 'The Youth of Nature' and 'Heine's Grave', pp. 532 and 552).

ll. 1-2. Fox How, the Arnold home in the Lake District, is beside the Rotha and in the shadow of Loughrigg.

ll. 4-6. See headn. to 'Stanzas in Memory of Edward Quillinan' (p. 530)—a poem written out in Rotha Quillinan's 'book'.

217 ll. 7-8. 'Charlotte Brontë and Harriet Martineau' (footnote in *1877*). The meeting took place on 21 December 1850, the date of Arnold's letter to Frances Lucy: 'talked to Miss Martineau (who blasphemes frightfully) about the prospects of the Church of England . . . talked to Miss Brontë (past thirty

and plain, with expressive grey eyes though) of her curates, of French novels, and her education in a school at Brussels, and sent the lions roaring to their beds at half-past nine' (*L* i. 13). For Charlotte Brontë's impression of Arnold see Introduction, p. xii).

ll. 9–12. *Jane Eyre* (1847).

ll. 13–17. Harriet Martineau (1802–76) known by 1855 for her unorthodox views in *Letters on the Laws of Man's Nature and Development* (1851) and *Philosophy of Comte* (1853). Arnold admired her energy, talent, and independence but found her antipathetic (letters to 'K', 7 May 1855 and to G. W. Boyle, 11 March 1877 (*L* i. 44, ii. 137).

ll. 19–25. See headnn. to 'Youth and Calm' and 'Stanzas in Memory of Edward Quillinan' (pp. 528 and 530). Scott's verses were written in the album on 22 September 1831.

ll. 30–4. See headnote (p. 548).

218 l. 65. Followed in *Fraser's Magazine* (1855) by sixteen lines referring to the desolation of the Brontës' father Patrick Brontë, and Charlotte's husband Arthur Nicholls.

ll. 79–100. Emily died at Haworth on 19 December 1848, Anne at Scarborough on 28 May 1849. For the mistake about their burial, see headn. Lines 90–2 refer to Anne, ll. 92–6 to Emily, whose poem 'No coward soul is mine' was at the time wrongly thought to be the last she wrote, hence 'dying song' (l. 99). It was 'too bold' perhaps because of its uncompromising statement that 'the thousand creeds / That move men's hearts' are 'unutter-ably vain' (ll. 9–10).

219 ll. 101–11. Written before the publication of Mrs Gaskell's *Life of Charlotte Brontë* (1857).

ll. 112–24. Recalling the celebrated closing paragraph of *Wuthering Heights* (1848).

220 *Rugby Chapel. November 1857.* Arnold's elegiac celebration of his father Dr Thomas Arnold was published in 1867, ten years after the date of inception indicated in the title. He may have conceived the poem when reading Thomas Hughes's *Tom Brown's Schooldays* (published April 1857), which celebrates Dr Arnold as a Carlylean 'Hero', and completed it in early 1860, since ll. 153 ff. closely resemble 'The Lord's Messengers' (written by May 1860: *Poems*, p. 490). The interval between completion and publication is indicated by Arnold's letter to his mother of 8 August 1867; 'I know . . . that the Rugby Chapel Poem would give you pleasure; often and often it had been in my mind to say it to you' (*CL*, p. 164). His earlier letter to her of 27 February 1855 shows that he had already formed the view of his father expressed in the poem: 'this is just what makes him great—that he was not only a good man saving his soul by righteousness, but that he carried so many others along with him in his hand, and saved them . . . along with himself' (*L* i. 42). The poetic pattern is that of Arnold's other elegies, notably

'Haworth Churchyard' and 'Heine's Grave' (pp. 216, 230); for the metre see headn. to 'The Youth of Nature' (p. 532).

ll. 1–13. Cf. Arnold on his schooldays at Rugby in his letter to his brother Tom of December 1886: 'What a long way back it is to the school field at this season, and the withered elm leaves, and the footballs kicking about, and the November dimness over everything' (W. T. Arnold, 'Thomas Arnold the Younger', *Century Magazine*, 1 (1903), 118).

l. 10. Similar groupings of epithets at ll. 35, 43, 139, 158, 160. See also the Miltonic positioning of adjectives at ll. 22, 41, 70.

221 ll. 12–13. Dr Arnold died on 12 June and was buried in the school chapel at Rugby on 17 June 1842.

ll. 37–43. The tentative intimation of immortality is characteristic; cf. Arnold's attempt to console his sister 'K' on the death of her father-in-law (27 February 1854): 'However, with them the pure in heart one feels—even I feel—that for their purity's sake . . . they shall undoubtedly, in some sense or other, see God' (*L* i. 35).

l. 41. 'Haworth Churchyard', l. 136 (p. 220): 'In the never idle workshop of nature'.

l. 47. 'Heine's Grave', ll. 214–17 (p. 236): 'What are we all, but a mood, / A single mood, of the life / Of the Spirit in whom we exist, / Who alone is all things in one?'

222 ll. 58–68. Freely adapting Milton's *Samson Agonistes*, notably ll. 667–70, 674–7; and cf. 'Epilogue to Lessing's "Laocoön"', ll. 169–70 (p. 251): 'But ah! how few, of all that try / This mighty march, do aught but die!'

l. 60. *eddy*: used again for purposeless activity at l. 77 below; and 'Resignation', l. 277, 'The Buried Life', l. 43, and 'Epilogue to Lessing's "Laocoön"', l. 157 (pp. 46, 154, 251).

l. 63. Wordsworth's sonnet 'The world is too much with us', l. 2: 'getting and spending we lay waste our powers'.

ll. 67–72. Suggested by Goethe, 'Grenzen der Menschheit' (*Werke* ii. 85): 'We by a billow are lifted—a billow engulfs us—we sink, and are heard of no more.'

l. 83. *Paradise Lost* iii. 259 (Christ ruining his enemies): 'Death last, and with his Carcass glut the Grave'.

223 ll. 124–7. See Arnold's February 1855 letter to his mother (headn.) and *Culture and Anarchy* (1869): 'The individual is required under pain of being stunted, and enfeebled in his own development if he disobeys, to carry others along with him in his march to perfection' (*CPW* v. 90).

224 ll. 128–33. A. P. Stanley records Dr Arnold's walks with his family in the Lake District: 'his delight in those long mountain-walks, when they would start with their provisions for the day, himself the guide and life of the party, always on the look out how best to break the ascent by gentle stages, comforting the little ones in their falls, and helping forward those who were

tired, himself always keeping with the laggers, that none might strain their strength' (*Life of Thomas Arnold* (London, 1844), i. 23–5).

ll. 145–208. Affected by the Carlylean idea of the 'great man' as reflected in the portrait of Dr Arnold in *Tom Brown's Schooldays* (headn., p. 549): cf. Carlyle: 'Universal history, the history of what man has accomplished in this world, is at bottom the History of the Great Men who have worked there. They were the leaders of men, these great ones . . . all things that we see standing and accomplished . . . are properly the outer material . . . of thoughts that dwelt in the Great Men sent into the world' (*On Heroes and Hero-Worship* (1841): Traill, v. 1).

ll. 162–5. J. H. Newman, *Apologia pro Vita Sua* (London, 1864), pp. 79–80: 'I say "The Gospel is a Law of Liberty. We are treated as sons, not as servants, not subjected to a code of formal commandments, but addressed as those who love God, and wish to please Him." ' Arnold was reading the *Apologia* in 1864 (see *CPW* iii. 244–5, 250 and cf. John 15:15).

225　ll. 166–7. Matthew 18:14; 'it is not the will of your Father which is in heaven, that one of these little ones should perish'.

l. 174. *A God*: some power, as in 'To Marguerite—Continued', l. 22 (p. 61): 'A God, a God, their severance ruled!'.

l. 177. Like the Israelites before reaching the Promised Land.

226　l. 208. *the City of God*: a poetical way of saying 'righteousness', as in *Literature and Dogma* (1873): 'The world's chief nations have now all come . . . to . . . profess themselves *born in Zion*,—born, that is, in the religion of Zion, *the city of righteousness*' (*CPW* vi. 398).

A Southern Night. The second of Arnold's two elegies for his younger brother William: his explanatory note to l. 33 of the first, 'Stanzas from Carnac', runs: 'William Delafield Arnold (1828–59), Director of Public Instruction in the Punjab and author of *Oakfield, or Fellowship in the East*, died at Gibralter on his way home from India on April the 9th, 1859.' William was just 31 and had already been invalided home from India in 1852 after serving as an ensign in the forces of the East India Company; he returned in 1855 to organize public education in the Punjab, carrying on throughout the Indian Mutiny. His wife Frances Ann had died of dysentery at Kangra in the Punjab on 24 March 1858 while her husband was away on a tour of duty; she left four young children who were sent to England ahead of their father in January 1859 and brought up by Arnold's sister 'K' and her husband W. E. Forster. The news about William reached Arnold in Paris on 13 April, while he was in France for the Newcastle Commission; his feelings found their first outlet on 6 May at Carnac and their second in this more reflective poem, begun about 19 May 1859 at the outset of his journey from Carcassonne through Narbonne and Cette to Montpellier, and expanding upon his sorrowful reflections about his brother in his letter to their mother of 14 April 1859 (*L* i. 79–80). The poem was first published in *Victoria Regia* in 1861 and reprinted in *1867* with 'Stanzas from Carnac'.

ll. 13-16. 'See the poem, "A Summer Night" ' (footnote in *1869*: see p. 150).

227   l. 27. *teen.* See 'The Scholar-Gipsy', l. 147 n. (p. 546).

ll. 50-2. At Dharmsala in the Punjab.

228   ll. 76-92. From E. Burnouf's *Introduction à l'histoire de Buddhisme indien* (Paris, 1844), quoted in 'On the Modern Element in Literature' (November 1857, Arnold's inaugural lecture as Professor of Poetry at Oxford), and included in his reading-list for May 1857 as 'Le Bouddhisme' (*Note-books*, p. 561); 'the pure goal of being' (l. 85) is the Nirvana of the Buddhist and Mukti (liberation) of the Hindu.

229   l. 94. Louis IX of France (1214-70), admired by Dr Arnold.

ll. 97-100. Recalling Arnold's lecture on the Troubadours (delivered 12 March 1859).

230   *Heine's Grave.* Heine died on 17 February 1856; Arnold visited his grave in Montmartre Cemetery on 14 September 1858, doubtless prompted by reading in the previous July Heine's 'Die Harzreise' (*Reisebilder*, 1824) from which many details in the elegy are taken; Arnold may thus have begun the poem in 1858 but the final stages of its composition overlap with the preparation for his Oxford lecture on Heine, delivered 13 June 1863 and published in the *Cornhill Magazine* the following August. In the November he offered the poem to the editor George Smith as 'a sort of poetic pendant to my prose article . . . it is in an irregular metre, in which several of my poems are composed' (see headnn. to 'The Youth of Nature' and 'Haworth Churchyard' (pp. 532 and 548); the poem did not appear in the *Cornhill*). Besides 'Die Harzreise' and other works by Heine, Arnold also draws on Saint-René Taillandier's description of his life, work, and prolonged affliction in 'Poètes contemporains de l'Allemagne: Henri Heine', *Revue des deux mondes*, NP 14 (1852), 5-36, which was illustrated by an arresting engraving of the dying poet. The views about poetry in ll. 103-20, echoing his *1853* Preface (pp. 172-83 above), reveal the distance between Arnold's delight in Heine's wit and irony, expressed in his essay, and his feeling here that these qualities are unsuitable for serious poetry.

231   ll. 13-14. Affected by C. Gleyre's portrait of the suffering Heine and the comments on it by Saint-René Taillandier (see headn. above; also Arnold on Heine's 'force of spirit . . . activity of mind, even gaiety amid all his suffering' (*CPW* iii. 117)).

ll. 46. *Bitter spirits.* Probably Aristophanes and Voltaire, with both of whom Heine was often compared.

232   ll. 62-4. Poetic licence (Shakespeare's grave is in Holy Trinity Church, Stratford-upon-Avon).

ll. 77-96. Arnold praises in *Culture and Anarchy* (1869): 'the England of Elizabeth . . . a time of splendid spiritual effort' above 'the England of the last twenty years' (*CPW* v. 97). He used ll. 87-96 to complete his 1866 essay 'My Countrymen' (collected in *Friendship's Garland* (1871)).

233 ll. 97–100. Goethe meant the modern antiromantic German poet and dramatist Platen: 'he has many brilliant qualities, but he is wanting in—*love*' (*Conversations with Eckermann*, Bohns Library, revised edn. 1883, pp. 164–5: Arnold's 1837 edition did not give the name).

ll. 116–17. Heine is a 'child of light' because he was at war with the Philistine, the 'strong, dogged, unenlightened opponent of . . . the children of the light' ('Heinrich Heine', *CPW* i. 112).

234 ll. 129–39. See 'Die Harzreise' (*Werke*, ed. H. Friedmann (n.d.), vii. 76: 'only go and turn over that pretty "Luneburg Chronicle" in which the good old lords live again . . . in the pleasant bearded faces you will plainly read how often they have yearned from foreign lands for the kind heart of their Hartz princess and the lively rustle of the Hartz woods—aye even from the citron-bearing and venomous southern realm into which they and their successors were so often drawn by the desire to be called Roman Emperors' (R. McLintock's 1881 translation).

ll. 152–90. All the details are from 'Die Harzreise', with the omission of Heine's humour and enjoyment of chance company (vividly described in the source), so that he is seen primarily as a Romantic solitary. With ll. 186–90 cf. Heine's less portentous account of his giddiness on the Ilsenstein cliff: 'I surely should have fallen . . . but that . . . I held fast to the iron cross . . . no one will think the worse of me for doing so in such a critical moment' (op. cit. vii. 76).

236 ll. 206–21. Cf. the expression of pantheism in 'Empedocles on Etna' I. ii. 287–301 (pp. 89–90).

ll. 227–30. Recalls the prayer in 'Lines Written in Kensington Gardens', ll. 37–40 (p. 147).

237 *The Terrace at Berne*. Arnold published this final 'Marguerite' poem in *1867* (see headn. to 'Switzerland', pp. 514–15); in *1869* and subsequent collections it became the last poem in the sequence, with the heading 'Composed Ten Years after the Preceding'. His dating indicates that the poem was conceived at the end of June 1859 when he visited Berne, very nearly ten years after the separation from 'Marguerite' at Thun; but the main period of composition was 1863, as shown by his diary entries and reading-lists during April to June of that year (e.g. 'work at Marguerite', 17, 24, 31 May; 'finish Marguerite', 14 June). The poem has been described as a 'mélange d'émotion encore vive et de ranceur puritain' (Bonnerot, p. 71), but it is emotion rather than puritanism that makes itself strongly felt. For the vividness of the topographical details associated with 'Marguerite's' dwelling place at Thun see headn. to 'Parting' (p. 515), and for the probable link by association with Rachel the headn. on p. 554.

ll. 9–11. The 'twin lakes' are those of Thun and Brienz; the 'garden-walk' was at the Hotel Bellevue, Thun.

l. 31. Cf. 'Parting', ll. 17–18: 'But on the stairs what voice is this I hear, / Buoyant as morning, and as the morning clear' (p. 57).

238 ll. 45–7. On the 'sea of life' in Arnold see headn. to 'To Marguerite—Continued' (p. 516).

From Rachel. This and two further sonnets on Rachel were printed in sequence as the second, third, and fourth of fourteen Italian sonnets printed together in *1867*. Rachel, the celebrated French tragedienne, died on 3 January 1858 aged only thirty-six; Arnold was probably inspired by reading in 1863 A. de Barréra's *Memoirs of Rachel* (1858)—the present sonnet was written on 28 June 1863, the other two probably in July 1863 (evidence from his diary and notebooks). His enthusiasm, however, dated from his early days when he saw her in *Andromaque* and *Phèdre* in London in July 1846, and from his visit to Paris (29 December following to 11 February 1847), when he saw her ten times in plays by Racine, Corneille, and other dramatists (see his much later essay 'The French Play in London' (*Irish Essays* (1882): *CPW*). His allusions to Switzerland and the Rhine (ll. 10–14) have a strong personal flavour, the former being associated with 'Marguerite' and the latter with his courtship of his wife (see headnn. to 'Switzerland' and 'Faded Leaves', pp. 514 and 528). Recollections of 'Marguerite' were recently in his mind, since he had finished 'The Terrace at Berne' only a week before (14 June; see preceding headn.). The memories of Rachel in association with memories of 'Marguerite' support the conjecture that he and 'Marguerite' may first have met in France in his enthusiastic Francophile days, recorded by Clough after his return to Oxford (see Introduction, p. xii; see also 'A Memory-Picture', p. 55). The quality of feeling in the sonnet lifts it above its companions and also the other sonnets in *1867*. The biographical details here as in the other Rachel sonnets are close to Barréra's *Memoirs*.

239 *Austerity of Poetry.* Written after reading in 1864 A. F. Ozanam's *Les Poètes franciscains en Italie au treizième siècle* (Paris, 1852) which records (pp. 170–1) the story used here. Published in *1867*.

ll. 1–2. 'Giacapone di Tode' (Arnold's footnote in *1867*). Ozanam describes Giacapone (*c.*1230–1306), who became a Franciscan after his wife's death, as a precursor of Dante.

l. 6. See 'To a Friend', l. 1 n. (p. 513).

ll. 12–14. Cf. Arnold's 'duality' (Introduction, pp. xiii–xv).

Palladium. Published in *1867* as a late expression of Arnold's abiding concern with the Stoic notion of a self withdrawn from the struggle of warring impulses; cf. his note in Yale MS (probably 1849–52): 'Our remotest self must abide in its remoteness awful and unchanged, presiding at the tumult of the rest of our being, changing thoughts contending desires etc. as the moon over the agitations of the Sea.' The Palladium was the archaic wooden image of Pallas preserved in the Trojan citadel as a pledge of the safety of Troy; the use of this image for the poem may be owed to Edmond Scherer's *Alexandre Vinet* (Paris, 1853): 'ce palladium de l'humanité, de la verité, de la vie, l'individualité' underlined in Arnold's copy with marginal comment: 'Yes, l'individualité is all this.' The setting derives from

Homer, with perhaps a memory of the battle tranquilly overlooked from Lucretius, *De Rerum Natura* ii. 5–6 ('Pleasant is it also to behold great encounters of warfare arrayed over the plains, with no part of yours in the peril'—cf. 'Resignation', ll. 164 ff., p. 43).

240 ll. 9–11. Arnold's ideal landscape to represent the soul's purity (with l. 9 cf. 'Empedocles on Etna' II. [i.] 392–3, p. 106).

l. 12. Cf. 'The Buried Life', ll. 77–90, 'Thyrsis', l. 31 (pp. 155, 241).

l. 16. *Iliad* iii. 184 ff.

l. 17. Cf. Tennyson's 'Ulysses', l. 3: 'To rust unburnished, not to shine in use'.

l. 19. See 'The Scholar-Gipsy', l. 167 n. (p. 546).

Thyrsis. A MONODY, *to commemorate the author's friend*, ARTHUR HUGH CLOUGH, *who died at Florence*, 1861. Arnold's heading for this elegy on Clough and companion poem to 'The Scholar-Gipsy' (pp. 208–15) is designed to recall Milton's heading for 'Lycidas': 'In this Monody the Author bewails a learned Friend, unfortunately drown'd in his Passage from *Chester* on the *Irish* Seas, 1637'. Arnold probably began to shape his poem in March or May 1862 at Oxford, as planned in his letter to Clough's widow of 22 January 1862, two months after Clough's death: 'Oxford, where I shall go alone after Easter,—and there, among the Cumnor hills where we have so often rambled, I shall be able to think him over as I could wish' (*CL*, p. 160), but his diary entries suggest that he composed the poem at intervals during 1864–5. He first published it in *Macmillan's Magazine* for April 1866, reprinting it in *1867* with the note: 'Throughout this Poem, there is reference to another piece, *The Scholar-Gipsy*': the two poems share the same 'Keatsian' stanza and are both stylistically influenced by Keats (see headn. to 'The Scholar-Gipsy', p. 546); in *1867–8* only Arnold added this motto:

> *Thus yesterday, to-day, to-morrow come,*
> *They hustle one another and they pass;*
> *But all our hustling morrows only make*
> *The smooth to-day of God.*
> > From LUCRETIUS, *an unpublished Tragedy*
> > [see headn. to 'Empedocles on Etna', p. 520]

Thyrsis is the shepherd in Virgil's *Eclogues* vii, whose rivalry with Corydon is mentioned in ll. 80–1; the name is also found in Theocritus' *Idylls* i. and Arnold claimed that he modelled his diction on Theocritus (*L* i. 325), but his poem seems to owe more to the 'Lament for Bion', then attributed to Moschus, whom Arnold read along with Theocritus and Bion in 1863. Arnold's 'lament' celebrates in Clough the scrupulous, melancholy, and Romantic side of his nature, omitting the wit, energy, and originality of the poet of 'The Bothie of Toper-na-Fuosich', 'Amours de Voyages', and 'Dipsychus'; see Arnold's views on wit in serious poetry cited in headn. to 'Heine's Grave' (p. 552) and the reference to 'a smoother reed' (l. 78 n., p. 556).

l. 4. Sybella Curr kept the Cross Keys in South Hinksey; she died in 1860.

l. 11. *Childsworth Farm*: usually Chilswell Farm.

241 l. 16. Arnold visited Oxford in November 1863 to lecture on Joubert and may have worked on the poem then.

ll. 19–21. 'Beautiful city! So venerable, so lovely, so unravaged by the fierce intellectual life of our century, so serene! . . . Adorable dreamer whose heart has been so romantic' (Arnold on Oxford, Preface to *E in C* I (1865): *CPW* iii. 290).

ll. 26–30. The 'single elm-tree bright / Against the west' is the poem's central symbol, as the scholar-gipsy is that of the earlier elegy.

l. 31. Repeated, l. 231.

ll. 35–7. *assayed*: essayed. Pastoral convention as in Milton's 'Lycidas'.

ll. 41–50. Religious scruple, not social conscience, led Clough to resign his Oriel fellowship in October 1848; see 'Religious Isolation' (p. 52).

242 ll. 51–80. 'The cuckoo on the wet June morning I heard in the garden at Woodford . . . those three stanzas you like are reminiscences of Woodford' (Arnold to his mother, 7 April 1866: *L* i. 325).

l. 57. The cuckoo in fact 'changes his tune' in June and migrates in August.

ll. 61–70. Keatsian in feeling.

l. 78. *a smoother reed*. Arnold speaks in his letter to Clough of 24 February 1848 of his 'growing sense of the deficiency of the *beautiful* in your poems' (*CL*, p. 66); see headn. (p. 555).

l. 80. Corydon conquers Thyrsis in the singing-match in Virgil's *Eclogues* vii.

ll. 82–100. Inspired by the plea for Bion's return from the underworld in 'Lament for Bion' 115–26. 'Bion's fate' (l. 84) alludes to the tradition that Bion died by poison. Orpheus' journey to the underworld to rescue Eurydice and Persephone's sojourn with Pluto there are both referred to in 'Lament for Bion' (115–26).

243 l. 106. 'The Scholar-Gipsy', ll. 82–3 (p. 210 above): 'Maidens, who from the distant hamlets come / To dance around the Fyfield elm in May'.

l. 109. *Ensham*: Eynsham. The spelling reflects local pronunciation of the name.

244 l. 137. *pausefully*. A neologism suggesting both 'purposefully' and 'so as to induce a pause'.

ll. 153–5. 'The Scholar-Gipsy', ll. 72–3 (p. 210): 'Oxford riders blithe / Returning home on summer-nights'.

245 l. 167. *Arno-vale*. Clough is buried in the Protestant cemetery at Florence.

l. 175. *boon*: gracious, benign. See 'A Farewell', l. 81 n. (p. 516).

l. 177. *Great mother*: Demeter, who rescued her daughter Persephone (Proserpine) from the underworld. See ll. 82–100 n. above.

ll. 182–90. Arnold's lengthy note, first added in *1869*, explains that

Lityerses, King of Phrygia, made strangers try a contest with him in reaping corn and killed them if he won; Daphnis became a contestant when seeking to rescue his mistress Piplea from Lityerses and was rescued by Hercules, who took over the contest successfully and killed the king. Arnold's note continues: 'The Lityerses-song connected with this tradition was, like the Linus-song, one of the early plaintive strains of Greek popular poetry, and used to be sung by corn-reapers. Other traditions represented Daphnis as beloved by a nymph who exacted from him an oath to love no one else. He fell in love with a princess, and was struck blind by the jealous nymph. Mercury, who was his father, raised him to Heaven, and made a fountain spring up in the place from which he ascended. At this fountain the Sicilians offered yearly sacrifices. See Servius, *Comment. in Virgil. Bucol.*, v 20 and viii 68.'

246 l. 201. Arnold's scholar-gipsy is like his Joubert who ardently seeks for truth and has 'clearly seized the fine and just idea that beauty and light are properties of truth' ('Joubert', *E in C* I (1865): *CPW* iii. 193)—Arnold was working on 'Joubert' in 1863 (l. 16 n., p. 556).

ll. 203–5. Arnold praises Clough's integrity in his obituary salute to him (*On Translating Homer: Last Words* (1862): *CPW* i. 215–16): 'In the saturnalia of ignoble personal passions, of which the struggle for literary success, in old and crowded communities, offers so sad a spectacle, he never mingled. He had not . . . traduced his friends, nor flattered his enemies, nor disparaged what he admired, nor praised what he despised.'

l. 224. Keats's 'Ode to a Nightingale', l. 24: 'here, where men sit and hear each other groan'.

l. 226. After 1847 Clough published no new poems in England; 'Amours de Voyage' appeared only in the American *Atlantic Monthly* (February–May 1858).

247 *Epilogue to Lessing's "Laocoön".* Arnold first read *Laokoön* (1766), the famous work by Gotthold Ephraim Lessing (1729–81), in the later 1840s ('a little mare's nesty—but very searching', *CL*, p. 97 and see p. 280) but probably composed this poem, published in *1867*, in the early 1860s, judging by his interest in a critical theme and the allusions at ll. 43–8 to Theocritus and the 'Lament for Bion', which he was reading when writing 'Thyrsis' in 1864–5 (headn., p. 555). Lessing is Arnold's source only in that he too rejects Horace's 'ut pictura poesis' but he takes the idea that poetry differs from the other arts (he adds music to Lessing's poetry and painting) in being a progressive imitation of an action and develops this until it becomes the poet's duty to convey the movement of life itself (ll. 129–42).

l. 16. Pausanias was a Greek of the second century AD and author of the *Description of Greece* which Arnold used for background detail in *Merope* (1858).

248 l. 29. Associated in 'Obermann', ll. 49–56 and 'Memorial Verses' (pp. 67, 137).

ll. 43–8. Polyphemus, the Cyclops, is from Theocritus, *Idylls* xi. 7–16; the cattle are from 'Lament for Bion' 58–63.

ll. 57–60. *Laokoön*, ch. iii.: 'If . . . the Painter . . . can only use this single moment with reference to a single point of view . . . then . . . this single moment, and the single point of view of this single moment, must be chosen which are most fruitful of effect' (R. Phillimore's translation, Morley's New Universal Library edn., London, 1905, p. 70).

l. 62. Over the Serpentine in Hyde Park.

249   l. 70. Westminster Abbey.

ll. 95–106. Presumably the Gloria in Beethoven's *Missa Solemnis* in D, op. 123 (if so, the phrase should be 'miserere nobis').

250   l. 107. *the Ride*: Rotten Row.

ll. 129–42. *Laokoön*, ch. iv: 'Nothing . . . constrains the poet to concentrate his picture upon a single moment. He takes each of his actions as he likes from their very beginning and carries them through all possible changes to the very end' (ed. cit., pp. 74–5).

ll. 141–2. Goethe, *Faust*, Part I, ll. 1938–9: 'Then are the parts in his hand, there's only lacking the thread of spirit.'

251   ll. 177–8. The secret stream of 'The Buried Life' (p. 153 above): moments of creative inspiration are glimpses of the buried stream.

252   ll. 193–8. See the *1853* Preface on the poet's duty to add to men's happiness as well as their knowledge (pp. 172–3).

Obermann Once More. Arnold's last major poem, published in *1867* and from *1869* onwards placed after the first 'Obermann' (pp. 66–71) with the sub-heading 'Composed Many Years After the Preceding', was almost certainly first inspired by his visit to Vevey in September 1865 when in Switzerland for the Taunton Commission. He worked on it at intervals between October 1865 and January 1866, his diaries and reading-lists showing that he was busy with the poem and reading *Obermann* when also composing 'Thyrsis' (pp. 240–7); he had still not finished on 13 March 1867, when he told A. Macmillan he had 'nearly done a poem I want to conclude my new volume with' (Buckler, p. 33); and he revised it considerably between *1867* and *1877*. The poem seeks a more positive response to Senancour than in the 1840s and reflects Arnold's constant concern in the 1860s with the problem of how to retain religious warmth outside the old religious forms. His answer to the question posed in the poem's epigraph from *Obermann* (Lettre XLI. i. 159) is 'the service of reason and nature warmed by Christian feeling': see his conviction in *God and the Bible* (1875) that 'at the present moment two things about the Christian religion must be clear . . . One is, that men cannot do without it; the other, that they cannot do with it as it is' (*CPW* vii. 378). The parallel he draws between the late Roman and the modern world derives from the characteristic view of history expressed in his 'On the Modern Element in Literature' (1857: *CPW* i. 18–37) and his Yale MS note (*c.*1847–52).

The Roman world perished for having disobeyed reason and nature. The infancy of the world was renewed with all its sweet illusions.

But infancy and illusions must for ever be transitory, and we are again in the place of the Roman world, our illusions past, debtors to the service of religion and nature.

O let us beware how we are again false to them: we perish and the world will be renewed: but we shall leave the same question to be solved by a future age.

I cannot conceal from myself the objection which really wounds and perplexes me from the religious side is that the service of reason is freezing to feeling, chilling to the religious mood.

And feeling and the religious moods are eternally the deepest being of man, the ground of all joy and greatness for him.

For the note on Senancour, first attached to this poem in *1868*, see headn. to 'Obermann' (pp. 517–18), and for the identification of Empedocles and Senancour headn. to 'Empedocles on Etna' (p. 520).

ll. 1–8. 'Probably all who know the Vevey end of the Lake of Geneva will recollect Glion, the mountain-village above: the castle of Chillon. Glion now has hotels, *pensions*, and villas: but twenty years ago it was hardly more than the huts of Avant opposite to it—huts through which goes that beautiful path over the Col de Jaman, followed by so many foot-travellers on their way from Vevey to the Simmenthal and Thun' (footnote in *1868*; Arnold probably arrived at Thun in 1848 by the route mentioned. 'Twenty years' (l. 1) was a round figure in 1867).

253 ll. 21–2. 'The blossoms of the Gentia lutea' (footnote in *1885*).

l. 24. 'Montbovon. See Byron's Journal, in his *Works* [1832], vol. iii, p. 258. The river Saane becomes the Sarine below Montbovon' (footnote in *1867*). Byron was with Hobhouse at Montbovon on 19 September 1816.

l. 31. *Jaman*: near Obermann's chalet ('Obermann', l. 114 n., p. 518).

254 ll. 53–6. 'Paganism and Christianity alike have tampered with man's mind and heart, and wrought confusion within' (Arnold, 1869 essay on *Obermann*, summarizing Senancour's argument in Lettre XLIV. i. 194 ff.).

ll. 65–6. *flower . . . book*. Indicating his poetic feeling for nature and probably his 'Manuel de Pseusophanes', much marked in Arnold's copy of *Obermann*.

255 ll. 79–80. Reflects Arnold's recent optimism about the nature of religious liberation in the Church of England; this yielded in the 1880s to pessimism about the advance of Ritualism.

ll. 85–104. See headnote above.

ll. 97–104. Derives from Lucretius, *De Rerum Natura* iii. 1060–7, 912–13.

256 l. 115. The joy of the Christian message: 'this . . . made the fortune of Christianity,—its gladness not its sorrow' ('Pagan and Religious Sentiment', *E in C* I (1865): *CPW* iii. 230).

ll. 121–40. In the Preface to *God and the Bible* (1875) it is the 'multitude's appetite for miracles . . . its inexact observation and boundless crudity', not its thirst for spiritual relief, which hastens the breakup of the old civilization (*CPW* vii. 385–6).

257 l. 136. The Egyptian desert, birthplace of Christian monasticism.

ll. 139–40. Shelleyan; see ll. 293–6 n. below.

ll. 146, 189. See *St Paul and Protestantism*, quoted in ll. 323–4 n. below.

258 l. 173. 'That Christ is alive is language truer to my own feeling' (Arnold to H. Dunn, 12 November 1867: see *Commentary*, pp. 271–2).

ll. 189–90. Echoing his own 'Dover Beach', ll. 21–8 (p. 136).

259 ll. 201–4. Cf. 'the fiery storm of the French Revolution' in the note on Senancour (headn. to 'Obermann', pp. 517–18), and ll. 65–6 (pp. 66, 68).

ll. 221–8, 283–8. Draws on currently familiar scientific ideas of the cooling of the earth from a molten mass and of the ensuing great ice age: the 'glow of central fire' is the unifying belief of a time of faith; extinction of the fire is succeeded by the ice age; the breakup of the ice in the French Revolution is the removal of what is merely a semblance of unity. Only at the appointed time (ll. 283–6) will the sun melt the ice and the green earth re-appear, the sun being the intelligence at the service of the *Zeitgeist* which 'irresistibly changes the ideas current in the world' ('Dr Stanley's Lectures on the Jewish Church' (1863): *CPW* iii. 77).

260 l. 232. Revelation 21:5.

ll. 237–40. Senancour's doctrine: see *Obermann*, Lettre XXXVIII. i. 143, on 'le sentiment de la joie', copied out by Arnold in his 1866 notebook.

ll. 245–6. Cf. 'wandering between two worlds' in 'Stanzas from the Grande Chartreuse', ll. 85–6 (p. 161) and n. (p. 536).

261 ll. 268–72. See headn. to 'Obermann' (p. 517).

ll. 283–6. See ll. 221–8 n. above and Arnold's belief in *Culture and Anarchy* (1869) that Culture 'needs times of faith and ardour . . . the intellectual horizon is opening and widening . . . the iron force of exclusion of all which is new has wonderfully yielded' (*CPW* v. 8–9).

l. 286. Revelation 21:1.

262 ll. 293–6. Reworking Shelley's chorus in 'Hellas' 1060–5 ('The world's great age begins anew, / The golden years return').

ll. 307–10. See 'Stanzas from the Grande Chartreuse', ll. 73–156 (pp. 161–3). The 'weeds of our sad time' echo 'Hellas' 1063–4: 'The earth doth like a snake renew / Her winter weeds outworn'.

l. 316. Milton has 'wounds immedicable' (*Samson Agonistes*, l. 620).

ll. 323–4. Arnold writes in *St Paul and Protestantism* of the 'wonder-working power of attachment' found by St Paul in the life of Christ: 'The struggling stream of duty . . . was suddenly reinforced by the immense tidal wave of sympathy and emotion' (*CPW* vi. 43).

263 l. 348. Recalling the birth of Christianity (see ll. 113–16).

Persistency of Poetry. Used, untitled, as the prefatory poem for *1867* (in *1869* for vol. ii); given its present title in *1877*.

264 *Bacchanalia; or, The New Age*. Arnold's feeling in the 1860s that a 'new era' was dawning (reflected in 'Obermann Once More', ll. 281–312, pp. 261–2) is again expressed in this poem, probably composed between 1864 and 1867 (ll. 19–20 recall 'the battle in the plain' of 'Palladium', l. 13 (p. 240) also written in the 1860s), and ll. 1–28 seems a variation on 'what we mean . . . by fame' in 'Joubert' (*National Review*, January 1864: *CPW* iii. 209–10); but l. 1–19 may have been written earlier judging by its resemblance to the description of early morning in 'Resignation', ll. 170–85 (p. 44) and Arnold's account of his evening walk near Louth in July 1853: 'I have been shaking off the burden of the day by a walk tonight along the Market Rasen road, over the skirts of the wolds between hedges full of elder blossoms and white roses' (*L* i. 131).

l. 30. *Iacchus*. See 'The Strayed Reveller', l. 38 n. (p. 510).

266 ll. 60. Ironic; a current expression at the time, deriving from Goethe's *Wilhelm Meister* (1795), ch. vi, 'Confessions of a Beautiful Soul'.

267 ll. 67–8. Adapting Terence's 'Homo sum; humani nil a me alienum puto'.

Growing Old. The title suggests initial impetus from a sardonic response to Browning's 'Rabbi Ben Ezra' (*Dramatis Personae*, 1864): 'Grow old along with me / The best is yet to be', but ll. 16–18 indicate Wordsworth's consolatory reflections on old age in *The Excursion* ix. 50–2, 55–8. The unrhymed five-line stanza with lines alternating of three and five stresses is Arnold's own. Lionel Trilling noted of this and several other poems in *1867*, including 'The Progress of Poesy. A Variation', 'The Last Word', and 'Below The Surface-Stream' (pp. 268 and 269), that they 'do not question but reply, do not hint but declare. A note of finality, even dismissal, is here' (Trilling, p. 293). See *Poems*, p. 583: ' "Growing Old" . . . is perhaps the best illustration of Henry James's remark that Arnold's poetic style shows "a slight abuse of meagreness for distinction's sake" ' (*English Illustrated Magazine*, January 1884, reprinted in *Literary Essays and Reviews by Henry James*, ed. A. Mordell (New York, 1957), p. 348).

ll. 16–18. Cf. Wordsworth's *The Excursion* ix. 50–2, 55–8 on 'Age / As . . . a final EMINENCE' and 'a place of power / A throne that may be likened unto his, / Who, in some placid day of summer, looks / Down from a mountain top'.

ll. 19–20. Echoing Tennyson's 'Tears, idle tears', ll. 1–5, which ends 'think of the days that are no more'.

268 ll. 23–5. Echoes his own images of imprisonment in 'A Summer Night', ll. 37–40 (p. 151).

l. 31. *As You Like It* ii. vii. 163–5; 'Last scene of all / That ends this strange eventful history / Is second childishness and mere oblivion'.

The Progress of Poesy. A Variation. The 'variation', presumably, is on Thomas

Gray's Pindaric Ode of the same title. If Arnold believed in a 'new age' in the 1860s he was none the less pessimistic about his own creativity: 'I am a scanty spring, and nearly choked now by all the rubbish that Mr Lowe's Revised Code (I am a school-inspector) causes to be shot into me' (Letter to George Smith, 17 March 1864: Buckler, p. 17). His development of the image may also have been affected by ll. 1367–71 of Browning's 'Mr Sludge, "The Medium" ' (published in *Dramatis Personae*, 28 May 1864) ending: 'you'd play off wondrous waterwork; / Only no water's left to feed their play'; l. 2 echoes Exodus 17: 6.

269 *The Last Word*. Arnold probably wrote this poem, published in *1867*, sometime between 1864 and 1867, to judge by its links with (*a*) 'Thyrsis', written 1864–5, at ll. 13–16 (see note below), and (*b*) the preoccupation with battle against the Philistines at the close of the Preface to *E in C* I (*CPW* iii. 290), which Arnold was still working on in January 1865. His idea in the poem is owed in part to Senancour's 'Périssons en resistant' quoted in 'The Function of Criticism at the Present Time' (*CPW* iii. 276) and to Heine's 'Enfant Perdu' in *Romanziero* (1851), also echoed in 'Heine's Grave' (pp. 230–6).

ll. 13–16. 'Thyrsis', ll. 146–8 (p. 244): 'Unbreachable the fort / Of the long-battered world uplifts its wall; / And strange and vain the earthly turmoil grows'.

'Below the Surface-Stream'. Another variation on the theme of the buried stream of life; see headn. to 'The Buried Life' (p. 153). The lines were published in the *Cornhill Magazine* for November 1869 as part of *St Paul and Protestantism* and reprinted in 1870 when the work was published as a book; Arnold told his mother on 21 February 1870: '[they] are my own, and I think them good' (*L* ii. 28), but did not reprint them among his poems.

Poor Matthias. This and the following poem, together with 'Geist's Grave' (*Poems*, pp. 592–5), each an elegy for a beloved family pet, draw on a gift for blending playfulness, wit, and tenderness of feeling which Arnold regrettably rarely used in his poetry (for his views about this see headnn. to 'Heine's Grave' and 'Thyrsis', pp. 552 and 555). Of this piece he told John Morley on 24 October 1882: 'The "dirge" is as good as done—a simple thing enough, but honest' (*L* ii. 207); it was published the following December in *Macmillan's Magazine* and reprinted in *1885*. Hardy paid tribute by emulating its style in 'Last Words to a Dumb Friend' (1904).

l. 6. Catullus iii. 6–7 on Lesbia's sparrow: 'honey-sweet he was, and knew his mistress as well as a girl knows her own mother'; Matthias's 'mistress' was Arnold's daughter Eleanor ('Nelly' in l. 191).

270 ll. 25–6. Rover and Toss appear in 'Kaiser Dead', ll. 65–6 (p. 277). 'Great Atossa' appears in Pope's 'Moral Essays' (Epistle ii, 115–50) but Arnold's Persian cat was named, appropriately, after Atossa, daughter of Cyrus the Great and wife of Darius.

271 ll. 47–55. Two of Arnold's sons, Basil and Thomas, died in 1868; a third, Trevenen William, in 1872.

l. 70. See 'Kaiser Dead', ll. 19–20 (p. 276).

272 l. 86. Echoing *Macbeth* I. v. 44–5.

273 l. 131. *the Grecian*: Aristophanes: 'See *The Birds* of Aristophanes, 465–85' (Arnold, footnote to ll. 153–4).

275 l. 201. Tributary of the Thames.

l. 209. The 'iter tenebracosum' travelled by Lesbia's sparrow (Catullus iii. 11).

Kaiser Dead. April 6, 1887. Arnold's last poem, written between April and June 1887. Kaiser, his mongrel dachsund mentioned in 'Poor Matthias' (see preceding headn.), died on 6 April 1887 and the poem was published in the *Fortnightly Review* the following July. Arnold took his stanza from Burns, imitating his elegy in 'The Death and Dying Words of Poor Mailie, the Author's only pet Yowe' (see ll. 11–12 n. below).

l. 3. *Farringford*. Tennyson's home near Freshwater, Isle of Wight, where he usually spent the winter.

l. 5. *Pen-bryn's bold bard*. Sir Lewis Morris (1833–1907), popular author—under the name 'Lewis Morris of Penbryn'—of *Songs Unsung* (1883). Penbryn was the name of his house near Carmarthen.

ll. 11–12. Burns's 'Poor Mailie' elegy, ll. 45–6: 'Come, join the melancholious croon / O' Robin's reed!'

276 ll. 40–3. 'All the world over, I will back the masses against the classes' were Gladstone's words in a campaign speech at Liverpool on 28 June during the General Election of 1886 (fought and lost on his policy of Irish Home Rule, which Arnold opposed).

277 ll. 65–6. See 'Poor Matthias', ll. 43–58 (pp. 270–1).

l. 73. *Burwood*: estate near Arnold's cottage at Cobham.

PROSE

279 *Letters to Arthur Hugh Clough*. The publication in 1932 of Arnold's letters to Arthur Hugh Clough over the period from 1845 to Clough's death in 1861 was comparable to the publication in 1848 of Keats's letters, an event on which Arnold comments with some passion here. Both are the frank, enthusiastic outpourings to intelligent and sympathetic friends of young poets as they gradually mature in their thinking about their own work and the nature of the poetic art. Arnold's letters provide a running commentary on many of the poems published in the present volume, and on the Preface to the *Poems* of 1853. It is interesting also to find in them seeds of ideas and even details that Arnold developed in his later prose writings. The text of the letters here is that of Howard Foster Lowry's original edition, the very few transcriptional errors corrected from the manuscripts by the kind assistance

and permission of the Yale University Library. A few of Arnold's slips of the pen (e.g. 'speroid' for 'spheroid') are also corrected; a few other misspellings (e.g. 'chalf' for 'chaff') are left as perhaps intended.

Shakspeare says: *Midsummer Night's Dream* v. i. 19–20.

Tennyson's dawdling. Arnold here cannot have been referring to anything later than Tennyson's two-volume *Poems* (1842).

Béranger. Pierre-Jean de Béranger (1780–1857) was author of light-hearted, witty *Chansons* that much appealed to Arnold during his university years.

Arthur's Bosom. The hostess of the Boar's Head Tavern, describing the death of Falstaff in Shakespeare's *Henry V* II. iii., was sure he went to Arthur's bosom (presumably she meant Abraham's bosom, Luke 16: 22).

New Zealand. Arnold's next younger brother Thomas sailed from Greenwich for New Zealand on 24 November 1847 in the face of some severe winter storms. He remained in New Zealand and Tasmania until 1856, then returned to England.

the Calf Poem: Clough's 'When Israel Came Out of Egypt', later published in *Ambarvalia* (1849). It dealt with nineteenth-century scientific materialism, by means of the story of Moses' forty-day conversation with God on Mt. Sinai and the Israelites' erecting an idol in the form of a golden calf while he was gone (Exodus 19–34).

about Burbidge: Thomas Burbidge (1816–92), a fellow Rugbeian and co-author with Clough of *Ambarvalia*, a joint collection of poems.

au reste: besides.

as Nelson said of Mack. The Austrian general Karl Mack von Leiberich, commander of the Neapolitan army against the French under Napoleon, was reputed to be a superb military man. But Nelson, the naval commander, after watching him blunder in action, commented, 'I have formed my opinion. I heartily pray I may be mistaken.' (Robert Southey, *Life of Nelson*, ch. vi).

280 *tourmenté*: laboured.

under a little gourd. Doubtless as Jonah sat under a gourd outside the city of Nineveh (Jonah 4: 5–6).

to unite matter. The words come at the end of a sheet, without punctuation, and the next sheet is missing. One might suppose Arnold wrote 'to unite matter with style' or 'with form'.

Keats' Letters: R. M. Milnes, *Life, Letters and Literary Remains of John Keats*, published about the beginning of September 1848.

Browning. By the date of this letter, Browning had published very few of the poems we know best. Arnold was referring to works like *Paracelsus* and *Sordello*.

added unto them. 'Seek ye first the kingdom of God, and his righteousness; and all these things shall be added unto you' (Matthew 6: 33).

the Laocoon of Lessing: Gotthold Ephraim Lessing, *Laokoön, oder über die Grenzen der Malerei und Poesie* (1766), a treatise on the necessary distinctions between the literary and the plastic arts. See Arnold's poem 'Epilogue to Lessing's "Laocoön" ' (pp. 247–52) and headn. (p. 557). Arnold himself, in the Preface to First Edition of *Poems* (1853) (pp. 172–83), was somewhat caught up in the same mare's nest. See Martin Corner, 'Arnold, Lessing, and the Preface of 1853', *Journal of English and Germanic Philology*, 72 (1973), 223–35.

281 *et id genus omne*: and all that sort of people.

ὁπλίτης: the common soldier in the heavy infantry.

στάσιμοι: steady sort of men.

Trench: Richard Chenevix Trench (1807–86), later Archbishop of Dublin, but at this time known as a minor poet.

cheeper: the chick of a partridge or grouse.

Collins, Green: William Collins (1721–59), best known for 'To Evening' and other odes (1747), and Matthew Green (1696–1737), author of *The Spleen* (1737).

the Poet sees, but wide: Arnold's soon-to-be-published poem 'Resignation', l. 214 (p. 45).

Cumner Hill. A spot west of Oxford much loved by Arnold and Clough as undergraduates; the countryside at Cumnor was the setting for 'The Scholar-Gipsy' and 'Thyrsis'. The towns of Cirencester and Cheltenham are west of Cumnor.

282 *The Iliad translation.* Arnold and Clough shared an interest in translating Homer, and Clough made many attempts at a hexameter version of the *Iliad*. Clough later discussed the problem in the second of his 'Letters of Parepidemus' (*Putnam's Monthly Magazine*, 2 (1853), 138–40) and Arnold devoted four lectures as Professor of Poetry at Oxford to translating Homer (1860–1). Clough also used free hexameters in some of his longer poems such as 'The Bothie of Toper-na-Fuosich' and 'Amours de Voyage'.

Voss. Johann Heinrich Voss (1751–1826) translated the *Odyssey* (1781) and *Iliad* (1793) into German hexameters. Clough mentioned Voss's 'clumsy' hexameter poem *Louise* in the article in *Putnam's* (see previous note).

Carlyle's Dante. John Aitken Carlyle, Thomas Carlyle's younger brother, in 1849 published an English prose translation of Dante's *Inferno*.

my Antigone. Arnold in his newly-published volume had a 'Fragment of an "Antigone" ' (see headn., pp. 508–9)—whether originally intended as part of a larger undertaking is not clear.

Brodie . . . John Coleridge: Benjamin Collins Brodie (1817–80) and John Duke Coleridge (1820–94), both friends of Arnold's at Balliol.

Shairp is δεξιά. John Campbell Shairp (1819–85), another Balliol contemporary, 'is favourable'.

283 *Thun*: town at the western tip of Lake Thun in the Bernese Oberland, about twenty miles south-east of Berne. See headn. to 'Switzerland' (p. 514).

faussé: strained, distorted.

born again: John 3: 7.

could not tame: Arnold's 'Stanzas in Memory of the Author of "Obermann"', ll. 143–4 (p. 70). The poem is dated 'November 1849' and was published in Arnold's 1852 volume (see headn., pp. 517–18).

dear Tom: Arnold's brother.

besonnenheit: thoughtfulness, reflectiveness.

still, considerate mind. In 'Fragment of Chorus of a "Dejaneira"' (first published in 1867 but probably written much earlier) Arnold used the expression 'purged, considerate minds' (l. 17, p. 23).

senses of the word. 'intuition', being 'insight' or vision, cannot be 'palpable' or touchable.

not of my fold: John 10: 16.

foolishness to me. Thomas Arnold after Matthew's death recalled how uninterested his older brother was in the abstract reasoning of geometry: see *Passages in a Wandering Life* (London, 1900), p. 10.

Blümlis Alp: a group of peaks reaching 12,000 feet, with spectacular scenery, about twenty-five miles south of Thun.

284 *I come*: Arnold's poem 'Parting', ll. 25–34, published in the 1852 volume (p. 58 above).

desperadoes: men reckless with despair.

ὡς ὁ φρόνιμος διορίσειεν: 'as the man of practical wisdom would define it', the standard for judging virtue in Aristotle's *Nicomachean Ethics* II. vii. 1107a.

convicts to the Cape. Henry, third Earl Grey, the Colonial Secretary, became the centre of a lively controversy in England when he attempted to establish a penal colony at the Cape of Good Hope.

Thou fool: Luke 12: 20 or 1 Corinthians 15: 36.

Maskelyne: M. H. Nevil Story-Maskelyne (1823–1911), at Wadham College, Oxford, while Arnold was at Balliol, and a friend of Brodie's (note to p. 282).

Forster: William Edward Forster, husband of Arnold's older sister Jane.

at Quillinan's sollicitation [sic]. At the request of Wordsworth's son-in-law Edward Quillinan Arnold wrote his 'Memorial Verses' on the death of William Wordsworth (23 April 1850) and published them at the beginning of June (see headn., p. 531).

F. Newman's book. Francis Newman (1805–97), brother of John Henry Newman, had just published *Phases of Faith*. It was Dogberry in Shakespeare's *Much Ado About Nothing* (IV. ii.) that was written down an hass.

285 *zu ekelhaft*: 'One must merely leaf through this work; if one tried to read it through, it would be much too nauseous.'

that article: money.

the poems: *Empedocles on Etna, and Other Poems* was published about 27 October 1852.

from America: Clough went out to Boston at the beginning of November 1852.

the Solitary: in Wordsworth's *Excursion* iii. 861–3.

unwilling wind: *Prometheus Unbound* II. i. 147.

286 *magister vitae*: director or teacher of life.

Miss Smith . . . Fanny Lucy: Blanche Smith, Clough's fiancée, and Fanny Lucy Arnold ('Flu'), Arnold's wife.

Church of England place. In general, schools and training colleges conducted by the Church of England had their own clerical inspectors; Arnold as a lay inspector dealt with schools conducted by Nonconformists.

me font l'effet: give me the impression.

wedding garment . . . highways and hedges. Arnold alludes to the parable told by Jesus in Matthew 22: 1–14. The parable ends with the verse 'For many are called, but few are chosen'.

287 *in contemplating*. Arnold published this quatrain in 1867 with the title 'A Caution to Poets' (p. 159).

Walrond: Theodore Walrond (1824–87), a Balliol contemporary of Arnold and Clough, and the close friend also of young Thomas Arnold at University College.

Guardian: *The Guardian* (a London-based Church of England newspaper), vol. 7 (8 December 1852), p. 823; anonymous. In the reviewer's opinion, 'A.', like Longfellow, expressed his philosophy too nakedly and did not fuse it with his poetry.

the April Westminster: actually not until the *Westminster Review* for January 1854 (pp. 146–59, anonymous). Arnold's name was not on the title-pages of his 1849 and 1852 volumes; it did appear on the 1853 volume. James Anthony Froude (1818–94) was Fellow of Exeter College, Oxford, while Arnold was Fellow of Oriel, and subsequently editor of *Fraser's Magazine*, where he published the earliest of the *Essays in Criticism*.

mollis et exspes: 'weak and hopeless' (Horace, *Epodes* xvi. 37).

Werter, René: introspective heroes of Goethe's *Die Leiden des jungen Werthers* (1774) and Chateaubriand's *René* (1802, 1805).

ὑγρὰ κέλευθα: 'the watery ways' (*Odyssey*, iii. 71 and *passim*).

ἀνάλογον: an equivalent.

recueillement: concentration of thought.

288 *Gordon Square*: the site of University Hall, of which Clough had been Principal from 1849 to 1852.

the Bothie: Clough's 'long-vacation pastoral' poem 'The Bothie of Toperna-Fuosich', published in 1848.

289 *demi-mot*: hint.

assiette: established position, firm attitude.

resolve to be thyself: Arnold's 'Self-Dependence', l. 31 (p. 73).

qui ne vous valaient pas: who mean nothing to you.

raffinements: extreme subtleties.

290 *aperçus*: ideas.

Combe Hurst: the home of Clough's fiancée Blanche Smith.

vacuas sedes et inania arcana. When the Roman general Pompey conquered Jerusalem and entered the Temple, he found no image of the god; 'the place was empty and the secret shrine contained nothing' (Tacitus, *Histories* V. ix. 1).

Miss Martineau. Harriet Martineau (1802–76), a writer in many fields, was a neighbour of the Arnold family in Ambleside. See headn. to 'Haworth Churchyard' (p. 548).

Halstead in Essex: Arnold's diary confirms that his school inspecting took him to Halstead on 18 March 1853.

the thinnest paper: thin in order to save postage.

the Italian poem: Clough's 'Amours de Voyage', the fruit of visits to Italy in 1849 and 1850.

incolore: colourless.

Margaret Fuller: Boston bluestocking and transcendentalist (1810–50). Her *Memoirs*, with notices of her life by Emerson and W. H. Channing, were published in three volumes in London, 1852. Clough met her in Italy in 1849.

oblate spheroid: a foreshadowing of Arnold's· dealing with the relation between science and culture nearly thirty years later in 'Literature and Science' (pp. 456–71).

291 *Miss Bronte*. Charlotte Brontë published *Villette* in 1852. Arnold's remark here is markedly different from the tone of 'Haworth Churchyard', the poem he published on her death two years later (pp. 216–20).

Thackeray. Thackeray crossed the Atlantic to Boston in the same ship as

Clough in November 1852, and Clough saw something of him in Boston during the next few weeks, but not after the date of Arnold's letter. 'I liked Esmond', Clough wrote to Blanche Smith about this time, '—but I don't know that it's much, nor much in it.'

Waverley and Indiana. Early novels by Sir Walter Scott (1814) and George Sand (1831).

the woman Stowe: Harriet Beecher Stowe (1811–96), whose *Uncle Tom's Cabin* (1852) was a current sensation.

Alexander Smith's poems. Clough took Arnold's suggestion promptly and reviewed Smith's *Poems* (1853) along with Arnold's *Empedocles on Etna, and Other Poems* (1852) in the *North American Review* for July 1853.

article on Wordsworth. It was actually Whitwell Elwin, not John Gibson Lockhart, who reviewed Wordsworth in the *Quarterly Review*, 92 (1852), 182–236.

Moore's: Thomas Moore (1779–1852), best known for his *Irish Melodies*.

Ampère's. Jean-Jacques Ampère published a series of nine articles on his travels in the United States in the *Revue des deux mondes*, January–June 1853. Like Arnold thirty or more years later, he became weary of the Americans' 'perpetual glorification of their nation . . . the conviction of the superiority of their country is at the bottom of everything they say' (1 January, p. 7). See above, pp. 500 ff.

désabusé: disillusioned.

toujours Homer: 'perpetually Homer'—presumably to set the tone for his work on 'Sohrab and Rustum'.

Susy: Arnold's younger sister Susan. Her future father-in-law was a wealthy Liverpool paper manufacturer.

what Curran said. The Irish lawyer John Philpot Curran, at the trial of A. H. Rowan on 29 January 1794, said: 'There are certain fundamental principles which nothing but necessity should expose to public examination; they are pillars, the depth of whose foundation you cannot explore, without endangering their strength.'

292 *philisterey*: Philistinism, thoughtless acceptance of convention.

Mrs Lingen. Emma Hutton in 1852 married R. R. W. Lingen, who had been Arnold's tutor at Balliol and was now his superior in the Education Office.

your friend's: presumably Blanche Smith's.

North's Plutarch: Sir Thomas North's 1579 version of Plutarch's *Lives*, done into English from Jacques Amyot's French translation. Clough was commissioned by a Boston publisher to revise John Langhorne's eighteenth-century translation. But he found Langhorne so dull and heavy that in fact he based his new version (published 1859) on Dryden's.

Cotton's Montaigne. Charles Cotton published his translation of Montaigne's *Essays* in 1685.

Long. Arnold praised George Long's notes on Plutarch (1844–8) and his translation of Marcus Aurelius (1862) in the essay on 'Marcus Aurelius' (1863).

article of yours on me: see note to p. 291.

my version of Tristram and Iseult: see headn. to 'Tristram and Iseult' (pp. 524–5).

got through a thing: 'Sohrab and Rustum' (pp. 186–207).

293   *or [thought me] to be other*. Arnold clearly omitted part of his sentence as he turned his sheet of paper. The suggested emendation combines Lowry's and the present editor's.

my father's journals: *Thomas Arnold's Travelling Journals*, ed. A. P. Stanley (London, 1852).

Tennyson's Morte d'Arthur: published in 1842.

the two last lines. Lowry parallels the last two lines of 'Sohrab and Rustum' (p. 207) with 'Morte d'Arthur', ll. 191–2:

> And on a sudden, lo! the level lake,
> And the long glories of the winter moon.

Hill's criticism. Herbert Hill, writer (presumably) of one of the flattering letters Arnold forwarded to Clough, was Southey's son-in-law and had been tutor of both Arnold and his brother Thomas before they went to Winchester.

295   *Democracy*. The Newcastle Commission, appointed on 30 June 1858 'to consider and report what Measures, if any, are required for the Extension of sound and cheap elementary Instruction to all Classes of People', resolved to send two assistant commissioners to the Continent to see how the problem was solved there; to Arnold fell the mission to study the elementary schools of France, the French-speaking cantons of Switzerland, and Holland. He was abroad for most of the period from 15 March to 26 August 1859. He visited schools, made the acquaintance of the men most concerned with the educational problems of the State, and read widely in the books they had written. When his official report was completed in late March 1860 he received permission to publish it separately as a book, *The Popular Education of France* (4 May 1861), and for the book wrote an Introduction showing the relation of educational matters to the whole theory of the responsibility of the State for the well-being of its people. The Introduction gave him much trouble, but in the end he was sufficiently pleased with it to think of republishing it in *Essays in Criticism* (1865)—a thought he then rejected; he did include it, with the title 'Democracy', in *Mixed Essays* (1879).

The close acquaintance he made with France, and especially with its leading intellectuals, was of the greatest importance to the development of his thinking on social and governmental matters, and indeed on literary affairs also: the report contained the seed of his essay on 'The Literary Influence of Academies'. In many ways the Introduction is the keystone of his subsequent thinking about politics and education. 'It is one of the things I have taken most pains with', he commented.

Burke (1770): 'Thoughts on the Cause of the Present Discontents', nearly two-thirds through.

plus le même. 'I have often said that one ought to change the manner of governing, when the government is no longer the same.'

Burke's Works: 'Thoughts on . . . the Present Discontents' (1770), eighth paragraph.

297 *in the grand style*: in the lectures *On Translating Homer* (1860–1: *CPW* i. 116 and *passim*).

298 *réduit là*: 'To administer is to govern; to govern is to reign; everything comes down to that' (*Correspondance [avec] de La Marck* ii. 75).

one's own essence: Spinoza, *Ethics*, part III, Propositions 6–7.

300 *la dignité humaine*: Arnold has translated the greater part of the passage in his text.

302 *life-peerages*: The House of Lords in 1856 declined to let Sir James Parke take his seat when created a life peer, on the ground that the crown had lost by disuse the power of creating life peerages. A century or so later, life peerages became the usual creation.

304 *Chesterfield's Letters.* For the Wood anecdote, see *On Translating Homer* (*CPW* i. 107–8). Chesterfield's letters to his son insist upon the necessity of knowing Greek and Latin if one is to live at ease in the best society.

posse videntur. 'They can because they think they can' (Virgil, *Aeneid* v. 231).

305 *Look at France.* The Second Empire of Napoleon III was the consequence of the Revolution of 1848 that overthrew the last of the French kings.

306 *no mind to*: Samuel Butler, *Hudibras* I. i. 215–16.

309 *Act of Uniformity.* These were measures passed by the first Restoration Parliament in 1661–5 to suppress the Nonconformist sects.

full and abounded: see Philippians 4: 12.

who were sick: see Matthew 9: 12.

310 *the old colleges.* These were secondary schools, largely supplanted in the French Republic by the new 'lyceums' (see next sentence).

Classical and Commercial Academy. This term was used to describe the multitude of private-venture schools set up primarily for middle-class patrons who did not have access to the expensive endowed 'public' schools like Eton and Harrow or to the 'free' grammar schools, of which there were in fact very few.

314 *corporate character.* Arnold's definition is apparently a brief summary of what Burke argues in *An Appeal from the New to the Old Whigs*, a little more than two-thirds through.

316 *Be ye perfect*: Matthew 5: 48.

to have apprehended: Philippians 3: 13. The verse continues: 'but this one thing I do, forgetting those things which are behind, and reaching forth unto those things which are before'.

317 *The Function of Criticism at the Present Time.* When Arnold began his second five-year term as Professor of Poetry at Oxford (a post which required three lectures a year), he sought a new organizing principle for his series, one which very much reflects his experience in France in 1859. He had there met Sainte-Beuve, whose critical essays he regarded as a model of graceful style, good sense, and literary tact. And so, with a view to helping his British compatriots broaden their intellectual horizons and remedy their provincialism and complacency, he lectured on Continental writers, and especially on the French (even Heine had lived so long in France that he seemed almost more French than German). Many of his authors Sainte-Beuve had already treated, but Arnold commonly added to his subject a British author as a parallel—Keats to Maurice de Guérin, Coleridge to Joubert—or a comparison like that between British religion (both Roman Catholic and Protestant) and the engaging Catholic moderation of Eugénie de Guérin. When he set out to gather these lectures into a volume, the *Essays in Criticism*, he was 'struck by the admirable riches of human nature that are brought to light in the group of persons of whom they treat, and the sort of unity that as a book to stimulate the better humanity in us the volume has'. He determined upon collecting them about the time he delivered the lecture on 'The Literary Influence of Academies', with its central definition of 'provincialism', and he then saw the necessity of designing a lecture/essay which would pull all the others together thematically by showing the virtue of disinterested objectivity in evaluating one's own writings as well as the writings of others. 'The Function of Criticism at the Present Time' was delivered at Oxford on 29 October 1864, and stood first in the new volume. It is the central critical statement of Arnold's career, witty, keen-minded, and full also of the characteristic catchwords he loved to use.

Europe most desires,—criticism: *On Translating Homer* (1860), Lecture II, last paragraph. Arnold here exaggerates the unfavourable reception of his proposition.

Mr Shairp's excellent notice: [John Campbell Shairp,] 'Wordsworth: the Man and the Poet', *North British Review*, 41 (1864), 1–54. Shairp was a friend of Arnold's from their undergraduate days at Balliol College, Oxford.

as genuine poetry: Christopher Wordsworth, *Memoirs of William Wordsworth* (London, 1851) ii. 53.

a practice. Arnold follows this 'practice' himself: his essays on, e.g., Wordsworth, Byron and Keats were designed as introductions to his selections from their work.

318 *is quite harmless*: Christopher Wordsworth, *Memoirs*, ii. 439.

Irenes. One of Samuel Johnson's earliest literary works was a neoclassical drama *Irene*.

his celebrated Preface: Wordsworth's 'Preface to the Second Edition of *Lyrical Ballads*'.

319 *current at the time.* Arnold echoes Goethe's remarks to Eckermann, 3 May 1827. 'Current' is Goethe's 'in Kurs'.

320 *creative activity*: the work of the British 'Romantic' poets—a term Arnold does not use.

321 *thrown quiet culture back*: in the poem 'Four Seasons: Autumn', no. 62.

322 *The old woman*: Jenny Geddes, on 23 July 1637. The minister's wearing a surplice was to her Presbyterian eye the symbol of Episcopalianism.

decimal coinage. After the French Revolution decimal coinage was gradually adopted by the Continental nations, but not in Great Britain until 1971.

323 *en attendant le droit.* Joseph Joubert (1754–1824), a French critic, was the subject of one of Arnold's earlier Oxford lectures, published in *Essays in Criticism*. Arnold quotes from his *Pensées*, Titre xv. 2.

324 *Dr Price.* Richard Price (1723–91) was a Nonconformist minister whose sermon in praise of the French Revolution on 4 November 1789 stirred Edmund Burke to write his *Reflections on the Revolution in France*.

meant for mankind: Goldsmith's description of Burke in his poem 'Retaliation', l. 32.

he ever wrote. In fact Burke lived nearly six years longer and continued to write.

in your mouth: Numbers 22: 38.

325 *Lord Auckland*: William Eden, first Baron Auckland (1744–1814), British ambassador at The Hague, in a memorial to the States General there on 25 January 1793.

traveller in the fable. In Aesop's fable of the contest between the wind and the sun to see which was stronger and could more quickly force the cloak off a traveller, the wind merely made him button up tighter, but the sun made him loosen and discard his cloak.

a long peace. After the Battle of Waterloo in 1815 England was not involved in war in western Europe until 1914.

326 *disinterestedness.* The concept of 'disinterestedness' is a constant in Sainte-Beuve, but Arnold took it immediately from Ernest Renan, 'L'Instruction supérieure en France' (*Revue des deux mondes*, 2me période, 51 (1864), 81): the duty of a professor in the University was to teach 'sans aucune vue d'application immédiate, sans autre but que la culture désintéressée de l'esprit' (with no view of immediate application, with no other purpose than the disinterested cultivation of the mind). 'Disinterestedness' was a favourite word of many authors Arnold admired, including Burke, Bishop Joseph Butler, and Coleridge; the last used it in his 1818 lecture on Shakespeare, 'Shakspere's Judgment Equal to His Genius'.

327 *Home and Foreign Review.* This liberal Catholic quarterly edited by Lord Acton was discontinued after two years (1862–4) in the face of ecclesiastical censure. Arnold's brother Thomas was one of its contributors.

328 *Sir Charles Adderley.* Adderley, a Conservative Member of Parliament, was speaking to the Warwickshire Agricultural Association at Leamington on 16 September 1863.

Mr Roebuck. John Arthur Roebuck, a Benthamite Radical MP for Sheffield, was speaking to his constituents on 18 August 1864.

noch übrig bleibt: Goethe, *Iphigenie auf Tauris* I. ii. 91–2.

diminish local self-government. These were Liberal and Utilitarian proposals. The Reform Act of 1832 provided a uniform parliamentary franchise for householders rated at ten pounds or over; Lord Palmerston in 1860 introduced a bill (later withdrawn) to reduce that qualification to six pounds for voters in boroughs. Church rates were property taxes levied by the parish vestries for the support of the (Anglican) Church; attempts to abolish them were unsuccessful until 1868.

329 *Wragg is in custody*. Elizabeth Wragg committed this crime on Saturday 10 September 1864. The Mapperley Hills are to the north-west of the industrial city of Nottingham.

the Ilissus. A pleasant stream that flows through Athens.

Christian name lopped off. The newspapers customarily used only last names in their reporting of criminal activity, but were more polite in their other columns (Christian name, 'Mr', 'Mrs', 'Miss').

330 *Indian virtue of detachment*. Arnold early had an interest in the *Bhagavadgita*.

inadequate. Arnold takes the terms 'adequate' and 'inadequate' from the philosophical writings of Spinoza, to whom he had devoted one of the *Essays in Criticism*.

Lord Somers. John, Lord Somers (1651–1716), distinguished statesman and constitutional lawyer, presided over the drafting of the Declaration of Rights after the abdication of James II in 1688.

331 *Philistines*. Heinrich Heine traces his use of the term 'Philistines' to German student slang, where it stood for townspeople as opposed to academics in a German university town. Arnold develops the idea to represent the English middle class in his essay on Heine (*CPW* iii. 111–14).

Cobbett. William Cobbett (1762–1835) was a prolific journalist and propagandist for the rights of the common man.

Latter-day Pamphlets. Carlyle's *Latter-Day Pamphlets* were published in 1850; Ruskin's *Unto This Last* appeared in the *Cornhill Magazine* in 1860.

terrae filii: 'sons of the earth' (Persius vi. 59). Arnold had affirmed that a book on religion, to be justifiable, must either edify the multitude or add to the knowledge of the learned few. The *London Review*, in a lively article entitled 'The Few and the Many' (31 January 1863) criticized Arnold and proclaimed 'We are all alike *terrae filii*.'

as Goethe says: *Wilhelm Meisters Lehrjahre*, Book VII, ch. ix, 'Lehrbrief'.

en résistant. 'Let us die resisting' (Senancour, *Obermann*, Lettre XC, ninth paragraph).

332 *says Joubert*: *Pensées*, Titre xxiii. 54.

Dr Stanley: A month after his attack on Colenso, Arnold praised his friend

Arthur Penrhyn Stanley's *Lectures on the History of the Jewish Church*, also in *Macmillan's Magazine*.

eighty and odd pigeons. Colenso had been a mathematics teacher at Harrow, and when faced with the statements in Leviticus 12: 6–8, 10: 12–20, and Numbers 18: 8–10 that for every child born to the Israelites, Aaron or one of his two sons had to eat a sacrificial pigeon, he calculated that there were at least 264 children born each day and that no priest could keep up the pace; therefore the Book of Leviticus could not have been divinely inspired.

Dr Colenso's book. Arnold severely (and wittily) criticized John William Colenso's *The Pentateuch and Book of Joshua Critically Examined* in 'The Bishop and the Philosopher', *Macmillan's Magazine*, 7 (1863), 241–56.

It has been said: in an anonymous article by Fitzjames Stephen written in response to the present essay: 'Mr Matthew Arnold and His Countrymen', *Saturday Review*, 18 (1864), 685.

333 *a lady*. Frances Power Cobbe, *Broken Lights: an Inquiry into the Present Condition and Future Prospects of Religious Faith* (London, 1864), pp. 116, 104.

M. Renan. Ernest Renan's *Life of Jesus* (1863) and his other studies of the Hebrew and Christian religion were highly admired by Arnold.

Dr Strauss's book. David Friedrich Strauss's *Life of Jesus* was published in an English translation by Marian Evans (later 'George Eliot') in 1846.

ne l'entend pas: 'Whoever imagines he can write it better doesn't understand it' (Claude Fleury, *Histoire ecclésiastique* (Paris, 1722), i. [xxii] (Preface).

its insufficiency. Renan, *Études d'histoire religieuse* (7th edn., Paris, 1880), pp. 199–200. Renan first made this statement before he undertook his *Life of Jesus*, the fruit of a subsequent visit to the Holy Land.

dixit esse: 'No educated man ever has said that a change of opinion is inconsistency' (*Letters to Atticus* XVI. vii. 3).

Coleridge's happy phrase: 'In the Bible there is more that *finds* me than I have experienced in all other books put together' (*Confessions of an Inquiring Spirit*, end of Letter I and beginning of Letter II).

334 *Religious Duty*. Miss Cobbe's *Religious Duty*, published in 1864, tried to establish the moral and intellectual necessity of Theism.

British College of Health. The British College of Health was the name given by James Morison, 'the Hygeist' (1770–1840), to the building from which he distributed his patent medicine, a vegetable pill of universal efficacy. It was just off the part of the New Road now known as King's Cross Road.

335 *Bossuet*. Jacques-Bénigne Bossuet (1627–1704), in his *Discours sur l'histoire universelle* (1681) conceived of all human affairs as expressing the purpose of God regarding humanity.

Bishop of Durham. Charles Thomas Baring (1807–79), Bishop of Durham, was a strong evangelical.

336 *nascitur ordo*: 'The order of the ages is born anew' (Virgil, *Eclogues* iv. 5 (the 'Messianic Eclogue')).

338 *die in the wilderness*. As Moses was not permitted to reach the Promised Land, but was permitted to see it in the distance from the top of Mount Pisgah (Deuteronomy 32: 48-52; 34: 1-6).

339 *Preface to Essays in Criticism*. The wit that showed itself in 'The Function of Criticism' was given full rein as Arnold composed for his new volume a Preface, completed only days before the book was published on 11 February 1865. It was a brilliant piece of topical satire which, as one reviewer said, added a *Dunciad* to Arnold's *Essay on Criticism*. But the wit played on many subjects that were merely ephemeral, matters that required a fresh knowledge of recent newspapers for example, and for that reason Arnold excised a considerable portion of the Preface when a second edition of *Essays in Criticism* appeared in 1869. For the same reason, the shorter version is printed here; brilliant as the wit of the original Preface is, too many of the allusions now need meticulous explanation (and indeed, modern scholars are still baffled by some of them).

not in my power. 'He has hardly any power of argument', [Fitzjames Stephen], 'Mr Matthew Arnold and His Countrymen', *Saturday Review*, 18 (1864), 683. Arnold's subsequent references to the *Saturday Review* are to this article.

strive or cry: Matthew 12: 19.

Mr Wright. Arnold had criticized the translation of the *Iliad* by Ichabod Charles Wright in his first lecture *On Translating Homer* (1860), and Wright replied in a pamphlet *Letter to the Dean of Canterbury on the Homeric Lectures of Matthew Arnold, Esq.* (London, 1864); the quoted passage is on p. 6.

reason for existing: CPW i. 103.

340 *Professor Frickel, and others*. John Henry Pepper, John Henry Anderson, and Wiljalba Frikell were stage magicians who, as was the custom of the trade, called themselves 'Professor'.

341 *the murderer, Müller*. On 9 July 1864 Franz Müller murdered and robbed Thomas Briggs in a first-class railway carriage on a train that had just left Bow station on the North London Railway, threw the body from the train, and made good his escape at the next stop. The murder was discovered by a boarding passenger who laid his hand 'on a cushion steeped in gore', but Müller remained at large for some time. Arnold and his family spent the summer of 1864 at The Rectory, Woodford, Essex; the Great Eastern Railway from Woodford joined the North London Railway near the spot where the murder was committed, and proceeded to the same London terminus in Fenchurch Street.

Cheapside. A street in London's business district.

d'homme nécessaire: 'There's no such thing as an indispensable man.'

Exeter Hall. Exeter Hall, in the Strand, was used for religious and philanthropic assemblies, and came to stand for a certain type of fundamental evangelicalism.

Marylebone Vestry. The St Marylebone Vestry was the popularly elected governing body of the Parish of St Marylebone, London, in Arnold's day decidedly middle class and something of a laughing-stock for its outspoken opposition to such matters as public sanitation.

great, dissected master. Jeremy Bentham (1748–1832) is said to have wished he might awaken once a century to see how the happiness of the world was increasing from the application of his Utilitarian ideas. His skeleton, dressed in his favourite clothes, with hat and walking-stick, and his mummified head are preserved at his request in University College London.

342 *all at play*. 'There were his young barbarians all at play' (Byron, *Childe Harold's Pilgrimage*, IV. cxli. 5). In *Culture and Anarchy* Arnold adopted the term 'barbarians' for the English aristocracy, in contrast to the 'Philistines' or middle class, and the 'populace'.

Tübingen. The University of Tübingen, in Württemberg, was chiefly known for its rationalistic and historical criticism of Scripture.

das Gemeine: 'what ties all of us down, the commonplace' (Goethe, 'Epilog zu Schillers Glocke', l. 32).

343 *Culture and Its Enemies*. Arnold's final year as Professor of Poetry at Oxford was sadly disrupted by his having to write his long report on higher schools and universities on the Continent; instead of three lectures in 1866–7 he gave only this one, on 7 June 1867. It was a triumphant close to his professoriate, and marks also a transition between his discussions of literature and his embarking anew upon social criticism. At the time he wrote he had much more to say upon the latter subject than a single lecture could allow, and the public criticisms his lecture stimulated led him to elaborate his theme in five more essays in the *Cornhill Magazine* in 1868. When all these were collected into the book *Culture and Anarchy* in 1869, 'Culture and Its Enemies' appeared as the Introduction (two paragraphs) and Part I; to the latter subsequent editions gave the title 'Sweetness and Light'. The essay nicely gives the flavour of the book that grew out of it.

'Having a sincere affection for [Oxford], I wanted to make my last lecture as pleasing to my audience and as *Oxfordesque* as I could. I succeeded, and finished my career amidst a most gratifying display of feeling.' And therefore he appended to the lecture as published in the *Cornhill* a note of explanation: 'What follows was delivered as Mr Arnold's last lecture in the Poetry Chair at Oxford, and took, in many places, a special form from the occasion. Instead of changing the form to that of an essay to adapt it to this Magazine, it has been thought advisable, under the circumstances, to print it as it was delivered.'

Mr Bright. John Bright was speaking in the Parliamentary debate on the Elective Franchise Bill on 30 May 1866—a bill that proposed an education test for the franchise.

silliest cant of the day: Harrison, 'Our Venetian Constitution', *Fortnightly Review*, 7 (1867), 276−7. Harrison, a Comtist, later denied that he wrote with Arnold in mind.

Know thyself. 'Know thyself' and 'Avoid extremes' were inscribed on the temple of Apollo at Delphi, and were discussed by Socrates in Plato's *Charmides* 164d−167a and *Protagoras* 343b. Arnold refers to a statement he had made in his essay 'My Countrymen' (1866): *CPW* v. 28.

Daily Telegraph. Leading article, 8 September 1866, pp. 4−5. It was written by the young journalist James Macdonell in response to Arnold's remarks in the Preface to *Essays in Criticism* (pp. 339−42).

344 *I have before now pointed out*: in 'The Function of Criticism at the Present Time' (p. 325).

345 *In the Quarterly Review*: [F. T. Marzials], 'M. Sainte-Beuve', *Quarterly Review*, 119 (January 1866), 80−108. The article refers to Arnold as Sainte-Beuve's disciple and calls the *Essays in Criticism* 'graceful but perfectly unsatisfactory'.

Montesquieu says: 'Discours sur les motifs qui doivent nous encourager aux sciences', *Œuvres complètes*, ed. Édouard Laboulaye (Paris, 1879) vii. 78.

words of Bishop Wilson: Thomas Wilson (1663−1755; Bishop of Sodor and Man), 'Sacra Privata', *Works* (London, 1796) ii. 303. Though Wilson was a favourite of both Arnold and his father, he was so little known in the mid-nineteenth century that Arnold was accused of inventing him.

346 *intellectual horizon*. Arnold elsewhere described these periods as 'epochs of expansion' and 'epochs of concentration'. See 'The Function of Criticism at the Present Time' (p. 324).

347 *and unprofitable*. At this point the manuscript of the essay contains the following passage, crossed out and never published by Arnold, but perhaps delivered from the lecture platform at Oxford: 'Bishop Wilson has the strongest and most present sense of the dangers of curiosity and vanity. . . . But knowledge and light are indispensable, and he sets his face against all disparagement of them. "After all," is his conclusion, "the better a man *knows* the grounds of his duty, the better he is prepared to practise it.". . .

'Religion has been apt so to lay stress upon emotion and affection as often to lose sight of the thought and reason which are necessary in order to guide emotion and affection and to give them their true object. But religion being undoubtedly the greatest, the deepest, the most important of the efforts by which the human race has manifested its impulse to perfect itself, and to draw near to the divine and incomprehensible life in which it exists and to be at one with it; religion being an endeavour through which the human race has on a far wider scale and with a far more intense earnestness sought to satisfy this impulse, than it has through art, or science, or poetry, all of them endeavours directed to the self-same end,—there is an advantage and a satisfaction in getting from religion, speaking by the witness of one of her truest children, sanction for the authenticity and necessity of the aim which is the great aim of culture, the aim of first seeing things as they are, of gaining a sense of how the universal order tends, what the will of God and reason prescribe, with the hope of giving to this knowledge, when gained, a practical influence for the direction of human life.'

within you: Luke 17: 21.

said on a former occasion: in *A French Eton* (*CPW* ii. 318).

348 *one's own happiness*: Wilson, 'Sacra Privata', *Works* ii. 176.

harmonious expansion. 'Civilisation must be grounded in cultivation, in the harmonious development of those qualities and faculties that characterise our humanity' (S. T. Coleridge, *On the Constitution of Church and State*, ch. v).

349 *Faith in machinery*. Thirty-eight years earlier Carlyle had remarked in an essay on 'Signs of the Times': 'Were we required to characterise this age of ours by any single epithet, we should call it . . . the Mechanical Age. It is the Age of Machinery.'

Mr Roebuck's stock argument: See 'The Function of Criticism at the Present Time' (p. 328).

350 *Philistines*: ibid. (p. 331).

their mouths. Arnold deliberately adopts the cadence of the scriptural and liturgical language affected by the evangelicals and Nonconformists (e.g. 'my mouth shall show forth thy praise').

351 *the Times*. 'When Marriages are many and Deaths are few it is certain that the people are doing well' commented the *Times* (3 February 1866, p. 9, col. 1) on the Registrar General's last quarterly report for 1865.

among the sheep. For the sheep and the goats, see Matthew 25: 31–46.

Epistle to Timothy: 1 Timothy 4: 8.

Franklin says. The first of the 'Rules of Health and Long Life' at the end of *Poor Richard, An Almanack* for 1742.

Epictetus: *Encheiridion* xli, somewhat modified.

Swift . . . calls them. Arnold alludes to Aesop's moralizing on the quarrel between the spider and the bee in Swift's *Battle of the Books*.

352 *to save us*. In place of this sentence there is another of the passages Arnold cancelled in manuscript and never published, but may have delivered in his lecture at Oxford: 'and if to some people I have seemed to try and use this chair too much for the general purposes of culture and not enough for the particular criticism of poetry, this is my answer: that culture and poetry have in view the same standard of human perfection, and that in order to give poetry its full reach and to enable men to feel its real power, it is necessary to get this standard of human perfection clear; for which end the varied and manifold workings of culture are the best of disciplines. It is by culture, by studying, with the disinterested aim of perfection in view, all the great manifestations of our human nature—morality, philosophy, history, art, religion, as well as poetry,—that we arrive at the clearest sense of the worth

of poetry, of its use for man's development, and of its function in the future. I have called religion the greatest and the most important manifestation which has yet been seen of human nature, and so it is; but it is so, religion conceived as something wholly separate from poetry is so, only because the rawness of the mass of mankind made them incapable of following religion, conceived as something in harmony with poetry, incapable of meeting the full demands on their spirit and powers which religion, conceived as something in harmony with poetry, would impose; they were not ripe for religion as a conscious study of perfection, pursued with devout energy, for the whole of their being; they could receive it but as a code of rules, and for one side of their being only. In no other way could the animality of the mass of mankind be broken down, and the first beginnings of perfection be made by them.'

353 *resist . . . wicked one*: James 4: 7; 1 John 2: 13, 14.

Independents. The Independents are more commonly known as the Congregationalists, who were Calvinists. 'Dissent' is of course opposition to the established Church of England.

St Peter: 1 Peter 3: 8.

355 *Professor Huxley*. The Darwinian scientist T. H. Huxley was a proponent of Agnosticism.

children of God. 'Ye are all the children of God by faith in Christ Jesus' (Galatians 3: 26).

builded for us to dwell in. See Psalms 122: 3; 107: 4.

privatim opulentia: 'public poverty and private opulence' (Sallust, *Catilina* lii. 22).

muscular Christianity. Fitzjames Stephen, in an anonymous review of *Tom Brown's Schooldays*, coined this phrase to describe the doctrine of Charles Kingsley and his followers on account of their stress on the sacredness, 'the great importance and value of animal spirits, physical stength, and a hearty enjoyment of all the pursuits and accomplishments which are connected with them' (*Edinburgh Review*, 107 (1858), 190).

356 *a speech at Paris*. Gladstone was speaking at a dinner of the Society of Political Economy in Paris, 31 January 1867. See *The Times*, 1 February, p. 10, col. 1.

357 *Mr Beales and Mr Bradlaugh*. Edmond Beales (1803–81), president of the Reform League, and Charles Bradlaugh (1833–91), one of its most influential members, organized the monster meeting on behalf of reform of the franchise that led to the Hyde Park riots of 23 July 1866.

movement which shook Oxford. The Tractarian movement, of which Newman gives an account in his *Apologia*. He frequently refers to 'liberalism' in that work, but especially in his Note A, 'Liberalism'.

plena laboris: 'What region of the earth is not full of [tales of] our hardship?' (Virgil, *Aeneid* i. 460).

358 *Mr Lowe.* Arnold alludes to the speech of Robert Lowe in the parliamentary debate on the Borough Franchise Extension Bill, 3 May 1865.

A new power. A reference to the Conservatives' assumption of power in 1866, and to their sponsoring the Reform Act of 1867, which greatly widened the urban franchise to include the working class, the 'democracy'.

359 *throughout all the world.* John Bright, speaking at Leeds on 8 October 1866 and reported in the *Morning Star* next day (p. 2, col. 5). The first sentence is not Bright's own, but from a leading article in the paper (p. 4, col. 2).

tabernacle. An allusion to C. H. Spurgeon's huge Metropolitan Tabernacle.

a wedding garment. See Matthew 22: 11–14.

the Journeyman Engineer. [Thomas Wright], *Some Habits and Customs of the Working Classes.* By the Journeyman Engineer (London, 1867).

Jacobinism. 'Jacobinism' was the term used for the extreme party in the first French Revolution. Arnold explains how he himself uses the term in his next sentence.

360 *Mr Congreve.* Richard Congreve (1818–99) was a few years ahead of Arnold at Rugby, and was converted to Positivism by a meeting with Auguste Comte in Paris in 1848. He was shortly thereafter tutor to Frederic Harrison at Oxford.

current in people's minds. This 'current' is what Arnold elsewhere calls the Zeitgeist.

Preller: Ludwig Preller, *Römische Mythologie* (3rd edn., Berlin, 1881) i. 21–23, 147.

version of the Book of Job: 'Bagatelles' (*Works*, ed. Jared Sparks (Boston, 1840) ii. 167). Franklin's version was clearly intended to be comic; perhaps Arnold read it out of context. See Job 1: 9.

361 *the Deontology:* Jeremy Bentham, *Deontology; or, the Science of Morality*, arranged and edited by John Bowring (London, 1834) i. 39–40. In his essay on Bentham (1838; republished 1859), J. S. Mill speaks regretfully of this passage.

Mr Buckle. Henry Thomas Buckle (1821–62) wrote the *History of Civilisation in England* (1857, 1861), a book hailed for applying scientific and empirical principles to the treatment of historical problems. 'Mr Mill' is John Stuart Mill.

called Rabbi: Matthew 23: 8.

professor of belles-lettres: see p. 343.

362 *envying and strife:* James 3: 16.

The pursuit of perfection. In the Oxford lecture the first sentence of this paragraph is replaced by a passage designed for the occasion: 'On this the last time that I am to speak from this place, I have permitted myself, in justifying culture and in enforcing the reasons for it, to keep chiefly on ground where I am at one with the central instinct and sympathy of Oxford. The pursuit of perfection is the pursuit of sweetness and light. Oxford has worked with all the bent of her nature for sweetness, for beauty; and I have allowed myself to-day chiefly to insist on sweetness, on beauty, as necessary characters of perfection. Light, too, is a necessary character of perfection; Oxford must not suffer herself to forget that! At other times, during my passage in this chair, I have not failed to remind her, so far as my feeble voice availed, that light is a necessary character of perfection. I shall never cease, so long as anywhere my voice finds any utterance, to insist on the need of light as well as of sweetness. To-day I have spoken most of that which Oxford has loved most. But he who works for sweetness works in the end for light also; he who works for light works in the end for sweetness also.'

363 *Saint Augustine*: *Confessions* xiii. 18.

364 *A Psychological Parallel*. Having published three books on Christianity and the Christian Church—*St Paul and Protestantism*, *Literature and Dogma*, and *God and the Bible*—Arnold felt impelled to write one more essay, 'my last theological paper, I hope', which would sum up his convictions, especially as regarded miracles, with an illustrative parallel aimed to remove the grounds from all arguments that supported them by appeal to the presumed authority of scriptural writers or to the integrity and intelligence of believers. 'I regard the belief in miracles as on a par, in respect of its inevitable disappearance from the minds of reasonable men, with the belief in witches and hob-goblins', he wrote to a friend. And by the same token, the belief that any 'God' was a person with human affections and human reason was bound to disappear from the mind of the future. The essay, published in the *Contemporary Review* for November 1876, was so explicitly conceived as Arnold's last word upon the whole large subject of religion and ecclesiastical affairs that it has, for so short a piece, uncommon comprehensiveness; it was given the first place in his new volume, *Last Essays on Church and Religion*, and bears much the same relation to Arnold's religious writings that the essay on 'The Function of Criticism at the Present Time' bears to the *Essays in Criticism*.

I have formerly spoken: in *St Paul and Protestantism* (1869: *CPW* vi. 50–6).

365 *St Paul's vision*. St Paul was converted to Christianity by a vision of a great light from heaven, and by hearing the voice of Jesus saying 'Saul, Saul, why persecutest thou me?' (Acts 9: 3–18; 26: 12–20).

366 *end of the world*. See 1 Thessalonians 4: 13–17.

and good humour: Burke's characterization in 'A Letter to a Noble Lord' (1796) three-fifths through.

Addison: *Spectator*, 110 (Friday, 6 July 1711) on ghosts; 117 (Saturday, 14 July 1711) on witchcraft.

hanged at Huntingdon. Arnold's principal source for his discussion of witchcraft is a collection of six tracts of 1619–82, separately reprinted in type facsimile in 1837–8 and then gathered with a common title-page and frontispiece. Charles Clark's Appendix to one of these reprints (*A Trial of Witches . . . at Bury St Edmunds (1682)*, p. 28), is the source of Arnold's statement about events of 1716–36.

367 *Grizzel Greedigut.* These and others are named in the 1837 reprint of [Matthew Hopkins,] *The Discovery of Witches* (1647), pp. 2–3.

atheist and infidel: Clark's Appendix to his reprint of *A Trial of Witches*, p. 22.

minister of vengeance: reprint of *The Wonderful Discoverie of the Witchcrafts of Margaret and Philip Flower* (1619), p. 7.

witch to live: Exodus 22: 18, quoted in the 1837 reprint of *A Prodigious & Tragicall History of the Arraignment . . . of Six Witches at Maidstone (1652)*, p. 8.

persons of great knowledge: Reprint of *The Wonderful Discoverie*, p. 11 and of *A Trial of Witches*, p. 16.

Burnet. Gilbert Burnet, Bishop of Salisbury, was Hale's contemporary and so described him in the Preface to his *Life and Death of Sir Matthew Hale*.

368 *and learned persons*: reprint of *A Trial of Witches*, p. 3.

trial of Rose Cullender and Amy Duny: ibid., pp. 8–12, 16–18, 20–21.

372 *fit to undertake.* Simon Patrick's sermon is printed at the end of Smith's *Select Discourses*, ed. H. G. Williams (Cambridge, 1859); Arnold quotes from p. 512. For biographical details Arnold has used Williams's prefatory memoir.

Queen's College. Properly 'Queens' College', and also on p. 374.

Burnet has well described: Gilbert Burnet, *History of His Own Time* (Oxford, 1833) i. 340: *anno* 1661, in Book II.

Principal Tulloch: John Tulloch (1823–86), Principal of St Mary's College, St Andrews.

373 *Laudian clergy*: followers of the high-church doctrines of William Laud (1573–1645), Charles I's Archbishop of Canterbury.

Bishop Wilson: see p. 345 and n. (p. 578).

Owen: John Owen (1616–83).

Cudworth: Ralph Cudworth (1617–88).

the ever memorable. Tulloch begins his chapter on Hales with the remark that Hales was often so designated.

Whichcote: Benjamin Whichcote (1609–83).

by some hand. Arnold himself planned to edit such a volume, but never did so.

374 *Bossuet, Pascal, Taylor, Barrow.* Jacques-Bénigne Bossuet (1627–1704), Blaise Pascal (1623–62), Jeremy Taylor (1613–67), and Isaac Barrow (1630–77) are frequently quoted in Arnold's pocket diaries and his published works.

Smith's great discourse: Smith's *Discourse* IX (*Select Discourses* (Cambridge, 1859), pp. 387–459).

M. Reuss's History. Édouard Reuss, *Histoire de la théologie chrétienne au siècle apostolique* (Strasbourg and Paris, 1852, 1860), which seemed to Arnold to contain the most perceptive analysis of St Paul's doctrine. Reuss was a Protestant theologian at the University of Strasbourg.

Lady-day: 25 March, then regarded as the beginning of the new year.

preach this sermon: *Select Discourses* (1859), pp. 461–87.

375 *flee from you*: James 4: 7.

376 *experimental and real.* Arnold conceived that religion, rightly understood, was fully verifiable by experience and hence truly 'scientific'.

377 *Sir Robert Phillimore.* Phillimore, as Dean of Arches, the highest ecclesiastical judge in England, in giving judgment on a case on 16 July 1875, affirmed that 'we must receive [the] doctrine [of the existence and personality of the devil] unless we impute error and deceit to the writers of the New Testament'.

John the Baptist: Matthew 14: 2 (Mark 6: 14, 16; Luke 9: 7).

378 *one of the old prophets*: Mark 6: 15 (Luke 9: 8).

appeared unto many: Matthew 27: 52–3.

his own coming resurrection. See Luke 24: 6–7.

If One died. The translation is presumably Arnold's own.

379 *promotion of goodness.* This was Arnold's published definition of the Church of England (*CPW* viii. 65, ll. 2–3).

Formerly. The prior form of these oaths dates from the Act of Uniformity of 1662; their form was modified by the Clerical Subscription Act of 1865.

the Three Creeds. '*Nicene* Creed, *Athanasius's* Creed, and that which is commonly called the *Apostles'* Creed'. 'They may be proved by most certain warrants of Holy Scripture', the Eighth Article continues.

380 *marriage with a deceased wife's sister.* The persistent attempt of some Liberal politicians to remove the ecclesiastical barrier to a man's marriage with his deceased wife's sister was a recurrent subject of Arnold's ridicule for its triviality; see the conclusion of *Culture and Anarchy* and Letter VIII of *Friendship's Garland* (*CPW* v. 205–8, 313–18).

Dr Newman. Probably an allusion to Newman's irony in *Lectures on the Present Position of Catholics in England* (Dublin, 1857), pp. 27–8.

Oath of Supremacy. A new and elaborate form of the Oath of the Queen's Supremacy was introduced in 1858. But by the Clerical Subscription Act seven years later, oaths were not to be administered during the services of ordination. Every person about to be ordained Priest or Deacon was required before ordination to subscribe the new declaration of assent and the oath of allegiance.

381 *offendiculum of scrupulousness*: Joseph Butler, *Works*, ed. W. E. Gladstone (Oxford, 1896) ii. 424. *Offendiculum* is a stumbling-block.

which are before: Philippians 3: 13.

an approximative rendering: An echo of *Literature and Dogma* (*CPW* vi. 170–71 and *passim*).

382 *Brady and Tate's metrical version*. In 1696 Nicholas Brady and Nahum Tate published their *New Version of the Psalms*, which by an Order in Council of William III was authorized for use in churches; it was so used almost universally down to the early nineteenth century. The Prayer Book version of the psalms was Miles Coverdale's translation of 1539.

their most eloquent spokesman: Frederic Harrison, 'The Religious and Conservative Aspects of Positivism', *Contemporary Review*, 26 (1875), 1010–11.

383 *bringer-in of righteousness*. The prophet Daniel is told that his people are 'to make an end of sins, and to make reconciliation for iniquity, and to bring in everlasting righteousness' (Daniel 9: 24).

advesperascit. 'Abide with us: for it is toward evening, and the day is far spent' (Luke 24: 29). Arnold inserted (from the context) the word 'Domine'—'O Lord'.

384 *I have elsewhere called*: *Literature and Dogma* (*CPW* vi. 340–4).

to Hegel's philosophy. The intuitionist philosopher Arthur Schopenhauer (1788–1860) had little good to say about the systematic rationalism of his predecessor G. W. F. Hegel (1770–1831)—'a colossal mystification', a 'pseudophilosophy', 'mere parody of scholastic realism and at the same time of Spinozism'.

through Jesus Christ. 'God hath not appointed us to wrath but to obtain salvation by our Lord Jesus Christ'(1 Thessalonians 5: 9).

385 *When Jesus talked*: Matthew 25: 31–46.

when Jesus spoke: Luke 22: 30; Matthew 26: 29.

sitting upon thrones: Matthew 24: 30; 26: 64 (Mark 13: 26; 14: 62).

386 *their own lifetime*. See Matthew 24: 2–51 (Mark 13: 2–37), especially 'This generation shall not pass, till all these things be fulfilled.'

an inward change. 'Behold, the kingdom of God is within you' (Luke 17: 21).

handful of leaven: Matthew 13: 31–3 (Mark 4: 30–2; Luke 13: 18–21).

So is the kingdom. The version is presumably Arnold's own.

to the whole world: Matthew 24: 14 (Mark 13: 10; 14: 9). Also 'Go ye into all the world, and preach the gospel to every creature' (Mark 16: 15).

signs of that time. 'Can ye not discern the signs of the times? (Matthew 16: 3).

387 *his new covenant*. The Old Testament expression 'new covenant' becomes 'new testament' in the Authorized Version of the Gospels: Luke 22: 20; Matthew 26: 28; Mark 14: 24.

Book of Enoch. The Book of Enoch, written partly in Hebrew, partly in Aramaic, probably dates from the second century BC. The English traveller James Bruce brought home in 1771 the Ethiopic version of the book in an eighteenth-century manuscript which was translated into English by

Richard Laurence, Archbishop of Cashel (Oxford, 1821) and into German by August Dillmann (Leipzig, 1853). A more recent English version is by R. H. Charles (Oxford, 1912).

388 *Bishop of Gloucester and Bristol.* Arnold became more closely acquainted with Charles John Ellicott, Bishop of Gloucester and Bristol, after having ridiculed him in *Literature and Dogma*; hence the compliment.

children of light. 'Children of light' occurs in Luke 16: 8; John 12: 36; Ephesians 5: 8; 1 Thessalonians 5: 5. See Enoch 108: 11.

Socrates talked of: in Plato's *Phaedo*, 111c–114c.

remembering the Book of Enoch: Matthew 26: 24 (Mark 14: 21); Luke, 10: 10; Matthew 18: 10; 10: 23; 13: 43. See Enoch 38: 2; 108: 7; 60: 2; 100: 1–2; 58: 3.

Tell it to the church: Matthew 18: 17.

he said to Peter: Matthew 16: 18.

congregation . . . of the just: See Enoch 38: 1.

389 *ecclesiam meam*: Matthew 16: 18.

390 *shall live*: Mark 1: 15; John 3: 15; 5: 24–5. Throughout this long italicized passage Arnold frequently modifies slightly the language of the Authorized Version.

him that sent me: John 6: 38–9, 44; 8: 47; 7: 16; 13: 20 (Luke 9: 48).

shall see God: Luke 6: 46; John 13: 17; Matthew 23: 26; 15: 18; 7: 3 (Luke 6: 41); Matthew 16: 6 (Luke 12: 1); Luke 16: 15; Matthew 5: 8.

my burden light: Matthew 11: 28–30.

shall never perish: John 6: 35, 51, 57, 63; 8: 51; 10: 27–8.

391 *a ransom for many*: John 12: 26; Luke 14: 27; 9: 23; Mark 8: 35–6; John 10: 17; 13: 34; Matthew 20: 28 (Mark 10: 45).

abode with him: John 11: 25–6; 10: 10; Luke 13: 32; John 14: 19; 15: 10; 14: 21; 14: 23.

give you the kingdom: John 10: 11; 10: 16; Luke 12: 32.

shall the end come: John 18: 36; Luke 17: 20–1; 13: 18–21; Mark 4: 26–7; Matthew 24: 14.

Scribes and Pharisees. The Scribes were the expounders of Jewish law, the Pharisees a severe sect of Jewish religionists who in the New Testament become the type of literal observance of religious duty without spiritual or moral commitment. See, for example, Matthew 5: 20.

392 *Dr Mozley*: J. B. Mozley, 'The Influence of Dogmatic Teaching on Education', *Sermons Preached before the University of Oxford and on Various Occasions* (London, 1876), pp. 332–3.

miracle and metaphysics. Arnold devotes a chapter of *God and the Bible* to 'The God of Miracles' and another to 'The God of Metaphysics'.

what they will be: Joseph Butler, *Works* (ed. Gladstone) ii. 134.

have them in derision: Psalm 2: 4.

393　*before the Son of Man*: John 4: 13; Luke 21: 36.

which hath foundations: Hebrews 11: 10.

394　*A French Critic on Milton.* As early as 12 October 1865—only eight months after the *Essays in Criticism* were published, with their overt praise of Sainte-Beuve—the elder Henry James remarked in *The Nation* (New York): 'M. Scherer is a solid embodiment of Mr Matthew Arnold's ideal critic.' Arnold met Scherer on his official mission for the Schools Inquiry Commission in April 1865, and found him 'one of the most interesting men in France'. Not until he read the fourth and fifth series of Scherer's literary essays in 1876, however, was he moved to write upon him, an anonymous article in the *Quarterly Review* for January 1877. It is a much fuller illustration of the contrast between valid and invalid criticism of literature than anything else Arnold wrote, more explicit even than his contrast of the 'personal estimate', the 'historic estimate', and the 'real estimate' in the essay on 'The Study of Poetry'. The high value Arnold sets upon strong good sense and human sympathy over rigorous theoretical consistency shows especially in his warm feeling towards Dr Johnson, despite Johnson's wrong-headedness about 'Lycidas'.

Mr Trevelyan says: George Otto Trevelyan, *The Life and Letters of Lord Macaulay* (New York, 1876) i. 116–17. Macaulay was 24 when the essay appeared. Robert Hall (1764–1831), a Baptist divine much admired for the eloquence of his sermons, was at that time a minister in Leicester. Pope refers to 'this long disease, my Life' in 'Epistle to Dr Arbuthnot', l. 132.

Jeffrey. Francis Jeffrey was editor of the *Edinburgh Review*.

real truth about his object. See 'The Function of Criticism at the Present Time' (p. 317).

395　*like Hezekiah's.* The Lord prolonged the life of King Hezekiah by fifteen years in response to the sick man's prayers (Isaiah 38: 5).

396　*Oromasdes and Arimanes*: the opposing powers of good and evil in the Zoroastrian religion.

said Chillingworth: William Chillingworth, first sermon at Oxford, para. 16 (*Works* (Oxford, 1838) iii. 14).

397　*Milton treats an opponent.* Milton, 'Colasterion: a Reply to a Nameless Answer against The Doctrine and Discipline of Divorce', halfway through, and sixth paragraph from the end.

398　*traveller in Australia*: John Morley, 'Macaulay', *Fortnightly Review*, 25 (1876), 494–5, reprinted in his *Critical Miscellanies*, Second Series (London, 1877), p. 373. Arnold himself in 'Joubert' (1864) called Macaulay 'the great apostle of the Philistines . . . a born rhetorician; a splendid rhetorician doubtless . . .; still, beyond the apparent rhetorical truth of things he never could penetrate' (*CPW* iii. 210).

Addison's Miltonic criticism. Addison's criticism of *Paradise Lost* appeared in the Saturday numbers of *The Spectator* from 5 January to 3 May 1712.

401   *Mr Trevelyan says*: Trevelyan, *Macaulay* i. 117.

M. Edmond Scherer. Edmond Scherer (1815–89), Paris born but of Protestant Swiss family, wrote literary essays for the newspaper *Le Temps* under the general title of 'Variétés'. His essay on 'Milton et le Paradis Perdu' appeared there on 10, 17, and 24 November 1868 and was reprinted in his *Études critiques de littérature* (Paris, 1876), pp. 151–94.

Alexandre Vinet. Vinet (1797–1847) held chairs of both theology and French literature in his native Lausanne.

402   *Sainte-Beuve.* Charles Augustin Sainte-Beuve (1804–69), whose critical essays or *Causeries* also appeared initially as newspaper articles, was one of the writers Arnold thought of as most influential in forming his own career. (The other two were Newman and George Sand.) He wrote a commemorative article on Sainte-Beuve for *The Academy* on 13 November 1869 and the *Encyclopaedia Britannica* article on him in 1886. Arnold and Scherer dined as guests of Sainte-Beuve in a Paris restaurant on 11 May 1866.

Voltaire's sheer disparagement. Count Pococurante, in ch. xxv of Voltaire's *Candide*, disparages Milton.

the man and the milieu. Scherer uses the word *milieu*, probably in allusion to Hippolyte Taine's rigorous theory in his *Histoire de la littérature anglaise* that writers can be explained as products of their race, milieu, and moment. See Arnold's own dictum in 'The Function of Criticism at the Present Time' (p. 319).

403   *art long*: *Ars longa, vita brevis est*, a Latinized aphorism of the Greek physician Hippocrates.

408   *prophets old*: *Paradise Lost* iii. 36.

and fiery arms: ibid. xii. 644.

famous inn: *Prelude* iii. 17. Arnold misses the humour of Wordsworth's line, the grown poet's tolerant smile at the awe with which as a country lad he first saw the great university town.

409   *says St Paul.* 2 Corinthians 6: 6.

410   *passage of another stamp*: 'An Apology for Smectymnuus', one-fourth through. Scherer translates this passage in his essay.

411   *and formidable critic*: *Spectator*, 2 February 1712.

412   *A French Critic on Goethe.* In his very earliest critical essay, the Preface to the 1853 edition of his own *Poems*, Arnold referred to Goethe as the man 'of strongest head' among all the critics of modern literature and culture; from that time onward, Goethe continued to be his touchstone intellectually. But the task of dealing directly and comprehensively with Goethe was too formidable for him. Edmond Scherer's critique gave Arnold the opportunity to say what was most important to him about Goethe within the framework

of what Scherer had written. The title he gave his essay is somewhat misleading and reductive: this is an essay, not on a French critic, but on critical method, and an essay also on a writer of very great stature whose influence on Arnold's world was powerful. Arnold had learned to know Goethe's work through Carlyle, but Carlyle wrote while Goethe was still alive, and indeed before the second part of *Faust* was published, to say nothing of the vast library of what Arnold refers to as 'the immense Goethe-literature of letter, journal, and conversation'; Arnold's essay, therefore, in part sets out to adjust an imbalance in Carlyle's evaluation. As for critical method, Arnold has nowhere summarized his views so concisely as on pp. 413–15 and 430–1 of this essay. It might be added that Arnold's translations of Scherer's French are patterns of style for translators. Like 'A French Critic on Milton', this essay was published anonymously in the *Quarterly Review*; it appeared in January 1878.

Joseph de Maistre comments thus: Joseph de Maistre, *Lettres et opuscules inédits* (3rd edn., Paris, 1853) ii. 399. Arnold published (anonymously) an article on de Maistre in the *Quarterly Review* of October 1879.

Joseph de Maistre has given: ibid. ii. 208–9, 211. Thomas Newton (1704–82), Bishop of Bristol after 1761, published his edition of *Paradise Lost* in 1749. The 'Fortieth Article' is a hypothetical addition to the Thirty-nine Articles that form the basis of the Established Church of England.

413 *Tieck, in his introduction*: J. M. R. Lenz, *Gesammelte Schriften*, ed. Ludwig Tieck (Berlin, 1828), i. p. cxxxix. Goethe made the statement at the beginning of his review of *Manfred* in his periodical *Über Kunst und Alterthum*, vol. ii (1820).

Herman Grimm ... lectures on Goethe: H. Grimm, *Goethe: Vorlesungen gehalten an der Kgl. Universität zu Berlin* (Berlin, 1877) ii. 296 (Lecture 25). Herman Grimm (1828–1901), professor in Berlin from 1873, was the eldest son of Wilhelm and the nephew of Jacob, the philologist brothers Grimm. The Franco-German war of 1870–1 ended in a decisive German victory and a political unification of the independent German states into a single nation.

415 *series of articles*. Scherer's articles on Goethe appeared in *Le Temps* on 21 and 28 May and 7 June 1872; his article on 'Dante et Goethe' appeared on 30 October 1866. All were collected in his *Études critiques de littérature* (Paris, 1876).

inane ... in German literature. See pp. 417–18 and 425. Scherer also describes *Werther* as 'niais' (*Études*, p. 299).

work on English schools: Ludwig Wiese, *German Letters on English Education*, tr. Leonhard Schmitz (London, 1877), p. 168. Arnold reviewed the German edition in the *Pall Mall Gazette*, 3 May 1877.

from Dr Wiese himself: ibid., p. 11 (in Letter II). 'The English belong to a great nation that occupies a position commanding respect in all corners of the earth.' Arnold's point is not the substance of the sentence, but the incredibly convoluted word order.

416 *how Mr Carlyle*: Carlyle, 'Goethe' (1828: *Works* (New York, 1899)) xxvi. 211–12, 217–18. Carlyle was 82 when Arnold's essay was published. The phrase 'old man eloquent' is from Milton's Sonnet X, 'To the Lady Margaret Ley', l. 8.

417 *gone so far*: *Études critiques*, pp. 335–6.

says Mr Hutton: R. H. Hutton, 'Goethe and His Influence', *Essays in Literary Criticism* (Philadelphia, 1876), pp. 40, 7.

at M. Scherer's hands: *Études critiques*, pp. 336–8. The poet Friedrich Gottlob Klopstock (1724–1803) was much admired by the young Goethe but in the later nineteenth century was considered a bore; Scherer reads Charlotte's remark as bathetic. *La nouvelle Héloïse* (1761) is the romantic novel by Rousseau. Johann Gottfried Herder (1744–1803) was a contemporary of Goethe's; Goethe knew both him and his fiancée in the years immediately before he wrote *Werther*.

418 *Kruppism and corporalism*. George Sand, in her *Journal d'un voyageur pendant la guerre* (4th edn., Paris, 1871), pp. 119–20, exclaimed bitterly, 'But what an awakening awaits you, if you pursue the stupid and vulgar ideal of "corporal-ism", or let us say rather of *Kruppism*! Poor Germany of scholars, of philosophers, of artists, Germany of Goethe and of Beethoven! What a fall, what a disgrace.' The firm of Krupp, in Essen, was Prussia's principal manufacturer of armaments.

a Dr Zimmermann. 'There was a Doctor Zimmermann who had been to Berlin to be operated on for a hernia, and who returned home to Hanover. "It was with thousands of tears of joy, he said, that I was received by my son and by my friends, male and female. Some were speechless with happiness, others swooned and still others fell down in convulsions" ' (*Études critiques*, pp. 337–8).

M. Scherer says truly: *Études critiques*, p. 313.

while he was in Rome: A. Mézières, *W. Goethe: les œuvres expliquées par la vie* (Paris, 1872) i. 320–1, quoted by Scherer (*Études critiques*, p. 317).

told the Chancellor von Müller: *Goethes Unterhaltungen mit dem Kanzler Friedrich von Müller*, ed. C. A. H. Burkhardt (Stuttgart, 1904), p. 8 (30 May 1814), quoted by Scherer (*Études critiques*, p. 320).

419 *the Paradiso tiresome*: Mézières, *Goethe* i. 310–11, quoted by Scherer (*Études critiques*, p. 318).

says M. Scherer very truly: *Études critiques*, pp. 339–40.

the late Mr Lewes: G. H. Lewes, *The Life and Works of Goethe* (London, 1855) ii. 13, 100 (Book V, chs. ii, ix). Lewes died on 28 November 1878, eleven months after Arnold's essay was first published, and 'the late' was introduced into all subsequent editions.

420 *he speaks of it*: *Études critiques*, pp. 347–9.

421 *to the formless Germany*: ibid., pp. 314, 321, 313, 319.

Of Egmont M. Scherer says: ibid., p. 344.

422 *hear Mr Lewes*: *Goethe* ii. 65 (Book V, ch. vi).

judicious would determine. This is Aristotle's way of determining the 'mean' which is virtue (*Nicomachean Ethics* 1107a).

writes Schiller: to Johann Heinrich Meyer, quoted by Lewes, *Goethe* ii. 237 (Book VI, ch. iv).

for M. Scherer: *Études critiques*, pp. 340–1.

turn to Mr Lewes: *Goethe* ii. 234–5 (Book VI, ch. iv).

423 *Carlyle on Wilhelm Meister*: 'Goethe' (1828: *Works* (1899) xxvi. 229, 224–5).

Schiller, too, said: letter to Goethe, quoted by Lewes, *Goethe* ii. 213–14 (Book VI, ch. ii).

of our French critic: *Études critiques*, pp. 331–2. *The Elective Affinities* (*Die Wahlverwandtschaften*) is a novel of Goethe's. Paul de Saint-Victor (1825–81) was a French critic and man of letters, highly regarded in his day. Jupiter Pluvius is the Roman god of rain.

424 *menagerie of tame animals*: quoted by Scherer (*Études critiques*, p. 299). Barthold Georg Niebuhr (1776–1831) was author of a history of Rome.

M. Scherer passes judgment. See *Études critiques*, pp. 299–301, 332–3. Like Arnold above (pp. 415–16), Scherer says flatly 'La prose allemande n'existe pas' ('There is no such thing as German prose').

Mr Lewes declares: *Goethe* ii. 342 (Book VI, ch. viii).

says Mr Carlyle: 'Goethe's Helena' (1828): *Works* (1899) xxvi. 195–6. When Carlyle wrote this, the second part of *Faust* had not yet been published as a whole and Carlyle could not foresee how different it would be from this prediction.

Goethe's last manner: *Études critiques*, pp. 344–6.

425 *of the second Faust*: 'Dante et Goethe', *Études critiques*, pp. 91–2. The ironic 'cabbage-leaf' allusion is introduced from *Werther*; see p. 418.

M. Scherer concludes: *Études critiques*, p. 346.

426 *vertraut ist*: *Die natürliche Tochter*, ll. 199–200 (I. iv. 2–3).

Every one has laughed: *Études critiques*, pp. 341–2. The verses are the close of Racine's *Mithridate*.

Goethe's artistic egotism. Lewes, for example, uses the expression 'the egoism of genius' (*Goethe* ii. 50 (Book V, ch. iv)).

427 *has some direct blame*: *Études critiques*, pp. 330–1. Scherer draws upon *Goethes Unterhaltungen mit . . . Müller*, pp. 153–4 (30 August 1827) and Johann Peter Eckermann, *Gespräche mit Goethe*, ed. Eduard Castle (Berlin, 1916) ii. 165 (14 February 1830).

the women of Europe: F. W. Riemer, *Mittheilungen über Goethe* (Berlin, 1841) ii. 707 n. (or ed. Arthur Pollmer (Leipzig, 1921), p. 308).

who loved Lola Montes. Ludwig I (1786–1868), King of Bavaria from 1825, was a warm patron of the arts who was largely responsible for Munich's fine buildings and collections. Not until fourteen years after Goethe's death did Ludwig meet and become infatuated with the 'Spanish' dancer 'Lola Montez' (an Irish girl named Marie Gilbert), whose influence over him was one of the causes that forced his abdication in 1848.

Dogberry: the pompous constable in Shakespeare's *Much Ado about Nothing*.

428 *the summing-up begins*: *Études critiques*, p. 351. Arnold had quoted Goethe to much the same effect, as 'liberator', in his essay on 'Heinrich Heine' (1863: *CPW* iii. 109).

article on Shakspeare: 'Shakspeare et la critique' (1869: *Études critiques*, pp. 149–50).

429 *Mr Hayward's*: Abraham Hayward's prose translation of *Faust*, Part I, first appeared in 1833.

Milton saw and said: Milton, 'Of Education', two-thirds through: poetry 'ought to be simple, sensuous, and impassioned'. Scherer cites this rule in his essay on Milton (*Études critiques*, p. 194).

Carlyle heaps such praise: Carlyle's Introduction to his translation (1832) of Goethe's *Das Märchen* ('The Tale'): *Works* (1899) xxvii. 448–9.

430 *Carlyle has a sentence*: 'Goethe' (1828): *Works* (1899). xxvi. 253.

431 *are what they are*. 'Things and actions are what they are, and the consequences of them will be what they will be; why then should we desire to be deceived?' (Joseph Butler, Sermon VII, 'Upon the Character of Balaam', para. 16: *Works*, ed. Gladstone (Oxford, 1896) ii. 134). This passage was a favourite of Arnold's, which he quoted in 'A Psychological Parallel' (p. 392), 'Bishop Butler and the Zeit-Geist', and 'Numbers'.

bleibende Verhältnisse sind: Riemer, *Mittheilungen über Goethe* (1841) ii. 281.

432 *Equality*. Like 'Democracy', this essay draws heavily on Arnold's observation of the social conditions of France, and it was paired with that essay at the beginning of his volume of *Mixed Essays* (1879), both to set the tone for the volume and to testify for how many years he had held these doctrines. It is a remarkable corrective for those who fancy *Culture and Anarchy* savours too much of the élite. One of the most perceptive statements of his social doctrine, it serves above all to show the inadequacy of nineteenth-century liberal ideology to cope with the problems of the modern world. It should be noticed that Arnold bases his argument, not, as the French were inclined to do, upon the doctrine of 'natural rights', but on the Englishman's ground of expediency and on the sound awareness that 'rights' are created—and taken away—by law.

The 'Royal Institution of Great Britain for the Promotion, Diffusion, and Extension of Science and of Useful Knowledge' was chartered by George III in 1800. It provides laboratories for scientific research—Davy and Faraday worked there—sponsors courses of lectures on a variety of subjects, and holds weekly meetings in the winter to hear papers in the arts and literature as well as science. Arnold read his paper at the Institution's buildings in Albemarle Street at the weekly meeting on the evening of Friday 8 February 1878; it was published in the *Fortnightly Review* for March.

the Burial Service. The Lesson read in the Church of England Burial Service was 1 Corinthians 15: 20–58. The Nonconformists chafed against being required to use the Book of Common Prayer ceremonials for burial in the churchyard—in many communities the only cemetery available.

Corinthians: 1 Corinthians 15: 33. Whether in fact the line comes from Menander's (lost) play *Thais* is a matter of dispute, but the tradition is older than St Jerome.

a Father: Tertullian (*De praescriptionibus adversus haereticos* ch. 7).

flee greed: *Menandri et Philemonis Reliquiae*, ed. A. Meineke (Berlin, 1823), p. 336.

Pleonexia. In nearly every instance, the word in the New Testament which the Authorized Version translated 'covetousness' is represented by *pleonexia* in Greek.

Lord Beaconsfield: Benjamin Disraeli, *Inaugural Address . . . as Lord Rector of the University of Glasgow* (Glasgow, 1873), pp. 15–23; delivered 19 November 1873.

433 *History of Democracy*: *Democracy in Europe: a History* (New York, 1878), end of ch. xvii. The French Revolution of 1789 was dedicated to 'Liberty, Equality, Fraternity'.

Mr Froude: J. A. Froude, 'On the Uses of a Landed Gentry', *Short Studies on Great Subjects*, 3rd ser. (New York, 1877), p. 299.

Mr Lowe and Mr Gladstone. The debate was carried on in the *Fortnightly Review* (Lowe), October and December 1877, and the *Nineteenth Century* (Gladstone), November 1877 and January 1878.

Mr Lowe declared: *Fortnightly Review*, 28 (1877), 451.

Mr Gladstone replied: *Nineteenth Century*, 2 (1877), 547–8.

434 *says George Sand*: *Impressions et souvenirs* (Paris, 1896), pp. 252, 256–7 (ch. xv, 'La Révolution pour l'idéal').

She calls equality: *Journal d'un voyageur pendant la guerre* (Paris, 1871), pp. 162, 114–15.

Real Estates Intestacy Bill. The most recent Real Estate Intestacy Bill was introduced on 18 January 1878; it was defeated on 10 July. Since such bills proposed to control the distribution of the estates of men who died intestate, but imposed no limitation on the power of making a will, they were, in Arnold's view, aiming at a trifle.

Turgot. See A. N. de Condorcet, *Vie de M. Turgot* (*Œuvres* (Paris, 1847) v. 187–8).

436 *The United States of America*. The laws of inheritance in the United States are within the jurisdiction of the several States, not the federal government.

to Mr Lowe: Lowe, 'A New Reform Bill', *Fortnightly Review*, 28 (1877), 448, 451.

437 *letter from Clerkenwell*. Arnold's lecture of course had been announced in advance in the newspapers. Clerkenwell is a part of the London borough of Finsbury, lying north of St Paul's Cathedral and Smithfield Market; it had something of a history of radicalism in the nineteenth century.

I have roughly divided: in *Culture and Anarchy (CPW* v. 143-6).

Plato says: *Phaedrus* 274a.

Mr Charles Sumner says: Edward L. Pierce, *Memoir and Letters of Charles Sumner, 1811-1845* (Boston, 1877) ii. 215 (letter of 8 July 1842).

438 *Rousseau says*. Jean Jacques Rousseau was author of a discourse on 'The Origin and Bases of Inequality among Men' (1754) and the treatise *On the Social Contract* (1762).

says Mr Lowe: *Fortnightly Review*, 28 (1877), 451.

I have often said: in *Culture and Anarchy* (1868), 'Endowments' (1870), *Literature and Dogma* (1871), and *God and the Bible* (1874): *CPW* v. 201; vi. 134, 188; vii. 145.

Sir Henry Maine and Mr Mill: Maine, *Ancient Law* (6th edn., London, 1876), p. 177 (ch. vi); J. S. Mill, *Principles of Political Economy*, Book II, ch. ii, sect. 3.

439 *Sir Erskine May . . . continues thus*: *Democracy in Europe* ii. 348.

says Lucan: *Civil War* ii. 381-2.

says Burke: *Reflections on the Revolution in France*, nearly two-fifths through. *extricate him from it*. Arnold echoes the delicately ironic tone of Socrates in Plato's *Republic*.

440 *by their lawgiver*: Moses, in Deuteronomy 4: 8, 6.

Isocrates could say: *Panegyricus* 50.

social life and manners. Arnold discusses these powers more fully in 'Literature and Science' (pp. 463-71).

Cardinal Antonelli. Arnold was presented to Cardinal Antonelli, the papal Secretary of State, on 6 June 1865, during his tour of the Continent to gather data on higher education.

441 *knowing scientifically*. See 'Literature and Science' (p. 459).

Voltaire, in a famous passage: *Siècle de Louis XIV*, ch. i.

Yet Burke says: *Reflections on the Revolution in France*, nearly one-third through.

442 *Obermann says*: E. P. de Senancour, *Obermann*, Lettre VII.

M. de Laveleye: Émile de Laveleye, 'Le Socialisme contemporain en Allemagne', *Revue des deux mondes*, 3rd pér., 18 (1876), 882.

from Mr Hamerton: Philip Gilbert Hamerton, *Round My House: Notes of Rural*

Life in France in Peace and War (3rd edn., London, 1876), pp. 229–30. Hamerton (1834–94) was an English artist and art critic who married a French wife and settled near Autun.

443 *Mr Gladstone says.* On 23 March 1866 Gladstone, during the Reform Bill debate, responded obliquely to Lowe's assertion of the venality, ignorance, drunkenness, and violence of the working class by pointing out that these were 'our fellow-subjects, our fellow-Christians, our own flesh and blood'.

just now, in France. Arnold alludes to the revival of the *noblesse* under Napoleon III, whose imperial reign came to an end in 1871.

Alsace. France lost Alsace to Germany after the military defeat of 1870–1, and ultimately regained it as a result of World War I.

444 *Michelet . . . gives us.* Jules Michelet (1798–1874), historian of France, perhaps made this remark in conversation with Arnold in Paris on 14 April 1859.

445 *that war-song.* This song by G. W. Hunt became popular in the London music-halls in 1878 when the British government was threatening to intervene on behalf of Turkey in the Russo-Turkish war.

Moody and Sankey meetings. Dwight Lyman Moody and Ira David Sankey were American revivalists who toured Great Britain in 1873–5.

446 *more than once said*: first in the lecture on 'Heinrich Heine' (1863: *CPW* iii. 121).

Mr Goldwin Smith: Smith, 'Falkland and the Puritans: In Reply to Mr Matthew Arnold', *Contemporary Review*, 29 (1877), 925–43. The two men were personal friends of long standing, but Smith's attack was savage.

journals of the House of Commons: cited by Horace Walpole, *Anecdotes of Painting in England* (2nd edn. Strawberry Hill, 1765) ii. 68.

447 *Christianity and its Founder*: an event that occurred in a violent debate on clericalism in the French Chamber of Deputies on 4 May 1877.

rejoins Milton: "Colasterion: a Reply to a Nameless Answer against The Doctrine and Discipline of Divorce', halfway through and two-thirds through. See 'A French Critic on Milton' (p. 397).

says Mr Goldwin Smith: *Contemporary Review*, 29 (1877), 933.

Mrs Hutchinson relates: *Memoirs of the Life of Colonel* [John] *Hutchinson*, written by his widow Lucy (Everyman's Library edn., pp. 238–9), *anno* 1646. John Tombes (1603–76) and Henry Denne (died *c.* 1660) were vigorous polemicists against infant baptism.

448 *Toronto.* Goldwin Smith went from England to Ezra Cornell's new university in Ithaca, New York, in 1868, then in 1871 moved to Toronto, where he lived until his death in 1910. Arnold visited him there in 1884.

says Bossuet. This expression, often attributed to Bossuet, is in fact from de La Mothe-Fénelon, 'Sermon pour la fête de l'Épiphanie (1685): Sur la vocation des gentils', para. 11. 'Man is restless, God guides him.'

that shall stand: Proverbs 19:21. Arnold substitutes 'the Eternal' for 'the Lord' to avoid the suggestion of a divine person.

449 *Bright would call*: in the debate on the Reform Bill, 26 March 1867.

agree with Mr Cobden: at the end of a speech to his constituency at Rochdale on 24 November 1863.

450 *virtus verusque labor*: Virgil, *Aeneid* xii. 435, translated in the words immediately preceding.

the wise man: Proverbs 14: 6.

fas et nefas: 'right or wrong' (Livy, *History* VI. xiv. 10).

Burke calls them: *Reflections on the Revolution in France*, two-fifths through.

451 *I have elsewhere said*. 'Even the Roman governor has his close parallel in our celebrated aristocracy, with its superficial good sense and good nature, its complete inaptitude for ideas, its profound helplessness in presence of all great spiritual movements' (*Literature and Dogma* (1873: *CPW* vi. 399)).

comment of Pepys: Diary for 3 November 1662.

453 *The Times itself*. The leading article in *The Times*, 22 November 1877, upon the marriage of the Duke of Norfolk and Lady Flora Hastings is a most extravagant paean of the British aristocracy.

journals of a new type. The popular 'journals of society', weekly gossip newspapers, of which the most popular were the *World*, founded in 1874, and *Truth*, founded in 1877.

the young lion. See Preface to *Essays in Criticism* (p. 340).

454 *Mr Mill . . . shown*: *Principles of Political Economy*, Book II, ch. ii, sect. 4. Arnold's essay is in part a reply to this section.

Lord Hartington. Spencer Cavendish, Marquis of Hartington, heir to the Duke of Devonshire, was leader of the Liberal party.

456 *Literature and Science*. When Arnold delivered the Rede Lecture at Cambridge to a crowded audience in the Senate House on 14 June 1882, he accepted the challenge his friend T. H. Huxley had thrown out to him on 1 October 1880 at the opening of Sir Josiah Mason's new Science College in Birmingham: 'Literature and Science' is a reply to Huxley's 'Science and Culture'. But it is not merely an occasional piece written from a love of debate: it is the epitome of Arnold's writing on education over nearly three decades—'in general my doctrine on Studies as well as I can frame it', he wrote to his sister. He closed his report on Continental education in 1868 with the assertion: 'The ideal of a general, liberal training is, to carry us to a knowledge of ourselves and the world. We are called to this knowledge by special aptitudes which are born with us; the grand thing in teaching is to have faith that some aptitudes of this kind everyone has. This one's special aptitudes are for knowing men,—the study of the humanities; that one's special aptitudes are for knowing the world,—the study of nature. The circle of knowledge comprehends both, and we should all have some notion, at any

rate, of the whole circle of knowledge. The rejection of the humanities by the realists, the rejection of the study of nature by the humanists, are alike ignorant' (*CPW* iv. 300). In one of his reports as a school inspector eight years later he wrote: 'To have the power of using, which is the thing wished, these data of natural science, a man must, in general, have first been in some measure *moralised*; and for moralising him it will be found not easy, I think, to dispense with those old agents, letters, poetry, religion. So let not our teachers be led to imagine, whatever they may hear and see of the call for natural science, that their literary cultivation is unimportant. The fruitful use of natural science itself depends, in a very great degree, on having effected in the whole man, by means of letters, a rise in what the political economists call *the standard of life*' (*Reports on Elementary Schools, 1852–1882*, ed. F. Sandford (London, 1889), p. 200). Indeed, some of the illustrations he used to support his argument in 'Literature and Science' were the direct fruit of his close observation of his country's schools. By the time he delivered the address he was already thinking of making a lecture tour of the United States, and having been assured on all sides (quite correctly, as it turned out) that the question was of perhaps even greater interest in America, he made this one of the three discourses he would offer to his American audiences, even though it had already been published in periodicals on both sides of the Atlantic more than a year before he began his tour in the autumn of 1883. He was required to deliver it twenty-nine times there (out of sixty-five lecture engagements), and became thoroughly weary of the chore. British audiences also flocked to hear it on at least two occasions after his return. It is the second of the three *Discourses in America* he collected as a book in 1885—the book by which, he said, of all his prose writings, he should most wish to be remembered.

The American version of the lecture is the version normally reprinted, but there is a consistency in the original Cambridge version which is worthy of notice. There the following five paragraphs replace the opening six paragraphs of the lecture printed in the text:

LITERATURE AND SCIENCE

No wisdom, nor counsel, nor understanding, against the Eternal! says the Wise Man. Against the natural and appointed course of things there is no contending. Ten years ago I remarked on the gloomy prospect for letters in this country, inasmuch as while the aristocratic class, according to a famous dictum of Lord Beaconsfield, was totally indifferent to letters, the friends of physical science on the other hand, a growing and popular body, were in active revolt against them. To deprive letters of the too great place they had hitherto filled in men's estimation, and to substitute other studies for them, was now the object, I observed, of a sort of crusade with the friends of physical science—a busy host important in itself, important because of the gifted leaders who march at its head, important from its strong and increasing hold upon public favour.

I could not help, I then went on to say, I could not help being moved with a desire to plead with the friends of physical science on behalf of letters, and in deprecation of the slight which they put upon them. But from giving effect to this desire I was at that time drawn off by more pressing matters. Ten years have passed, and the prospects of any pleader for letters have certainly not mended. If the friends of physical science were in the morning sunshine of popular favour even then, they stand now in its meridian radiance. Sir Josiah Mason founds a college at Birmingham to exclude 'mere literary instruction and education'; and at its opening a brilliant and charming debater, Professor Huxley, is brought down to pronounce their funeral oration. Mr Bright, in his zeal for the United States, exhorts young people to drink deep of 'Hiawatha'; and the *Times*, which takes the gloomiest view possible of the future of letters, and thinks that a hundred years hence there will only be a few eccentrics reading letters and almost every one will be studying the natural sciences—the *Times*, instead of counselling Mr Bright's young people rather to drink deep of Homer, is for giving them, above all, 'the works of Darwin and Lyell and Bell and Huxley', and for nourishing them upon the voyage of the 'Challenger'. Stranger still, a brilliant man of letters in France, M. Renan, assigns the same date of a hundred years hence, as the date by which the historical and critical studies, in which his life has been passed and his reputation made, will have fallen into neglect, and deservedly so fallen. It is the regret of his life, M. Renan tells us, that he did not himself originally pursue the natural sciences, in which he might have forestalled Darwin in his discoveries.

What does it avail, in presence of all this, that we find one of your own prophets, Bishop Thirlwall, telling his brother who was sending a son to be educated abroad that he might be out of the way of Latin and Greek: 'I do not think that the most perfect knowledge of every language now spoken under the sun could compensate for the want of them'? What does it avail, even, that an august lover of science, the great Goethe, should have said: 'I wish all success to those who are for preserving to the literature of Greece and Rome its predominant place in education'? Goethe was a wise man, but the irresistible current of things was not then manifest as it is now. *No wisdom, nor counsel, nor understanding, against the Eternal!*

But to resign oneself too passively to supposed designs of the Eternal is fatalism. Perhaps they are not really designs of the Eternal at all, but designs—let us for example say—of Mr Herbert Spencer. Still the design of abasing what is called 'mere literary instruction and education', and of exalting what is called 'sound, extensive, and practical scientific knowledge', is a very positive design and makes great progress. The Universities are by no means outside its scope. At the recent congress in Sheffield of elementary teachers—a very able and important body of men whose movements I naturally follow with strong interest—at Sheffield one of the principal speakers proposed that the elementary teachers and the Universities should come together on the common ground of natural science. On the ground of the dead languages, he said, they could not possibly come together; but if the Universities would take natural science for their chosen and chief ground instead, they easily might. Mahomet was to go to the mountain, as there was no chance of the mountain's being able to go to Mahomet.

The Vice-Chancellor has done me the honour to invite me to address you here to-day, although I am not a member of this great University. Your liberally conceived use of Sir Robert Rede's lecture leaves you free in the choice of a person to deliver the lecture founded by him, and on the present occasion the Vice-Chancellor has gone for a lecturer to the sister University. I will venture to say that to an honour of this kind from the University of Cambridge no one on earth can be so sensible as a member of the University of Oxford. The two Universities are unlike anything else in the world, and they are very like one another. Neither of them is inclined to go hastily into raptures over her own living offspring or over her sister's; each of them is peculiarly sensitive to the good opinion of the other. Nevertheless they have their points of dissimilarity. One such point, in particular, cannot fail to arrest notice. Both Universities have told powerfully upon the mind and life of the nation. But the University of Oxford, of which I am a member, and to which I am deeply and affectionately attached, has produced great men, indeed, but has above all been the source or the centre of great movements. We will not now go back to the middle ages; we will keep within the range of what is called modern history. Within this range, we have the great movements of Royalism, Wesleyanism, Tractarianism, Ritualism, all of them having their source or their centre in Oxford. You have nothing of the kind. The movement taking its name from Charles Simeon is far, far less considerable than the movement taking its name from John Wesley. The movement attempted by the Latitude men in the seventeenth century is next to nothing as a movement; the men are everything. And this is, in truth, your great, your surpassing distinction: not your movements, but your men. From Bacon to Byron, what a splendid roll of great names you can point to! We, at Oxford, can show nothing equal to it. Yours is the University not of great movements, but of great men. Our experience at Oxford disposes us, perhaps, to treat movements, whether our own, or extraneous movements such as the present movement for revolutionising education, with too much respect. That disposition finds a corrective here. Masses make movements, individualities explode them. On mankind in the mass, a movement, once started, is apt to impose itself by routine; it is through the insight, the independence, the self-confidence of powerful single minds that its yoke is shaken off. In this University of great names, whoever wishes not to be demoralised by a movement comes into the right air for being stimulated to pluck up his courage and to examine what stuff movements are really made of.

says Plato: *Republic* vi. 495d–e.

He draws for us: *Theaetetus* 172–3.

Emerson declares: *Nature, Addresses, and Lectures*: 'Literary Ethics', three-fourths through.

457   *says Plato*: *Republic* ix, 591b–c.

458   *practical scientific knowledge.* Huxley, in 'Science and Culture', a lecture given at the opening of the new Science College in Birmingham on 1 October

1880, praised its founder, Sir Josiah Mason, for setting these aims for his college. See *Science and Culture and Other Essays* (London, 1881), pp. 5–6.

a phrase of mine: 'The Function of Criticism at the Present Time' (pp. 320, 337–8; also *CPW* v. 175); quoted by Huxley, 'Science and Culture', pp. 8–9.

Professor Huxley remarks: 'Science and Culture', pp. 9–10. Arnold first used the expression 'criticism of life' in his essay on 'Joubert' (1864: *CPW* iii. 209).

459 *M. Renan talks of*: *Souvenirs d'enfance et de jeunesse* (Paris, 1883), pp. 195, 180. Ernest Renan (1823–92) was a French critic and biblical scholar whom Arnold much admired.

Wolf, the critic of Homer: quoted in [Mark Pattison], 'F. A. Wolf', *North British Review*, 42 (1865), 262.

460 *to know, says Professor Huxley*: 'Science and Culture', pp. 8–9, 14–15.

Our ancestors learned: ibid., pp. 11–12, 15–16.

461 *Levites of culture*. Huxley speaks of 'the classical scholars, in their capacity of Levites in charge of the ark of culture and monopolists of liberal education', and later refers to them as 'the Levites of culture'. (ibid., pp. 3, 6). By command of the Lord, the Levites were guardians of the sacred tabernacle of the Israelites and all the vessels thereof, and likewise of the ark of the covenant (Numbers 1: 50; 3: 31; Exodus 38: 21).

its Nebuchadnezzars. When Nebuchadnezzar king of Babylon conquered Judah, he carried out of Jerusalem 'all the treasures of the house of the Lord, and the treasures of the king's house, and cut in pieces all the vessels of gold which Solomon king of Israel had made in the temple of the Lord' (2 Kings 24: 13). It is tempting also to see an allusion to Nebuchadnezzar's losing his reason, so that he 'did eat grass as oxen'. (Daniel 4: 33).

462 *tell us, if he likes*. All three propositions, Arnold would say, are myths.

lay it down that: 'Science and Culture', p. 7.

a certain President: Charles Watkins Merrifield, FRS, speaking at the Glasgow meeting of the British Association for the Advancement of Science in September 1876. 'Esaias is very bold', wrote St Paul (Romans, 10: 20).

463 *enumerate the powers*. Arnold first listed these powers in the essay on 'Equality' (1878: p. 440), and referred to them frequently thereafter.

464 *Diotima . . . explained to Socrates*: Plato, *Symposium* 206a, in the language of Shelley's translation.

instinct of self-preservation. 'M. [Émile] Littré . . . traces up . . . all our impulses into two elementary instincts, the instinct of self-preservation and the reproductive instinct', remarked Arnold in the first chapter of *Literature and Dogma* (1871: *CPW* vi. 174).

Professor Sylvester. James Joseph Sylvester (1814–97) was, as Arnold says, recognized as one of the foremost mathematicians of his day. He became professor of mathematics at the Johns Hopkins University when that university was founded in Baltimore in 1877, and was still there when Arnold lectured in Baltimore in December 1883, but immediately thereafter he took a chair at Oxford.

I once ventured. In the original version of the present lecture, in fact. Cambridge was noted for its predominance in mathematics and science, Oxford in the humanities.

465 *Mr Darwin's famous proposition*: *The Descent of Man*, Part III, ch. xxi, para. 7.

Professor Huxley delivers: 'Science and Culture', p. 15.

of physical science: ibid.

not very long ago. Darwin died on 19 April 1882, less than two months before Arnold first delivered this lecture in Cambridge.

466 *Faraday.* Michael Faraday (1791–1867), the physicist and chemist, was brought up in and remained for life a member of the sect founded by the Scottish religious leader Robert Sandeman (1718–71).

holds up to scorn: 'Science and Culture', p. 11.

467 *mediaeval thinking.* 'Scholarly and pious persons, worthy of all respect, favour us with allocutions upon the sadness of the antagonism of science to their mediaeval way of thinking' (ibid., p. 16).

Spinoza: *Ethics* IV. xviii, scholium.

say with the Gospel: Luke 9: 25.

469 *passage in Macbeth*: *Macbeth* v. iii. 40. Arnold commented on this paraphrase in his general report as inspector of schools for 1876.

member of our British Parliament: H. Hussey Vivian, *Notes of a Tour in America* (London, 1878), especially p. 233. Vivian (1821–94) developed his family copper-smelting business into one of the largest metallurgical plants in the world. He sat as a Liberal in the House of Commons from 1852 to 1893, when he was made Baron Swansea.

470 *every kind of excellence*: Huxley, 'Science and Culture', p. 19.

Lady Jane Grey. Lady Jane Grey (1537–54) at the age of thirteen was described by her tutor Roger Ascham as incredible in her accomplishment in writing and speaking Greek, to which accomplishment she added a knowledge of Latin, French, Italian, and Hebrew. She was beheaded at sixteen after an abortive attempt to place her on the throne of England in the place of her cousin Mary Tudor.

our English universities. At Cambridge the women's colleges were Girton (founded 1869) and Newnham (founded 1871); Anne Jemima Clough, sister of Arnold's friend, was first head of the latter. At Oxford, Lady Margaret Hall was founded in 1878, Somerville Hall in 1879. Bedford College, London, was founded as early as 1849. 'Amazons' are female warriors of Greek mythology.

here in America. Arnold delivered this lecture at Smith College on

11 December 1883, and his lecture on 'Emerson' at Wellesley College on 6 December, at Vassar College on 7 January 1884.

Leonardo da Vinci: An epitaph made for him in his lifetime, quoted in Charles Clément, *Michel-Ange, Léonard de Vinci, Raphaël* (Paris, 1861), p. 191.

471 *Mr Ruskin's province.* Ruskin delivered the Rede Lecture at Cambridge fifteen years earlier, 24 May 1867. When Arnold spoke, Ruskin was soon to resume his professorship at Oxford.

strive nor cry. 'He shall not strive, nor cry' (Matthew 12: 19).

472 *Emerson.* No essay of Arnold's is quite so personal as his lecture on Emerson. He first met Emerson, through Clough, in London in 1848. His earliest volume of poems, in 1849, contains a sonnet 'Written in Emerson's *Essays*' (p. 8 above). When Clough went out to Boston in 1852 in the hope of finding a career for himself, one persistent question Arnold's letters to him asked was 'What does Emerson say to my poems? [*Empedocles on Etna, and Other Poems*] Make him look at [them]'. Thereafter there was occasional interchange of letters and books between Arnold and Emerson, and very constant mutual regard. When Emerson died on 27 April 1882, Arnold proposed to write an introduction for an English edition of his works and embarked on an extensive rereading of them. (He had to abandon the edition, however, because his friend John Morley had already undertaken it.) 'I have a strong sense of his value which I am glad to say has deepened instead of diminishing on re-reading him. I always found him of more use than Carlyle, and I now think so more than ever.' And so he resolved to lecture on Emerson on his American tour; he carried with him the *Essays* and the newly published two-volume *Correspondence* of Emerson and Carlyle, wrote as much as he could on shipboard, and finished the writing after arrival in America. Literary America was pleased: Charles Eliot Norton described the lecture to James Russell Lowell as 'a piece of large, liberal, genuine criticism' and William Dean Howells remarked that as he listened to it he was constantly thinking, 'Ah! that is just what I should have liked to say!' It is a fine example of the practical application of Arnold's critical principles, warmed by a genuine sense of affection and literary indebtedness. Arnold delivered the lecture eighteen times on his tour, chiefly in the eastern cities of the United States (first in Boston on 1 December 1883), and on his return delivered it to 'a large and distinguished audience' at the Royal Institution (21 March 1884), where six years earlier he had lectured on 'Equality'. It was published third in *Discourses in America* (1885) after an earlier appearance in *Macmillan's Magazine* for May 1884.

at Oxford. Arnold was an undergraduate at Balliol College, Oxford, from the autumn of 1841 to the autumn of 1844.

to him for ever. Thucydides asserted that his history was composed as 'a possession for ever' (I. xxii. 4).

hear him still, saying: 'Peace in Believing' (1839), *Parochial Sermons* (London and Oxford, 1842) vi. 400–1.

Littlemore. The chapel at Littlemore, a village three miles south-east of

Oxford, was in Newman's care as Vicar of St Mary's, Oxford. He retired to his house at Littlemore as he drew closer to a determination to give up his Anglican titles at St Mary's and Oriel College.

tent-making at Ephesus. See Acts 18: 3. But the place was Corinth, not Ephesus.

seem to hear him: 'Worship, a Preparation for Christ's Coming' (1838), *Parochial Sermons* (London and Oxford, 1840) v. 2–3, with omission. Both sermons Arnold quotes were preached before Arnold matriculated at Oxford.

473 *I have spoken*: in the Preface to *Essays in Criticism* (1865); see p. 342.

Carlyle upon Edward Irving: 'Death of Edward Irving' (1835), in *Critical and Miscellaneous Essays* (*Works* (New York, 1899) xxviii. 319–20). Irving, a friend of Carlyle's and a native of the same part of Scotland, was an immensely popular Presbyterian minister in London from 1822; in 1832 he founded the Holy Catholic Apostolic Church, which still survives in Gordon Square, London.

dirge over Mignon: Goethe's *Wilhelm Meister's Apprenticeship*, Book VIII, ch. viii; translated in Carlyle's *Works* (1899) xxiv. 157.

Mr Lowell has well described: James Russell Lowell, *My Study Windows* (Boston, 1871), pp. 179–81 ('Emerson the Lecturer', halfway through). 'Here', of course, is 'in America'.

474 *thousand thousand men*: *Nature, Addresses and Lectures*, 'Literary Ethics', second-last paragraph.

he can understand: *Essays, First Series*, 'History', first paragraph.

Chaos and the Dark: ibid., 'Self-Reliance', third paragraph. Arnold uses Emerson's earlier version, as it was published in London in 1841.

a German critic. See 'A French Critic on Goethe' (pp. 413–14).

notice of Emerson: E. C. Stedman, 'Emerson', *Century Magazine*, 25 (1883), 880.

475 *personal sort of estimate*. The 'personal estimate' is the fallacious kind of judgment about poetry which occurs when we allow such matters as personal affection or patriotism to govern our valuation of a work. Arnold distinguishes it from the 'real estimate' in his essay on 'The Study of Poetry' (1880): *CPW* ix. 163–4).

Patience on a monument: *Twelfth Night* II. iv. 117.

Darkness visible: *Paradise Lost* i. 63.

ignorance is bliss: 'Ode: On a Distant Prospect of Eton College', l. 99.

Milton says that poetry: 'Of Education', two-thirds through. Arnold alluded to this definition also in 'A French Critic on Goethe' (p. 429).

on the Concord Monument. 'Concord Hymn' is inscribed on the monument to the Battle of Concord; the next poem to which Arnold alludes is 'Ode Sung in the Town Hall, Concord, July 4, 1857'.

476 *replies, I can*: 'Voluntaries', iii. 13–16.

he ought to die: the quatrain 'Sacrifice'.

Eden's balmier spring: 'May-Day', ll. 98–103.

477 *his transcendentalist friends*. The 'transcendentalists' were a group or club of Emerson's friends who espoused the doctrine of 'idealism'; the best description of them is Emerson's lecture on 'The Transcendentalist'. *The Dial* was their quarterly journal.

Stanley used to relate. Stanley (later Dean of Westminster) was quarantined for five days at Malta in January–February 1841.

478 *Emerson so admirably says*: Emerson to Carlyle, 30 October 1840 and 8 August 1839 (*The Correspondence of Thomas Carlyle and Ralph Waldo Emerson, 1834–72*, ed. Charles Eliot Norton (London, 1883) i. 308, 255).

description is Carlyle's: Carlyle to Emerson, 13 May 1835 (*Correspondence* i. 65).

well-beloved John Sterling: Carlyle to Emerson, 8 December 1837 (*Correspondence* i. 54).

Emerson to London: Carlyle to Emerson, 15 November 1838 (*Correspondence* i. 199).

a London Sunday: 29 September 1844 (Correspondence ii. 74).

for his histories. 'I think you have written a wonderful book [*The French Revolution*], which will last a very long time' (Emerson to Carlyle, 13 September 1837; *Correspondence* i. 129).

479 *Mr Charles Norton*. The correspondence was published in Boston and London about the middle of February 1883, some eight months before Arnold left England for America. Arnold had known Norton personally for more than a decade.

not by his works. Arnold would certainly exempt *The Lives of the Poets*, of which he edited a selection, from this sweeping censure of Johnson's works.

says of the 'Dial': Carlyle to Emerson, 26 September 1840 (*Correspondence* i. 304).

orations he says: Carlyle to Emerson, 8 February 1839 (*Correspondence* i. 217). He alludes to 'The American Scholar', the Divinity School Address, and the Dartmouth College oration ('Literary Ethics') of 1837–8.

Emerson himself formulates: Emerson to Carlyle, 30 October 1841 and 10 May 1838 (*Correspondence* i. 345, 161).

480 *English Traits*: published in 1856.

Our Old Home. An account of England, published in 1863, based largely on Hawthorne's observation as American consul in the commercial city of Liverpool from 1853 to 1857.

481 *in reply to Carlyle*: 25 April 1839 (*Correspondence* i. 238).

He deprecates: Emerson to Carlyle, 31 July 1841 (*Correspondence* i. 340–2).

he writes to Carlyle: 15 October 1870 (*Correspondence* ii. 334).

to envious Time. 'Love, friendship, charity, are subjects all / To envious and calumniating time' (Shakespeare, *Troilus and Cressida* III. iii. 173–4).

live in the spirit: an echo of Arnold's evaluation of the Stoic emperor-philosopher in his essay on 'Marcus Aurelius': 'He remains the especial friend and comforter of all clear-headed and scrupulous, yet pure-hearted and upward-striving men' (1863: *CPW* iii. 156).

482 *Character is everything*: *Essays, Second Series*, 'Politics', fourth para. from end.

that iron string: *Essays, First Series*, 'Self-Reliance', third para.

a not ourselves. Arnold, affirming the notion of a transcendental force in which we participate, but seeking to avoid all suggestion that God is a person, in *Literature and Dogma* defined God as 'a power, not ourselves, which makes for righteousness' (*CPW* vi. 189, 196, and *passim*).

of its communications. See *Essays, Second Series*, 'New England Reformers', fourth para. from end.

whole scene changes: *Essays, Second Series*, 'New England Reformers', halfway through.

are miserably dying: *Essays, Second Series*, 'The Poet', three-fourths through.

in obscure duties: *Essays, First Series*, 'Heroism', second-last paragraph.

whole pleasure for us: *Essays, Second Series*, 'Experience', two-fifths through.

that thyself is here: *Essays, First Series*, 'Heroism', two-thirds through.

in two classes: *Essays, Second Series*, 'New England Reformers', immediately following previous quotation from that essay.

something unique: *Essays, First Series*, 'Spiritual Laws', nearly one-third through.

483 *lose your own*: *Essays, First Series*, 'Compensation', halfway through.

important benefit: *Essays, Second Series*, 'New England Reformers', three-fourths through.

suffering men: *Essays, First Series*, 'Heroism', fourth para. from end.

to have done it: *Essays, Second Series*, 'New England Reformers', third para. from end.

your own debt: *Essays, First Series*, 'Compensation', three-fifths through.

shall have it: *Essays, First Series*, 'Spiritual Laws', two-fifths through.

inopportune or ignoble: *Essays, Second Series*, 'The Poet', last sentence.

renowned as any: *Essays, First Series*, 'Spiritual Laws', two-fifths through.

Moody and Sankey. See note to p. 445.

Mr Howells. William Dean Howells' novel *The Lady of the Aroostook* was published in volume form in 1879.

484 *for ever safe*: *Essays, First Series*, 'Heroism', final para.

The Democrats: *Essays, Second Series*, 'Politics', halfway through.

withdrawal in New England: For example, *Essays, Second Series*, 'New England Reformers', para. 4 and one-third through. Brook Farm was a co-operative community established in 1841 at West Roxbury, near Boston, by George Ripley, at whose house the first meeting of the Transcendentalist Club had been held five years earlier. The community was dissolved in 1847. 'The Dissidence of Dissent and the Protestantism of the Protestant Religion' was the motto of the *Nonconformist*, weekly newspaper of the English Congregationalists; see p. 353 above.

485 *manhood to withhold*: *Essays, First Series*, 'Self-Reliance', para. 7.

do not arrive: *Essays, First Series*, 'Spiritual Laws', para. 8.

my little Sir: ibid., para. 7.

have their being. 'For in him we live, and move, and have our being' (Acts 17: 28).

an immortal youth: *Essays, First Series*, 'Spiritual Laws', para. 11.

cut out my tongue: Emerson to Carlyle, 31 July 1841 (*Correspondence* i. 341–2).

on which these draw: Emerson to Carlyle, 15 October 1870 (*Correspondence* ii. 337–8).

by which it lives: *Essays, Second Series*, 'New England Reformers', final para. 'By [faith] he being dead yet speaketh' (Hebrews 11: 4).

486 *always of eternity*: Emerson to Carlyle, 17 September 1836 (*Correspondence* i. 95).

truth and justice: From Emerson's diary, October 1847 (*Correspondence with Carlyle* ii. 148).

own account of it: Carlyle to Emerson, 8 February 1839 (*Correspondence* i. 214–15).

in Sartor. Book II, chap. ix, 'The Everlasting Yea', para. 14.

Epictetus. This Stoic philosopher (*c*.50–120 AD) was a constant favourite of Arnold's from about 1848, when he was praised in Arnold's sonnet 'To a Friend' (p. 53).

487 *piece of poetry*: N. P. Willis, 'Saturday Afternoon', ll. 26–8.

488 *Wordsworth well says*: Sonnet, 1811 ('Here pause: the poet claims at least this praise'), ll. 5–6.

habit as he lived: *Hamlet* III. iv. 135.

avaricious America: Emerson to Carlyle, 31 July 1841 (*Correspondence* i. 342). The rhetoric of Arnold's conclusion is much like that with which he ended his essay on 'Marcus Aurelius' (1863).

interesting lines: 'At the Saturday Club', ll. 141–50. The poem was actually published in the *Atlantic Monthly* in January 1884, while Arnold was still in America.

489 *Civilisation in the United States*. After a second visit to the United States in

1886, where he not only had many friends but now also had a daughter who had married an American, Arnold set out to evaluate the quality of the life he had seen there. Having expressed his social doctrines in 'Democracy' and 'Equality', he here examined their working in practice: 'America is so deeply interesting to me, and to its social conditions we [in England] must more and more come.' The essay uses the comparative method he had so often used in writing about literature, and illuminates not only American but British society; obviously it was directed primarily at a home audience. He delivered his remarks as a lecture in provincial cities of England on 31 January, 3 February, and 8 March 1888. The essay was published in the *Nineteenth Century* for April, delighted his British readers and offended the Americans. 'After all,' wrote Theodore Roosevelt (still thirteen years short of the presidency), 'taming a continent is nobler work than studying *belles lettres.*' This was Arnold's last essay; on 15 April he died suddenly of a heart attack.

in this Review: 'A Word More about America' (1885): *CPW* x. 198–217). Arnold concluded that essay by citing the opinion expressed by Sir Lepel Griffin, 'A Visit to Philistia', *Fortnightly Review* 41 (1884), 50.

Theophrastus. In the Prologue to the *Characters*, Theophrastus speaks of having reached his ninety-ninth year (an exaggeration: he was closer to fifty-five).

never reach it. About a fortnight after this essay was published Arnold died suddenly at the age of sixty-five.

490 *humanisation of man in society.* Arnold thus defined 'civilisation' in the essay on 'Equality' (1878); see p. 439.

says Plato: *Gorgias* 512e or *Republic* I. 352d.

491 *I have often insisted.* Arnold first listed these powers in 'Equality'; see p. 440 and also p. 463.

fifteen hundred: pounds sterling, of course. Arnold's annual income at this time was about £1,400–£1,600 (about $7,000–$8,000).

horse-cars. Electric trams were still a few years away.

493 *greatest number.* The axiomatic goal of society among the Benthamite Utilitarians, and thence among the Liberals in general.

Pericles and Camillus. Pericles was the great leader of Athens at the height of its military, literary, and artistic glory in the fifth century BC. Camillus restored the city of Rome and its government after the invasion by the Gauls in 387–386 BC.

absurd Esquire. The custom fell out of use in England only after the Second World War. Arnold himself followed it when he addressed letters, even to Americans.

495 *Professor Norton.* Charles Eliot Norton, a long-time American friend of Arnold's, severely criticized J. A. Froude's *Thomas Carlyle, a History of His Life* (4 vols., 1882–84) in the Preface to his edition of *Early Letters of Thomas Carlyle* (2 vols., London, 1886). Froude's biography was too frank for

contemporary taste, especially as to Carlyle's treatment of his wife, but it is notwithstanding a superb piece of work.

Carlyle dissuades him: Thomas to Alexander Carlyle, 22 February 1822 (*Early Letters* ii. 51–2). Despite this advice, Alexander Carlyle emigrated.

Amiel. Henri-Frédéric Amiel's *Journal intime* was translated by Arnold's niece Mrs Humphry Ward in late 1885; Arnold wrote an essay on her translation and quoted at length the passage from which these remarks are drawn (*Amiel's Journal*, tr. Ward (London, 1885), ii. 116–17: *CPW* xi. 279).

das Gemeine. See note to p. 342.

496 *in English society*. Arnold had little awareness of the non-British origins of vast numbers of American families in his day.

St Pancras . . . Somerset House. The façade of the hotel at St Pancras Station in London is a remarkable neo-Gothic red brick structure designed by Sir George Gilbert Scott in 1865. Somerset House, a government office building on The Strand, was principally designed in the Palladian style by Sir William Chambers and constructed in 1777–86.

Richardson. Henry Hobson Richardson (1838–86) liked best of all his works the courthouse in Pittsburgh, begun in 1884. A somewhat earlier work of his was Trinity Church, in Copley Square, Boston.

497 *Jacksonvilles*. In 1874 there were 2 Briggsvilles, 5 Higginsvilles, and 21 Jacksonvilles in the United States.

classical dictionary. Arnold's lecture tour in January 1884 took him to or through Utica, Rome, Syracuse, and Aurora. There are at least two dozen other town names from the classical dictionary in upstate New York, including Marcellus and Camillus.

bestes Theil: *Faust*, Part II, act I, l. 6272.

498 *by a foreigner*. The founder of the weekly New York *Nation*, E. L. Godkin, was born in Ireland of an English family. The *Saturday Review* is the London weekly newspaper with which Arnold came into amused conflict, especially in 'The Function of Criticism at the Present Time' and 'My Countrymen'.

New York newspaper. Arnold's memory is only slightly inaccurate. A headline in the *Chicago Tribune* for 19 January 1884 (the day of Arnold's arrival in Chicago) reads: 'A Skeleton's Bride. Two Days of Wedded Bliss Make Her a Maniac'. The marriage was reported to have taken place in Philadelphia the preceding Tuesday. The German emperor Frederick III, whose wife was Queen Victoria's daughter, was seriously ill with cancer of the throat and died only some three months after he ascended the throne on 9 March 1888. Timothy Daniel Sullivan, MP for Dublin and Lord Mayor of that city, was imprisoned for two months for publishing in the *Nation* (Dublin) the proceedings of the suppressed branches of the National League.

499 *an elderly gentleman*. Medill was three and a half months younger than Arnold, who was a few weeks past his sixty-first birthday when he reached Chicago.

a New York paper. About a month after Arnold returned to England from his lecture tour, the *New York Tribune* on 6 April 1884 published under a London dateline a long article on culture in Chicago, purporting to be the first of a series on America being written by 'Mr Arnold' for the *Pall Mall Journal*, a non-existent newspaper. The hoax was taken seriously throughout America, and entrapped many of Arnold's American 'friends' into indignant attacks upon him.

brass and iron: Deuteronomy 28: 23 and Leviticus 26: 19, quoted by Arnold in 'Heinrich Heine' (1863: *CPW* iii. 113).

500 *national historian*: George Bancroft, *History of the United States from the Discovery of the American Continent* (Boston, 1858), vii, 21 (ch. i: May 1774), quoted in Arnold's 'A Word More about America' (1885: *CPW* x, 196).

Roe. Edward Payson Roe (1838–88), a Presbyterian minister in upstate New York, discovered the road to success and fortune through writing novels at the rate of one a year and quickly gave up his pastorate.

501 *type of mankind was born.* This resembles, but does not quite reproduce, several passages in the writings of Thomas Wentworth Higginson.

dinner-party of authors: *The Expedition of Humphry Clinker*, para. 7 of the first letter of 10 June.

Our Country. Josiah Strong, a Congregational minister in Cincinnati, became instantly famous with his book *Our Country: Its Possible Future and Its Present Crisis* (New York, 1885), with its warnings against the perils of immigration, Romanism, Mormonism, intemperance, socialism, wealth, and the city.

502 *American faults*: Strong, *Our Country*, pp. 219, 170, 168, 173, 168.

Congregationalist instructor: ibid., p. 169.

503 *lower class brutalised*: a description Arnold first used in 'Equality' (1878), and frequently thereafter; see p. 450.

not as they are. 'To see things as they really are' is a favourite expression of Arnold's, which he used as early as his second Oxford lecture *On Translating Homer* (8 December 1860: *CPW* i. 140). See 'The Function of Criticism at the Present Time' (p. 317).

des Américains du Nord. In a letter to Clough about 24 February 1848 Arnold attributed this expression to Jules Michelet.

light and leading. Burke, in *Reflections on the Revolution in France*, two-thirds through, uses the expression 'the men of light and leading in England'.

504 *are truly excellent*: a pastiche of expressions from 2 Corinthians 9: 3, Romans 1: 21, 1 Peter 3: 11, and Philippians 1: 10.

well-known one: John 3: 3, a passage central to very many of Arnold's writings throughout his career.

FURTHER READING

MAJOR EDITIONS

Arnold. The Complete Poems, ed. Kenneth Allott (1965). Second edn. by Miriam Allott (London and New York, 1979).
Arnold. The Complete Prose Works, ed. R. H. Super (Ann Arbor, Mich., 1960–77), 11 vols.

BIOGRAPHY AND CRITICISM

K. Allott, 'A Background for "Empedocles on Etna" ', *Essays and Studies*, NS 21 (1968), 80–100.
K. Allott, 'Matthew Arnold (and A. H. Clough)', in *The Victorians*, ed. Arthur Pollard (London, 1970), pp. 41–73.
Kenneth Allott (ed.), *Matthew Arnold*, 'Writers and their Background' Series (London, 1975). Includes: Fraser Neiman, 'A Reader's Guide to Arnold'; William A. Madden, 'Arnold the Poet: (i) Lyric and Elegiac Poems'; Kenneth and Miriam Allott, 'Arnold the Poet: (ii) Narrative and Dramatic Poems'; David J. DeLaura, 'Arnold and Literary Criticism: (i) Critical Ideas'; R. H. Super, 'Arnold and Literary Criticism: (ii) Critical Practice'; James Bertram, 'Arnold and Clough'; Peter Keating, 'Arnold's Social and Political Thought'; Basil Willey, 'Arnold and Religion'; Warren Anderson, 'Arnold and the Classics'; James Simpson, 'Arnold and Goethe'.
Miriam Allott, 'Matthew Arnold: "All One and Continuous" ', in *The Victorian Experience: The Poets*, ed. Richard A. Levine (Athens, Ohio, 1982), pp. 67–93.
Warren D. Anderson, *Matthew Arnold and the Classical Tradition* (Ann Arbor, Mich., 1965).
Ruth apRoberts, *Arnold and God* (Berkeley and London, 1983).
Paull Franklin Baum, *Ten Studies in the Poetry of Matthew Arnold* (Durham, NC, 1958).
Louis Bonnerot, *Matthew Arnold—Poète: Essai de biographie psychologique* (Paris, 1947).
Douglas Bush, *Matthew Arnold: A Survey of His Poetry and Prose* (New York and London, 1971).
Joseph Carroll, *The Cultural Theory of Matthew Arnold* (Berkeley and London, 1982).
Merton A. Christensen, 'Thomas Arnold's Debt to German Theologians: a Prelude to Matthew Arnold's *Literature and Dogma*', *Modern Philology*, 55 (1957), 14–20.
William F. Connell, *The Educational Thought and Influence of Matthew Arnold* (London, 1950).
Sidney M. B. Coulling, *Matthew Arnold and His Critics: a Study of Arnold's Controversies* (Athens, Ohio, 1974).
A. Dwight Culler, *Imaginative Reason: The Poetry of Matthew Arnold* (New Haven and London, 1966).

David J. DeLaura, 'Arnold and Carlyle', *PMLA* 79 (1964), 104–29.

David J. DeLaura, *Hebrew and Hellene in Victorian England: Newman, Arnold, and Pater* (Austin, Texas, and London, 1969).

David J. DeLaura, 'Matthew Arnold', in *Victorian Prose, a Guide to Research*, ed. David J. DeLaura (New York, 1973), pp. 249–320.

Robert A. Donovan, 'The Method of Arnold's *Essays in Criticism*', *PMLA* 71 (1956), 922–31.

Fred A. Dudley, 'Matthew Arnold and Science', *PMLA* 57 (1942), 275–94.

John P. Farrell, 'Matthew Arnold's Tragic Vision', *PMLA* 85 (1970), 107–17.

Leon Gottfried, *Matthew Arnold and the Romantics* (Lincoln, Nebr., and London, 1963).

Frank J. W. Harding, *Matthew Arnold the Critic and France* (Geneva, 1964).

Park Honan, *Matthew Arnold: A Life* (New York, 1981).

Walter E. Houghton, 'Arnold's "Empedocles on Etna" ', *Victorian Studies*, 1 (1958), 311–36.

D. G. James, *Matthew Arnold and the Decline of English Romanticism* (Oxford, 1961).

James C. Livingston, *Matthew Arnold and Christianity* (Columbia, SC, 1986).

Howard Foster Lowry (ed.), *The Letters of Matthew Arnold to Arthur Hugh Clough* (London and New York, 1932). Introductory Chapters, pp. 1–53.

Howard Foster Lowry, *Matthew Arnold and the Modern Spirit* (Princeton, NJ, 1941).

William A. Madden, *Matthew Arnold, a Study of the Aesthetic Temperament in Victorian England* (Bloomington, Ind., 1967).

William Robbins, *The Ethical Idealism of Matthew Arnold: a Study of the Nature and Sources of His Moral and Religious Ideas* (London, 1959).

William Robbins, *The Arnoldian Principle of Flexibility* (Victoria, BC, 1979).

James Simpson, *Matthew Arnold and Goethe* (London, 1979).

G. Robert Stange, *Matthew Arnold, The Poet as Humanist* (Princeton, NJ, 1967).

R. H. Super, 'Emerson and Arnold's Poetry', *Philological Quarterly* 33 (1954), 396–403.

R. H. Super, 'Vivacity and the Philistines', *Studies in English Literature, 1500–1900*, 6 (1966), 629–37.

R. H. Super, *The Time-Spirit of Matthew Arnold* (Ann Arbor, Mich., 1970).

R. H. Super, 'The Humanist at Bay: The Arnold–Huxley Debate', in *Nature and the Victorian Imagination*, ed. U. C. Knoepflmacher and G. B. Tennyson (Berkeley and London, 1977) pp. 231–45.

R. H. Super, 'The Epitome of Matthew Arnold', in *The Victorian Experience: The Prose Writers*, ed. Richard A. Levine (Athens, Ohio, 1982). pp. 175–202.

Lionel Trilling, *Matthew Arnold* (New York and London, 2nd edn., 1949).

Fred G. Walcott, *The Origins of Culture and Anarchy: Matthew Arnold and Popular Education in England* (Toronto, 1970).

INDEX OF TITLES AND FIRST LINES

(Titles are indicated in italic, first lines in roman.)